Handbook of Research on Solving Societal Challenges Through Sustainability-Oriented Innovation

Luísa Cagica Carvalho
Instituto Politécnico de Setúbal, Portugal & CEFAGE, University of Évora, Portugal

Paulo Bogas
Instituto Politécnico de Setúbal, Portugal

Jordana Kneipp
Federal University of Santa Maria, Brazil

Lucas Avila
Federal University of Santa Maria, Brazil

Elis Ossmane
Universidade Aberta, Portugal

A volume in the Practice, Progress, and Proficiency in Sustainability (PPPS) Book Series

Published in the United States of America by
IGI Global
Engineering Science Reference (an imprint of IGI Global)
701 E. Chocolate Avenue
Hershey PA, USA 17033
Tel: 717-533-8845
Fax: 717-533-8661
E-mail: cust@igi-global.com
Web site: http://www.igi-global.com

Copyright © 2023 by IGI Global. All rights reserved. No part of this publication may be reproduced, stored or distributed in any form or by any means, electronic or mechanical, including photocopying, without written permission from the publisher. Product or company names used in this set are for identification purposes only. Inclusion of the names of the products or companies does not indicate a claim of ownership by IGI Global of the trademark or registered trademark.

Library of Congress Cataloging-in-Publication Data

Names: Carvalho, Luísa Cagica, 1970- editor. | Bogas, Paulo, 1970- editor.
 | Kneipp, Jordana, 1986- editor. | Avila, Lucas, 1988- editor. |
 Ossmane, Elis, 1980- editor.
Title: Handbook of research on solving societal challenges through
 sustainability-oriented innovation / Luisa Carvalho, Paulo Bogas,
 Jordana Kneipp, Lucas Avila, and Elis Ossmane, editors.
Description: Hershey, PA : Engineering Science Reference, [2023] | Includes
 bibliographical references and index. | Summary: "This book is intended
 to be an essential reference source that emphasizes the importance of
 innovation and sustainability as a possible solution for the big
 challenges of our society, leveraging the articulation between
 innovation and sustainability with the big challenges that our existence
 are facing today and in the near future"-- Provided by publisher.
Identifiers: LCCN 2022035916 (print) | LCCN 2022035917 (ebook) | ISBN
 9781668461235 (hardcover) | ISBN 9781668461259 (ebook)
Subjects: LCSH: Sustainable development. | Technological innovations. |
 Economic policy.
Classification: LCC HC79.E5 H318539 2023 (print) | LCC HC79.E5 (ebook) |
 DDC 338.9/27--dc23/eng/20220816
LC record available at https://lccn.loc.gov/2022035916
LC ebook record available at https://lccn.loc.gov/2022035917

This book is published in the IGI Global book series Practice, Progress, and Proficiency in Sustainability (PPPS) (ISSN: 2330-3271; eISSN: 2330-328X)

British Cataloguing in Publication Data
A Cataloguing in Publication record for this book is available from the British Library.

All work contributed to this book is new, previously-unpublished material. The views expressed in this book are those of the authors, but not necessarily of the publisher.

For electronic access to this publication, please contact: eresources@igi-global.com.

Practice, Progress, and Proficiency in Sustainability (PPPS) Book Series

Ayman Batisha
International Sustainability Institute, Egypt

ISSN:2330-3271
EISSN:2330-328X

Mission

In a world where traditional business practices are reconsidered and economic activity is performed in a global context, new areas of economic developments are recognized as the key enablers of wealth and income production. This knowledge of information technologies provides infrastructures, systems, and services towards sustainable development.

The **Practices, Progress, and Proficiency in Sustainability (PPPS) Book Series** focuses on the local and global challenges, business opportunities, and societal needs surrounding international collaboration and sustainable development of technology. This series brings together academics, researchers, entrepreneurs, policy makers and government officers aiming to contribute to the progress and proficiency in sustainability.

Coverage

- Socio-Economic
- Innovation Networks
- Strategic Management of IT
- Sustainable Development
- Knowledge clusters
- Green Technology
- Outsourcing
- Global Content and Knowledge Repositories
- ICT and knowledge for development
- Intellectual Capital

IGI Global is currently accepting manuscripts for publication within this series. To submit a proposal for a volume in this series, please contact our Acquisition Editors at Acquisitions@igi-global.com or visit: http://www.igi-global.com/publish/.

The Practice, Progress, and Proficiency in Sustainability (PPPS) Book Series (ISSN 2330-3271) is published by IGI Global, 701 E. Chocolate Avenue, Hershey, PA 17033-1240, USA, www.igi-global.com. This series is composed of titles available for purchase individually; each title is edited to be contextually exclusive from any other title within the series. For pricing and ordering information please visit http://www.igi-global.com/book-series/practice-progress-proficiency-sustainability/73810. Postmaster: Send all address changes to above address. Copyright © 2023 IGI Global. All rights, including translation in other languages reserved by the publisher. No part of this series may be reproduced or used in any form or by any means – graphics, electronic, or mechanical, including photocopying, recording, taping, or information and retrieval systems – without written permission from the publisher, except for non commercial, educational use, including classroom teaching purposes. The views expressed in this series are those of the authors, but not necessarily of IGI Global.

Titles in this Series

For a list of additional titles in this series, please visit: www.igi-global.com/book-series/practice-progress-proficiency-sustainability/73810

Handbook of Research on Socio-Economic Sustainability in the Post-Pandemic Era
Jozef Oleński (High School Technology and Economics, Jaroslaw, Poland) Jeffrey Sachs (Columbia University, USA) Masayuki Susai (Nagasaki University, Japan) Yannis Tsekouras (University of Macedonia, Greece) and Arjan Gjonça (London School of Economics and Political Science, UK)
Information Science Reference • © 2023 • 400pp • H/C (ISBN: 9781799897606) • US $270.00

Handbook of Research on Promoting Sustainable Public Transportation Strategies in Urban Environments
Zafer Yilmaz (TED University, Turkey) Silvia Golem (University of Split, Croatia) and Dorinela Costescu (Polytechnic University of Bucharest, Romania)
Engineering Science Reference • © 2023 • 401pp • H/C (ISBN: 9781668459966) • US $295.00

Energy Transition in the African Economy Post 2050
Olayinka Ohunakin (Covenant University, Nigeria)
Engineering Science Reference • © 2023 • 300pp • H/C (ISBN: 9781799886389) • US $215.00

Implications of Industry 5.0 on Environmental Sustainability
Muhammad Jawad Sajid (Xuzhou University of Technology, China) Syed Abdul Rehman Khan (Xuzhou University of Technology, China) and Zhang Yu (ILMA University, Pakistan)
Business Science Reference • © 2023 • 328pp • H/C (ISBN: 9781668461136) • US $250.00

Climatic and Environmental Significance of Wetlands Case Studies from Eurasia and North Africa
Abdelkrim Ben Salem (Faculty of Sciences, Mohammed V University, Rabat, Morocco) Laila Rhazi (Faculty of Sciences, Mohammed V University, Rabat, Morocco) and Ahmed Karmaoui (Moulay Ismail University of Meknes, Morocco & Moroccan Center for Culture and Science, Morocco)
Engineering Science Reference • © 2023 • 208pp • H/C (ISBN: 9781799892892) • US $195.00

Climate Change, World Consequences, and the Sustainable Development Goals for 2030
Ana Pego (Nova University of Lisbon, Portugal)
Engineering Science Reference • © 2023 • 306pp • H/C (ISBN: 9781668448298) • US $250.00

Positive and Constructive Contributions for Sustainable Development Goals
Cristina Raluca Gh. Popescu (University of Bucharest, Romania & The Bucharest University of Economic Studies, Romania)
Engineering Science Reference • © 2023 • 281pp • H/C (ISBN: 9781668474990) • US $245.00

701 East Chocolate Avenue, Hershey, PA 17033, USA
Tel: 717-533-8845 x100 • Fax: 717-533-8661
E-Mail: cust@igi-global.com • www.igi-global.com

Editorial Advisory Board

Thiago Antonio Beuron, *UNIPAMPA, Brazil*
Daniel Henrique Dario Capitani, *State University of Campinas, Brazil*
Vanderli Correia, *Federal University of Acre, Brazil*
Crisomar Lobo de Souza, *Pontifical Catholic University of São Paulo, Brazil*
Clandia Maffini Gomes, *Federal University of Santa Maria, Brazil*
Muriel de Oliveira Gavira, *State University of Campinas, Brazil*
Izabela Rampasso, *Universidad Católica del Norte, Brazil*
Roberto Schoproni Bichueti, *Federal University of Santa Maria, Brazil*
Tiago Zardin Patias, *Federal University of Santa Maria, Brazil*

List of Contributors

Agarwal, Juhi / *IIT Roorkee, India* 243
Anderluh, Alexandra / *St. Pölten University of Applied Sciences, Austria* 159
Antunes, André / *Escola Superior de Tecnologia de Setúbal, Portugal* 41
Barbieri da Barbieri da Rosa, Luciana Aparecida / *Pontifical Catholic University of Rio de
    Janeiro, Brazil* 74
Borges, Rui Pedro / *Instituto Politécnico de Setúbal, Portugal* 17
Both, Thiago Paulo / *Federal Institute Sul-rio-Grandense, Brazil* 261
Braga Blanco, Gloria / *Universidad de Oviedo, Spain* 328
Caetano, João / *Escola Superior de Tecnologia do Barreiro, Portugal* 41
Cagica Carvalho, Luisa / *Instituto Politécnico de Setúbal, Portugal & CEFAGE, Universidade
    de Évora, Portugal* 116
Calvo-Salvador, Adelina / *Universidad de Cantabria, Spain* 328
Campos, Waleska Yone Yamakawa Zavatti / *Pontifical Catholic University of Rio de Janeiro,
    Brazil* 74
Carriço, Nelson / *Instituto Politécnico de of Setúbal, Portugal* 159
Carriço, Nelson Gaudêncio / *Escola Superior de Tecnologia do Barreiro, Portugal* 41
Channi, Harpreet Kaur / *Chandigarh University, India* 56
Chowdhary, Chiranji Lal / *Vellore Institute of Technology, Vellore, India* 56
Coelho, Dulce Matos / *Polytechnic Institute of Setúbal, Portugal* 183
Cohen, Marcos / *Pontifical Catholic University of Rio de Janeiro, Brazil* 74
De Coninck, Sarah / *UC Leuven-Limburg, Belgium* 159
Dias, Nilmara Braz / *Instituto Politécnico de Setúbal, Portugal* 17
Duarte, Raquel Galamba / *Instituto Politécnico de Setúbal, Portugal* 17
Ferreira, Bruno / *Escola Superior de Tecnologia do Barreiro, Portugal* 41
Fueyo Gutiérrez, Aquilina / *Universidad de Oviedo, Spain* 328
Galatanu, Sergiu Valentin / *Poitehnica University Timisoara, Romania* 159
Gomes, Mónica Filipa Nunes Carvalho / *Instituto Superior de Educação e Ciências, Portugal* ... 215
Kneipp, Jordana Marques / *Federal University of Santa Maria, Brazil* 261
Leščevica, Maira / *Vidzeme University of Applied Sciences, Latvia* 159
Machado Jr., Celso / *Universidade Municipal de São Caetano do Sul, Brazil* 284
Mantovani, Daielly M. N. / *Universidade de São Paulo, Brazil* 284
Mesbahi, Zahra / *St. Pölten University of Applied Sciences, Austria* 159
Nascimento, Thaisa Barcellos Pinheiro / *Universidade de São Paulo, Brazil* 284
Nevado Gil, Teresa / *University of Extremadura, Spain* 116
Nolz, Pamela / *St. Pölten University of Applied Sciences, Austria* 159

Pache Durán, María / *University of Extremadura, Spain* 116
Paula, Fábio de Oliveira / *Pontifical Catholic University of Rio de Janeiro, Brazil* 74
Pereira, Marcelo / *Polytechnic Institute of Setubal, Portugal* 307
Pereira, Raquel / *Institute Polytechnic of Setúbal, Portugal* 183
Pinto, Andreia Raquel / *Institute Polytechnic of Setúbal, Portugal* 183
Pontelli, Greice Eccel / *Federal University of Santa Maria, Brazil* 261
Radványi, Dalma / *Hungarian University of Agriculture and Life Sciences, Hungary* 159
Raichande, Elis Shaida / *Instituto Politécnico de Setúbal, Portugal* 215
Reis, Leonilde / *Instituto Politecnico de Setubal, Portugal* 93, 307
Risbud, Mrudula / *Vishwakarma University, Pune, India* 201
Rodrigues, Leon Maximiliano / *Universidade Estadual do Rio Grande do Sul, Brazil* 215
Rodrigues, Maria Carolina Martins Martins / *Universidade do Algarve, Portugal* 74
Rodríguez-Hoyos, Carlos / *Universidad de Cantabria, Spain* 328
Rodríguez-Maillard, Carlos / *Universidad Cristobal Colón, Mexico* 1
Saini, Damini / *Indian Institute of Management, Raipur, India* 243
Santos, Beatriz Sara / *Instituto Politécnico de of Setúbal, Portugal* 159
Santos, Kleber Rodrigues / *Universidade de São Paulo, Brazil* 284
Sardinha, Boguslawa M. B. / *Instituto Politécnico de Setúbal, Portugal* 116
Semedo, Mirian Benair / *Universidade de Santiago, Cape Verde* 215
Serralha, Maria de Fátima Nunes / *Instituto Politécnico de Setúbal, Portugal* 17, 159
Silva, Jean Marcos da / *Federal Institute Sul-rio-Grandense, Brazil* 261
Silveira, Clara / *Polytechnic of Guarda, Portugal* 93, 307
Sousa, Maria J. / *University Institute of Lisbon, Portugal* 74
Stefanidou, Antonia / *Technical University of Crete, Greece* 135
Teixeira, Cristiano / *Polytechnic of Guarda, Portugal* 93
Vargas-González, Omar C. / *Tecnológico Nacional de México, Ciudad Guzmán, Mexico* 1
Vargas-Hernandez, José G. / *Instituto Tecnológico Mario Molina, Mexico* 1
Waghmare, Rahul Baburao / *Vishwakarma University, Pune, India* 201

Table of Contents

Preface xx

Section 1
Technological Innovations for Sustainability

Chapter 1
Green Technological Innovation and Environmental Regulation 1
    José G. Vargas-Hernandez, Instituto Tecnológico Mario Molina, Mexico
    Carlos Rodríguez-Maillard, Universidad Cristobal Colón, Mexico
    Omar C. Vargas-González, Tecnológico Nacional de México, Ciudad Guzmán, Mexico

Chapter 2
Contribution of Sustainable Fuels for the Future of the Energy Sector 17
    Maria de Fátima Nunes Serralha, Instituto Politécnico de Setúbal, Portugal
    Nilmara Braz Dias, Instituto Politécnico de Setúbal, Portugal
    Raquel Galamba Duarte, Instituto Politécnico de Setúbal, Portugal
    Rui Pedro Borges, Instituto Politécnico de Setúbal, Portugal

Chapter 3
The Challenge of the Digitalization of the Water Sector 41
    Nelson Gaudêncio Carriço, Escola Superior de Tecnologia do Barreiro, Portugal
    Bruno Ferreira, Escola Superior de Tecnologia do Barreiro, Portugal
    André Antunes, Escola Superior de Tecnologia de Setúbal, Portugal
    João Caetano, Escola Superior de Tecnologia do Barreiro, Portugal

Chapter 4
Blockchain-Based IoT E-Healthcare 56
    Harpreet Kaur Channi, Chandigarh University, India
    Chiranji Lal Chowdhary, Vellore Institute of Technology, Vellore, India

Chapter 5
Mapping Research Trends in Eco-Innovation: A Citespace-Based Scientometric Analysis 74
    Waleska Yone Yamakawa Zavatti Campos, Pontifical Catholic University of Rio de Janeiro, Brazil
    Fábio de Oliveira Paula, Pontifical Catholic University of Rio de Janeiro, Brazil
    Maria J. Sousa, University Institute of Lisbon, Portugal
    Luciana Aparecida Barbieri da Barbieri da Rosa, Pontifical Catholic University of Rio de Janeiro, Brazil
    Marcos Cohen, Pontifical Catholic University of Rio de Janeiro, Brazil
    Maria Carolina Martins Martins Rodrigues, Universidade do Algarve, Portugal

Chapter 6
Education for Sustainability: Promoting the Sustainable Development Goals in the Development of Mobile Applications 93
    Clara Silveira, Polytechnic of Guarda, Portugal
    Cristiano Teixeira, Polytechnic of Guarda, Portugal
    Leonilde Reis, Instituto Politecnico de Setubal, Portugal

Section 2
Governance for Sustainability

Chapter 7
Entrepreneurship and Local Government: A Study of the Information Available on Web Pages and Its Evolution Over Time... 116
    Teresa Nevado Gil, University of Extremadura, Spain
    María Pache Durán, University of Extremadura, Spain
    Luisa Cagica Carvalho, Instituto Politécnico de Setúbal, Portugal & CEFAGE, Universidade de Évora, Portugal
    Boguslawa M. B. Sardinha, Instituto Politécnico de Setúbal, Portugal

Chapter 8
The Challenges Cities Face on Their Way Towards Creativity: Indicative Case Studies of Creative Cities 135
    Antonia Stefanidou, Technical University of Crete, Greece

Chapter 9
The Contribution of Urban Domestic Waste Management to the Circular Economy: The Perspective of Six European Countries 159
    Maria de Fátima Nunes Serralha, Instituto Politécnico de Setúbal, Portugal
    Alexandra Anderluh, St. Pölten University of Applied Sciences, Austria
    Beatriz Sara Santos, Instituto Politécnico de of Setúbal, Portugal
    Dalma Radványi, Hungarian University of Agriculture and Life Sciences, Hungary
    Maira Leščevica, Vidzeme University of Applied Sciences, Latvia
    Zahra Mesbahi, St. Pölten University of Applied Sciences, Austria
    Nelson Carriço, Instituto Politécnico de of Setúbal, Portugal
    Pamela Nolz, St. Pölten University of Applied Sciences, Austria
    Sarah De Coninck, UC Leuven-Limburg, Belgium
    Sergiu Valentin Galatanu, Poitehnica University Timisoara, Romania

Section 3
Entrepreneurship and Corporate Innovations

Chapter 10
Teleworking: New Challenges and Trends 183
    Andreia Raquel Pinto, Institute Polytechnic of Setúbal, Portugal
    Dulce Matos Coelho, Polytechnic Institute of Setúbal, Portugal
    Raquel Pereira, Institute Polytechnic of Setúbal, Portugal

Chapter 11
Sustainability Through Innovation: The Case of Indian Startup Thaely 201
    Mrudula Risbud, Vishwakarma University, Pune, India
    Rahul Baburao Waghmare, Vishwakarma University, Pune, India

Chapter 12
Corporate Governance and Ethics for Sustainability: The Case of the Company Mercur S.A. in Brazil 215
    Leon Maximiliano Rodrigues, Universidade Estadual do Rio Grande do Sul, Brazil
    Elis Shaida Raichande, Instituto Politécnico de Setúbal, Portugal
    Mónica Filipa Nunes Carvalho Gomes, Instituto Superior de Educação e Ciências, Portugal
    Mirian Benair Semedo, Universidade de Santiago, Cape Verde

Chapter 13
Symbiosis of Humanistic Leadership, Sustainability, and Circular Economy 243
    Damini Saini, Indian Institute of Management, Raipur, India
    Juhi Agarwal, IIT Roorkee, India

Section 4
Social Engagement and Inclusion

Chapter 14
Challenges and Perspectives of Pinhão Production Considering the Dimensions of Sustainability:
A Study in a City in Southern Brazil 261
    Jean Marcos da Silva, Federal Institute Sul-rio-Grandense, Brazil
    Jordana Marques Kneipp, Federal University of Santa Maria, Brazil
    Thiago Paulo Both, Federal Institute Sul-rio-Grandense, Brazil
    Greice Eccel Pontelli, Federal University of Santa Maria, Brazil

Chapter 15
Engaging People on E-Participation Through Social Media Interactions 284
    Daielly M. N. Mantovani, Universidade de São Paulo, Brazil
    Kleber Rodrigues Santos, Universidade de São Paulo, Brazil
    Thaisa Barcellos Pinheiro Nascimento, Universidade de São Paulo, Brazil
    Celso Machado Jr., Universidade Municipal de São Caetano do Sul, Brazil

Chapter 16
The Social Challenge of Migrant Integration: The Role of Mobile Apps 307
    Leonilde Reis, Polytechnic Institute of Setubal, Portugal
    Marcelo Pereira, Polytechnic Institute of Setubal, Portugal
    Clara Silveira, Polytechnic Institute of Guarda, Portugal

Chapter 17
Youth Civic Engagement: New Strategies for Social Innovation 328
    Carlos Rodríguez-Hoyos, Universidad de Cantabria, Spain
    Adelina Calvo-Salvador, Universidad de Cantabria, Spain
    Aquilina Fueyo Gutiérrez, Universidad de Oviedo, Spain
    Gloria Braga Blanco, Universidad de Oviedo, Spain

Compilation of References 352

About the Contributors 411

Index 417

Detailed Table of Contents

Preface xx

Section 1
Technological Innovations for Sustainability

Chapter 1
Green Technological Innovation and Environmental Regulation 1
    José G. Vargas-Hernandez, Instituto Tecnológico Mario Molina, Mexico
    Carlos Rodríguez-Maillard, Universidad Cristobal Colón, Mexico
    Omar C. Vargas-González, Tecnológico Nacional de México, Ciudad Guzmán, Mexico

This study intends to analyze the impact of environmental regulations and incentives in socio-ecological and green technological innovation. It departs from the assumption that environmental regulation affects the performance of socio-ecological and green technological innovation in organizations. The method employed is the critical analytical and reflective, based on the theoretical, conceptual, and methodological literature. It is concluded that the analysis confirms that there is a direct relationship between the incentives of environmental regulation and the socio-ecological and green technological innovation in organizations.

Chapter 2
Contribution of Sustainable Fuels for the Future of the Energy Sector 17
    Maria de Fátima Nunes Serralha, Instituto Politécnico de Setúbal, Portugal
    Nilmara Braz Dias, Instituto Politécnico de Setúbal, Portugal
    Raquel Galamba Duarte, Instituto Politécnico de Setúbal, Portugal
    Rui Pedro Borges, Instituto Politécnico de Setúbal, Portugal

The main driving forces of the development of alternative energy are growing energy demand combined with the search for energy independence and environmental issues, such as global warming. Throughout this chapter, the sustainability of the currently most used alternative fuels, their characteristics, applications, global consumption, and demand data will be discussed. The different strategies and policies for the adoption of renewable energies also will be discussed. Fuels are compared by their contributions to the development of the circularity of the energy sector, by the feedstock and process efficiency. The advantages, disadvantages, and barriers that each one presents are evaluated to better understand which are the most promising and how their production and consumption can be increased. The aim of this chapter is to present the potential alternative fuels within their applications and analyze their contribution to make the energy sector more circular and sustainable.

Chapter 3
The Challenge of the Digitalization of the Water Sector 41
    Nelson Gaudêncio Carriço, Escola Superior de Tecnologia do Barreiro, Portugal
    Bruno Ferreira, Escola Superior de Tecnologia do Barreiro, Portugal
    André Antunes, Escola Superior de Tecnologia de Setúbal, Portugal
    João Caetano, Escola Superior de Tecnologia do Barreiro, Portugal

The digitalization of the water sector is of utmost importance for improving the efficiency and sustainability of the managed systems. The digitalization process, however, can be seen as a ladder with several steps that the water utility must climb to become a smart utility. The reality is that many water utilities worldwide have not realized yet the benefits of digital transformation and, thus, the digitalization of the water sector lags behind other industries. This chapter presents the major challenges and the promising future that water utilities face in the journey of digitalization. Guidelines on how to choose the most adequate digital solution are also presented, as well as the trends for a smarter water utility.

Chapter 4
Blockchain-Based IoT E-Healthcare 56
    Harpreet Kaur Channi, Chandigarh University, India
    Chiranji Lal Chowdhary, Vellore Institute of Technology, Vellore, India

Numerous industries, including e-healthcare, are capitalizing on and using blockchain and internet of things (IoT) technology. IoT devices may collect patient vitals and other sensory information in real-time, which medical professionals can then examine. All information gathered from the internet of things is stored, processed, and computed in one place. Such concentration raises concerns since it increases the likelihood of a catastrophic failure, distrust, tampering with data, and even the circumvention of privacy protections. By offering decentralized processing and storage for IoT data, blockchain has the potential to address these critical issues. As a result, designing a decentralized IoT-based e-healthcare system that incorporates IoT and blockchain technology might be a viable option. First, the authors provide some context about blockchain in this essay. The viability of blockchain systems for the internet of things-based e-healthcare is then assessed.

Chapter 5
Mapping Research Trends in Eco-Innovation: A Citespace-Based Scientometric Analysis 74
    Waleska Yone Yamakawa Zavatti Campos, Pontifical Catholic University of Rio de Janeiro, Brazil
    Fábio de Oliveira Paula, Pontifical Catholic University of Rio de Janeiro, Brazil
    Maria J. Sousa, University Institute of Lisbon, Portugal
    Luciana Aparecida Barbieri da Barbieri da Rosa, Pontifical Catholic University of Rio de Janeiro, Brazil
    Marcos Cohen, Pontifical Catholic University of Rio de Janeiro, Brazil
    Maria Carolina Martins Martins Rodrigues, Universidade do Algarve, Portugal

Eco-innovation contributes to a more sustainable environment through the development of green technologies. In this context, eco-innovations are capable of improving the performance of organizations in order to reduce environmental and social impacts. Knowing the behavior of studies on eco-innovation can help researchers in understanding the scientific and intellectual structure of the field. Therefore, the general objective of this research is to evaluate the literature on eco-innovation on the world stage from bibliometric and Scientometric analysis, with the help of CiteSpace software. The results show a highly fruitful, dynamic, and rapidly expanding field of studies, marked by a high degree of interdisciplinarity and multidisciplinarity. The conclusions of the research reveal the trends for studies in the area, from the understanding of the behavior of the field of studies in eco-innovation.

Chapter 6
Education for Sustainability: Promoting the Sustainable Development Goals in the Development of Mobile Applications 93
    Clara Silveira, Polytechnic of Guarda, Portugal
    Cristiano Teixeira, Polytechnic of Guarda, Portugal
    Leonilde Reis, Instituto Politecnico de Setubal, Portugal

Information and Communication Technologies enhance human progress, bringing value to people and society. The role of software in society requires a paradigm shift for software development. The Karlskrona Manifesto reflects this change by establishing a focus on sustainability education. The objective of the chapter is to present the development of an Android mobile application inspired by the Sustainable Development Goals to promote sustainability. The methodology adopted used agile development integrated with the Software Engineering Method and Theory - SEMAT approach. SEMAT and agile development are two complementary initiatives, and perfectly aligned, both are structured and non-prescriptive that help to think and improve software development capability. The developed mobile application, Android, thus allows learning more about sustainability by answering questionnaires, thus contributing for the target audience to apply knowledge in environmental and social domains enhancing human progress, bringing value to people and society.

Section 2
Governance for Sustainability

Chapter 7
Entrepreneurship and Local Government: A Study of the Information Available on Web Pages and Its Evolution Over Time 116
    Teresa Nevado Gil, University of Extremadura, Spain
    María Pache Durán, University of Extremadura, Spain
    Luisa Cagica Carvalho, Instituto Politécnico de Setúbal, Portugal & CEFAGE, Universidade de Évora, Portugal
    Boguslawa M. B. Sardinha, Instituto Politécnico de Setúbal, Portugal

Entrepreneurship is one of the main drivers of social development, innovation, global competitiveness, and the economy growth. Because of that, local governments around the world, carry out initiatives focused on the promotion of entrepreneurship in order to support an economic growth and social development. This paper has two objectives. In the first place, the authors analyse the degree and type of information that local governments of the Alentejo region offer to entrepreneurs through their web pages, and calculate the index of information disclosure. Then the authors analyse the evolution of this index between 2015 and 2019. The results show a generalized increase in the information offered using the content analysis technique.

Chapter 8
The Challenges Cities Face on Their Way Towards Creativity: Indicative Case Studies of Creative Cities 135
    Antonia Stefanidou, Technical University of Crete, Greece

The chapter highlights how important it is for modern cities to invest in the field of creative economy, which has dominated the international scene in recent decades. Promoting innovation, with emphasis on competitiveness and cultural diversity, creative economy's sectors are developing dynamically. The chapter aims to define the concept of the city, to present cities' categorization nationally (in Greece), as well as in European and global level. It highlights the relationship between culture and development, creative cities' characteristics, as well as examining the creative sectors, with emphasis to the challenges that cities face in their effort to benefit from the creative economy. For the empirical analysis of the theoretical part, indicative case studies of cities, such as Nantes and Medellin, models of urban revival at European and world level, are used. The challenges faced by these cities and the policies they have implemented, aimed at development, are presented.

Chapter 9
The Contribution of Urban Domestic Waste Management to the Circular Economy: The Perspective of Six European Countries 159
    Maria de Fátima Nunes Serralha, Instituto Politécnico de Setúbal, Portugal
    Alexandra Anderluh, St. Pölten University of Applied Sciences, Austria
    Beatriz Sara Santos, Instituto Politécnico de of Setúbal, Portugal
    Dalma Radványi, Hungarian University of Agriculture and Life Sciences, Hungary
    Maira Leščevica, Vidzeme University of Applied Sciences, Latvia
    Zahra Mesbahi, St. Pölten University of Applied Sciences, Austria
    Nelson Carriço, Instituto Politécnico de of Setúbal, Portugal
    Pamela Nolz, St. Pölten University of Applied Sciences, Austria
    Sarah De Coninck, UC Leuven-Limburg, Belgium
    Sergiu Valentin Galatanu, Poitehnica University Timisoara, Romania

In line with the European community's goal, each EU Member State should recycle at least 60% of municipal waste or prepare them for reuse. In this chapter, the authors intend to show the waste management strategies implemented in six European countries, namely, Austria, Belgium, Hungary, Latvia, Portugal, and Romania. The methodology used was to analyse reports and publications on the management of urban waste and dialogue with some technicians of the municipalities. This knowledge of what is done in each country allows others to learn from the best and most innovative solutions and reflect on the various waste management forms implemented, according to environmental, economic, and social perspectives. The analysis identifies several challenges to bring up in further research and projects, with the contribution of the different countries and the synergies that might be obtained. The authors intend to promote a decrease in consumption and an increase in reuse, separately collected waste and recycling, contributing to circular economic growth and the sustainability of the planet.

Section 3
Entrepreneurship and Corporate Innovations

Chapter 10
Teleworking: New Challenges and Trends 183
    Andreia Raquel Pinto, Institute Polytechnic of Setúbal, Portugal
    Dulce Matos Coelho, Polytechnic Institute of Setúbal, Portugal
    Raquel Pereira, Institute Polytechnic of Setúbal, Portugal

In light of new trends and realities, today's society promotes and values the importance of innovation and sustainability as a solution to the challenges ahead, where companies contribute to the attraction and retention of young talent through Telework. This study addresses Telework as a solution to new challenges. The focus of the study is five small and medium-sized companies. It was concluded that companies opted for Telework at home, with advantages of cost reduction, conciliation of personal and professional life, and reduction of pollutant gases; and as disadvantages, the investment in equipment, social isolation, and precariousness of the labor market. The impacts of telework were verified at the level of productivity, commitment, and performance, where it was concluded that telework can be partially maintained. The contribution of this work focuses on the importance that virtual experiences have in Generation Z and the flexibility of companies in adopting pre-existing work models.

Chapter 11
Sustainability Through Innovation: The Case of Indian Startup Thaely 201
    Mrudula Risbud, Vishwakarma University, Pune, India
    Rahul Baburao Waghmare, Vishwakarma University, Pune, India

The highlighting aspect of the Indian startup ecosystem is the inclusion of startups and SMEs in the various sectors. The Indian startups vary from traditional business sectors to technology-based businesses, and from traditional to social entrepreneurs. A significant number of Indian startups have focused on solving social and environmental problems through creativity and innovation. This chapter discusses one of the social entrepreneurs, Mr. Ashay Bhave, who recently got international recognition for his innovative startup, Thaely. Thaely is involved in manufacturing sneakers from recycling plastic bags with an innovative process. This chapter used secondary data to discuss the case of innovative startup Thaely. The chapter focuses on understanding the problem identification, the innovative process of sneaker manufacturing from plastic bags & bottles, and the social and sustainable impact this startup is making on society.

Chapter 12
Corporate Governance and Ethics for Sustainability: The Case of the Company Mercur S.A. in Brazil.. 215
    Leon Maximiliano Rodrigues, Universidade Estadual do Rio Grande do Sul, Brazil
    Elis Shaida Raichande, Instituto Politécnico de Setúbal, Portugal
    Mónica Filipa Nunes Carvalho Gomes, Instituto Superior de Educação e Ciências, Portugal
    Mirian Benair Semedo, Universidade de Santiago, Cape Verde

This study analyzed, in the form of a case study, the experience of a large company (Mercur S.A) in the implementation of a new vision focused on sustainability. The target company is located in the south of Brazil, has almost a century of existence, and has been managed by the same family since its foundation.

The study is based on the notions and concepts of corporate governance, ethics for sustainability, and social-ecological systems. The study shows that the changes implemented in the company are catalyzed by an initial change of the 'inner change' type, and are disseminated by the company and the community through social and cultural innovations. Important changes in the economic vision and social and environmental responsibility related to the structure/architecture and operation of the company, as well as trade-offs between financial and social and environmental aspects, were identified.

Chapter 13
Symbiosis of Humanistic Leadership, Sustainability, and Circular Economy 243
    Damini Saini, Indian Institute of Management, Raipur, India
    Juhi Agarwal, IIT Roorkee, India

Corporate sustainability is presumed to be a business strategy that creates long-term value by focusing on innovative measures aimed towards the natural environment and aligning it with its external environment. Continuous measures at global, institutional, and individual level have to be taken up to ensure sustainability paving the way to bring concepts like circular economy, sustainable practices into limelight among the scholars, academicians, and even corporate houses. This chapter tries to establish a link between a leader's role in an organization and creating spaces for circular economy and sustainability in their already existing cycles. On the basis of the steps taken for creating a sustainable business by TATA rganization, the chapter explores their leadership style and suggests that the values such as responsibility, humanity, and empathy in leadership became more important and goes well with the vision of circular economy and how well it has been dwelled up with the business models of TATA.

Section 4
Social Engagement and Inclusion

Chapter 14
Challenges and Perspectives of Pinhão Production Considering the Dimensions of Sustainability:
A Study in a City in Southern Brazil 261
    Jean Marcos da Silva, Federal Institute Sul-rio-Grandense, Brazil
    Jordana Marques Kneipp, Federal University of Santa Maria, Brazil
    Thiago Paulo Both, Federal Institute Sul-rio-Grandense, Brazil
    Greice Eccel Pontelli, Federal University of Santa Maria, Brazil

The activity of collecting non-timber forest products (NTFPs) is a secular activity in Brazil, more specifically the sertão [Brazilian backlands]. Many of these activities have ceased to exist over the years, although pinhão (Araucaria angustifolia seeds) production still persists as an income generator. Given this context, this study sought to answer the following question: "What are the challenges and perspectives of pinhão production considering the context of the community of Barro Preto in the city of Arvorezinha (Rio Grande do Sul State, southern Brazil) from the dimensions of sustainability?" Using thematic analysis, this interpretative qualitative study employed conversational interviews by Boje and Rosile (2021). In mapping the challenges and perspectives of the extractive activity of pinhão, these findings showed that inserting actors from the base of the pinhão productive chain to induce them to tell their stories is not enough to build a narrative that contemplates all the dimensions of sustainability.

Chapter 15
Engaging People on E-Participation Through Social Media Interactions 284
　Daielly M. N. Mantovani, Universidade de São Paulo, Brazil
　Kleber Rodrigues Santos, Universidade de São Paulo, Brazil
　Thaisa Barcellos Pinheiro Nascimento, Universidade de São Paulo, Brazil
　Celso Machado Jr., Universidade Municipal de São Caetano do Sul, Brazil

The chapter aims to analyze how social media engages citizens in issues related to municipal management in Brazilian capital cities (27 cities). For that, Twitter data was collected, and descriptive analysis, text mining, and social network analysis were carried out. Results show the most frequent interactions regarded sharing posts, replies, and reactions were less frequent. Text mining suggested behavior on Twitter is related on the hot news, so discussions tend to be superficial; network analysis showed mayor accounts have more connections with users than the cities' official accounts, which suggests a necessity for personification on the conversation. Interactions are both centralized (started by the city) and decentralized (start by the citizen), but consist merely of information transmission and opinion sharing, and more complex kinds of participation, such as co-creation and decision-making were not observed. These findings show the potential of social media communication for public management and give insights on how to develop a successful policy to participate in social media.

Chapter 16
The Social Challenge of Migrant Integration: The Role of Mobile Apps 307
　Leonilde Reis, Polytechnic Institute of Setubal, Portugal
　Marcelo Pereira, Polytechnic Institute of Setubal, Portugal
　Clara Silveira, Polytechnic Institute of Guarda, Portugal

Information and communication technologies can be a driving factor towards the resolution of social challenges and assertively contribute to the development of people and regions. The objective of this chapter is to present a mobile application designed to contribute to the inclusion of Migrants in the city of Viseu in Portugal. The methodology used to support the study was a design science research of creating the artifact to involve the various stakeholders in the process. The main results emphasize the relevance in organizational context of the use of information and communication technologies to bridge the gaps underlying the various social challenges and to develop innovative solutions so that the access to information is global and ubiquitous. In this sense, it is considered that the IntegraBrasil mobile application contributes to the well-being and integration of migrants, given the difficulties they face when they arrive in Portugal.

Chapter 17
Youth Civic Engagement: New Strategies for Social Innovation 328
　Carlos Rodríguez-Hoyos, Universidad de Cantabria, Spain
　Adelina Calvo-Salvador, Universidad de Cantabria, Spain
　Aquilina Fueyo Gutiérrez, Universidad de Oviedo, Spain
　Gloria Braga Blanco, Universidad de Oviedo, Spain

This chapter analyses the scientific bibliography on the civic engagement of young people published in English between 2014 and 2019. Sixty-nine articles published in international scientific journals were analyzed. The main objective of this meta-research is to understand what problems have been investigated and the data collection techniques used. The authors found that the most researched problem was the analysis and understanding of the impact that participation has on plans and programs designed by diverse socio-educational actors with regard to the civic engagement of young people. Based on the meta-research carried out, they propose three socially innovative intervention strategies aimed at improving youth civic engagement. Each strategy is designed to enhance young people's agency in order to increase their levels of social participation.

Compilation of References 352

About the Contributors 411

Index 417

Preface

Considering deeply dynamic and challenging contexts, innovation as a driver of sustainability could provide recommendations and solutions to achieve a better world. Sustainability-driven innovations such as new technologies or good governance practises are critical aspects of modern innovation, as they must consider natural capital, social inclusion, and development beyond economic growth.

Handbook of Research on Solving Societal Challenges Through Sustainability-Oriented Innovation emphasizes the importance of sustainability-driven innovations as a possible solution for the societal challenges. This book covers many different sustainability dimensions such as entrepreneurship, technological and green innovations, social innovation, circular economy, digitalization, blockchain, governance, education, and corporate social responsibility.

This reference source is ideal for industry and government leaders and officials, policymakers, researchers, scholars, entrepreneurs, practitioners, instructors, and students.

The book presents seventeen chapters organized in four sections: "Technological Innovations for Sustainability," "Governance for Sustainability," "Entrepreneurship and Corporate Innovations," and "Social Engagement and Inclusion." Despite the chapters may be somehow interrelated it is considered that this structure will render the book's consultation more effective considering the main related dimensions.

This book has an interdisciplinary nature appealing to lectures, academic management and regional development scholars, students, and practitioners. Contributions received from international authors focusing on a more practical approach for managerial education, at different levels of education and highlighting its importance for local communities and companies, provided a broader geographic overview.

SECTION 1: TECHNOLOGICAL INNOVATIONS FOR SUSTAINABILITY

The first chapter, "Green Technological Innovation and Environmental Regulation," intends to analyze the impact of environmental regulation and incentives in socio-ecological and green technological innovation.

The second chapter, "Contribution of Sustainable Fuels for the Future of the Energy Sector," discusses sustainability of the currently most used alternative fuels, aiming to present the potential alternative fuels within their applications and analyse their contribution to make the energy sector more circular and sustainable.

The third chapter, "The Challenge of the Digitalization of the Water Sector," presents the major challenges and the promising future that water utilities face in the journey of digitalization. Guidelines on

Preface

how to choose the most adequate digital solution are also presented, as well as the trends for a smarter water utility.

The fourth chapter, "Blockchain-Based IoT E-Healthcare," provides some context about blockchain systems in the context of data security in the healthcare industry, assessing its viability for the Internet of Things-based e-healthcare.

The fifth chapter, "Mapping Research Trends in Eco-Innovation: A Citespace-Based Scientometric Analysis," evaluates the literature on eco-innovation on the world stage from bibliometric and Scientometric analysis, with results showing a highly fruitful, dynamic, and rapidly expanding field of studies, marked by a high degree of interdisciplinarity and multidisciplinarity.

The sixth chapter, "Education for Sustainability: Promoting the Sustainable Development Goals in the Development of Mobile Applications," presents the development of an Android mobile application, based on SEMAT approach, which allows the target to learn more about sustainability and apply knowledge in environmental and social domains enhancing human progress.

SECTION 2: GOVERNANCE FOR SUSTAINABILITY

The seventh chapter, "Entrepreneurship and Local Government: A Study of the Information Available on the Web Pages and Its Evolution Over Time," considering the importance of local government support on entrepreneurship activities, analyses the degree and type of information that local governments of the Alentejo region offer to entrepreneurs through their web pages, and calculate the index of information disclosure, analysing its evolution between 2015 and 2019.

The eighth chapter, "The Challenges Cities Face on Their Way Towards Creativity: Indicative Case Studies of Creative Cities," aims to define the concept of the city, to present cities' categorization nationally (in Greece), as well as in European and global level. It highlights the relationship between culture and development, creative cities' characteristics, as well as it examines the creative sectors, with emphasis to the challenges that cities face in their effort to benefit from the creative economy.

The ninth chapter, "The Contribution of Urban Domestic Waste Management to the Circular Economy: The Perspective of Six European Countries," intends to show the waste management strategies implemented in six European countries, namely, Austria, Belgium, Hungary, Latvia, Portugal, and Romania.

SECTION 3: ENTREPRENEURSHIP AND CORPORATE INNOVATIONS

The tenth chapter, "Teleworking: New Challenges and Trends," addresses telework as a solution to new challenges related to attraction and retention of young talent but also bringing environmental, social, and economic advantages for both companies and employees.

The eleventh chapter, "Sustainability Through Innovation: The Case of Indian Startup Thaely," discusses one of the social entrepreneurs, Mr. Ashay Bhave, who recently got international recognition for his innovative startup, Thaely. Thaely is involved in manufacturing sneakers from recycling plastic bags with an innovative process.

The twelfth chapter, "Corporate Governance and Ethics for Sustainability: The Case of the Company Mercur S.A. in Brazil," analyzed, in the form of a case study, the experience of a large company (Mercur S.A.) in the implementation of a new vision focused on sustainability. The target company is located in the south of Brazil, has almost a century of existence, and has been managed by the same family since its foundation.

The thirteenth chapter, "Symbiosis of Humanistic Leadership, Sustainability, and Circular Economy," tries to establish a link between a leader's role in an organization and creating spaces for circular economy and sustainability in their already existing cycles. On the basis of the steps taken for creating a sustainable business by TATA organisation, the chapter explores their leadership style and suggest that the values such as responsibility, humanity and empathy in leadership became more important and goes well with the vision of circular economy and how well it has been dwelled up with the business models of TATA.

SECTION 4: SOCIAL ENGAGEMENT AND INCLUSION

The fourteenth chapter, "Challenges and Perspectives of Pinhão Production Considering the Dimensions of Sustainability: A Study in a City in Southern Brazil," sought to answer the following question: "what are the challenges and perspectives of the activity of collecting pinhão considering the context of the community of Barro Preto in Arvorezinha (Rio Grande do Sul State, southern Brazil) from the dimensions of sustainability?". The community of Barro Preto was chosen as it is inserted in the biome formed by the Atlantic Forest, a vegetation with a predominance of pinhão.

The fifteenth chapter, "Engaging People on E-Participation Through Social Media Interactions," aims to analyze how social media engages citizens in issues related to municipal management in Brazilian capital cities (27 cities). For that, Twitter data was collected and descriptive analysis, text mining and social network analysis were carried out.

The sixteenth chapter, "The Social Challenge of Migrant Integration: The Role of Mobile Apps," presents a mobile application designed to contribute to the inclusion of Migrants in the city of Viseu in Portugal.

The seventeenth chapter, "Youth Civic Engagement: New Strategies for Social Innovation," analyzes the scientific bibliography on the civic engagement of young people published in English between 2014 and 2019. 69 articles published in international scientific journals were analysed. The main objective of this meta-research is to understand what problems have been investigated and the data collection techniques used.

Preface

To conclude, we would like to thank the authors whose collaboration has made this project possible and express our hope that readers will find this publication inspiring and useful.

Luisa Cagica Carvalho
Instituto Politécnico de Setúbal, Portugal & CEFAGE, University of Evora, Portugal

Paulo Bogas
Instituto Politécnico de Setúbal, Portugal

Jordana Kneipp
Federal University of Santa Maria, Brazil

Lucas Avila
Federal University of Santa Maria, Brazil

Elis Ossmane
Universidade Aberta, Portugal

Section 1
Technological Innovations for Sustainability

Chapter 1
Green Technological Innovation and Environmental Regulation

José G. Vargas-Hernandez
Instituto Tecnológico Mario Molina, Mexico

Carlos Rodríguez-Maillard
https://orcid.org/0000-0003-2406-196X
Universidad Cristobal Colón, Mexico

Omar C. Vargas-González
https://orcid.org/0000-0002-6089-956X
Tecnológico Nacional de México, Ciudad Guzmán, Mexico

ABSTRACT

This study intends to analyze the impact of environmental regulations and incentives in socio-ecological and green technological innovation. It departs from the assumption that environmental regulation affects the performance of socio-ecological and green technological innovation in organizations. The method employed is the critical analytical and reflective, based on the theoretical, conceptual, and methodological literature. It is concluded that the analysis confirms that there is a direct relationship between the incentives of environmental regulation and the socio-ecological and green technological innovation in organizations.

INTRODUCTION

The rapid spread at global scope of innovative technologies has modernized the economic sectors that are contributors of environmental pollution and other health risk hazards with limiting findings in technological innovations and ecological solutions. The green transformation of the manufacturing industry is crucial, and environmental regulation and technological innovation may play key roles (Zhao et al., 2021)

In recent years, climate change and environmental sustainability have become some of the most pressing global economic issues (Chen et al., 2022), but the contradiction between economic and ecol-

DOI: 10.4018/978-1-6684-6123-5.ch001

ogy has become increasingly prominent (Liu et al., 2020), The research in population socio-ecology is accompanied by criticisms and organizational research regarding the abandonment of organizations (Amburgey & Rao, 1996). Research can be advanced at the intersection between sustainability and population socio-ecology (Salimath, & Jones, 2011).

In other hand, when the environmental regulation system is weak, firms tend to emphasize maximization of profits, environmental taxes and engage in treatments to expand production scale and balance the regulation cost and with the economic recession boosting the propose of Industry 4.0 and circular economy, innovation is generally accepted as the solution to the contradiction among environmental protection, pollution prevention, resource recovery, and economic growth (Zhou & Du, 2022). That why is important to identify the relationship between environmental regulation and green innovation.

This analysis on the impact of environmental regulation and incentives in socio-ecological and green technological innovation assumes that environmental regulation and market financing are important factors affecting enterprise green technological innovation (Wang et al., 2022) and affects the performance of socio-ecological in organizations. The study presents some implications of the organizational socio-ecology to develop a framework for the analysis followed by the organizational green technological innovation leading to determine the impact of the environmental regulation policies and incentives. Finally, some conclusions are presented based on the analysis.

The theoretical framework of organizational socio-ecology and sociological theory is a tool to formulate and implement socioecological principles to organizational strategies for turning around low-performance organizations, inducing the emergence of an economic and social environment and with an impact on non-linear development. The theory of organizational socio-ecology is based on a social Darwinism of organizational populations The organizational socio-ecology theory analyzes the alterations, difficulties, and restrictions of the organizational populations in application conditions (DiMaggio and Powell, 1983).

The organizational socio-ecology theory assumes that there is convergence between the paradigm of organizational socio-ecology and sociological perspectives leading to organizational research. The fundamental assumption of organizational socio-ecology is the differentiation of organizational populations. Organizational socio-ecology is linked to other areas with the incorporation of methodological innovations such as strategic simulation models. The socio-ecology of routines at the organizational level, and the notion organizational character (Birnholtz, Cohen & Hoch, 2007) is defined as the ability to regenerate a coherent socio-ecology of patterns of action. The theory of population socio-ecology is a tool for the theoretical and empirical analysis of organizational phenomena.

The research focuses on the socio-ecology of populations in coordination and convergence with organizational sustainability considering that there are several areas related to the sustainability of populations that potentially contribute to population sustainability (Salimath, & Jones, 2011). The variables gas emissions, energy consumption, green technology innovation, institutional quality, and economic growth and urbanization are all interdependent and cointegrated.

Finally, the research for green innovation focuses on product innovation and environmental management systems, the development of technological systems adopted by organizations, and the adoption of innovative technologies dealing with the integration of environmentally friendly practices. Environmental science and technology and strong normative and regulatory institutions support the transformation towards green innovation of firms.

LITERATURE OVERVIEW

Organizational Socio-Ecology

Although the term ecology is not used Kremser, Pentland, & Brunswicker (2017) they theorize the interdependence traced from the dynamics of social phenomena while sets of practices (Nicolini & Monteiro, 2017). The dynamic interplay of ecological and social systems leads to a socio-ecological systems (Berkes et al., 2003; Folke et al., 2002; Gunderson & Holling, 2002).

Organizational socio-ecology has implications for organizational theories. The emergence of models that value resources such as the resource-based theory, the stakeholder theory and theory of organizational socio-ecology that value external influences. Firms are restricted by internal factors such as capital and technology and external factors such as inclusiveness, participants, and bureaucratic factors in the context of environmental regulation policies (Hu et al., 2020; Chen et al., 2021). Strict environmental policies lead small firms to be inclined to reduce expenditures to enhance pollution control rather than green technologies development. Organizational socio-ecology theory studies organizational size, sustaining that small organizations have more problems to capitalize, so they are prone to dissolve.

Organizations compete for resources and resource positioning in organizational socio-ecology. The niche width theory is a category of organizational socio-ecology based on the survival opportunities of environmental changes of organizations (Hannan & Freeman, 1977). Ecological contingencies imposed by the environment require the fulfilment of components of organizational socio-ecology.

Organizational socio-ecology analyzes the characteristics of the vital manifestations of organizational populations, such as disappearance considering individual situations such as their size and age that are associated with survival (Aldrich & Auster, 1986). Population socio-ecology analyzes the sample as the entire population over time periods. Organizational socio-ecology as a social demand on its organizations that exceed internal limits and affect the influence of human beings on the climate and natural environment. The internal social responsibility of organizations is introduced in the fields of organizational socio-ecology with the application of socio-ecological and environmental measures (Oller Alonso, 2021).

Building an organizational socio-ecology to learn as a management model of professional development must be based more on guiding than on control (Le Boterf, 1999). The organizational socio-ecology framework uses its knowledge to formulate and implement turn around organizational strategies.

The practice of sustainability is related in convergence with the socio-ecology of the population. Population socio-ecology is a tool for the analysis of macro-organizational phenomena that traces the use in theoretical and empirical research (Salimath, & Jones, 2011). Population socio-ecology supports organizational sustainability and explores the convergence that links the theory and practice of the organizational socio-ecology of populations. Population socio-ecology has areas of convergence with organizational sustainability for the analysis of population sustainability (Salimath, & Jones, 2011).

The industrial structure has effects on regional economic growth and ecological welfare (Zhao et al., 2020; Huang et al., 2021). Technological innovation infrastructure system improves the innovation efficiency by protecting the property rights. For rapid economic growth, most developing countries are dependent on non-renewable energy consumption, leading to significant CO_2 releases with adverse effects on environmental quality.

Improving the well-being via development increases output accompanied by the substantial consumption of fossil fuel with harmful influence on the environment. Energy consumption releasing more carbon is the major contributor to environmental footprint. The organizations prefer to scale up production rather than use high-tech equipment exacerbating environmental pollution and weakening motivation (Bian et al., 2021). Many studies are used to evaluate environmental performance (Zhou et al., 2008; Sueyoshi & Goto 2012; Wu et al., 2019). The pollution emissions characteristics leading to the environmental performance affect the green development efficiency.

Organizational Green Technological Innovation

Organizations focus on technological innovation advances to implement green innovation characteristics that may affect the adoption and use (Sneideriene & Rugine, 2019; Kousar, Sabri, Zafar, & Akhtar, 2017; Weng & Lin, 2011; Lin & Ho, 2011). The government encourages organizations to adopt and use green technology innovation with technical assistance and provides financial support and training for adoption.

Organizations adopt technological innovation before use with elevated levels of environmental uncertainty that affects green technology innovation (Zailani, Iranmanesh, Nikbin, & Jumadi, 2014). The adoption of green technology innovation reduces negative organizational impacts, organizations may be able to try to adopt the green technology innovation several times before the final adoption (Baines, Brown, Benedettini, & Ball, 2012).

The green innovation of firms is related to the life cycles. Green organizational innovation adoption leads to offer ecological and reliable product and services environmentally friendly at competitive prices delivered on time (Thatte, 2007). Some inconsistencies may lie on the measures of green innovation performance, such as using granted-patents and pollutant reduction (Amore & Bennedsen, 2016; Fujii & Managi 2016). Triability is the organizational innovation characteristic to adopt any trial of innovation before the use and implementation because the degree of uncertainty (Rogers, 2003).

Research finds relative advantage in the green innovation adoption The relative advantage is the perception of the characteristic of organizational technology innovation in terms of price, quality, satisfaction, use and life span over its costs. Perception of organizational relative advantage is positively related to green innovation. If the organizational perception of relative advantage of innovation is greater than the cost and better that the current technology, the organization may adopt the innovation. (Lin & Ho, 2011; Rogers, 2003; Kousar et al., 2017; Grandon & Pearson, 2004).

Investments on research and development, human capital, and gross domestic product per capita have effects on the development of green innovation. Investments in research and development enhances the firm technological innovation capabilities. Green investment is the ratio between fiscal expenditure on energy conservation and environmental protection and total fiscal expenditure. Environmental protection is a social responsibility of governments (Li et al., 2015; Jiang et al., 2020; Wang & Jiang, 2021). Regional economic development is a macroeconomic operation related to the investment level of innovation (Grossman & Krueger, 1995). Local governments may invest and encourage firms to develop innovative activities through technology spillovers from foreign investments and competition incentives.

Some of the concepts related to Green Innovation are:

- Complexity of innovation is the degree of organizational perception of technology innovation as being difficult to use. If the complexity of innovation increases, the adoption decrease. The level of complexity increases with the difficulties of innovation diffusion and knowledge sharing. The perception of organizational complexity is negatively related to green innovation. Organizations tends to adopt technology innovation when innovation is diffused, and knowledge shared. Innovation requires skills and efforts increasing the level of complexity while the level of adoption decreases (Kousar et al., 2017; Etzion, 2007; Rogers, 2003; Tornatzky & Klein, 1982; Deng & Ji, 2015; Bradford & Florin, 2003). There is a lack of green technological innovations in developing countries compared to developed nations (Li, Pan, Kim, Linn, & Chiang, 2015). A systematic review by Taklo, Tooranloo, & Shahabaldini (2020) of multiple industrial sectors and regions on the emergence of green technologically innovative advances for enhancing technical systems and enrichment of the customer experience.

- Technological innovation is adopted after observing the impacts (Shahrul & Normah, 2015; Tan & Eze, 2008). Use of innovation that is visible to potential adoption reflects the result of the characteristic observability (Rogers, 2003). The perception of observability is positively related to green innovation. Observability is a significant characteristic of technological innovation because enables organizations to identify the benefits of innovation with high possibilities to be adopted by organizations (Hatimtai & Hassan, 2018; Rogers, 2003).

Green innovation is categorized in green management, green process, and green product (Chen, Shih, Shyur, & Wu, 2012; Chen & Hsu, 2009). Green innovation capabilities, green product innovation, green process innovation, and governance technology capabilities must have to be implemented based on the green technology innovation capabilities of firms (Wang et al., 2019). Organizational technological innovations compatible with capabilities of the organization have positive effects on the environment (Etzion, 2007). The perception of organizational compatibility is positively related to green technological innovation.

The innovation of green process is related to updates in systems, methods and processes and process innovation is aimed to produce eco-friendly goods and services to meet the ecological concerns and environmental targets (Antonioli, Mancinelli, & Mazzanti, 2013). The indicators of a green process innovation capabilities are the strengthening transformation and enhancing protection design processes, pollution elimination and green process review. Actions of government environmental regulation has effects on the investments of pollution elimination. Pollution control benefits are greater than those of reduction of production promoting firms to invest in green innovation activities

The green technology innovation capabilities of firms are related to the elements of green product innovation, process innovation and technology governance. Organizations must intensify green technology innovation to meet the technical specifications and standards of green emissions set by governments, improve the technology governance capabilities through wastes and the innovation capabilities of green products, processes, and develop green technology innovation (Li, 2019; Wang et al., 2018; Saud et al., 2019).

From the perspective of motivation, green innovation should be separated from technological innovation ability. Organizational behaviors are classified based on their motivations as symbolic strategic environmental actions for cooperation and substantive environmental actions, or substantive strategies and cooperation (Delmas & Montes Sancho, 2010; Neumann, 2021; Truong et al., 2021).

Environmental Regulation Policies and Incentives

The mechanism of environmental regulation policy has influence on the motivation of green innovation firms to differentiate between the green innovation behaviors in regulation and regulated objects of government enterprise. Government environmental support increases the adoption of green innovation (Lin & Ho, 2010). Government environmental support refers to the systems that take the form of environmental regulations, policies, standards, and interventions that tend to reduce the environmental effect and encourage firms to engage in environmental technology innovation (Eiadat, Kelly, Roche, & Eyadat, 2008). Government environmental support motivates firms to adopt environmental methods of production (Gadenne, Kennedy, & McKeiver, 2009).

Mechanisms of environmental regulations policies affect organizational green innovation in different green innovations behaviors. State-owned enterprises assume responsibilities than non-state-owned do not assume in implementing the green innovation development under constraints of environmental regulations (Zhou et al., 2017; Bai et al., 2019). State-owned enterprises improve after the implementing of environmental regulations since constrained environmental regulations policy, firms respond faster to policies on green innovation due to the responsibility and use of scarce resources.

Environmental regulation policies cause small organizations to enhance pollution control mora than implementing green technological innovation. Growing firms have access to abundant funds for R&D investments in green innovation to improve environment-friendly production and green technology (Cao & Cao, 2019). Environmental policy formulation must consider the specific measures for pollution control capacities of the organization.

Policy on environmental regulations considering the value of carbon trading, it has a strong incentive effect in improving firm green innovation. Openness has no significant impact on green innovation because there is an indirect effect between openness and green innovation openness. That technological innovation in organizations reduces carbon emissions and enhances environmental quality (Dauda et al., 2021; Ko et al., 2021). Green innovation significantly lessens energy consumption and CO_2 gas emissions.

Carbon emission trading market must be liberalized positioning the emissions of firms in price mechanisms to stimulate green technological innovation activities. Carbon finance is the ratio between carbon dioxide emissions and gross domestic product. A higher level of CO_2 emissions and environmental pollution as the causal links requires the establishment of effective institutional quality to green innovation. The causality link is established between the energy consumption and the institutional quality. Thus, there is a relationship between green innovation and institutional quality.

Research has found causal relationships among the variables are determined from CO_2 emissions to economic growth, CO_2 emissions to green innovation and institutional quality, urbanization and energy consumption to economic growth and institutional quality. From institutional quality to green innovation. Research also has found a bidirectional link between economic growth and gas emissions (Onofrei, Vatamanu, & Cigu, 2022).

The influence of environmental regulations on green technological innovation of growing and mature firms achieves the highest impact while the green technological innovation of startup firms has severe financial constraints, struggle to meet capital needs, face elevated level of risk-taking leading to achieve the lowest impact on investments in green innovation technologies (Shahzad et al., 2019).

The mechanism between capital misallocation, green development, and technological innovation leads to more efficient socio-ecology environment. Incorporating capital allocation into technological innovation in an analytical framework promotes regional green development efficiency. Regarding heterogeneity, the effect of capital allocation is significant. Capital allocation has a positive relationship with technological innovation enhancing the input of the firms with an impact on green development efficiency. Capital allocation enhances green development efficiency leading to reducing the level of technological innovation. The impact of capital allocation on technological innovation is positive (Schmidt et al., 2020; Gao et al., 2019).

Government intervention on capital misallocated may lead to deterioration of the environmental. Prices distortion of capital market, capital allocation cannot be optimized (Zhang & Lu., 2017). Firms have difficulties in obtaining the capital required for technological innovation of the firm leading to weaken the motivation to conduct research and development (Gao et al., 2019). If the price of capital controlled by government, firms tend to rent-seek to obtain low-cost factors crowing research and development funds and reducing innovation efficiency (Murphy et al., 1993). The intervention of external factors may destroy the price of capital mechanism lowering costs of production factors for polluting industries.

Technological innovation increases the use of clean processes reducing the environmental pollution (Hua et al., 2018). Technological innovation stimulates firms to use efficient resources, adopt equipment and promote the upgrading of industries (Sun et al., 2020b). Technological innovation tends to facilitate green economy development having a mediating role on capital allocation on green economic efficiency while capital allocation improves the green development efficiency. Technological innovation has a mediating role with capital allocation and enhances green development efficiency.

Green finance addressing the financial gap is a relevant financial instrument that contributes to green innovation development of firms (Luo et al., 2021). Green finance motivates firms to invest in green technological innovation and is the driving force for green transformation of polluted firms through the engagement in green innovation activities. Environmental regulation improves the effects of green finance in green innovation. Environmental regulations strengthen the impact of green innovation projects and the effects of green finance.

Firms tend to adopt conservative strategies to invest in green technological projects with low-risk and intense environmental regulations (Boubakri et al. 2013; Chiu and Lee 2020). Green finance and strict environmental regulations have positive effects on green innovation. The moderation effects of environmental regulation advance the effects of green finance on green innovation. Environmental regulations at low levels leads to increasing of innovation performance as green finance performance increases.

An emphasis on green innovation outcomes by the firms requires more efficient green finance in producing green innovation. Green innovation output can be achieved by urging firms to use green finance and investments and avoiding time-consuming projects. Environmental regulation improves the effects of green finance in green innovation. Green finances are contingent to the implementation of environmental regulations and can improve with more intense regulations.

Large firms with more available resources tend to improve the quality of green technological innovation. Large firms have more mature development of organizational systems achieving high innovation management (Yu et al., 2022b). Large firms count on access to abundant innovation resources such as capital, R&D, patents, technological talents (Gupeng & Xiangdong, 2012; Hu et al., 2017). The green patent application is a measure that has effects of green finance in driving green technological innovation. The use of green patent application to measure innovative behaviors is not comprehensive for the different motivations on green innovation choices.

Large organizations promote green technological innovation even if they are constrained by environmental regulations and small organizations that are more restricted by capital, technology, talent, etc., have more difficulties in promoting green innovation. The implementation of an intense environmental regulations improves green innovation considering constraints may not lead to transformation of technologically innovative production. Stringent environmental regulations tend to strength the motivation to pursue technological innovation.

Green innovation as dependent variable can be measured by the number of patent applications associated to energy intensity (Wurlod & Noailly, 2018). Green patent applications indicate the advance of green innovation (Tolliver et al., 2021) and they are leading to innovation activities without institutional preferences and time-lag, used as indicators to evaluate green innovation (Fujii & Managi 2016; Cho & Sohn, 2018; Fujii 2016).

Environmental regulation has a positive impact on green innovation patents and green utility model patents, although from the perspective of influence, the impact of environmental regulation on green innovation may decrease if firms have not eliminated strategic green innovation. Green technology innovation strategies must be formulated and implemented with sensitivity considering cultural differences that have an impact on innovation processes (Westwood & Low, 2003).

Environmental regulations influence government subsidies to eliminate compliance costs improving green innovation with minimal impact on green innovation behaviors. Government subsidies on green innovation must be classified in the diverse types to determine the effect on the choices of innovative behaviors with different motivations. Regulatory capture and government subsidies are moderating variables. Government subsidies supplying funds provide firms to conduct green innovation aligned to the government goals of technological innovation (Boeing, 2016; Bai et al., 2019). Government subsidies promote green innovation; however, regulatory capture has a negative role in the regulatory process of environmental regulation promoting green innovation. Increasing green subsidies lead to Organization to have more resources and funds to invest in green technology development enhancing green product, process innovation and technology governance capabilities (Saud et al., 2019; Lin et al., 2020).

Environmental regulations tend to influence green innovation of growing and mature firms at the highest level, while the startup stage has the lowest impact due to the available resources in capital, technology, and talent.

METHOD

To carry out this bibliometric review, electronic, literary, and scientific sources were used as the basis for recent research on environmental regulation, Technological innovation, and its impact on Socioecological model. The study was conducted by searching in the main scientific search engines. The consulted studies have been in English and most of them are design in Chinese universities in the recent years.

RESULTS

Environmental regulation has a moderating effect on environmental uncertainty and enterprises' green technological innovation (Chen et al., 2022), Although environmental regulation helps to control and solve the problem of environmental pollution, it is bound to increase enterprise production costs and squeeze profit space, thus hindering technological innovation and green transformation (Zhao et al., 2021), The socio-ecology of the organizational population has implications for sustainability. Institutional reforms are essential to reduce ecological emissions and achieve climate change goals.

Figure 1.

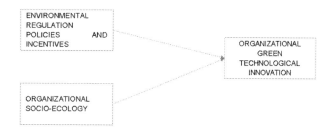

CONCLUSION

In general terms, this analysis concludes that there is a direct relationship between the incentives of environmental regulation with the Organizational Socio-ecological and green technological innovation in organizations. There is still much to be defined, we have some research primarily in China that will help us to have more data to support these impacts.

Firms realize high quality green innovation behavior to promote green technology to obtain a competitive advantage. Green innovation firms are higher during the growth and maturity periods when they improve green technological innovation and develop a keen sense of social responsibility for more active innovation activities. The role of technological innovation needs to be strengthened to improve green development efficiency. The organization must implement and scrutinize green product innovation capabilities implementation, green process innovation capabilities, and technology governance capabilities.

Environmental regulation affects the firm´s green innovation through the mechanisms of innovation compensation and compliance costs. The scientific design of environmental regulation policy tools must be strengthened to monitor the emission reduction and conduct green innovation. Environmental regulation intensity must be enhanced according to the specific situations of firms. Firms that bear excessive environmental costs must be inclined to regulatory capture rather than improve green technological environmental protection which may destroy the incentive mechanism with extended time and higher costs.

Governments in policymaking and stakeholders must facilitate organizational green strategies to invest in technological innovation projects aimed to reduce environmental pollution. Policymaking need to strengthen institutional quality and enact more effective policies to address the environmental challenges. The different impacts of institutional environments must take the research model. Institutional environment is relevant in factor allocation and environmental performance.

Differences in ownership, life cycle, size, etc., have a significant impact on the environmental regulation green innovation incentives. Environmental regulations and green finance have effects on green technological innovation. The green finance levels on green innovation have heterogeneous impacts with a moderating effect on environmental regulations. Green finance, green innovation and environmental regulations tend to be stable with fixed effects. Environmental regulation is a moderator is used to examine the effects. Regional green innovation can be measured by the level of contribution of green finance and the green patent application. Governments must support with fiscal incentives and financial subsidies the development of innovative firms encouraging more flows of capital into the green technology innovations and clean processes. Research findings suggest that subsidies do not explain the gap between the green technological innovation behaviors and

Future research in applying population socio-ecology principles and assumptions to examine organizational sustainability.

REFERENCES

Aldrich, H., & Auster, E. R. (1986). Even dwarfs started small: Liabilities of age and size and their strategic implications. In B. M. Staw & L. L. Cummings (Eds.), *Research in Organizational Behavior* (Vol. 8, pp. 165–198). JAI Press.

Amburgey, T., & Rao, H. (1996). Organizational ecology: Past, present, and future directions. *Academy of Management Journal*, *39*(5), 1265–1286. doi:10.2307/256999

Amore, M. D., & Bennedsen, M. (2016). Corporate governance and green innovation. *Journal of Environmental Economics and Management*, *75*, 54–72. doi:10.1016/j.jeem.2015.11.003

Antonioli, D., Mancinelli, S., & Mazzanti, M. (2013). Is environmental innovation embedded within high- performance organizational changes? The role of human resource management and complementarity in green business strategies. *Research Policy*, *42*(4), 975–988. doi:10.1016/j.respol.2012.12.005

Bai, Y., Song, S., Jiao, J., & Yang, R. (2019). The impacts of government R&D subsidies on green innovation: Evidence from Chinese energy-intensive firms. *Journal of Cleaner.*, *233*(9), 819–829. doi:10.1016/j.jclepro.2019.06.107

Baines, T., Brown, S., Benedettini, O., & Ball, P. (2012). Examining green production and its role within the competitive strategy of manufacturers. *Journal of Industrial Engineering and Management*, *5*(1), 53–87. doi:10.3926/jiem.405

Becker, M. C. (Ed.). (2008). Handbook of organizational routines. Cheltenham, UK: Edward Elgar Publishing Berkes, F., Folke, C., Colding, J. (eds.) (2003). Navigating Social-Ecological Systems: Building Resilience for Complexity and Change, Cambridge University Press. doi:10.4337/9781848442702

Bian, Y., Wu, L., Bai, J., & Yang, Y. (2021). Does factor market distortions inhibit green economic growth? *World Econ Papers*, *2*, 105–119.

Birnholtz, J. P., Cohen, M. D., & Hoch, S. V. (2007). Organizational character: On the regeneration of Camp Poplar Grove. *Organization Science*, *18*(2), 315–332. doi:10.1287/orsc.1070.0248

Boeing, P. (2016). The allocation and effectiveness of China's R&D subsidies—Evidence from listed firms. *Research Policy*, *45*(9), 1774–1789. doi:10.1016/j.respol.2016.05.007

Boubakri, N., Cosset, J. C., & Saffar, W. (2013). The role of state and foreign owners in corporate risk-taking: Evidence from privatization. *Journal of Financial Economics*, *108*(3), 641–658. doi:10.1016/j.jfineco.2012.12.007

Bradford, M., & Florin, J. (2003). Examining the role of innovation diffusion factors on the implementation success of enterprise resource planning systems. *International Journal of Accounting Information Systems*, *4*(3), 205–225. doi:10.1016/S1467-0895(03)00026-5

Cao, X., Deng, M., Song, F., Zhong, S., & Zhu, J. (2019). Direct and moderating effects of environmental regulation intensity on enterprise technological innovation: The case of China. *PLoS One*, *14*(10), e0223175. doi:10.1371/journal.pone.0223175 PMID:31589643

Chen, C. C., Shih, H. S., Shyur, H. J., & Wu, K. S. (2012). A business strategy selection of green supply chain management via an analytic network process. *Computers & Mathematics with Applications (Oxford, England)*, *64*(8), 2544–2557. doi:10.1016/j.camwa.2012.06.013

Chen, H. L., & Hsu, W. T. (2009). Family ownership, board independence, and R&D investment. *Family Business Review*, *22*(4), 347–362. doi:10.1177/0894486509341062

Chen, J., Wang, X., Shen, W., Tan, Y., Matac, L. M., & Samad, S. (2022). Environmental uncertainty, environmental regulation and enterprises' green technological innovation. *International Journal of Environmental Research and Public Health*, *19*(16), 9781. doi:10.3390/ijerph19169781 PMID:36011417

Chen, Z., Zhang, X., & Chen, F. (2021). Do carbon emission trading schemes stimulate green innovation in enterprises? Evidence from China. *Technological Forecasting and Social Change*, *168*, 120744. doi:10.1016/j.techfore.2021.120744

Chiu, Y. B., & Lee, C. C. (2020). Effects of financial development on energy consumption: The role of country risks. *Energy Econ*, *90*, 104833. doi:10.1016/j.eneco.2020.104833

Cho, J. H., & Sohn, S. Y. (2018). A novel decomposition analysis of green patent applications for the evaluation of R&D efforts to reduce CO2 emissions from fossil fuel energy consumption. *Journal of Cleaner Production*, *193*, 290–299. doi:10.1016/j.jclepro.2018.05.060

Dauda, L., Long, X., Mensah, C. N., Salman, M., Boamah, K. B., Ampon-Wireko, S., & Dogbe, C. S. K. (2021). Innovation, trade openness and CO2 emissions in selected countries in Africa. *Journal of Cleaner Production*, *281*, 125143. doi:10.1016/j.jclepro.2020.125143

Delmas, M. A., & Montes-Sancho, M. J. (2010). Voluntary agreements to improve environmental quality: Symbolic and substantive cooperation. *Strategic Management Journal*, *31*(6), 575–601.

Deng, Q., & Ji, S. (2015). Organizational green IT adoption: *Concept and evidence. Sustainability*, *7*(12), 16737–16755. doi:10.3390u71215843

DiMaggio, P., & Powell, W. W. (1983). The iron cage revisited: Collective rationality and institutional isomorphism in organizational fields. *American Sociological Review*, *48*(2), 147–160. doi:10.2307/2095101

Eiadat, Y., Kelly, A., Roche, F., & Eyadat, H. (2008). Green and competitive? An empirical test of the mediating role of environmental innovation strategy. *Journal of World Business*, *43*(2), 131–145. doi:. jwb.2007.11.012 doi:10.1016/j

Etzion, D. (2007). Research on organizations and the natural environment, 1992-present: A review. *Journal of Management*, *33*(4), 637–664. doi:10.1177/0149206307302553

Folke, C., Carpenter, S., Elmqvist, T., Gunderson, L., Holling, C. S., & Walker, B. (2002). Resilience and Sustainable Development: Building Adaptive Capacity in a World of Transformations. *Ambio*, *31*(5), 437–440. doi:10.1579/0044-7447-31.5.437 PMID:12374053

Fujii, H. (2016). Decomposition analysis of green chemical technology inventions from 1971 to 2010 in Japan. *Journal of Cleaner Production, 112,* 4835–4843. doi:10.1016/j.jclepro.2015.07.123

Fujii, H., & Managi, S. (2016). Research and development strategy for environmental technology in Japan: A comparative study of the private and public sectors. *Technological Forecasting and Social Change, 112,* 293–302. doi:10.1016/j.techfore.2016.02.012

Gadenne, D. L., Kennedy, J., & McKeiver, C. (2009). An empirical study of environmental awareness and practices in SMEs. *Journal of Business Ethics, 84*(1), 45–63. doi:10.100710551-008-9672-9

Gao, X., Lyu, Y., Shi, F., Zeng, J., & Liu, C. (2019). The impact of financial fac-tor market distortion on green innovation efficiency of high-tech industry. *Ekoloji, 28*(107), 3449–3461.

Grossman, G. M., & Krueger, A. B. (1995). Economic growth and the environment. *The Quarterly Journal of Economics, 110*(2), 353–377. doi:10.2307/2118443

Gunderson, L. H., & Holling, C. S. (Eds.). (2002). Panarchy; Understanding Transformations in Human and Natural Systems, Island, Washington, DC.

Gupeng, Z., & Xiangdong, C. (2012). The value of invention patents in China: Country origin and technology field differences. *China Economic Review, 23*(2), 357–370. doi:10.1016/j.chieco.2012.02.002

Hannan, M. T., & Freeman, J. (1989). *Organizational Ecology*. Harvard University Press. doi:10.4159/9780674038288

Hannan, M. T., & Freeman, J. H. (1977). The Population Ecology of Organizations. *American Journal of Sociology, 82*(5), 929–963. doi:10.1086/226424

Hatimtai, M. H., & Hassan, H. (2018). The relationship between the characteristics of innovation towards the effectiveness of ICT in Malaysia productivity corporation. *Malaysian Journal of Communication, 34*(1), 253–269. doi:10.17576/JKMJC-2018-3401-15

Hu, A. G. Z., Zhang, P., & Zhao, L. (2017). China as number one? Evidence from China's most recent patenting surge. *Journal of Development Economics, 124,* 107–119. doi:10.1016/j.jdeveco.2016.09.004

Hua, Y., Xie, R., & Su, Y. (2018). Fiscal spending and air pollution in Chinese cities: Identifying composition and technique effects. *China Economic Review, 47,* 156–169. doi:10.1016/j.chieco.2017.09.007

Huang, M., Ding, R., & Xin, C. (2021). Impact of technological innovation and industrial-structure upgrades on ecological efficiency in China in terms of spatial spillover and the threshold effect. *Integrated Environmental Assessment and Management, 17*(4), 852–865. doi:10.1002/ieam.4381 PMID:33325155

Jiang, Z., Wang, Z., and Lan, X. (2021). How Environmental Regulations Affect Corporate Innovation? the Coupling Mechanism of Mandatory Rules and Voluntary Management. *Technol. Soc., 65,* 101575. doi:. 2021.101575 doi:10.1016/j.techsoc

Ko, Y.-C., Zigan, K., & Liu, Y.-L. (2021). Carbon capture and storage in South Africa: A technological innovation system with a political economy focus. *Technological Forecasting and Social Change, 166,* 120633. . techfore. 2021. 120633 doi:10.1016/j

Kousar, S., Sabri, P. S. U., Zafar, M., & Akhtar, A. (2017). Technological factors and adoption of green innovation- moderating role of government intervention: A case of SMEs in Pakistan. *Pakistan Journal of Commerce and Social Sciences, 11*(3), 833–861.

Kremser, W., Pentland, B. T., & Brunswicker, S. (2017). *The Continuous Transformation of Interdependence in Networks of Routines*. Presented at European Group for Organizational.

Li, J., Pan, S. Y., Kim, H., Linn, J. H., & Chiang, P. C. (2015). Building green supply chains in eco-industrial parks towards a green economy: Barriers and strategies. *Journal of Environmental Management, 162*, 158–170. doi:10.1016/j.jenvman.2015.07.030 PMID:26241931

Li, N. B. (2019). Environmental Regulation and Corporate Green Technology Innovation - a Conditional Process Analysis. *Inn. Mong. Soc. Sci. Chin. Ed, 40*(6), 109–115. doi:10.14137/j.cnki.issn1003-5281.2019.06.016

Li, S., Song, X., & Wu, H. (2015). Political connection, ownership structure, and corporate philanthropy in China: A strategic-political perspective. *Journal of Business Ethics, 129*(2), 399–411. doi:10.100710551-014-2167-y

Li, Z., Liao, G., Wang, Z., & Huang, Z. (2018). Green loan and subsidy for promoting clean production innovation. *Journal of Cleaner Production, 187*, 421–431. doi:10.1016/j.jclepro.2018.03.066

Lin, C. Y., & Ho, Y. H. (2010). The influences of environmental uncertainty on corporate green behavior: An empirical study with small and medium-size enterprises. *Social Behavior and Personality, 38*(5), 691–696. doi:10.2224bp.2010.38.5.691

Lin, Y. X., Sha, K. C., & Wang, J. (2020). Practical Experience and Inspiration of Foreign Emission Permit System. *Environ. Impact. Assess., 42* (1), 14–16.

Liu, J., Zhao, M., & Wang, Y. (2020). Impacts of government subsidies and environmental regulations on green process innovation: A nonlinear approach. *Technology in Society, 63*, 101417. doi:10.1016/j.techsoc.2020.101417

Liu, Y., Wang, A., & Wu, Y. (2021). Environmental regulation and green innovation: Evidence from China's new environmental protection law. *Journal of Cleaner Production, 297*, 126698. doi:10.1016/j.jclepro.2021.126698

Liu, Y., Zhu, J., Li, E., Meng, Z., & Song, Y. (2020). Environmental regulation, green technological innovation, and ecoefficiency: The case of Yangtze river economic belt in China. *Technological Forecasting and Social Change, 155*, 1–21. doi:10.1016/j.techfore.2020.119993

Luo, Y., Salman, M., & Lu, Z. (2021). Heterogeneous impacts of environmental regulations and foreign direct investment on green innovation across different regions in China. *The Science of the Total Environment, 759*, 143744. doi:10.1016/j.scitotenv.2020.143744 PMID:33341514

Murphy, K. M., Shleifer, A., & Vishny, R. W. (1993). Why is rent-seeking so costly to growth? *The American Economic Review, 83*(2), 409–414. https://www.jstor.org/stable/2117699

Neumann, T. (2021). Does it pay for new firms to be green? An empirical analysis of when and how different greening strategies affect the performance of new firms. *Journal of Cleaner Production*, *317*, 128403. doi:10.1016/j.jclepro.2021.128403

Nicolini, D., & Monteiro, P. (2017). The practice approach: For a praxeology of organizational and management studies. In A. Langley & H. Tsoukas (Eds.), *The Sage Handbook of Process Organization Studies*. Sage.

Oller Alonso, M. (2021). *La Responsabilidad Social Corporativa de las empresas del mármol en la comarca del Almanzora, Almería (2019-2021): análisis de sus estrategias de comunicación integral.* [*The Corporate Social Responsibility of marble companies in the Almanzora region, Almería (2019-2021): analysis of their integral communication strategies.*] [Doctorate, Universidad de Murcia Escuela Internacional de Doctorado].

Onofrei, M., Vatamanu, A. F., & Cigu, E. (2022). The Relationship Between Economic Growth and CO2 Emissions in EU Countries: A Cointegration Analysis. *Frontiers in Environmental Science*, *10*, 934885. doi:10.3389/fenvs.2022.934885

Rogers, E. M. (2003). Elements of diffusion. *Diffusion of Innovations*, *5*, 1–38.

Salimath, M., & Jones, R. III. (2011). Population ecology theory: Implications for sustainability. *Management Decision*, *49*(6), 874–910. doi:10.1108/00251741111143595

Saud, S., Chen, S., Haseeb, A., Khan, K., & Imran, M. (2019). The Nexus between Financial Development, Income Level, and Environment in Central and Eastern European Countries: A Perspective on Belt and Road Initiative. *Environmental Science and Pollution Research International*, *26*(16), 16053–16075. doi:10.100711356-019-05004-5 PMID:30968296

SchmidtC.SchneiderY.SteffenS.StreitzD. (2020). Capital misal-location and innovation. doi:10.2139/ssrn.3489801

Shahrul, N. S., & Normah, M. (2015). Digital version newspaper: Implication towards printed newspaper circulation in Malaysia. *Malaysian Journal of Communication*, *31*(2), 687–701.

Shahzad, F., Lu, J., & Fareed, Z. (2019). Does firm life cycle impact corporate risk taking and performance? *Journal of Multinational Financial Management*, *51*, 23–44. doi:10.1016/j.mulfin.2019.05.001

Snaideriene, A., & Rugine, H. (2019). Theoretical approach on the green technologies development. *Regional Formation and Development Studies*, *2*(28), 124–134.

Studies: EGOS Le Boterf, G. (1999). *L'ingénierie des compétences*. París: Éditions d'Organisation.

Sueyoshi, T., & Goto, M. (2012). Efficiency-based rank assessment for electric power industry: A combined use of data envelopment analysis (DEA) and DEA-discriminant analysis (DA). *Energy Economics*, *34*(3), 634–644. doi:10.1016/j.eneco.2011.04.001

Sun, X., Zhou, X., Chen, Z., & Yang, Y. (2020b). Environmental efficiency of electric power industry, market segmentation and technologi-cal innovation: Empirical evidence from China. *The Science of the Total Environment*, *706*, 135749. doi:10.1016/j.scitotenv.2019.135749 PMID:31940733

Taklo, S. K., & Tooranloo, H. S. & Shahabaldini parizi, Z. (. (2020). Green Innovation: A Systematic Literature Review. *Journal of Cleaner Production, 2020*(7). Advance online publication. doi:10.1016/j.jclepro.2020.122474

Tan, K. S., & Eze, U. C. (2008). An empirical study of internet-based ICT adoption among Malaysian SMEs. *Communications of the IBIMA, 1*, 1–12.

Thatte, A. A. (2007). *Competitive advantage of a firm through supply chain responsiveness and SCM practices.* [Doctoral dissertation, The University of Toledo].

Tolliver, C., Fujii, H., Keeley, A. R., & Managi, S. (2021). Green innovation and finance in Asia. *Asian Economic Policy Review, 16*(1), 67–87. doi:10.1111/aepr.12320

Tornatzky, L. G., & Klein, K. J. (1982). Innovation characteristics and innovation adoption-implementation: A meta-analysis of findings. *IEEE Transactions on Engineering Management, 29*(1), 28–45. doi:10.1109/TEM.1982.6447463

Truong, Y., Mazloomi, H., & Berrone, P. (2021). Understanding the impact of symbolic and substantive environmental actions on organizational reputation. *Industrial Marketing Management, 92*, 307–320. doi:10.1016/j.indmarman.2020.05.006

Wang, F. Z., Jiang, T., & Guo, X. C. (2018). Government Quality, Environmental Regulation and Corporate Green Technology Innovation. *Sci. Res. Manage. 39* (1), 26–33. CNKI:SUN:KYGL.0.2018-01-004.

Wang, K., & Jiang, W. (2021). State ownership and green innovation in China: The con-tingent roles of environmental and organizational factors. *Journal of Cleaner Production, 314*, 128029. doi:10.1016/j.jclepro.2021.128029

Wang, Y., Sun, X., & Guo, X. (2019). Environmental regulation and green productivity growth: Empirical evidence on the Porter Hypothesis from OECD industrial sectors. *Energy Policy, 132*, 611–619. doi:10.1016/j.enpol.2019.06.016

Weng, M. H., & Lin, C. Y. (2011). Determinants of green innovation adoption for small and medium-size enterprises (SMES). *African Journal of Business Management, 5*(22), 9154–9163.

Westwood, R., & Low, D. R. (2003). The multicultural muse: Culture, creativity, and innovation. *International Journal of Cross Cultural Management, 3*(2), 235–259. doi:10.1177/14705958030032006

Wu, H. Y., Tsai, A., & Wu, H. S. (2019). A hybrid multi-criteria decision analysis approach for environmental performance evaluation: an example of the tft-lcd manufacturers in taiwan. [EEMJ]. *Environmental Engineering and Management Journal, 18*(3), 597–616. doi:10.30638/eemj.2019.056

Wurlod, J. D., & Noailly, J. (2018). The impact of green innovation on energy intensity: An empirical analysis for 14 industrial sectors in OECD countries. *Energy Econ, 71*, 47–61. doi:10.1016/j.eneco.2017.12.012

Yu, Z., Khan, S. A. R., Ponce, P., & Jabbour, A. B. L. (2022). Factors affecting carbon emissions in emerging economies in the context of a green recovery: Implications for sustainable development goals. *Technological Forecasting and Social Change, 176*, 121417. doi:10.1016/j.techfore.2021.121417

Zailani, S., Iranmanesh, M., Nikbin, D., & Jumadi, H. B. (2014). Determinants and environmental outcome of green technology innovation adoption in the transportation industry in Malaysia. *Asian Journal of Technology Innovation*, *22*(2), 286–301. doi:10.1080/19761597.2014.973167

Zhang, D. G., & Lu, Y. Q. (2017). Impact of market segmentation on energy efficiency. *China Popul Resour Environ*, *27*(1), 65–72.

Zhao, X., Ding, X., & Li, L. (2021). Research on Environmental Regulation, Technological Innovation and Green Transformation of Manufacturing Industry in the Yangtze River Economic Belt. *Sustainability*, *13*(18), 10005. doi:10.3390u131810005

Zhao, X., Shang, Y., & Song, M. (2020). Industrial structure distortion and urban ecological efficiency from the perspective of green entre-preneurial ecosystems. *Socio-Economic Planning Sciences*, *72*, 100757. doi:10.1016/j.seps.2019.100757

Zhou, K. Z., Gao, G. Y., & Zhao, H. (2017). State Ownership and firm innovation in China: An integrated view of institutional and efficiency logics. *Administrative Science Quarterly*, *62*(2), 375–404. doi:10.1177/0001839216674457

Zhou, P., Ang, B. W., & Poh, K. L. (2008). Measuring environmental performance under different environmental DEA technologies. *Energy Economics*, *30*(1), 1–14. doi:10.1016/j.eneco.2006.05.001

Zhou, X., & Du, J. (2021). Does environmental regulation induce improved financial development for green technological innovation in China? *Journal of Environmental Management*, *300*, 113685. doi:10.1016/j.jenvman.2021.113685 PMID:34517232

Chapter 2
Contribution of Sustainable Fuels for the Future of the Energy Sector

Maria de Fátima Nunes Serralha
Instituto Politécnico de Setúbal, Portugal

Nilmara Braz Dias
Instituto Politécnico de Setúbal, Portugal

Raquel Galamba Duarte
Instituto Politécnico de Setúbal, Portugal

Rui Pedro Borges
Instituto Politécnico de Setúbal, Portugal

ABSTRACT

The main driving forces of the development of alternative energy are growing energy demand combined with the search for energy independence and environmental issues, such as global warming. Throughout this chapter, the sustainability of the currently most used alternative fuels, their characteristics, applications, global consumption, and demand data will be discussed. The different strategies and policies for the adoption of renewable energies also will be discussed. Fuels are compared by their contributions to the development of the circularity of the energy sector, by the feedstock and process efficiency. The advantages, disadvantages, and barriers that each one presents are evaluated to better understand which are the most promising and how their production and consumption can be increased. The aim of this chapter is to present the potential alternative fuels within their applications and analyze their contribution to make the energy sector more circular and sustainable.

INTRODUCTION

The negative impact of anthropogenic CO_2 emissions has been established beyond any doubt. It is now clear that the energy consumption patterns of humanity will have to change very quickly and on a global scale if the worst effects of climate change are to be avoided.

The still growing energy demand, driven by the expansion of consumer societies, has been mostly satisfied by fossil sources, such as coal, oil, and natural gas. Fossil fuels are used intensively since the dawn of the industrial era and are the main sources of anthropogenic CO2 emissions, which has led to a search for suitable renewable energy alternatives. It is important to stress that there are other forces behind the drive for renewable energy besides climate change, such as geopolitical tensions, scarcity of resources or citizens awareness. Also, the search for renewable energy sources and technologies is not the sole area of activity in the fight against climate change. Ecosystem preservation, sustainable agricultural practices, or waste valorization to comply with circular economy objectives, are just a few examples.

The energy policy of countries in general, including Portugal, is based on security of energy supply, sustainability, and competitiveness. Thus, to produce renewable energy all those factors must be considered. Renewable energy, besides its environmental advantages, works towards a diversification of primary resources, the constitution of strategic reserves and the decreased of dependence on imports (Gulbenkian, 2022).

In 2015, the Paris Agreement (PA) defined that paradigm changes in society are necessary to contain the effects of climate change. The main goal of the Paris Agreement is to limit global warming to 2 or preferably to 1.5 degrees Celsius, compared to pre-industrial levels. The European Union (EU) and all its member states have signed and ratified the Paris Agreement and are committed to its implementation through some strategies as the reduction the greenhouse emissions. Thus, EU countries have agreed to set the EU on the path to becoming climate-neutral economy and society by 2050, through the decarbonization of the energy sector (UNFCCC- United Nation Climate Change, 2022).

The path for a carbon neutral economy is based on the electrification of energy use (with electricity increasingly coming from renewable energy sources) and the use of sustainable fuels in those sectors that are not prone to electrification (such as aviation or some branches of industry).

Finding suitable substitutes of fossil fuels that can be used in existing infrastructure and equipment and are free of CO2 emissions is the main aim of the search for sustainable renewable alternative fuels. Sustainable fuels will be a central factor for the decarbonization of the energy sector and compliance with the objectives of the Paris Agreement. They can simultaneously be used for power generation and as an energy carrier where the utilization of new technologies and energy sources is required and will have a more important role in the future as they are likely to be utilized on a greater scale (UNFCCC- United Nation Climate Change, 2022). The need for new types of fuels to substitute fossil fuels is particularly important in the aviation sector due to the greater difficulty of electrification of the sector. Currently there are a variety of alternative fuels the mostly used are Green Hydrogen, Ammonia, biodiesel, bioethanol, synthetic fuels (Stancin, Mikuleié, Wang, & Duié, 2020).

It is necessary to clarify the distinction between renewable, sustainable, and alternative fuel. In accordance with European Commission, renewable fuels are fuels produced from renewable energy sources. Renewable energy sources are non-fossil sources that are naturally replenished on a human timescale (wind, solar, geothermal, wave, tidal, hydropower, biomass). A fuel can be defined as renewable only when it is based on renewable sources. To be sustainable, a renewable fuel should not increase the concentration of CO2 in the atmosphere (Al-Breiki & Bicer, 2021) as well as other pollutants (European Commission, 2019b). Sustainable fuels are produced from renewable and/or alternative feedstocks, such as plant and vegetable waste, which are sustainably sourced. Alternative or non-conventional fuels are fuels such as electricity, hydrogen, biofuels, natural gas, synthetic fuels, ammonia, with the potential

Figure 1. The three pillars of sustainability aspect of the renewable fuels
Source: adapted from (IEA, 2021; Guimarães, 2019; Hasenheit, Gerdes, Kiresiewa, & Beekman, 2016)

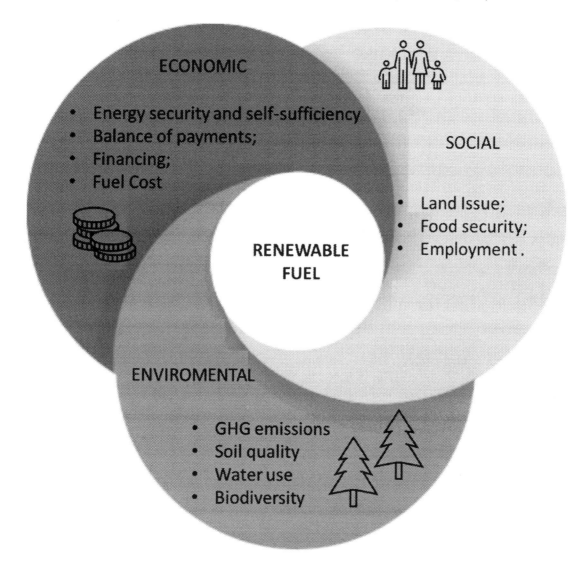

for long-term replacement of petroleum fuels regardless of the energy and feedstock used and possible fuel combinations. (Directives, 2014)

Developing fuels such as biodiesel, bioethanol and green hydrogen considering all the three pillars of sustainability (environment, economic and social) plus addressing properly the security of supply policies is a challenge that is globally increasing the interests of scientists and researchers. Figure 1 (adapted from (IEA, 2021; Guimarães, 2019; Hasenheit, Gerdes, Kiresiewa, & Beekman, 2016) represents the main features of the three pillars of sustainability of renewable fuels.

The expansion of the renewable energy matrix stimulates market added value in manufactured products and sustainable energy commodities with the decrease of fossil fuels dependence and contributes to the sustainable development goals (SDG) mainly SDG7, 12 and 13.

This work aims to make a bibliographic review on sustainable fuels showing their forms of production and by establishing the comparative advantages and disadvantages of each fuel, to analyse their contribution to make the energy sector more circular and sustainable.

RENEWABLE AND SUSTAINABLE: FUEL CONTRIBUTIONS

Green Hydrogen

Hydrogen gas (in the form of molecular hydrogen, H2) plays a significant industrial role in both energy and non-energy applications. Worldwide hydrogen demand in 2020 was approximately 90 Mt (IEA, 2021), corresponding to 10,80 EJ of energy (considering a net calorific value of 119,96 MJ/kg). The main current applications of hydrogen are in the production of ammonia for the fertilizer industry and as feedstock for hydro-cracking units in oil refineries (IEA, 2021). Its energy applications are also wide, from direct combustion to electricity production using fuel cells. The fact that it is possible to extract energy from the hydrogen molecule without producing CO2 is making this compound one of the centerpieces of the energy transition (IEA, 2021).

Most of the hydrogen consumed worldwide is produced from fossil fuels, with hydrogen being obtained mainly through steam methane reforming and coal gasification (IEA, 2019). These techniques all involve the emission of CO2 and in general using hydrogen produced from unabated fossil fuels (so called grey hydrogen) as an alternative to the fossil fuels themselves, offers very limited environmental benefits and can even lead to higher global emissions in most applications. If we consider electricity production in a fuel cell with an efficiency of 60% where hydrogen is obtained from steam methane reforming followed by a water gas shift reaction, the overall emissions are 0,65 kgCO2/kWh. For electricity produced in a CCGT plant with an efficiency of 50%, the emissions are 0,39 kgCO2/kWh.

The threat posed by global warming and the expectation that carbon prices will increase significantly in the future has led to a search for ways of decreasing the carbon emissions associated with energy products.

In the case of hydrogen there is a growing interest in production methods based on low carbon techniques. Low carbon techniques can still be based on fossil fuels, but emissions are reduced by carbon capture, storage, and utilization (CCUS) (Pahija, Golshan, Blais, & Boffito, 2022). The hydrogen obtained from such processes is called "blue hydrogen". CCUS is seen as one the main technologies for reducing emissions while maintaining the operation of existing infrastructures and methods. In 2021, there were six facilities worldwide where hydrogen was produced from methane reforming coupled to CCUS facilities (IEA, 2021). Blue hydrogen, although not yet produced on a mass scale is considered the most cost-effective alternative to grey hydrogen (IRENA, 2019).

The other low carbon option is "green hydrogen" that is obtained from renewable energy sources. Among these options there is water electrolysers powered by renewable electricity (usually solar PV or wind). There have been some significant advances in electrolysis technology and today alkaline electrolysis and proton exchange membrane electrolysis are commercially available for large scale applications (JRC Publication Repository, 2021). These types of electrolysers can be powered by dedicated renewable technologies, having no access to grid electricity and being subject to the intermittence of renewable production for their operation. Alternatively, they may be connected to the electricity grid, but in that case the hydrogen produced can only be labelled "green hydrogen" if the renewable origin of the electricity is certified through instruments such as the renewable electricity certificates used in the USA (EPA, s.d.) or the guarantees of origin used in the EU (Eu Directives, 2018). The Eu has also started issuing guarantees of origin for green or low carbon hydrogen (FCH, s.d.)

Production of hydrogen from electrolysis requires very large amounts of water that can be obtained from the treated wastewater or desalinated sea water (Fonseca, Carmargo, Commenge, Falk, & Gil, 2019), particularly in areas where fresh water is a scarce resource.

One of the advantages of connecting electrolysers to the grid is that they can use excess renewable electricity production and act as efficient energy storage systems. Hydrogen could be converted back onto electricity at times of lower renewable production although the overall conversion efficiency of the process electricity -> hydrogen -> electricity is rather low, of the order of 40%. Another option is the power-to-gas (P2G) path, where the hydrogen produced from excess renewable electricity could be used in other sectors such as mobility (in fuel cell vehicles), injection in natural gas networks or other industrial non-energy uses. This type of approach would promote sector coupling among different parts of the energy system and could provide a robust mechanism for reducing emissions across the economy (Frischmuth & Härtel,, 2022; Sorknaes, et al., 2022; Emonts, et al., 2019).

Blending in natural gas networks could allow hydrogen to be distributed to a variety of end users such as households and industries, resulting in reduced emissions in all combustion applications. The fraction of blending strongly depends on the technical characteristics of each network (Bard, et al., 2022; Eames, Austin, & Wojcik, 2022), related to issues as pipeline material or compression stations. Furthermore, it should be noticed that the energy content by unit of mass (the net calorific value is 119,96 MJ/kg) of hydrogen is approximately three times larger than that of natural gas, but its density is nine times smaller (Ma, et al., 2019), and therefore, the blending of hydrogen in natural gas networks effectively decreases the energy content of the gas. This limits the fraction of blending and might require some adaptation from end users.

Green hydrogen could also be used to produce a variety of fuels such as synthetic fuels produced from recycled carbon monoxide using a Fischer-Tropsch process. This is a gas-to-liquid conversion technology that transforms syngas (a mixture of carbon monoxide and hydrogen in variable fractions) in hydrocarbon chains that could substitute diesel and petrol (Okolie, et al., 2021). Other liquid fuels of organic origin, such as hydrogenated vegetable oil (HVO), could be made totally renewable by using green hydrogen in their composition.

Ammonia

Ammonia is too an entirely carbon-free chemical compound widely used as a fertilizer, which recently gained significant attention as a potential energy carrier or alternative fuel and can be produced from renewable energy, such as green hydrogen (Valera-Medina, Xiao, Owen-Jones, David, & Bowen, 2018).

However, ammonia, being the second most produced chemical product worldwide, with an annual production of 200 million tons per year, relies mainly on fossil fuels for its production (about 2% of world´s fossil fuel energy) and generates 420 million tons of CO_2 per year (over 1% pf global energy related emissions). At present, more than 80% of ammonia production is used as fertilizer and only 5% is used in other chemical commodities. Little ammonia production is used as energy carrier (Giddey, Badwal, Minnings, & Dolan, 2017).

Energy used in industrial and residential sectors is a relevant part of the total anergy accounting for more than 50% of total energy consumption, and ammonia can have an active role in the decarbonization of both sectors. (Valera-Medina, et al., 2021)

Ammonia can be used as power generation, from systems that produce low power energy to the ones that can potentially move transoceanic vessels and provide power to national grids. Also, due to its carbon free combustion reaction, it can be use as fuel in combustion systems, such as furnaces/boilers (industrial sector) and internal combustion engines (automotive application, marine sector and auxiliary power), gas turbines, and fuel cells, as direct feedstock or, due to easy cracking to produce H2, to feedstock fuel cells or other hydrogen applications (Osamah Siddiqui, 2018; Giddey et al., 2017; Aziz, 2017; Valera-Medina et al., 2018).

However, due to some of its physical and chemical properties, ammonia must be used (Shiazawa, 2020) and handled with some care. Ammonia can cause corrosion (Yapicioglu & Dincer, 2019) however, is not particularly corrosive to ferrous material used in storage and distribution equipment's (Valera-Medina, et al., 2021). NOX emissions also need to be taken in consideration when ammonia combustion is considered for high temperatures (Valera-Medina, et al., 2021; Stancin, Mikuleié, Wang, & Duié, 2020; Yapicioglu & Dincer, 2019). Another issue with ammonia use is its toxicity (Stancin, Mikuleié, Wang, & Duié, 2020) but, since a wide network of production, storage and transportation already exists, its use is highly regulated by health, safety, security transport and storage and environment guidelines (Valera-Medina, et al., 2021)

One of the main problems for using ammonia as a sustainable fuel is its production. The bibliography describes three possible routes to produce ammonia: the currently used, the Haber Bosch process (Matar & Hatchl, 2001), the renewable or CO2 free ammonia synthesis (Giddey et al., 2017; Giddey, Badwal, & Kulkarni, 2013) and the electrochemical synthesis (Giddey, Badwal, & Kulkarni, 2013; Amar, Petit, & Tao, 2011).

Ammonia is mainly produced by the Haber Bosch process (Giddey, Badwal, & Kulkarni, 2013; Yapicioglu & Dincer, 2019; Valera-Medina et al., 2018; Giddey, Badwal, Minnings, & Dolan, 2017) which is a highly energy and carbon intensive process. This production process requires temperatures of 450-500°C and pressures up to 200 bar. This process uses natural gas, coal or fuel oil and nitrogen from the air as feedstock (Amar, Petit, & Tao, 2011). The costs associated to the ammonia production using the Haber-Bosch process (mainly dependent of the natural gas feedstock) can be estimated around 200 $ per ton of NH3 (Shiazawa, 2020), compared to the gasoline equivalent of 0.96-2.83 US $ per gallon) (Valera-Medina, et al., 2021)

Therefore, there is a need to produce ammonia with a lower carbon footprint and some technologies are being developed. The 1st generation will produce blue hydrogen that will ensure carbon capture and storage of CO2 produced during the manufacturing of ammonia (Valera-Medina et al., 2021).

As alternative, the CO2 free ammonia, also called 2nd generation, (Valera-Medina, et al., 2021) can be produced using green hydrogen produced by electrolysis as feedstock (Amar, Petit, & Tao, 2011; Giddey, Badwal, & Kulkarni, 2013), as is showed in Figure 2 (Adapted from (Aziz, Wijayanta, & Nandiyanto, 2020)) or biomass-derived hydrogen (Valera-Medina, et al., 2021). The required electricity is generated from renewable sources and nitrogen is obtained from an air separation unit and compressed to the required synthesis pressure (200bar) (Giddey et al., 2017). The production cost from this process is mainly associated with the electrolysis capital costs and operating (energy input) costs (Giddey, Badwal, Minnings, & Dolan, 2017). It is estimated by the Ammonia Energy Association (AEA), the CO2 free ammonia can cost around 320 $ per ton in 2050 (considering a target production cost of 2 $ per kg of H2) (Shiazawa, 2020), being its production cost highly dependent of green hydrogen production costs.

Contribution of Sustainable Fuels for the Future of the Energy Sector

Figure 2. Simplified diagram of the production of ammonia and utilization as fuel. Adapted from (Aziz, Wijayanta, & Nandiyanto, 2020)

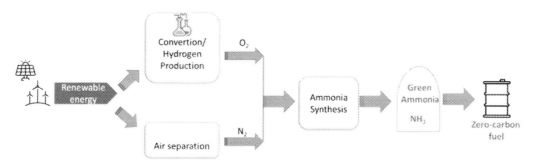

As an alternative to the Haber-Bosch process (or reactor), a so-called Electrochemical Synthesis (Amar, Petit, & Tao, 2011; Giddey, Badwal, & Kulkarni, 2013) has been reported (3rd generation). This ammonia synthesis process is still in a fundamental scale and consists of a single electrochemical reactor with water or steam and nitrogen as feedstocks. Depending on the electrolyte and electrodes used in the process, the reaction can occur at a temperature between room temperature and 800°C (Amar, Petit, & Tao, 2011). This process presents a very low reaction rate and needs to be improved at least 1 or 2 orders of magnitude to achieve commercial use (Amar, Petit, & Tao, 2011). However, this technology presents the advantage of the use of renewable sources to feedstock the reactor at significantly milder process conditions and in a single reactor. It has been estimated that electrochemical routes of ammonia production can save more than 20% of the energy consumption as compared to the conventional Haber-Bosch route (Giddey, Badwal, & Kulkarni, 2013).

For the 1st and 2nd generation, while the technology is feasible, the main barrier are the costs associated with it, being greatly dependent of the renewable energy cost feedstock to the process. For the third-generation technology, the technological feasibility towards commercial implementation needs to be improved.

The future of use of ammonia as fuel need to pass to:

- Ammonia fuel production system based on the Haber-Bosch process needs to reduce the production costs using renewable energy sources.
- Ammonia is seen to have an end use as a direct replacement for fossil fuels in transport systems, replacing fossil fuels in internal combustion engines, power turbines, jet engine turbines and marine propulsion power plants.
- The current cost of ammonia as fuel is generally higher than fossil fuels. Because of the environmental-based restrictions and the cost of carbon-based fuel are expected to increase in the future, there are several applications where fuel cost of ammonia will be lower.

Biofuels

Biofuels are alternative fuels, produced from renewable feedstocks (in contrast to fossil fuels that are exhaustible), that can be blended or fully substitute fossil fuels (such as gasoline, diesel, and heating oil) and are normally made from biomass such as corn, sugar cane, soy, switchgrass, agricultural waste, wood, waste cooking oil and animal fat. A variety of solid, liquid, and gaseous biofuels can be produced, such as ethanol, biodiesel, methane, methanol, and bio-oil (EIA- U.S. Energy Information Administration, 2022; EPA- U-S- Environemntal Protection Agency, s.d.).

There are currently three major transport biofuel types: bioethanol produced mainly by sugarcane or corn, biodiesel produced mainly from vegetable oils or residual raw material, and Hydrogenated vegetable oil (HVO) produced mainly from vegetable oil and animal fat (OECD, 2021).

Usually, biofuel can be classified, depending on the feedstock, in four groups: first, second, third and fourth generation. First generation biofuels (also called conventional biofuel) are produced from crops directly from the field (corn, sugar cane, cereals…); second-generation biofuels (also called advanced biofuel if the feedstock is listed in Part A of Annex IX (such as agricultural and forestry residues, wastes, energy crops, or aquatic biomass of the Renewable Energy Directive Recast (REDII)) are made with alternative, non-food crop biomass such as waste biomass; agriculture and forestry residues (lignocellulosic biomass) and other organic waste; third generation biofuels are made with improved non-food crop biomass, as algae, and finally the fourth generation, which are the biofuels made with non-food crop biomass plus microbes (Zubkova et al., 2019; Chandrasekhar et al., 2021; Gasparatos et al., 2022; Raj et al., 2022).

However, there is a huge concern about the first-generation biofuel, which is linked to many negative impacts as, among others, indirect land use change (ILUC) which is when the soil is used to produce feedstock to biofuels instead of food, deforestation, causing biodiversity loss and social conflicts, and most importantly, affecting food security. The use of food crops like corn and palm oil can reflect in the price of food, making the food, for some part of population, in some countries with low income, inaccessible (Rulli et al., 2015; Ahmed, Abubakari, & Gasparatos, 2019; Kline et al., 2016; Gasparatos et al., 2022; DRE - Diário da República Eletrónico, 2021).

Currently, the term "advanced biofuel" refer to the biofuel made from lignocellulosic feedstocks (i.e. agricultural and forestry residues, e.g. wheat straw/corn stover/bagasse, wood based biomass), non-food crops (i.e. grasses, miscanthus, algae), or industrial waste and residue streams, present low carbon dioxide emission or high Green House Gases (GHG) reduction, and reaches zero or low ILUC impact (Europex, 2018).

The production of second generation biofuel complies not only with the food security concept, but also with the Waste to Energy (WtE) concept transiting from a linear economy (exploitation-production-consumption-disposal) to a circular economy model where the waste is the feedstock to produce biofuel, which means that increase wasted valorization minimizes the quantity of waste incinerated or landfilled (Dias & Vieceli, 2018; Barros et al., 2020), plus contributing to reduce CO2 emissions. Howsoever, to comply with all the tree pillars of sustainability is still a challenge. The production of first-generation biofuel should be avoided, since it competes with the food sector, in other words doesn´t comply with the pillar of social sustainability.

In another hand, the production of the second-generation biofuel involves costs, related to the pre-treatment of the feedstock, even knowing that the waste feedstock can have a lower price than feedstock from second- generation, such as waste cooking oil and sugarcane bagasse, that are not competitive with conventional biofuels (Hasenheit et al., 2016). In other hand, the feedstock variability, availability, and affordability can increase the risks for the sustainability of the production of biofuels.

It is true that it is not easy to produce biofuel with a competitive market price, compared to fossil fuels. As an example, the average free on board (FOB) price for American biodiesel in later 2021 was around $1.23 per liter while the on-highway price for conventional diesel was about $0.82 per gallon.

Contribution of Sustainable Fuels for the Future of the Energy Sector

Figure 3. Global biofuel production in 2019 and forecast to 2023. Adapted (IEA, 2021)

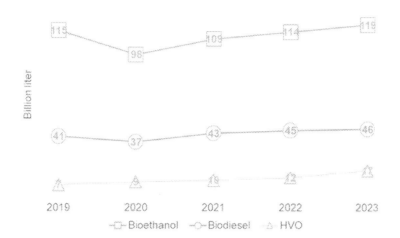

To minimize the risks and uncertainties for the future of the biofuel sector, stronger local policies should be applied and include tax exemptions, subsidies for biofuels and innovation to reduce production costs (IEA, 2021; OECD, 2021). In Europe, the Renewable Energy Directive (RED II), promotes biofuel produced from a list of material (such as algae, unsorted urban waste, biowaste, bagasse, peels…). This list is the basis for member states to define different levels of support depending on the types of biofuels (Europex, 2018; TE- Transport & Enviroment, 2020), the type of support will depend of the local policies.

Nevertheless, transition from conventional to advanced biofuel is still small. According to the Organization for Economic Co-operation and Development (OECD) review, only 20% of the biodiesel produced in 2020 come from waste cooking oil (WCO) and advanced technologies based on cellulosic feedstock to produce ethanol is still not significant compared to total biofuel production (OECD, 2021), mainly because the production of biofuel of second generation still need to be better developed to reduce the costs.

It is important to underline that the production of biofuel is strongly influenced by national regulations, which is based on greenhouse gas emissions, agriculture support, the necessity to increase the energy independence plus the global demand. Those policies will direct influence the demand of the type of biofuel. Improved polices mainly in the sector of climate change can help to propel sustainable biofuel growth. The International Energy Agency estimated that the global demand for biofuel is set to grow by 41 billion liter (28% grow) between 2021 and 2026 (IEA, 2021).

The transport biofuel demand contracted 8% from 2019 to 2020, largely due to impacts of the Covid-19 pandemic that cause a global transport disruption. While biofuel demand grew 5% per year on average between 2010 and 2019, thus the estimation made by IEA taking in account the recovery to pre-covid-19 demand levels (IEA, 2021; OECD, 2021).

Figure 3 (Adapted form (IEA, 2021)) shows the global biofuel production, composed by the currently main transport biofuels, in billion liter up to 2020, and forecast in 2021 to 2023. The estimation of the production between 2021-2023 taking in account the expected economic recovery, higher blend (depend on national policies) plus decarbonization initiatives (OECD, 2021).

Figure 3 shows that bioethanol is the most consumed biofuel, compared to biodiesel and Hydrogenated Vegetable Oil (HVO), also known as renewable diesel or green Diesel. This difference in production is related to demand that is highly influenced by the technologies and polices as mentioned before.

Figure 4. Distribution of the production of ethanol (A) and Biodiesel (B) by region. Adapted from (IEA, 2021; RFA, 2021; OECD, 2021)

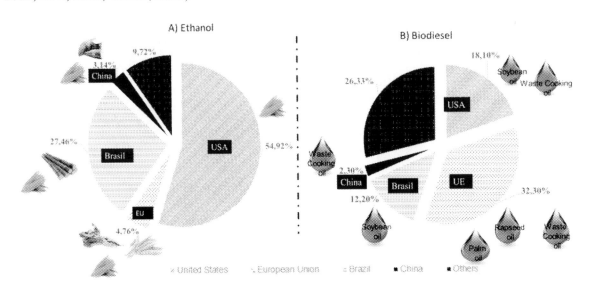

It is well known that biodiesel and HVO, can be both blended with diesel fuel. Renewable diesel has the same chemical composition as fossil diesel, both are hydrocarbons, and so it is fully compatible with existing diesel engines, thus it can complete substitute the fossil diesel. However, biodiesel, is an ester, that means it has a different chemical composition to fossil diesel, which limit the percentage of blending. In Europe, such as in Portugal for example, blending (Biodiesel and diesel) is limited to 11% in energetic value and in Brazil to 10%. It is important to stress that some countries present a variety of blending (fossil diesel + biodiesel), as USA, that have B2, B5, B20 (the number means the % of biodiesel) and the B100 (neat biodiesel) (PURE- European Renewable Ethanol, 2000; RESOLUÇÃO Nº 807, DE 23 DE JANEIRO DE 2020, 2020; EIA, 2022)

In term of ethanol, the blending in gasoline also varies from country to country, for example in Europe and USA is around 10% (E10), Brazil 25 – 27%, in Brazil there are vehicles moved 100% by ethanol. Both in Brazil and in the United States of America there are flexible-fuel vehicle (FFV) (commonly called flex-fuel vehicle) (PURE- European Renewable Ethanol, 2000; RESOLUÇÃO Nº 807, DE 23 DE JANEIRO DE 2020, 2020; EIA, 2022).

Currently, United States of America is the world´s largest producer of ethanol, with a production of around 14 billion gallons in 2020. Together, the United States and Brazil produced over 80% of the ethanol in the world. In terms of biodiesel, European Union is the largest producer followed by United States (18%) and Indonesia (15%), as can be seen at Figure 4 (IEA, 2021; RFA, 2021; OECD, 2021).

Since the first and second generation are discussed in this chapter, it is important to enhance the difference between both processes, to better understand the cost involved. Figure 5 shows the simplified process to produce bioethanol and biodiesel from first and second generation.

In the first generation, the production of ethanol, mainly from sugar biomass, where the fermentable sugar is completely available, thus the fermentation process can occur without special pre-treatment, as hydrolysis that should be present in the production of ethanol from starchy biomass. In other side, the production of ethanol from lignocellulosic biomass requires special pretreatment to release ferment-

Contribution of Sustainable Fuels for the Future of the Energy Sector

Figure 5. Simplified process of bioethanol and biodiesel. Main feedstocks, main simplified reactions, and processes
Source: own author

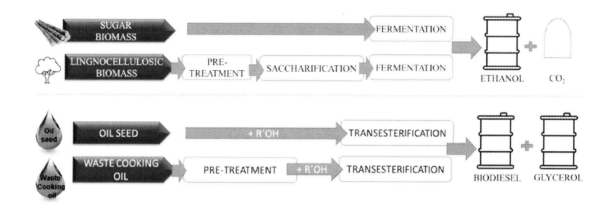

able sugars. The pre-treatment can be chemical (acid or basic hydrolysis); mechanical (ultrasonication, milling, supercritical CO2 exposure); thermal (steam explosion); biological (enzymatic hydrolysis) or a combination of any two or more processes.

The production of biodiesel from first generation doesn´t need a special pre-treatment of the feedstock since the oil present a well-known characteristic for its origin. In other side the production of biodiesel from waste feedstock, per example from waste cooking oil, need to be more complex, including a special pre-treatment line, since the characteristic of the feedstock is very heterogeneous, presenting a variety of contaminants such as solid organic matter, water and depending on the previous use, a high content in fatty free acid and sulfur content.

Those pre-treatment stages, needed to produce second-generation ethanol or biodiesel, increase the cost of the production. To increase the economic sustainability and increase waste valorization, besides the support that can be received from the government, the biofuel should be produced within the biorefinery concept. It is well known that the concept of biorefinery is globally discussed, and there are different definitions worldwide, however in accord to International Energy Agency (IEA), biorefinery is the sustainable processing of the biomass, valorizing it at its maximum resulting in a range of marketable products and energy (IEA, 2021). The production of biofuel complying with the three pillar of sustainability is still a challenge. The development of new technologies to optimize the process and valorize the feedstock at its maximum, producing coproducts with a market value, is the key to achieve economic sustainability

Hydrogenated vegetable oil (HVO), also called as Hydro-processed and Fatty Acids (HEFA) and green diesel, is a renewable diesel and as biodiesel, it can be produced from vegetable oils and animal fats. The difference between HVO and biodiesel, is how the feedstock is processed in HVO, with hydrogenation (hydrogenation under high pressure) or hydrocracking (breaking down larger molecules using hydrogen using hydrogen) of vegetable oil, and the final product is pretty similar, from a chemical point of view, to the fossil diesel, since, contrary to biodiesel, it doesn´t present oxygen in its chemical structure (HVO is a hydrocarbons (carbon + hydrogen) and biodiesel is an esters (carbon + hydrogen + oxygen). Overall, it has similar chemical properties as fossil diesel, plus, compared to biodiesel, HVO (can reach -40°C) has

Figure 6. Main differences between HVO and Biodiesel
Source: own author

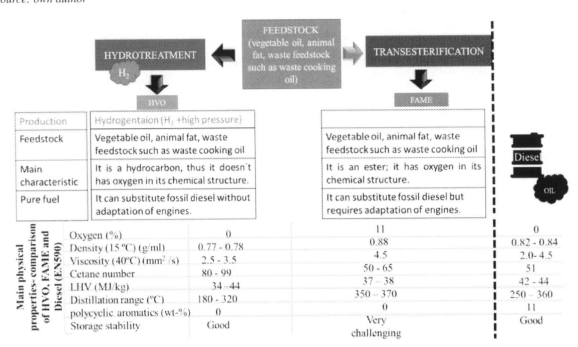

better cloud point than biodiesel (1 – 17°C depending on the feedstock, close to the cloud point of fossil diesel (-28 to -40°C depending on the diesel). The cloud point is the temperature of the fuel at which small, solid crystals can be observed as the fuel cools (FARM ENERGY, 2019; Zeman, et al., 2019).

Some differences, between HVO and fossil diesel, are that it has a lower density and energy content than fossil diesel. In accordance with Statista the global production of HVO increased to 6.2 million metric tons in 2020, the triple compared to the production of 2013 (2.2 million metric tons), which almost half of this production was produced by the European Union (3.4 million metric tons in 2020 (Statista, 2021; McCaffery, et al., 2022). HVO and biodiesel are different in terms of chemical structure. While biodiesel is an ester, that is, it has oxygen in its chemical structure, HVO is a hydrocarbon like fossil diesel. Figure 6 shows the main differences between HVO, Biodiesel and fossil diesel (based on the specification of EN590) (Garrain et al., 2014).

As can be seen at Figure 6, HVO, except for density (lower) and cetane number (higher than fossil), meet all the requirements (showed in the figure) diesel standard. The LHV in Biodiesel is lower than in HVO, especially due to the presence of oxygen in the chemical structure of biodiesel.

Synthetic Fuels

Synthetic fuel, also called synfuel, is obtained from syngas (composed mainly by a mixture of carbon monoxide and hydrogen), derived from gasification of solid feedstocks such as coal or biomass or by reforming of natural gas. Synfuel is fuel modified by anthropogenic processes from chemical reactions to any manufactured fuel with the approximate composition and comparable specific energy of a fossil fuel, with the aims of reducing the impurities of the substances for a more effective and clean combustion, lowering the emission of pollutants and reducing dependence on oil (Kobayashi, 2021).

In 1925, Franz Fischer and Hans Tropsch developed what is considered the first synthetic fuel process obtained from the gasification of coal to synthesis gas, and it can be converted into hydrocarbons of variable carbon number in a stepwise polymerization Fischer–Tropsch reaction. The products are like crude oil and are further refined into various fuels as for example gasoline or diesel. The products are light hydrocarbons (methane, ethane, propane, butane), naphtha, kerosene, Diesel fuel, low-molecular-weight wax, and high-molecular-weight wax. The chain length distribution of the product depends on the catalyst characteristics, the reaction conditions (pressure, temperature, H_2/CO ratio) (Towoju, 2021; Hanggi, et al., 2019).

Beyond the electrolysis of water to produce hydrogen, the life-cycle of a synthetic fuel considers the separation of carbon dioxide from the atmosphere or from other sources, chemical synthesis and purification of the desired fuel, transportation and storage, and finally the fuel oxidation in a fuel cell or combustion engine, with the release of gaseous water and carbon dioxide to the atmosphere (Hanggi, et al., 2019). In term of storage and transportation, these are very simplified because synthetic Diesel can be blended with conventional Diesel without any changes in the existing infrastructure (Klerk, 2016).

Synfuel can be classified, based on the main feedstock used as: Coal-To-Liquids (CTL), Gas-To-Liquids (GTL) and Biomass-To-Liquids (BTL).

More recently synthetic fuels can be produced using water or hydrogen and another carbon source that is atmosphere carbon dioxide. These fuels can be considered sustainable if they used renewable energies, such as wind and solar, to produce green hydrogen, as well as if the CO_2 used is captured from the atmosphere or from industries gaseous waste or from biomass.

When the synthetic hydrocarbons are consumed, carbon dioxide will be generated and can be used again by entering a cycle powered by renewable energy. In this extended carbon cycle, CO_2 remains part of the global biogeochemical cycle and the CO_2 concentration in the atmosphere will be reduced, contributing to carbonic neutrality, which is a strong contribution to the circular economy (Braun & Toth, 2020)

The most common process used to produce synfuel (converting gas, coal, biomass or natural gas into syngas) is the Fischer-Tropsch process (Anissimov, 2022), as demonstrated in Figure 7 (Adapted from (Anissimov, 2022; SYNGASCHEM, 2022). Currently, researchers and companies are trying to produce syngas from atmospheric CO_2 using Carbon Capture and Utilization (CCU), where de CO_2 is co-electrolyzed with water near room temperature. Through this technology water and CO_2 is converted into syngas as an intermediate process for syngas (SYNGASCHEM, 2022; NovaInnovation, 2020).

Several companies are working on creating a viable and feasible process of producing synthetic fuels, such Sasol, Indian Oil Corporation, Royal Dutch Shell, Phillips 66, ExxonMobil, Petrochina, Reliance Industries and Bosch. Now, the main challenge to the large-scale implementation of synthetic fuels is the associated costs because the processing facilities are expensive and there are not enough test plants. Despite this, the current results are already encouraging, and they become considerably more accessible when production capacities are expanded, and the cost of electricity generated from renewable sources is reduced. It is expected that by 2030 fuel costs will already be at more affordable prices and that by 2050 they will be close to fossil fuel prices.

The applications of synthetic fuels are transportation, heating, electricity production, fuel cells and chemical production and in transports as ships, trucks, or planes, which are hardly electrified.

Figure 7. Simplified diagram of the main process to produce Syngas and synfuels. Adapted from (Anissimov, 2022; SYNGASCHEM, 2022)

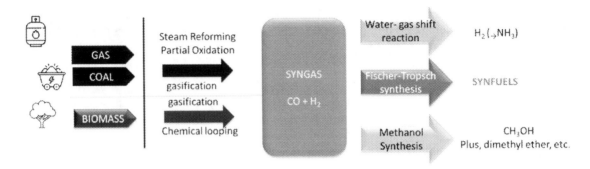

Synthetic fuels are important for the aviation sector and are considered sustainable aviation fuel (also known as SAF - Sustainable Aviation Fuel) and can be one of the solutions to ensure sustainable travelling, contributing enormously to the sector's emissions reduction strategy. In fact, it is one of the sectors with a high contribution to CO2 emissions. In 2019 it was responsible for burning 363 billion litres of jet fuel, representing 2% of all CO2 emissions, or 914 million tons of CO2. (Braun & Toth, 2020) (ATAG, 2020).

Due to limited decarbonisation options, the implementation of solutions for sustainable aviation fuels is crucial for reducing greenhouse gas emissions (Doliente, Narayan, Tapia, & Samsatli, 2020) The use of SAF was approved in 2011 (ATAG, 2020) but currently less than 1% of aviation fuels are sustainable, although it is expected to reach 2% by 2025 (equivalent to 7 billion liters) (ATAG, 2020). A SAF can emit about 80% less carbon during its life cycle compared to fossil fuels. Research is also underway on fuels that may have negative emissions, which means that they absorb more CO2 than they emit, using raw materials that are abundant in a particular region of the world, be it forest waste, agricultural waste, or municipal waste, what contributes strongly to the sustainability of the region. (Al-Breiki & Bicer, 2021).

H2 production from renewable energy (solar and wind) and biogenic CO2 are two critical elements that can contribute to the success of the industrial production of carbon-neutral synthetic fuels, with a view to the decarbonization of the aviation industry, and Portugal can benefit from the strong competitiveness it presents in the production of green hydrogen. The current challenge is to achieve sufficient fuel quantity to become economically viable. Therefore, there is such great interest in the development of these fuels (Hanggi, et al., 2019).

In conclusion, synthetic fuels are considered good alternatives to traditional fuels, because:

- they can contain climate change as they reduce the emission of pollutants and greenhouse gases into the atmosphere and then they are environmentally friendly alternative.
- they are compatible with the infrastructure and the technology of motor engines and industrial equipment, so they can be used in the transition between combustion engines and electric motors and other transport types.

DISCUSSION

To facilitate the comparison of the alternative fuels mentioned in this chapter, it was chosen to group in Figure 8, showing the main advantages and disadvantages of alternative fuels discussed in this chapter. Biodiesel and bioethanol both present a well-established production technology but are under considerable legislative pressure to be produced from non-food crops, which increases the production costs. In terms of quality, for example biodiesel is not good to be use in cold weather due to its high cloud point, but this is not a big issue if the biodiesel is blended with fossil fuel.

Biodiesel and HVO both are produced from the same feedstock, but since HVO uses hydrogen in its production, it has a higher price than biodiesel. HVO is very similar to fossil fuels and doesn't have the same problem related to cloud point as biodiesel does, plus it doesn't have oxygen in its structure, which means that HVO presents more stability under storage (Greena Team, 2014).

Hydrogen, despite having good features, has to be stored at very low temperature and high pressure, resulting in a high price of the final product, and it still doesn´t have a widespread infrastructure.

The decarbonization process requires multiple decarbonization measures, one of the alternatives is the use a combination of sustainable fuels such as HVO, bioethanol and synthetic fuels (ammonia and methanol), but as could be seen at table 1, mostly of then present a high cost production due to the energy consumption for the production or storage of the final product.

CONCLUSION

As is well known, the current global energy consumption patterns are unsustainable and will dramatically affect the planet's climate with consequences that are suspected to be catastrophic for the planet and for humanity. What can be done to reverse this situation? The solution relies on the adoption of a variety of measures and the use of sustainable fuels could be a very important contribution, as this chapter aims to demonstrate. However, the production of alternative fuels that contribute to decarbonize the transport sector, plus contribute to a circular economy, comply with all the tree pillars of sustainability and all the national and international regulations is a strong challenge.

This sector, even for the alternative fuels that have a well-established and developed technology, still need improvements in their technology to reduce costs, mainly when the feedstock is very heterogenous (such as lignocellulosic biomass, waste…), and be able to compete with fossil fuel prices. Government support can help to turn this sector more economical sustainable. In fact, it is necessary to increase environmental awareness of consumers, industrialists, and governments, so that there are incentives and investments for the implementation of the technological processes mentioned throughout the chapter. Legislation could also help promoting the use of more sustainable fuels, notably with greater fractions of incorporation of biodiesel in commercial diesel.

Figure 8. Main advantages and disadvantages of alternative fuels (Oddo & Masi, 2021; Kim, Koo, & Joung, 2020; Górniak et al., 2018; Hofstrand, 2009)

TECHNOLOGY	ADVANTAGE	DISADVANTAGE
BIODIESEL	Valorisation of waste cooking oil and animal fat Sub product (as glycerol) has a market (such as pharmacies, chemical industries) Increase energetic independence (in term of transport's fuels) CO_2 neutral	Doesn't have good stability to be storage for long period (more than 6 months) High cloud point (gel point) of biodiesel. It's difficult to start engine in cold weather High cost of the production
BIOETHANOL	Valorization of lignocellulosic biomass Old and consolidate technology Increase energetic independence (in term of transport's fuels) Good stability to be storage CO_2 neutral	Ethanol also absorbs water easily giving it a high tendency to corrode materials in gasoline-based engines, thus the vehicles must be prepared to it. Low yield of ethanol (due to yeasts poisoning). **High cost of the production when has as feedstock lignocellulosic biomass**
HVO	Valorization of waste cooking oil and animal fat High-performance drop-in biofuel Increase energetic independence (in term of transport's fuels) It can fully replace fossil diesel Good stability to be storage CO_2 neutral	Hydrogen required **High cost of the** production
GREEN HYDROGEN	Valorisation of waste cooking oil and animal fat Enable zero emission (with full cell) CO_2 neutral	High fuel price No available piston engine and infrastructure Very low storage temperature for liquefied hydrogen (-253 °C) Very high storage pressure for gas (245 atm)
RENEAWABLE AMONIA	Good stability to be storage Can be used in various combustion engines (with minor modifications) as well as fuel cells Can be stored relatively low pressure and high temperature (liquefied ammonia) The existing retail fuel dispensing infrastructure can be converted to ammonia	Toxicity and environmental impact when leaked Need to add hydrogen when used for internal combustion engines The use of Ammonia can negatively impact the use of ammonia as a nitrogen fertilizer
RENEAWABLE SYNTHETIC FUELS	High quality synthetic fuels Different cuts available (gasoline, diesel and naphtha per example) Well known with existing infrastructure Reduction of GHG through the consumption of Atmospheric CO_2	Purification is crucial for catalyst Catalyst may be costly High cost of the production.

REFERENCES

Ahmed, A., Abubakari, Z., & Gasparatos, A. (2019). Labelling large-scale land acquisitions as land grabs: Procedural anddistributional considerations from two cases in Ghana. *Geoforum, 105*, 191–205. doi:10.1016/j.geoforum.2019.05.022

Al-Breiki, M., & Bicer, Y. (2021, January 10). Comparative life cycle assessment of sustainable energy carriers including production, storage, overseas transport and utilization. *Journal of Cleaner Production, 279*, 1–16. doi:10.1016/j.jclepro.2020.123481

Amar, I. A., Petit, C. G., & Tao, S. (2011). *Solid-state electrochemical synthesis of ammonia: a review.* Solid State Electrochem. doi:10.100710008-011-1376-x

Anissimov, M. (2022). What are Different Types of Synthetic Fuels? *About Mechanics.* https://www.aboutmechanics.com/what-are-different-types-of-synthetic-fuels.htm

ATAG. (2020, 9). Aviation Benefits beyond borders. *Global Fact Sheet.*

Aziz, M., Oda, T., Morihara, A., & Kashiwagi, T. (2017). Combine4d nitrogen production, ammonia synthesis, and power generation for efficient hydrogen dtorage. *Energy Procedia, 143*, 674–679. doi:10.1016/j.egypro.2017.12.745

Aziz, M., Wijayanta, A. T., & Nandiyanto, A. D. (2020). Ammonia as Effective Hydrogen Storage: A Review on Production, Storage and Utilization. *Energies*, (2022).

Bard, J., Gerhardt, N., Selzam, P., Beil, M., Wiemer, M., & Buddensiek, M. (2022). *The limitations of hydrogen blending in the european gas grid.* IEE- Fraunhofer Institute for Energy Economics and Energy System Technology. https://www.iee.fraunhofer.de/content/dam/iee/energiesystemtechnik/en/documents/Studies-Reports/FINAL_FraunhoferIEE_ShortStudy_H2_Blending_EU_ECF_Jan22.pdf

Barros, M. V., Salvador, R., de Francisco, A. C., & Piekarski, C. M. (2020). Mapping of research lines on circular economy practices in agriculture: From waste to energy. *Renewable & Sustainable Energy Reviews, 131*, 131. doi:10.1016/j.rser.2020.109958

Bevill, K. (2008). Building the 'Minnesota Model. *Ethanol Producer Magazine*, 114-120.

Braun, A., & Toth, R. (2020, December 15). Circular economy:national and global policy -overview. *Clean Technologies and Environmental Policy, 23*(2), 301–304. doi:10.100710098-020-01988-8

Chandrasekhar, K., Kumar, A., Raj, T., Kumar, G., & Kim, S. (2021). *Bioelectrochemical system-mediated waste valorization. SMAB.* Systems Microbiology and Biomanufactturing. doi:10.100743393-021-00039-7

Costa, C. C., Cunha, M. P., & Guilhoto, J. M. (2011). *The role of ethanol in the brazilian economy: three decades of progress.* MPRA- Munich Personal RePEc Archive.

Dias, N. B., & Vieceli, N. (2018). Mechanical Biological Treatment. In Waste-to-Energy (WtE). NOVA.

Directives, E. (2014, outubro 22). Diretiva 2014/94/UE do Parlamento Europeu e do Conselho [Directive 2014/94/EU of the European Parliament and of the Council]. *Jornal Oficial da União Europeia.* https://eur-lex.europa.eu/legal-content/PT/TXT/?uri=CELEX:32014L0094

EU Directives. (2018). *Directive (eu) 2018/2001 of the european parliament and of the council.* EU. https://eur-lex.europa.eu/legal-content/EN/TXT/PDF/?uri=CELEX:32018L2001&from=EN

Doliente, S., Narayan, A., Tapia, J., & Samsatli, N. (2020, july 10). Bio-aviation Fuel: A Comprehensive Review and Analysis of the Supply Chain Components. *Fronties in Energy Research.*

DRE - Diário da República Eletrónico. (2021). Decreto-Lei 8/2021, de 20 de Janeiro [Decree-Law 8-2021 of January 20]. *Diário da República n.º 13/2021, Série I de 2021-01-20. [Diario da republica no. 13/2021, Series I of 2021-01-20.]* https://dre.tretas.org/dre/4390632/decreto-lei-8-2021-de-20-de-janeiro

Eames, I., Austin, M., & Wojcik, A. (2022). Injection of gaseous hydrogen into a natural gas pipeline. *International Journal of Hydrogen.*

EIA. (2022). *EIA- US- Energy Information Administration.* EIA. https://www.eia.gov/tools/faqs/faq.php?id=27&t=10#:~:text=The%20ethanol%20content%20of%20most,ethanol%20production%20capacity%20is%20located

EIA- U.S. Energy Information Administration. (2022). Biofuels explained. *Biodiesel, renewable diesel, and other biofuels.* EIA.https://www.eia.gov/energyexplained/biofuels/biodiesel-rd-other-use-supply.php

Emonts, B., Reuß, M., Stenzel, P., Welder, L., Knicker, F., Grube, T., Görner, K., Robinius, M., & Stolten, D. (2019). Flexible sector coupling with hydrogen: A climate-friendly fuel supply for road transport. *International Journal of Hydrogen Energy, 44*(26), 12918–12930. doi:10.1016/j.ijhydene.2019.03.183

EPA. (n.d.). *Renewable Energy Certificates (RECs) | US EPA.* EPA. https://www.epa.gov/green-power-markets/renewable-energy-certificates-recs

EPA- U-S- Environemntal Protection Agency. (n.d.). *Economics of Biofuels.* EPA. https://www.epa.gov/environmental-economics/economics-biofuels

European Commisision. (2022). *Sustainable development.* EC. https://policy.trade.ec.europa.eu/development-and-sustainability/sustainable-development_en

European Commission. (2022). *Energy.* EC. https://energy.ec.europa.eu/topics/renewable-energy/bioenergy/biofuels_en#:~:text=By%202030%2C%20the%20EU%20aims,the%20achievement%20of%20this%20target

European Parliament. (2015). *Circular economy: definition, importance, and benefits.* European Parliament. https://www.europarl.europa.eu/news/en/headlines/economy/20151201STO05603/circular-economy-definition-importance-and-benefits

Europex. (2018). Renewable Energy Directive (RED II) -Directive (EU) 2018/2001 (recast) on the promotion of the use of energy from renewable sources.

FARM ENERGY. (2019). Biodiesel Cloud Point and Cold Weather Issues. Retrieved 2022, from https://farm-energy.extension.org/biodiesel-cloud-point-and-cold-weather-issues/#:~:text=In%20cold%20climates%2C%20it%20can,from%20which%20it%20is%20made

FCH. (n.d.). Fuel Cells and Hydrogen - Green Hydrogen Guarantees of Origin (GO) now available on the market. Retrieved 2022, from https://www.fch.europa.eu/news/green-hydrogen-guarantees-origin-go-now-available-market

Filoso, S., Carmo, J. B., Mardegan, S. F., Lins, S. M., Gomes, T. F., & Martinelli, L. A. (2015). Reassessing the environmental impacts of sugarcane ethanol production in Brazil to help meet sustainability goals. *Renewable & Sustainable Energy Reviews, 52*, 1847–1856. doi:10.1016/j.rser.2015.08.012

Fonseca, J. D., Carmargo, M., Commenge, J.-M., Falk, L., & Gil, I. D. (2019). Trends in design of distributed energy systems using hydrogen as energy vector: A systematic literature review. Internaqtional Journal of hydrogen energy, 44, 9486-9504.

Frischmuth, F., & Härtel, P. (2022). Energy - Hydrogen sourcing strategies and cross-sectoral flexibility trade-offs in net-neutral energy scenarios for Europe. *Energy, 238*, 121598. doi:10.1016/j.energy.2021.121598

Garrain, D., Herrera, I., Lechón, Y., & Lago, C. (2014). *Well-to-Tank environmental analysis of a renewable diesel fuel from vegetable oil through co-processing in a hydrotreatment unit.* Biomassa and Bionergy. doi:10.1016/j.biombioe.2014.01.035

Gasparatos, A., Mudombi, S., Balde, B. S., Maltitz, G. V., Johnson, F. X., Romeu-Dalmau, C., ... Willis, K. J. (2022). Local food security impacts of biofuel crop production in southern Africa. *Renewable & Sustainable Energy Reviews, 154*, 154. doi:10.1016/j.rser.2021.111875

Giddey, S., Badwal, S. S., & Kulkarni, A. (2013). Review of electrochemical ammonia production technologies and materials. *International Journal of Hydrogen Energy, 38*(34), 14576–14594. doi:10.1016/j.ijhydene.2013.09.054

Giddey, S., Badwal, S. S., Minnings, C., & Dolan, M. (2017). Ammonia as Renewable Energy Transportation Media. *ACS Sustainable Chemistry & Engineering, 5*(11), 10231–10239. doi:10.1021/acssuschemeng.7b02219

Górniak, A., Midor, K., Kaźmierczak, J., & Kaniak, W. (2018). *Advantages and Disadvantages of Using Methane from CNG in Motor Vehicles in Polish Conditions.* MAPE- Multidisciplinary Aspects of Production Engineering. doi:10.2478/mape-2018-0031

Greena Team. (2014). Waste based biofuels, waste based feedstock. Retrieved from http://www.greenea.com/wp-content/uploads/2016/07/10.-HVO-market.pdf

Guimarães, D. (2019). Sustentabilidade. (Meio sustentável) Retrieved 2022, from https://meiosustentavel.com.br/sustentabilidade/

Gulbenkian. (2022). Fiundação Calouste Gulbenkian. Foresight Portugal 2030- 3 cenários para o futuro de Portugal. Retrieved 2022, from https://gulbenkian.pt/wp-content/uploads/2022/02/FCG_BROCHURA_ForesightPortugal2030_05as.pdf

Hanggi, S., Elbert, P., Butler, T., Cabalzar, U., Teske, S., Bach, C., & Onder, C. (2019). A review of synthetic fuels for passenger vehicles. *Energy Reports*, *5*, 555–569. doi:10.1016/j.egyr.2019.04.007

Hasenheit, M., Gerdes, H., Kiresiewa, Z., & Beekman, V. (2016). Summary report on the social, economic and environmental impacts of the bioeconomy. European Union´s Horizon 2020.

Hofstrand, D. (2009). Ammonia as a Transportation Fuel. AgMRC Renewable Energy Newsletter.

IEA. (2019). International Energy Agency - The Future of hydrogen - Seizing today's opportunities - Report prepared by the IEA for the G20, Japan. Retrieved 2020, from https://iea.blob.core.windows.net/assets/9e3a3493-b9a6-4b7d-b499-7ca48e357561/The_Future_of_Hydrogen.pdf

IEA. (2021). International Energy Agency - Hydrogen. Retrieved from https://www.iea.org/reports/hydrogen

IEA. (2021). Renewables 2021- Biofuels. (IEA – International Energy Agency) Retrieved 2022, from https://www.iea.org/reports/renewables-2021/biofuels?mode=transport&region=North+America&publication=2021&flow=Consumption&product=Ethanol

IEA. (2021). World Energy Outlook. Retrieved from https://www.iea.org/reports/world-energy-outlook-2021

IRENA. (2019). International Renewable Energy Agency - Hydrogen: a renewable energy perspective. Abu Dhabi.

IRENA-International Reneawable Energy Agency. (2014). IRENA- Global bionergy supply and deand projections: A working paper for REmap 2030. Retrieved 2022, from https://www.irena.org/publications/2014/Sep/Global-Bioenergy-Supply-and-Demand-Projections-A-working-paper-for-REmap-2030

JRC Publication Repository. (2021). Historical Analysis of FCH 2 JU Electrolyser Projects. (European COmission) Retrieved 2022, from https://publications.jrc.ec.europa.eu/repository/handle/JRC121704

Kim, H., Koo, K. Y., & Joung, T. (2020). A study on the necessity of integrated evaluation of alternative marine fuels. Journal of International Maritime Safety, Environmental Affairs, and Shipping.

Klerk, A. (2016). Chapter 10 - Aviation Turbine Fuels Through the Fischer-Tropsch Process. Biofuels for aviation, pp. 241-259.

Kline, K. L., Msangi, S., Dale, V. H., Woods, J., Souza, G. M., Osseweijer, P., . . . Mugera, H. K. (2016). Reconciling food security and bioenergy: priorities for action. GCB-Bioenergy- Bioproducts for a sustainable bioeconomy.

Kobayashi, M. (2021). Introduction. In *Dry Syngas Purification Processes for Coal Gasification Systems* (pp. 1–49). Elsevier. doi:10.1016/B978-0-12-818866-8.00001-X

Ma, M., Yang, G., Wang, H., Lu, Y., Zhang, B., Cao, X., Peng, D., Du, X., Liu, Y., & Huang, Y. (2019). Ordered distributed nickel sulfide nanoparticles across graphite nanosheets for efficient oxygen evolution reaction electrocatalyst. *International Journal of Hydrogen Energy*, *44*(3), 1544–1554. doi:10.1016/j.ijhydene.2018.11.176

Matar, S., & Hatchl, F. (2001). *Ammonia production (Haber Process)*. Chem Petrochemical Process.

McCaffery, C., Zhu, H., Ahmed, C. S., Canchola, A., Chen, J. Y., Li, C., ... Karavalaskis, G. (2022). Effects of hydrogenated vegetable oil (HVO) and HVO/biodiesel blends on the physicochemical and toxicological properties of emissions from an off-road heavy-duty diesel engine. *Fuel*, *323*, 323. doi:10.1016/j.fuel.2022.124283

NovaInnovation. (2020). Green syngas from carbon dioxide. Retrieved 2022, from https://novainnovation.unl.pt/2020/01/22/green-syngas-from-carbon-dioxide/

Oddo, E., & Masi, M. (2021). Roadmap to 2050: The Land-Water-Energy Nexus of Biofuels - Biofuels Technologies. Retrieved 2022, from https://roadmap2050.report/biofuels/biofuels-technologies

OECD. (2021). Biofuels. (OECD-Organisation for Economic Co-operation and Development-FAO Agricultural Outlook 2021-2030) Retrieved 2022, from https://www.oecd-ilibrary.org/sites/89d2ac54-en/index.html?itemId=/content/component/89d2ac54-en

Okolie, J. A., Patra, B. R., Mukherjee, A., Nanda, S., Dalai, A. K., & Kozinski, J. A. (2021). Futuristic applications of hydrogen in energy, biorefining, aerospace, pharmaceuticals and metallurgy. Journal of Hydrogen Energy, 46.

Osamah Siddiqui, I. D. (2018). A review and comparative assessment of direct ammonia fuel cells. *Thermal Science and Engineering Progress*, *5*, 568–578. doi:10.1016/j.tsep.2018.02.011

Pahija, E., Golshan, S., Blais, B., & Boffito, D. C. (2022). Chemical Engineering and Processing - Process Intensification.

PURE- European Renewable Ethanol. (2000). Fuel Blends. Retrieved 2022, from https://www.epure.org/about-ethanol/fuel-market/fuel-blends/

Raj, T., Chandrasekhar, K., Kumar, A. N., Banu, R. J., Yoon, J.-J., Bhatia, S. K., ... Kim, S.-H. (2022). Recent advances in commercial biorefineries for lignocellulosic ethanol production: Current status, challenges and future perspectives. *Bioresource Technology*, *344*, 126292. doi:10.1016/j.biortech.2021.126292 PMID:34748984

RESOLUÇÃO N° 807, DE 23 DE JANEIRO DE 2020. (2020). Imprensa Nacional_ Brasil_Diário oficial da União. Retrieved 2022, from https://www.in.gov.br/web/dou/-/resolucao-n-807-de-23-de-janeiro-de-2020-239635261

RFA. (2021). World Fuel Ethanol Production by Region. (RFA- Renewable Fuels Association) Retrieved 2022, from https://ethanolrfa.org/markets-and-statistics/annual-ethanol-production

Rulli, M. C., Bellomi, D., Cazzoli, A., Carolis, G. D., & D'Orico, P. (2016). The water-land-food nexus of first-generation biofuels. *Scientific Reports*, *6*(22521). PMID:26936679

Shiazawa, B. (2020). Ammonia Energy Association. Retrieved from https://www.ammoniaenergy.org/articles/the-cost-of-co2-free-ammonia/

Sorkhabi, R. (2015). The First Oil Shock. (GeoExPro) Retrieved 2022, from https://www.geoexpro.com/articles/2015/06/the-first-oil-shock

Sorknaes, P., Johannsen, R. M., Korberg, A. D., Nielsen, T. B., Petersen, U. R., & Mathiesen, B. V. (2022). Electrification of the industrial sector in 100% renewable energy scenarios. Energy.

Staista. (2022). Chemical & Resource- HVO biodiesel production volume worldwide from 2013 to 2020. *Staista*. https://www.statista.com/statistics/1297290/hvo-biodiesel-production-worldwide/

Stancin, H., Mikuleié, H., Wang, X., & Duié, N. (2020). A review on aalternative fules in futiure energy system. *Renewable & Sustainable Energy Reviews*, 128.

Statista. (2021). Biofuel productin form 2000-2020. *Statista*. https://www.statista.com/statistics/274163/global-biofuel-production-in-oil-equivalent/

SYNGASCHEM. (2022). Clean Coal to Liquids as a Transitional Technology. Syngachem. https://www.syngaschem.com/our-vision/

TE- Transport & Enviroment. (2020). *RED II and advanced biofuels - Recommendations about Annex IX of the Renewable Energy Directive and its implementation at national level.*

Towoju, O. A. (2021, April). Fuels for automobiles: The Sustainable Future. *Journal of Energy Research and Reviews,* pp. 8-13.

UNFCCC- United Nation Climate Change. (2022). *United Nation Climate Change- What is the Paris Agreement?* UN. https://unfccc.int/process-and-meetings/the-paris-agreement/the-paris-agreement

U.S. department of energy. (2019). Flexible Fuel Vehicles (FFV). *Energy Efficiency & Renewable Energy*. https://afdc.energy.gov/vehicles/flexible_fuel.html

Valera-Medina, A., Amer-Hatem, F., Joannon, M., Fernandes, R. X., Glarborg, P., Hashemi, H., & Costa, M. (2021). Review on Ammonia as Potential Fuel: From Synthesis to Economics. *Energy & Fuels, 35*(9), 6964–7029. doi:10.1021/acs.energyfuels.0c03685

Valera-Medina, A., Xiao, H., Owen-Jones, M., David, W., & Bowen, P. (2018). Ammonia for power. *Progress in Energy and Combustion Science, 69*, 63–102. doi:10.1016/j.pecs.2018.07.001

Yapicioglu, A., & Dincer, I. (2019). A review on clean ammonia as a potential fuel for power generatiors. *Renewable & Sustainable Energy Reviews, 103*, 96–108. doi:10.1016/j.rser.2018.12.023

Zeman, P., Honig, V., Kotek, M., Táborsky, J., Obergruber, M., Marik, J., Hartová, V., & Pechout, M. (2019). Hydrotreated Vegetable Oil as a Fuel from Waste Materials. *Catalysts, 9*(4), 337. doi:10.3390/catal9040337

Zubkova, V., Strojwas, A., Bielecki, M., Kieush, L., & Koverya, A. (2019). Comparative study of pyrolytic behavior of the biomass wastes originating in the Ukraine and potential application of such biomass. Part 1. Analysis of the course of pyrolysis process and the composition of formed products. *Fuel, 254*, 115688. doi:10.1016/j.fuel.2019.115688

KEY TERMS AND DEFINITIONS

Alternative Fuels: (also called non-conventional fuels): Fuels with the potential to replace long-term petroleum fuels such as electricity, hydrogen, biofuels, natural gas, synthetic fuels, ammonia, regardless of the energy and feedstock used and it is possible to use fuel combinations.

Biofuels: Liquid or gaseous transport fuels, that are made from biomass, that is a renewable energy that can replace or blend with fossil fuel. Since the biomass can absorb the carbonic gas emitted when the biofuel is burned, it is considerate that it is a carbon neutral.

Biomass: (also called organic matter): Material that comes from plants and animal used in energy production through the decomposition of a variety of renewable resources, such as plants, wood, agricultural waste, and food waste.

Circular Economy: economic model of closed loops where raw materials, components and products lose their value in a way that they still can be reused, repaired, or valorised to produce new products, saving primary raw material, reducing waste, and minimizing the greenhouse gases to contribute to a sustainable development.

Green Ammonia: Ammonia made by hydrogen that comes from water electrolysis powered by renewable energy.

Green Hydrogen: Hydrogen made by electrolysis from electricity produced with renewable energy, then it is an alternative fuel helps reducing emissions.

Hydrogenated Vegetable Oil: Commonly called HVO is a biofuel made from vegetable oil or animal fat. The HVO is produced by reacting the feedstock with hydrogen at high temperature and pressure.

Renewable Fuels: fuels produced from renewable energy sources, which are non-fossil sources that are naturally replenished on a human timescale (wind, solar, geothermal, wave, tidal, hydropower, biomass).

Sustainable development: Meet the needs of the present without compromising the needs of future generations. To not compromise the future the development should be based on the three pillars of the sustainability (social, environmental, and economic).

Synthetic Fuels: Fossil fuels are made from petrol and their basic chemical structure is composed by hydrogen and carbon atoms (hydrocarbons). The synthetic fuels imitate fossil fuels, and it can be produced by hydrogen from water and carbon from the air, thus is helps to offset the greenhouses emitted when the synfuel is burned.

Chapter 3
The Challenge of the Digitalization of the Water Sector

Nelson Gaudêncio Carriço
https://orcid.org/0000-0002-2474-7665
Escola Superior de Tecnologia do Barreiro, Portugal

Bruno Ferreira
https://orcid.org/0000-0002-2863-7949
Escola Superior de Tecnologia do Barreiro, Portugal

André Antunes
Escola Superior de Tecnologia de Setúbal, Portugal

João Caetano
https://orcid.org/0000-0002-8537-3169
Escola Superior de Tecnologia do Barreiro, Portugal

ABSTRACT

The digitalization of the water sector is of utmost importance for improving the efficiency and sustainability of the managed systems. The digitalization process, however, can be seen as a ladder with several steps that the water utility must climb to become a smart utility. The reality is that many water utilities worldwide have not realized yet the benefits of digital transformation and, thus, the digitalization of the water sector lags behind other industries. This chapter presents the major challenges and the promising future that water utilities face in the journey of digitalization. Guidelines on how to choose the most adequate digital solution are also presented, as well as the trends for a smarter water utility.

DOI: 10.4018/978-1-6684-6123-5.ch003

INTRODUCTION

The fourth industrial revolution, also called Industry 4.0 (Water 4.0, in the case of the water industry), brought extraordinary technological advances to all economic sectors of activity, including the water sector. This digital transformation integrates different types of technology in all areas of a business, changing its operation profoundly by optimizing processes and rethinking how organizations create value. Any sector that wants to jump into Industry 4.0 needs first to conclude the process started with the third industrial revolution, also known as the digital revolution, by investing in adequate digital technology and the digitization of existing data and information. Only after the digitization of papers and processes, an organization is in the condition to start the digitalization process. Note that digitization and digitalization, despite being different, are closely related and as a consequence, some literature uses them interchangeably. The first term can be simply defined as converting or representing physical data or information (i.e., non-digital assets) into a digital format that can be used by specific information systems. The second term is a step forward, which leverages digitized data and processes by improving or transforming organizations' operations, functions, models, processes, or activities. So, the digitalization process starts with a huge amount of data that is continuously collected and stored by the water utilities in several information systems. However, such data needs to be combined and processed aiming to extract useful information. When this information is used with physics-based or data-driven models (for instance by simulating the behaviour of the whole system or its assets), then knowledge is generated. Knowledge about the managed systems supports water utilities in the decision-making process (Karmous-Edwards et al., 2022). This hierarchy of Data-Information-Knowledge (DIK) can be found in much information science literature as the DIKW hierarchy in which the W stands for Wisdom (see Figure 1).

Figure 1. The Data-Information-Knowledge-Wisdom hierarchy

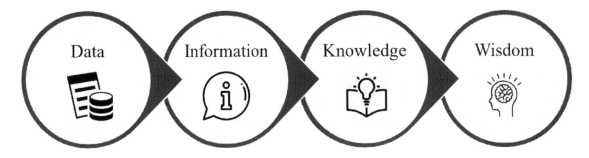

Data need to be put together in just one place to help the water utilities address their problems, namely, related to ageing assets, non-revenue water, water stress and scarcity, and variabilities of hydrologic events from climate change, among others. This process is typically called data integration and depends on the technology used to do it. For example, data can be linked to a third-party application using interoperability web services (this technology may be called middleware data integration) or by an Extract, Transform, Load (ETL) process which extracts data from a source system and loads the data after transformation into a target destination. As a large amount of assets and data involved makes data and information management a challenging task, new tools and processes are often needed to collect, gather, manage, analyse and use asset data (Carriço & Ferreira, 2021). The use of smart Internet of

The Challenge of the Digitalization of the Water Sector

Things (IoT) devices, to measure variables (e.g., water quality parameters, hydrometeorological data, flood data, and water consumption) and to continuously monitor network conditions in urban water systems, generates considerable amounts of data that needs to be stored and processed. This big data soon becomes unmanageable by humans and requires data mining and machine learning methods to extract relevant information and present it in a more user-friendly format.

Water utilities need to climb a ladder with several steps in order to become smart utilities, with each step being an advance toward full digitalization. For example, the first step in the ladder consists of just having the physical assets managed without any digital tool, being the second step having control and measuring devices and sensors to collect data from the physical systems. The third step may consist of having communication technologies connected to the sensors and automation to control the system, and so on. This means that water utilities are at quite different digital maturity levels. Regardless of their maturity, the utilities have to solve similar daily problems.

Many water utilities worldwide have not realized yet the benefits of digital transformation and this is challenging because it slows the digitalization of the sector. There are many benefits of digital transformation, such as water resources management, energy efficiency, operational efficiency, anomalous event management, and rehabilitation action schedules, among many others.

This chapter gives the authors' reflection on the challenges that water utilities face in achieving full digitalization.

ADVANTAGES OF DIGITALIZATION IN THE WATER SECTOR

Operation and management of urban water systems have become challenging and backbreaking tasks, not only having to deal with old and obsolete assets reaching the end of their service life and limited budgets that do not allows the satisfaction of rehabilitation needs, but also having to take short and medium-term decisions based on big and disorganised data generated by many separate information systems (IS). Nowadays, most water utilities collect a multitude of data time signals (e.g. pressure, flow, levels) with different time frequencies, aiming to get useful insights into the current and future status of the system. However, most data are acquired and stored but not used, since very few available tools can process and analyse them in real-time.

Water utility managers aim to have all information in a single platform linked to all IS that collects and processes data in real-time and that can virtually replicate the urban water systems' behaviour and help them in the daily decision-making process. This type of solution could allow the real-time detection and location of anomalous events (e.g., bursts, illegal connections), to help to make real-time decisions on the operation of pumps, tanks, and valves, as well as reduce water losses and energy consumption. Such a solution is called a digital twin (DT) and constitutes a virtual copy of the real urban water system, describing its behaviour through real-time monitoring and serving as a basis for experimentation and simulations (Bonilla et al., 2022; Pesantez et al., 2022). Nowadays, several definitions of DT exist in literature and Karmous-Edwards et al. (2022) developed a unified definition in the water sector which is as follows "a DT is a digital, dynamic system of real-world entities and their behaviours using models with static and dynamic data that enable insights and interactions to drive actionable and optimized outcome". According to Cooper et al. (2022), there are four digital twin maturity levels in water utilities: the lowest (Level 0) is a DT as the result of the digitization process, providing a virtual representation of the physical asset, enhancing operations modelling, design support, and traditional planning capabilities.

The last level (Level 3) connects and updates the DT using real-time data analytics to coordinate functions across domains. In short, there is no specific path to the digitalization process that fits optimally to all water utilities. In fact, the most appropriate digitalization process should be established depending on the water utilities' degree of digital maturity.

Although hydraulic models are often erroneously considered "simplified" versions of digital twins of WSS, currently, these are not fed with real-time data nor connected to existing IS, and are usually used for running offline analyses (Conejos Fuertes et al., 2020; Pedersen et al., 2021). Ideally, a hydraulic model should be automatically updated with processed real-time data and continuously calibrated to accurately describe the hydraulic system behaviour, being essential to efficiently anticipate, detect and manage anomalous behaviours and to assist in real-time operation.

A DT of urban water systems may also measure the performance state of a system by calculating key performance indicators (KPI) and, based on the overall information of the system, simulations of operation scenarios (e.g., valve settings) can be run to support operational decisions. For that, several tools should be coupled to enhance this capability, such as artificial intelligence tools.

In the author's opinion a DT of urban water systems should, at least, process and integrate real-time data from field sensors, detect and locate anomalous events with those data, and update and calibrate the hydraulic model to simulate operation scenarios (e.g., predictive control) for resource savings (e.g., energy), water quality testing and alerts (e.g., alerts on water contamination), or performance management (e.g., to minimize carbon footprint).

Several challenges need to be overcome, such as data assimilation from different data sources, data quality and uncertainty, the need to forecast systems malfunction based on accurate and real-time data, and computation time for some simulation scenarios. In the next section, some barriers to the digitalization process are presented.

BARRIERS TO DIGITALIZATION

It is unanimous that digitalization in the water sector will bring enormous benefits and the global research on smart water solutions is accelerating to meet industry and government demand. Nonetheless, there is still a large gap between the existence of technological solutions and their massive implementation in water utilities. As noted by Müller-Czygan et al. (2021), the conceptual, technical, and practical gaps between providers and customers are still not sufficiently close since there is a particular lack of solutions that can be integrated into everyday working life and which do not present additional barriers to the daily-life activities.

The water sector, especially in systems managed by municipalities, is where the digitalization process is proving to be extremely difficult. According to Alzamora et al. (2019), the development of a DT (and the overall digitalization process) is a challenge as it consists of a continuous process of adjustments and learning, which enriches the model with day-to-day knowledge. Municipal structures are extremely complex and hierarchical with ageing human resources and where technological literacy is far from desirable. In a broad sense, the barriers to digitalization can be related to five main areas (see Figure 2) (Savić, 2021), namely, 1) insufficient funds, 2) legacy data collection and management systems, 3) lack of a clear and plan for the digital transformation, 4) the "human factor", and 5) cybersecurity. Each area is further described.

Insufficient funds are the main obstacle seen by water utilities in the digitalization process. External funding is needed in most situations to start the digitalization process and it is not always easy to obtain (Brianne Nakamura, Williams, 2017). SCADA systems require a high first investment, continuous operation and maintenance costs and specialized human resources. Water utility managers can apply for European funds for the first investment, but the operation and maintenance are part of the water users' current costs. As noted by Moreira et al. (2021), and directly related to collective irrigation systems, the price that the irrigators pay for the water must cover the expenses associated with the provided service. If the water prices rise, the crops' economic sustainability may be put at risk, jeopardizing their production. On the other hand, there is a lack of know-how in water utilities' internal staff for planning and implementing the digitalization process (Stoffels & Ziemer, 2017). This can be cooped by improving water utility boards with specialized personnel (in digital transformation) but is often not done due to a lack of financial resources, leaving the already existing water utility personnel with the arduous task of dealing with (and prioritizing) the everyday tasks (that ensure the quality of the provided service), plus exploring the existing digital possibilities necessary for the digital transformation (Müller-Czygan et al., 2021).

Legacy data collection and management systems and inadequate digital solutions can be seen as major barriers to the digitalization process. Water utilities already have some sort of digitalization degree implemented, often using old software that was created and expanded to address specific everyday tasks. However, such solutions have ambiguous IT specifications and there is a lack of technical standardization (amongst digital solutions) (Stoffels & Ziemer, 2017). This can be due to the missing demand for digital solutions from customers and suppliers as it is a mostly niche market (in software relatable terms and when compared to other markets like health and wellbeing or financial). As noted by Savić (2021), the water industry still lags behind the automotive and aircraft industries in terms of the number of sensors and their density; an increased level of redundancy is required and will improve data quality and reduce the uncertainty associated with sensing data. As a result, the digitalization process can be challenging as there is a need for an adequate technical infrastructure that allows easy integration of smart solutions, whilst minimizing potential technological constraints (Brianne Nakamura, Williams, 2017). On the other hand, digital solutions may not yet meet the water utilities' needs. For instance, conventional hydraulic solvers (e.g., EPANET) were originally developed for scenario analysis and are not readily adopted for real-time hydraulic modelling (Mesquida, 2021; Shafiee et al., 2020). Furthermore, network model simplifications are often carried out (e.g., skeletonization) but, as noted by Pesantez et al. (2021), may create some barriers to implementing a DT as the pipe flows and hydraulic performance cannot be simulated at a finer resolution with the existing model. The development and use of advanced Artificial Intelligence techniques for predictive and optimal control of water treatment stations have also proved to be a challenge, specifically the macro intelligence model and decision scheme for entire treatment plants to support the overall management of the water supply systems (Mondejar et al., 2021).

The **lack of a clear plan for digital transformation** can be seen as another relevant barrier. The overall digital transformation process may seem incomprehensible to some, in particular advanced concepts like building a DT which resort to several technologies that must be previously available (Pedersen et al., 2021). This is a common reality in municipal water utilities that have not yet installed sensors extensively in their systems (and such sensors are required for the overall digital transformation). A common understanding of the multiple purposes of the digitalization process must emerge between the company's various functions and staff members. There is thus a need to discuss the requirements and difficulties for the digital transformation openly and transparently and by involving all partners. As noted by Stoffels & Ziemer (2017), water utility staff is often unclear about the overall benefits of the digital transformation and, as result, the novel digital technologies (when acquired) function below the full efficiency level (Banerjee et al., 2022).

The **"human factor"** is also a real barrier to the digitalization process and water utilities' staff often see the digitalization process as an unexpected high effort with unclear benefits (Müller-Czygan et al., 2021). The staff might also fear that the additional involvement may be useless as not everyone will commit to the additional tasks in the same way, or that the market is not yet ready. As noted by Nakamura and Williams (2017), human resistance due to fear of replacement or redundancy can prevent fluid transition and implementation of new methods and innovative digital solutions. The implementation of such technologies also requires the training of human resources, with an associated learning curve.

Cybersecurity can be seen as the last major barrier to digital transformation. As noted by Müller-Czygan et al., (2021), many water utilities report unresolved IT security issues and there is an overall lack of standardized interfaces. Data security is one of the major concerns on big-data sources' wide accessibility and openness as, being a sensitive topic, generates debate associated with safety risks and network integrity of these digitalized services (Mondejar et al., 2021). In short, water utilities fear data theft (Stoffels & Ziemer, 2017). Consider the example of a data streaming of end-user water consumption. On the one hand, such a stream of data can be malicious consulted to assess if water is being used, thus indicating that activity is happening (or not) inside the house. On the other hand, the end-users might maliciously alter the readings, for instance, to simulate a smaller water consumption. The overall data connectivity and security issues must be solved as new platforms must integrate and communicate with existing technologies (Nakamura and Williams, 2017). Despite the best efforts that a water utility makes to prevent cyberattacks, some may happen. Hence, to minimize the consequences of a cyberattack the water utility should have backup and recovery systems.

The above-mentioned barriers are enhanced when considering the digital transformation of the water sector in developing countries. In this regard, Yusuf et al. (2021) highlighted financial constraints, technological barriers, data security, lack of qualified personnel, lack of government support, resistance to change, lack of awareness, and poor internet facility as the major challenges of adopting digital technology in developing countries.

In the authors' view, for digitalization in the water sector to be effectively achieved, more than technological solutions are needed. The harmony between the technical, human, and financial aspects is the key to success. A plan for the digital transformation is also of utmost importance; its development should be the first step and should indicate the desired objectives and advantages, information on the current information processes inside the water utility, and establish the steps in a given timeframe for the digital transformation.

DATA QUALITY

Reliable data is the basis of an effective digital transformation. In water utilities, such data include cadastral (e.g., location, components' technical specifications,), operational information (e.g., valves and pumps operating conditions, water quality parameters), maintenance (e.g., condition, information about interventions and failures in the various assets), as well as billing and account information (i.e., network-related maintenance and operational costs, energy and personnel costs, amortizations and interest, revenues, among others). Such diverse data is usually collected, stored, and manipulated using specific

The Challenge of the Digitalization of the Water Sector

Figure 2. Barriers to digitalization of the water sector

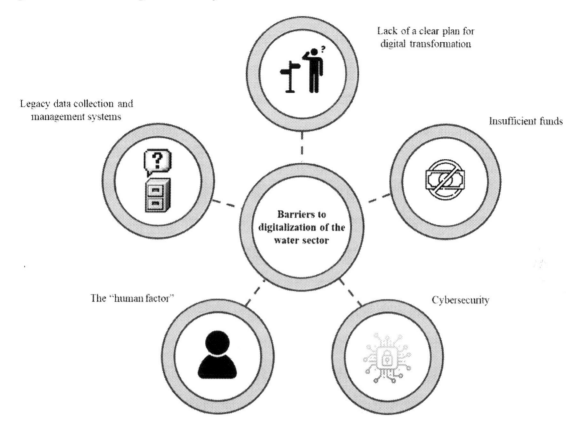

information systems which are, generally, in different departments or divisions. A strategic vision of the technological environment of the water utility is usually inexistent, leading to a vast ecosystem of software that is accessed and updated by numerous technicians. This results in many water utilities being "data rich" and "information poor".

In this reality, it is important to understand that the problem may be not technology-based but people-based. Data of quality can only be achieved with a data-driven philosophy among the water utility personnel which requires, amongst other practices, the optimization of data and information processes and flows, the definition of a strategic vision of the technological environment of the water utility, and the coordination between the various users of information within the utility.

It is, thus, essential that stakeholders are engaged to define what type of information is needed and what data will directly (or indirectly) satisfy its needs for the system's management. Such stakeholders may include accountants, maintenance, water quality or design technicians, amongst others. Although each stakeholder may have its distinct data needs for the system's management, some overlaps can be found in essential types of data, for instance, in the needs for consumers' location and information or the geographical information of physical assets. Coordination between the multiple departments must occur in such situations to avoid duplication of information in multiple databases and information systems.

Field collected data must be done systematically, in which the same type of data is collected using a standardized procedure and equipment (e.g., a form that is filled on-site using an application on a Tablet or smartphone).

However, most of the collected data come from specific instrumentation that must be installed, operated, and maintained properly. Such instrumentation usually represents a large capital expenditure, and its location and characteristics must be thoroughly assessed to ensure that suitable data is generated. Furthermore, such measuring equipment must be periodically maintained and calibrated to guarantee that collected data and insights are based on accurate fundamentals and that both tactical and strategic views are not compromised.

Regardless of the acquisition and transmission settings, the collected raw data may contain measurement errors, such as missing, repetitive or even false readings. These errors can have their origin in sensor or logger malfunctioning, faulty transmission system due to battery failure, inadequate acquisition range or data storage limitations. These measurement errors, typically known as outliers, should be detected and corrected before they can be used in engineering applications (Ferreira et al., 2022).

As noted by Moreira et al. (2021), there is a danger that system managers may consider that all collected data are correct and represent the process being measured. Thus, the more progress is done by a water utility in data acquisition (and the more dependent it is on collected data), the greater the need to check the quality of such data. Such data quality has a direct impact on future uses of such data, for instance, by directly influencing the results of water and energy balances.

In this regard, distinct approaches have been used, either by pre-processing the collected data before using or by assuming that data contains errors and analysis methods must take that into account. In Pesantez et al. (2021), analytical tools are connected to the data source in order to process the acquired data by filling in missing values and by filtering out corrupted or missing data. Ferreira et al. (2022) developed a tool to automatically process flowrate evenly and unevenly spaced time series data by using a set of tests that identify the main anomalies in collected flowrate time series. Valverde-Pérez et al. (2021) removed outliers by using statistical tests and historical records, whilst Shafiee et al. (2020) used analytical tools to ensure the quality of user consumption data before updating the hydraulic simulation model. In a distinct approach, Romano et al., (2020) used machine learning and Artificial Intelligence algorithms that can efficiently deal with the vast amount of often imperfect sensor collected data.

Hydraulic simulation models are often used, in addition to the already collected sensor data, in order to provide enhanced insights into the system's behaviour. Such models include simulation of hydraulic behaviour in distribution or drainage networks, water quality parameters, and material degradation, amongst others. The calibration and validation of such models are essential to guarantee that reliable results are being obtained. Such calibration and validation are often done using sensor-collected data which, if done incorrectly, may jeopardize the model's outputs. Additionally, the overall calibration process of models is cumbersome in many water utilities as it is a time-consuming process, especially compared with the many other duties of a water utility. Thus, calibration is often carried out for short periods and with few measuring points (Pedersen et al., 2021).

Keeping data quality is also important if the water utility wants to use the collected data with artificial intelligence algorithms, such as machine learning or deep learning techniques. Such techniques require a vast amount of data regarding the system's behavior, both in the base operation scenario and during abnormal. In short, Data Analytics is always hungry for data which must be up to date.

In summary, several criteria should be taken into consideration regarding data quality, namely, accuracy (e.g., is it acceptable?), precision (e.g,., is the granularity sufficient?), availability (e.g., is the right data available to the right people within the water utility?), relevancy (i.e., irrelevant information was filtered out?), completeness (e.g,., is the data enough for the analysis?), timeliness (e.g.,., the analysis is made in batch or in real-time?), consistency (e.g,., is the data in the expected format?), validity (e.g.., the data model which connects the data has a sound structure?) and ability to be understood (e.g., were the data cleaned and are they more readable? is data presented to the user in an adequate dashboard?).

HOW TO CHOOSE THE BEST DIGITAL SOLUTION?

Digital Transformation Plan

Digital solutions are essential in the various areas of activity of a water utility and their acquisition should be planned given the technological environment of the water utility as they usually represent relevant costs in acquisition or subscriptions. The vast choice in digital systems and technologies can be, however, overwhelming (Pub, 2020). The water utility should elaborate a strategic digital transformation plan, with a long-term focus on the implementation of digital systems and technologies that fits better the water utility's needs. The first step to starting the plan is to identify the status quo situation (i.e., which is the current situation of the digitalization process?). Then, clear and specific objectives should be defined, for which strategies to achieve the targets defined with the objectives are also established. A water utility may adopt digital solutions with different goals, such as asset management (e.g., inventory of the assets), customer engagement (e.g., water use and service), and predictive analytics (e.g., asset failure prediction), among others. The plan should also be developed with the support of the water utility's highest-level decision-makers (i.e., the executive board) to ensure the overall viability of the plan.

Figure 3 depicts the five main stages of a digital transformation plan aiming at obtaining a digital twin of a certain physical infrastructure (e.g., the water distribution network). The plan includes obtaining cadastral information of the assets in a first stage, which will support the construction of the geographic information system model. By using such systems, it is thus possible to create and easily update the hydraulic simulation model. This simulation model can finally be updated with real-time measurements to develop a digital twin.

Digitalization Process

The digitalization process should follow the guidelines previously established in the digital transformation plan. It is not expected that the digitalization process occurs simultaneously with all the information and management processes inside a water utility and, as such, priorities must be followed (as established in the digital transformation plan). In practical terms, the techniques, equipment, and investment involved in digitalizing the management of a treatment station are very distinct from the ones used to optimize the operation of a pumping station or even for the remote measurement of end-user consumption. Depending on the overall objective, a thorough assessment of the required data must be carried out. This may involve the acquisition of new sensors and changing internal practices like adopting a common format for data acquisition. Common industry standards should be used when possible, for instance, common date and time formats, and common acquisition file formats, among others. The software already being used should be fully explored to assess if it has the capabilities to address the new needs. The acquisition of new software should be done carefully as it usually represents high investments and involves personnel training. Moreover, the digitalization of water utilities may encounter internal resistance from the employees due to fears that they will lose their jobs to automation. Is important to involve and engage

Figure 3. Main stages of a digital transformation plan for the development of a digital twin

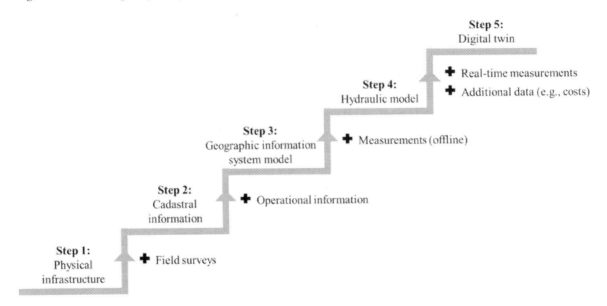

them in the elaboration of the digitalization plan by hearing its feedback and also communicating effectively the advantages of digitalization.

Acquisition or Development of Digital Solutions?

Water utilities with a greater digitalization maturity level may have the required resources (i.e., financial, human, and technological) to develop their solutions for a specific need. These solutions integrate data gathered from different sources (i.e., sensors, GIS, equipment automation signals, among others), as well as results from hydraulic simulation models, aiming ultimately at predictive models of network hydraulic behaviour and decision support tools. Some water utilities choose to integrate data through GIS-driven solutions, in which it is possible to easily access data related to infrastructures, fleets, water demands, and billing, among others). Most water utilities simply lack the resources to resort to the former solution and frequently give in to the commercial pressure for reactively acquiring a solution that addresses a specific end-use. As such, the complete IS capabilities are usually not fully explored since it was not acquired in a long-term philosophy. Different commercial software claims to provide comprehensive, actionable analytics to assist water utilities with specific efficiency issues and key business outcomes (Carriço et al., 2020). However, the cost of such solutions is usually a barrier to many water utilities.

Regarding the wireless infrastructure needed to transmit the data collected by the sensors, there are plenty of solutions in the market, such as 4G, 5G, LoRa, and SigFox, among others. In the choice of one of these network communication technology to use, the water utility should give attention to some important aspects, such as range and transmission rate, size of data to be transmitted, the network's power consumption and latency, and security, among other factors (Pub, 2020).

It is recommended that water utilities, especially those struggling to manage ageing infrastructure with minimal staff, digitalize incrementally and at their pace. Water utilities may not need a massive digital renovation, but step-wise solutions tailored to their capacity and needs. Nonetheless, the best solutions

The Challenge of the Digitalization of the Water Sector

may not be affordable for everyone. Since the digital transformation will not happen overnight, it is of utmost importance that water utilities prepare broader systems that can be aligned with newer (or to be acquired) digitalized services. Although technology advances fast every year, water utilities mustn't pursue exciting solutions whilst disregarding the ultimate goal to improve the provided service (Pokhrel, 2022).

TRENDS FOR A SMARTER WATER UTILITY

In the future, water needs to be managed more sustainably and effectively. Nowadays, water management is done separately at a system level by function (e.g., drinking water, wastewater). Nonetheless, water circulates in a loop through different functions and uses and is scarce in several locations on the planet. Therefore, water management should be done in an integrated mode. Integrated Water Resources Management (IWRM) should be the future of water-wise management. Additionally, water circularity should be intensified by using and reusing water in the best way possible. Most of the existing cities in the world prioritized the building of concrete infrastructures, usually called grey infrastructures, which ultimately increase city floods. In the future, cities should include green infrastructures in their design, namely based on nature-based solutions. Of course, this bright future brings challenges that the digital transformation of water utilities may help to overcome.

The future generation of water utilities will be much more tech-savvy than the present ones due to the youngsters which will be employed by the water utilities, and which constitute the most digital generation ever existed until now. These water experts will have digital skills and adapt themselves more quickly to new technological advances. The most advanced water utilities will handle technologies like artificial intelligence, digital, twin, fintech (e.g., blockchain, cryptocurrencies, smart contracts), metaverse, cybersecurity, 5G/6G, robotics, nanotechnology, augmented and virtual realities, and new imaging technologies.

As mentioned above, Digital Twins are also a technological solution that can revolutionize the operation and management of assets in a water utility. A digital twin of an urban water system combines a multitude of real-time data signals (e.g., pressure, flow, levels) with different time frequencies, aiming to get useful insights into the current and future status of the system. In the last few years, the water sector has given attention to this solution. A digital twin aggregates the advantage of several technologies, namely, monitoring of variables and information, data science applications, predictive analytics, direct visualization, scenario planning, and mapping futures (Meschengieser, 2022).

Blockchain technology is a shared, immutable ledger that facilitates the process of recording transactions and tracking assets in a business network. Virtually anything of value can be tracked and traded on a blockchain network, thus reducing risk and cutting costs for all involved (Gupta, 2018). An example of the use of Blockchain in the water sector is through the use of smart contracts. A smart contract is an application executed in a blockchain network that constitutes a digital twin of the contract, and which is self-executing. This means that the water utility needs to have an Advanced Metering Infrastructure (AMI) in which the blockchain performs smart contract functions concerning monthly water consumption determination and billing (EMA, 2018). Of course, this type of transaction requires that the utility pays special attention to issues like cybersecurity. Artificial Intelligence chatbots may also be used by water utilities to answer users' questions instead of leaving them on hold with a call center. The use of such systems will ultimately will the relationship between the water utility and the user.

The most advanced water utilities are already using virtual reality (VR) and augmented reality (AR) in their activities. Both technologies are distinct types of data and information visualization, but VR goes a step further allowing managers to explore different scenarios. These technologies can be used to train new water experts without exposing them to a risky environment and any mistakes made will not have harmful consequences and allow them to teach their staff. Of course, these technologies provide a myriad of advantages.

The use of data outside the scope of the water industry, for instance, social data available through social media platforms (e.g., Twitter and Facebook) and transport applications (e.g., Waze or Google Maps), may provide important information for the system's management, namely, by being integrated into water demand forecasting models for more accurate estimations.

CONCLUSION

This chapter briefly presented the enormous challenges and the promising future that water utilities face in the journey of digitalization. The huge amount of data that is collected by a water utility in every day's activity tends to increase and digital transformation should aim for open data and data standardization (von Ditfurth et al., 2021). To transform data into useful insights (i.e., wisdom in the DIKW hierarchy) the water utilities need to verify the quality of their collected data aiming to avoid misinformation and always bearing in mind that "garbage in, garbage out". Ultimately, effective digitalization requires the cultivation of good metadata management practices amongst the water utility personnel (Grievson et al., 2022).

Notwithstanding the fast digital solutions advances in recent years, the success of digitalization transformation in a water utility relies fundamentally upon the human element. So, employees should be involved in the process and motivated to embrace this hard journey. Though enthusiasm is essential for digitalization, so is having knowledge and capabilities. In many water utilities, employees simply make spreadsheets and compile data for reports (Pub, 2020). A water utility will be more efficient if its employees can explore the available data by using more advanced tools, for instance, machine learning or forecasting models. However, many water utilities face several difficulties in achieving their digital transformation, namely, ageing assets, ageing employees, and a lack of financial resources to invest.

Finally, water will only be efficiently and sustainably managed if all water stakeholders and users (i.e., urban, industry, and agriculture) are digitally connected into a comprehensive community of shared knowledge, from the basin to the city. That is, for sure, the biggest challenge to overcome in the future. The journey is long but promising.

ACKNOWLEDGMENT

The authors acknowledge the Fundação para a Ciência e a Tecnologia for supporting this chapter, under the WISDom project (grant number DSAIPA/DS/0089/2018), and the project's team for their contribution.

REFERENCES

Alzamora, F. M., Carot, M. H., Carles, J., & Campos, A. (2019). Development and Use of a Digital Twin for the Water Supply and Distribution Network of Valencia (Spain). *Water Quality Models for Water Distribution Systems*. Pilar Conejos Aguas de Valencia.

Banerjee, C., Bhaduri, A., & Saraswat, C. (2022). Digitalization in Urban Water Governance: Case Study of Bengaluru and Singapore. *Frontiers in Environmental Science*, *10*(March), 1–12. doi:10.3389/fenvs.2022.816824

Bonilla, C. A., Zanfei, A., Brentan, B., Montalvo, I., & Izquierdo, J. (2022). A Digital Twin of a Water Distribution System by Using Graph Convolutional Networks for Pump Speed-Based State Estimation. *Water (Basel)*, *14*(4), 514. doi:10.3390/w14040514

Carriço, N., & Ferreira, B. (2021). Data and Information Systems Management for the Urban Water Infrastructure Condition Assessment. *Frontiers in Water*, *0*, 670550. Advance online publication. doi:10.3389/frwa.2021.670550

Carriço, N., Ferreira, B., Barreira, R., Antunes, A., Grueau, C., Mendes, A., Covas, D., Monteiro, L., Santos, J., & Brito, I. S. (2020). Data integration for infrastructure asset management in small to medium-sized water utilities. *Water Science and Technology*, *82*(12), 2737–2744. Advance online publication. doi:10.2166/wst.2020.377 PMID:33341766

Conejos Fuertes, P., Martínez Alzamora, F., Hervás Carot, M., & Alonso Campos, J. C. (2020). Building and exploiting a Digital Twin for the management of drinking water distribution networks. *Urban Water Journal*, *17*(8), 704–713. doi:10.1080/1573062X.2020.1771382

Cooper, J. P., Jackson, S., Kamojjala, S., Owens, G., Szana, K., & Tomić, S. (2022). Demystifying Digital Twins: Definitions, Applications, and Benefits. *Journal - American Water Works Association*, *114*(5), 58–65. doi:10.1002/awwa.1922

EMA. (2018). *Blockchain and the water industry—Smart contracts*. https://www.ema-inc.com/news-insights/2018/10/blockchain-and-the-water-industry-smart-contracts/

Ferreira, B., Carriço, N., Barreira, R., Dias, T., & Covas, D. (2022). Flowrate Time Series Processing in Engineering Tools for Water Distribution Networks. *Water Resources Research*, *58*(6), 1–20. doi:10.1029/2022WR032393 PMID:35813986

Grievson, O., Holloway, T., & Johnson, B. (Eds.). (2022). *A Strategic Digital Transformation for the Water Industry*. IWA Publishing., doi:10.2166/9781789063400

Gunn, G., & Stanley, M. (2018). Harnessing the Flow of Data: Fintech opportunities for ecosystem management (p. 16).

Gupta, M. (2018). Blockchain for dummies (2nd IBM limited edition). John Wiley & Sons, Inc.

Karmous-Edwards, G., Tomić, S., & Cooper, J. P. (2022). Developing a Unified Definition of Digital Twins. *Journal - American Water Works Association*, *114*(6), 76–78. doi:10.1002/awwa.1946

Meschengieser, G. (2022). *Waterfuturism, a new perspective that is here to stay*. IWA. https://iwa-network.org/waterfuturism-a-new-perspective-that-is-here-to-stay/

Mesquida, M. (2021). *Digital Twin in Water Distribution Networks*. Issue February.

Mondejar, M. E., Avtar, R., Diaz, H. L. B., Dubey, R. K., Esteban, J., Gómez-Morales, A., Hallam, B., Mbungu, N. T., Okolo, C. C., Prasad, K. A., She, Q., & Garcia-Segura, S. (2021). Digitalization to achieve sustainable development goals: Steps towards a Smart Green Planet. *The Science of the Total Environment*, *794*(June), 148539. doi:10.1016/j.scitotenv.2021.148539 PMID:34323742

Moreira, M., Mourato, S., Rodrigues, C., Silva, S., Guimarães, R., & Chibeles, C. (2021). Building a Digital Twin for the Management of Pressurised Collective Irrigation Systems. In *Proceedings of the 1st International Conference on Water Energy Food and Sustainability (ICoWEFS 2021)* (Vol. 3, Issue ICoWEFS 2021, pp. 785–795). Springer International Publishing. 10.1007/978-3-030-75315-3_83

Müller-Czygan, G., Tarasyuk, V., Wagner, C., & Wimmer, M. (2021). How does digitization succeed in the municipal water sector? The waterexe4.0 meta-study identifies barriers as well as success factors, and reveals expectations for the future. *Energies*, *14*(22), 1–21. doi:10.3390/en14227709

Nakamura, B., Williams, P. (US) Inc. (2017). Intelligent Water System: The Path to a SMART Utility. *Joint Knowledge Development Forum (KDF) in Conjunction with the International Society of Automation's (ISA) Water/Wastewater and Automation Controls Symposium*.

Pedersen, A. N., Borup, M., Brink-Kjær, A., Christiansen, L. E., & Mikkelsen, P. S. (2021). Living and prototyping digital twins for urban water systems: Towards multi-purpose value creation using models and sensors. *Water (Switzerland)*, *13*(5), 592. doi:10.3390/w13050592

Pesantez, J. E., Alghamdi, F., Sabu, S., Mahinthakumar, G., & Berglund, E. Z. (2021, July). Using a digital twin to explore water infrastructure impacts during the COVID-19 pandemic. *Sustainable Cities and Society*, *77*, 103520. doi:10.1016/j.scs.2021.103520 PMID:34777984

Pesantez, J. E., Alghamdi, F., Sabu, S., Mahinthakumar, G., & Berglund, E. Z. (2022). Using a digital twin to explore water infrastructure impacts during the COVID-19 pandemic. *Sustainable Cities and Society*, *77*, 103520. doi:10.1016/j.scs.2021.103520 PMID:34777984

Pokhrel, N. (2022). *Five Lessons For Digitizing Asia's Water Systems* [Text]. Asian Development Bank. https://blogs.adb.org/blog/five-lessons-digitizing-asia-s-water-systems

Pub, S. N. W. A. (2020). *Digitalising Water – Sharing Singapore's Experience*. doi:10.2166/9781789061871

Romano, B. Y. M., Boatwright, S., Mounce, S., & Nikoloudi, E. (2020). *AI – BASED EVENT MANAGEMENT AT UNITED UTILITIES*. *4*, 104–109.

Savić, D. (2021). Digital Water Developments and Lessons Learned from Automation in the Car and Aircraft Industries. *Engineering*. doi:10.1016/j.eng.2021.05.013

Shafiee, M. E., Rasekh, A., Sela, L., & Preis, A. (2020). Streaming Smart Meter Data Integration to Enable Dynamic Demand Assignment for Real-Time Hydraulic Simulation. *Journal of Water Resources Planning and Management, 146*(6), 06020008. doi:10.1061/(ASCE)WR.1943-5452.0001221

Stoffels, M., & Ziemer, C. (2017). Digitalization in the process industries – Evidence from the German water industry. *Journal of Business Chemistry, 14*(3), 94–105. doi:10.17879/20249613743

Valverde-Pérez, B., Johnson, B., Wärff, C., Lumley, D., Torfs, E., Nopens, I., & Townley, L. (2021). *Digital Water in the urban water*. International Water Association.

von Ditfurth, H., Weisbord, E., Danielsen, T., Zutari, L. F.-J., Hafemann, A. C., Hima, J., & Oraeki, T. C. (2021). *Digital Water: An overview of the future of digital water from a YWP perspective*. International Water Association. https://iwa-network.org/publications/digital-water-an-overview-of-the-future-of-digital-water-from-a-ywp-perspective/

Yusuf, S., Musa, M. A., Diugwu, I., Adindu, C., & Afeez, B. (2021). A Systematic Literature Review Approach on the Role of Digitalization in Construction Infrastructure and Sustainable City Development in Developing Countries. *ZEMCH International Conference*, (pp. 1075–1093).

Chapter 4
Blockchain-Based IoT E-Healthcare

Harpreet Kaur Channi
Chandigarh University, India

Chiranji Lal Chowdhary
　https://orcid.org/0000-0002-5476-1468
Vellore Institute of Technology, Vellore, India

ABSTRACT

Numerous industries, including e-healthcare, are capitalizing on and using blockchain and internet of things (IoT) technology. IoT devices may collect patient vitals and other sensory information in real-time, which medical professionals can then examine. All information gathered from the internet of things is stored, processed, and computed in one place. Such concentration raises concerns since it increases the likelihood of a catastrophic failure, distrust, tampering with data, and even the circumvention of privacy protections. By offering decentralized processing and storage for IoT data, blockchain has the potential to address these critical issues. As a result, designing a decentralized IoT-based e-healthcare system that incorporates IoT and blockchain technology might be a viable option. First, the authors provide some context about blockchain in this essay. The viability of blockchain systems for the internet of things-based e-healthcare is then assessed.

INTRODUCTION

The decentralized ledger, source provenance, and tamper-proof features of blockchain (BC) technology have made it an integral part of several applications, including healthcare and the internet of things (IoT) (Abu-Elezz et al., 2020; Yoon, 2019, p. 3; Ben Fekih & Lahami, 2020). Health systems may now increase their scalability and stability on a decentralized platform because to IoT and BC. As a result, numerous scientists have investigated the potential of BC technology in many areas of eHealthcare, leading to the development of BC-enabled IoT eHealth systems. Because of its potential to improve the safety, dependability, and sturdiness of distributed systems, blockchain technology has recently come

DOI: 10.4018/978-1-6684-6123-5.ch004

into the spotlight. Studies based on this technique have benefitted several fields, including healthcare, finance, remote sensing, and data analysis (Srivastava, Parizi, & Dehghantanha, 2020; Anjum et al., 2020; Chanchaichujit et al., 2019). The primary benefits of blockchain technology are its immutability of data, privacy, transparency, decentralization, and distributed ledgers. This system is already complex due to the need to protect the privacy of individuals whose information is contained in healthcare records (Hussien et al., 2021).

Simply said, a blockchain is a shared digital record that can be accessed by any computer connected to the internet. Since information is stored digitally, a blockchain may be compared to an online database (Wong, Yee, & Nøhr, 2018). Blockchains serve an important purpose in cryptocurrency networks like Bitcoin's by providing a secure and decentralized ledger of transactions. Novel blockchain technology guarantees the authenticity of each recorded transaction without the need for a neutral third party (Siyal et al., 2019).

One definition of a blockchain is "a distributed database that maintains a continuously expanding list of ordered items, called blocks." The encrypted connection between these blocks. Each block contains transaction data, a timestamp, and a cryptographic hash of the previous block (Dash, Gantayat, & Das, 2021; Hussien et al., 2019). Blockchains are distributed digital ledgers that are public and decentralized, and which are used to record transactions across several computers in a way that precludes modifications to the ledger from being made in the past without also modifying the subsequent blocks and getting network permission (Angraal, Krumholz, & Schulz, 2017). In comparison to conventional databases, blockchains have a fundamentally different approach to data organization. Data on a blockchain is organized into blocks, which are collections of records. In a blockchain, each completed block is cryptographically linked to the one before it. After a new block is uploaded to the chain, all subsequent data is merged into a single block and then added to the chain as well. If you're familiar with databases, you'll notice that blockchain data is presented in the form of linked records rather than the tabular format more characteristic of databases (Zubaydi et al., 2019; Engelhardt, 2017; Shahnaz, Qamar & Khalid, 2019). When used in a decentralized setting, this data format creates a permanent record of events in the data without any human intervention. As each empty spot is filled, a permanent entry is made in this history book. An accurate timestamp is appended to each block in the chain whenever a new one is added.

The whole chapter has been organized as follows: Firstly we discussed the need problem formulation and objectives of using Blockchain-based IoT e-Healthcare. Then it is followed by features of blockchain, its uses, Internet of things, health information technology (HIT), future scope of the technology, and in the last section - result and discussion is introduced with conclusion.

NEED, PROBLEM FORMULATION, AND OBJECTIVES

Several issues prohibit the end-to-end security of IoT devices. Because networking appliances and other things is relatively new, product designers haven't always prioritised security. Because IoT is a new industry, many designers and manufacturers are more focused in bringing their goods to market fast than building in security from the outset. Hardcoded or default passwords may compromise IoT security. Passwords are frequently not strong enough to prevent intrusion, even when updated.

IoT devices frequently lack the CPU power needed to enable good security. As a result, many gadgets lack sophisticated security. Humidity and temperature sensors can't handle modern encryption or security. As many IoT devices are "set it and forget it," they seldom get security upgrades or fixes. From a manufacturer's perspective, integrating security in from the start may be expensive, hinder development, and cause device malfunction. Legacy assets not built for IoT are another security problem. Many assets will be retrofitted with smart sensors instead of replacing traditional infrastructure with linked technologies. The attack surface expands if outdated assets aren't upgraded or secured against contemporary threats.

Since decades, the worldwide healthcare system has deteriorated. With bed shortages, lengthy waiting lists, rising expenses, and worldwide pandemics, the global healthcare system has never been more complex. Global healthcare underuses blockchain technology. It can enable efficient, transparent, safe, and successful data and information communication in healthcare. Tokenization and smart contracts may minimise or eliminate healthcare pre-authorization.

Blockchain-based health documentation systems enable strong encryption to protect patient data. Tokenization, smart contracts, and blockchain network encryption will eliminate pre-authorization, speeding up care. Before, healthcare providers depended on patients or papers received or emailed from local doctors, laboratories, etc. Looking at current research in the subject, this chapter provides scholars with a balanced perspective on the development of Blockchain Technology, the Internet of Things, and healthcare.

IMPORTANT FEATURES OF BLOCKCHAIN

Leading firms have realised that the only way to acquire a competitive advantage is to use time compression in supply chain process design and operation. Companies must deliver more value for less money in less time in order to thrive, i.e. raise the proportion of time spent "adding value" in the whole supply chain process.

Information on a blockchain, which is a sort of distributed database, is stored in blocks that are cryptographically linked to one another. When new information is received, it is stored in a new block. Once a block is completely filled, it is connected to the one before it, producing a chain that stores information in chronological order (Saeed et al., 2022; Radanović, & Likić, 2018; Haleem et al., 2021).

- Blockchains may be used to record other sorts of data, although thus far they have mostly been used as a distributed ledger for recording financial transactions.
- Blockchain is employed in a decentralized fashion in Bitcoin, meaning that no central authority or group of people has control over the system and instead all Bitcoin users have equal voting rights.
- By its very nature, the data placed into a decentralized blockchain cannot be changed after it has been recorded. For Bitcoin, this implies that all trades are publicly accessible and forever recorded.

BLOCKCHAIN TECHNOLOGY

The term "blockchain" is used to refer to the digital ledger (or list of records) used to keep track of information as shown in Figure 1. It's organized as a series of blocks, and the cryptographic principle is built into the whole thing. Data and conversational exchanges can be protected through the use of cryptography. Supported by mathematical ideas and a collection of rule-based calculations (algorithms), it alters communications in a way that is difficult to decode (Leeming, Cunningham & Ainsworth, 2019; Chakraborty, Aich & Kim, 2019):

Figure 1. Blockchain technology

- Blockchain technology is decentralized, thus it may be used by anybody. In other words, no one can claim ownership.
- The level of transparency is too high and protects individual privacy.
- Because information saved on a blockchain cannot be altered after it has been recorded, it is considered immutable.

Blockchain technology's core purpose is to keep sensitive information safe from prying eyes while making it available to a select, pre-approved audience. Following are the are the three major ideas that make up blockchain development (Yaeger et al,, 2019; Khan et al., 2020; Massaro, 2021; Azogu et al., 2019):

Blocks

Numerous blocks make up each chain, with the following three providing the bulk of its functionality:

- Information alone constitutes the building component.
- Each time a new block is formed, an encrypted one is generated using a random 32-bit nonce (number only used once). And so, a block header hash is born.
- Connected to the nonce, the hash is a 256-bit integer. The first digits should be a bunch of zeroes.
- In addition, the nonce will be used to produce the cryptographic hash whenever a new block is created. The nonce effectively makes it unusable until it is mined.

Miners

It's also crucial to use Miners. It's a lot of work to mine. Every new block generates a new hash and nonce, but always uses the hash from the prior block as its starting point. In order to solve the extremely difficult mathematical problem of producing a decent hash, miners employ specialized software. For a 256-bit hash to be successfully cracked using a 32-bit nonce, roughly four billion nonce-hash permutations must be tested. This signifies that the miners have found the "golden nonce" and that their block has been successfully put to the network. Whenever a block in the chain is modified, mining must begin over from the very first block (Kumar et al., 2018; Hasselgren et al., 2020; Daniel et al., 2017; Wang, 2020; Clauson et al., 2018). This makes it extremely hard to tamper with blockchain data. Since it takes a lot of time and processing power to obtain a golden nonce, they are called "secure in math."

Nodes

Blockchain technology relies heavily on its ability to be decentralized. The blockchain cannot be controlled by a centralized server or organization. On the other side, it is a blockchain platform shared by all of the nodes in the network. To store copies of the blockchain, nodes are computerized devices (Laroiya, Saxena, & Komalavalli, 2020; Saha et al., 2019; Tanwar, Parekh & Evans, 2020). There is a copy of the blockchain on every computer, and the network as a whole must approve each new block that is mined before it can be added to the chain and verified as authentic. Since all transactions in a blockchain are public, the entire ledger can be checked and audited with ease. For the purpose of keeping tabs on everyone's dealings, an alphanumeric participant ID is provided to each person (Faisal et al., 2022).

HOW BLOCKCHAIN TECHNOLOGY IS TRANSFORMING THE CYBERSECURITY

Fintech, healthcare, eCommerce, etc. all stand to benefit from the cutting-edge tools made possible by blockchain technology. Businesses in the modern day are dependent on digitization because of the countless benefits it provides in terms of expansion, return on investment (ROI), longevity, and the capacity to bounce back from setbacks (McBee & Wilcox, 2020; Sharma et al., 2021). But there are always two sides to a tale. As beneficial as technology and digitalization may be, they are not without their drawbacks. You have enemies who are just waiting to cash in on your hard work. The prevalence of cybercrime and security holes increases as technology advances (Govindan et al., 2022). The blockchain's large variety of trustworthy and secure applications can help us meet these difficulties head-on. For instance, in the banking industry, hackers pose a serious threat to clients' digital banking accounts and the sensitive information stored inside. They have a greater chance of having their financial information compromised, including their account number, cash flow, and balance (Li et al., 2021; Mallikarjuna, Shrivastava, and Sharma, 2022; Sharma, Kaur & Singh, 2021; Dutta et al., 2020).

Both public and private blockchains exist. In a public blockchain, anybody may participate by reading, writing, or verifying the ledger's records. The fact that there is no central authority in charge of a public blockchain makes it incredibly difficult to reverse previously recorded transactions (Idrees et al., 2021; Srivastava et al., 2021; Kodali, Swamy & Lakshmi, 2015). A private blockchain, on the other hand, is one that is only accessible within a certain network, such as a company or a club. They are the only ones who can oversee a blockchain ledger that is exclusively accessible internally. It's the only thing that can tell who can crack the code. They have the ability to retroactively make changes to the blockchain. This private blockchain implementation is similar to a corporate intranet database, except the data is spread out among many servers for added safety (Kashani et al., 2021; Darshan, & Anandakumar, 2015; Azzawi, Hassan & Bakar, 2016).

USES OF BLOCKCHAIN TECHNOLOGY

There is a growing number of mandatory and expanding use cases for blockchain technology. The process of protecting digital accounts to facilitate asset transfers is only one example of the many sophisticated processes that are encouraged (Yeole & Kalbande, 2016; Selvaraj & Sundaravaradhan, 2020; Tyagi,

Figure 2. Uses of Blockchain technology

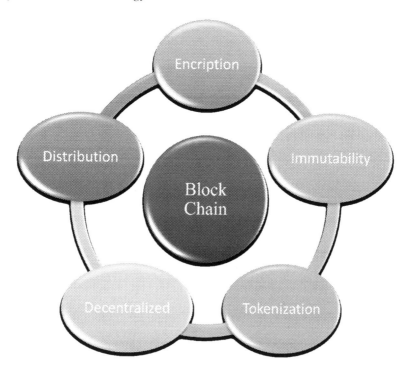

Agarwal & Maheshwari, 2016; Mathew, Pillai, & Palade, 2018). Figure 2 shows the application of Blockchain technology.

Cryptocurrency

Blockchain technology has emerged because of the rise of cryptocurrency. These days, Bitcoin and Ethereum are built on blockchain, making it the most well-known use case for this technology. There is a public ledger called a "blockchain" that keeps track of every bitcoin transaction ever made. If more people start using cryptocurrencies, blockchain might get a wider audience (Tyagi, Agarwal & Maheshwari, 2016).

Banking Sector

Fiat currencies like dollars and euros, as well as cryptocurrencies, may all be transacted using Blockchain in the banking sector. This method may be quicker than using a bank or other financial institution to send money abroad because overseas banking transactions are processed outside of regular business hours (Mathew, Pillai, & Palade, 2018).

Assets Transfer

Asset ownership may be recorded and transferred using Blockchain technology as well. With the rise of NFTs and other forms of digital currency, this is currently a popular trend. New Forms of Ownership (NFT) serves to validate and promote digital creative works. The transactions involving physical assets used to back up financial claims are also verified. By employing this method, they wouldn't even have to change the deed on file with the county government in order to transfer ownership of the property (Elhoseny et al., 2018).

Healthcare Sector

Cybersecurity tools that integrate blockchain technology with healthcare infrastructure help limit hackers' access to sensitive data and systems. The use of blockchain technology helps them save time and money by reducing the likelihood of these types of costly complications. Information is the ultimate double-edged sword; it has the power to destroy whole nations and the ability to change your life for the better by illuminating what lies ahead (Wu et al., 2017; Javaid & Khan, 2021; Bharadwaj et al., 2021; Hamza et al., 2020).

Supply Chain Monitoring

When dealing with international trade, keeping track of the supply chain became a major problem. Tracking the movement of commodities from one end of the world to the other may be a daunting task. Blockchain technology is used in businesses to keep records and locate older documents that may have been lost (Thakar & Pandya, 2017; Aktas, Ceken, & Erdemli, 2018; Ray et al., 2020).

INTERNET OF THINGS (IoT)

A component of the Future Internet, the internet of things is defined by the seamless integration of "things" with unique identities, physical traits, and digital personas through smart interfaces. The internet of things relies on many different kinds of software, middleware, and hardware to function. Constraints on memory, computing power, and battery life are all well-known issues for IoT devices. Plus, they are dispersed and may be found in open areas in a wide range of locales. Not only that, but they represent cutting-edge technology (Trayush et al., 2021; Tao et al., 2018). Therefore, they are susceptible to assaults, and they must be taken into consideration while designing security measures for them. Inter-device communication may also employ ad hoc IP protocols as NFC, Bluetooth, IEEE 802.15.4, Wi-Fi, ZigBee, and 6LoWPAN. Ad hoc communication can occur via the internet, mobile networks, satellite networks, or wireless networks.

Possible risks include infiltration and modification of data as a result of such communications reaching computer systems. On top of that, there are architectural differences amongst them because there is no generally accepted layering approach for the internet of things. IoT devices have three layers: application, network, and perception. Smart energy, healthcare, and smart cities are application fields. Routers, switches, gateways, and firewalls are network layer devices. Perception includes embedded systems and sensors. Some add a physical layer, a service management layer, or a middleware layer (Li et al., 2020; Krishnamoorthy, Dua & Gupta, 2021; Subramaniyaswamy et al., 2019). Layering discrepancies are caused by the IoT device's release version, the supplier's standard, the device's functionality, and complexity, etc. This discrepancy leaves Internet of Things devices open to assault and compromises their ability to communicate securely. In addition, there is no single dominant operating system used by a sizable percentage of IoT devices, in contrast to conventional electronics. There are also not enough standardized formats for storing information. Thus, it is difficult to ensure compatibility between them. Middleware has been deployed to help integrate IoT devices, but this has its own security challenges. While many other initiatives have been launched to solve these issues, blockchain technology has lately been proposed as a potential solution (Hou & Yeh, 2015).

HEALTH INFORMATION TECHNOLOGY (HIT)

Several studies have focused on HIT (health information technology) for many years. Diverse structures have been developed and put into place. Electronic health records (EHRs), computerized provider order entry (CPOE), electronic medical records (EMRs), clinical decision support (CDS), electronic result reporting (ERRs), electronic prescribing (EPSRs), mobile computing (MC), telemedicine (remote monitoring), electronic health communication (EHC), and telehealth (TeH) are examples. eMAR and PACS are more examples (PACS). Embedded medical equipment, sensors, and IoT-enabled wearable devices are examples of HIT utilized in health IoT. (HIoT) (Verdejo Espinosa et al., 2021). As with any IoT setup, sensors are used in HIoT, however they may be either worn or implanted. There are five types of wearable sensors: pulse, respiration rate, body temperature, blood pressure, and pulse oximetry monitors. HIT's overarching objective is to boost hospital value, efficiency, and quality of treatment all while decreasing the number of preventable medical mistakes and improving patients' health outcomes. Health IoT supports triage, patient monitoring, staff monitoring, disease transmission modelling and control, practitioner assistance through real-time health status and prediction information, and pandemic data. Overall, the evidence supports the conclusion that HIT is most useful when used in a group setting (Said & Tolba, 2021). There has been a lengthy period of time since fragmentation in use was first recognized, and legislation has been introduced to facilitate the exchange of health information across providers. Defining precisely what data should be kept in a given system has proven difficult in this setting. Specifically, many doctors and hospitals still consider EHR to be a private database. One may also argue that EHR is relevant to PHR since PHR uses data from many different sources, including EHR. It is also found that one system can motivate the implementation of another. In this regard, eMAR and PACS are two examples that aid in the implementation of EMR and CPOE. There must be integration because of the occurrence of redundant data and interdependencies of various kinds. This need must be met, yet system integration is unusual. In addition, despite efforts, technological and non-technical obstacles arise on a variety of levels. Incorrect system-level workflow design and integration, a discrepancy between the rate of HIT innovation and the complexity of security and privacy problems, and a lack of consensus on how to solve these issues inhibit integration (de Morais Barroca Filho et al., 2021). Lack of distinct patient identification, messaging that allows syntactic and semantic compatibility between systems, and data encoding standards impede data integration between systems. Although rare, the integration approach takes into consideration healthcare stakeholders. Primary stakeholders include healthcare providers, data purveyors, and patients. Secondary stakeholders include insurers, health authorities, clinical researchers, and technology suppliers. HIT needs to consider the varying priorities and ever-evolving information needs of these parties. For instance, patients have evolved over time from passive recipients of healthcare information to engaged participants in HIT networks. As a vital aspect of human existence, healthcare is always being challenged by technological advancements. Using IoT solutions, we can improve hospital infrastructure and medical practice overall. Health care services and outcomes can benefit from the use of IoT systems and smart devices in several ways (Aujla & Jindal, 2020).

IoT health devices may be able to monitor a patient in real time and send that data to a doctor's office at a remote location. Internet-of-Things devices can take readings from a patient's vitals (blood pressure, heart rate, blood glucose level, etc.), upload them to the cloud, and make them available to the treating physician as necessary as shown in Figure 3 (Techskill Brew, 2021). This innovation has the potential to reduce the number of needless clinic visits, improve the quality of treatment given to patients, and even enable remote consultations regardless of location or time.

Devices like this have the potential to empower individuals to take charge of their health at all times, while also easing the burden on healthcare providers and infrastructure. Connected health gadgets also allow doctors to remotely assess serious conditions like asthma and heart attacks as shown in Figure 4 (Techskill Brew, 2021). Even before a patient arrives at the hospital, doctors and nurses may look up their information and prepare to give them the treatment they need. Costs are mitigated and quality of care for medical emergencies is enhanced (Djenna & Saïdouni, 2018).

BLOCKCHAIN AND IoT

However, there are obstacles that IoT devices must surmount, and Blockchain technology can assist in doing so. Here's an example:

Improves Security

Smart IoT devices collect data in real time and save it to the cloud. Data security and privacy concerns are major obstacles to adopting a cloud-based data storage model as shown in figure 5. In this case, a hacker may get access to a patient's data and potentially damage his medical records. After obtaining this information, it is possible to forge the patient's identity in order to obtain pharmaceuticals illegally or to submit false insurance claims. Since data saved on Blockchain is invulnerable to hacking and manipulation, it can provide the necessary answer to overcome security challenges encountered by IoT devices (Techskill Brew, 2021; Yang et al., 2018).

Because of inherent differences in data transfer protocols and computing power, integrating various IoT devices can be difficult, limiting the usefulness of IoT for healthcare applications as shown in Figure 5 (Techskill Brew, 2021). Although Blockchain offers a solution to this problem, Internet of Things devices are not yet powerful enough to communicate with Blockchain. These Internet of Things (IoT) smart gadgets require an intermediary to link them to the Blockchain. Cloud computing provides a powerful intermediary system that can gather data from sensors and upload it to a distributed ledger.

Each user of the Blockchain-based system will be assigned a cryptographically-secured, cryptographically-verifiable ID tied directly to the distributed ledger. He has the ability to control who has access to the data his smart gadgets collect. If he has more than one Internet of Things device, he may select the ones whose information he wants to share with his doctor.

Better Integration of Data of IoT Devices with Electronic Healthcare Records

Internet of Things (IoT) smart health wearables such as fit bits, health bands, watches, and blood glucose monitors capture a user's daily data and activities such as calories burned, steps taken, kilometres travelled,

Figure 3. IoT devices sharing information to the doctors through cloud (Techskill Brew, 2021)

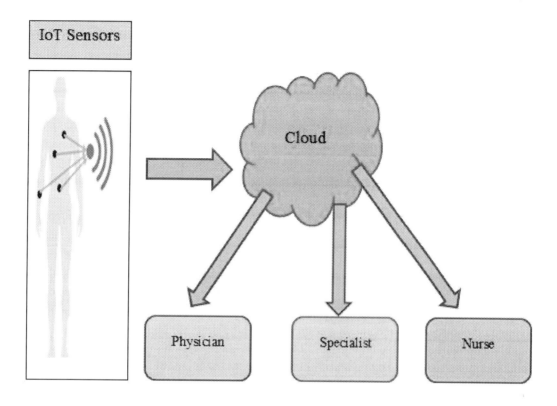

Figure 4. IoT gadgets allow doctors to remotely assess emergencies (Techskill Brew, 2021)

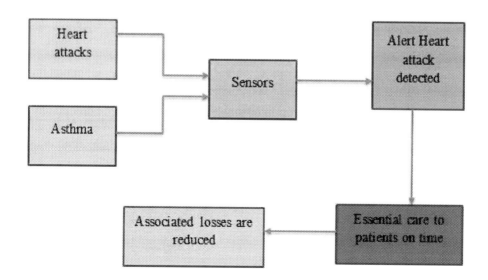

Figure 5. Cloud gathers IoT data and transfers it to Blockchain (Techskill Brew, 2021)

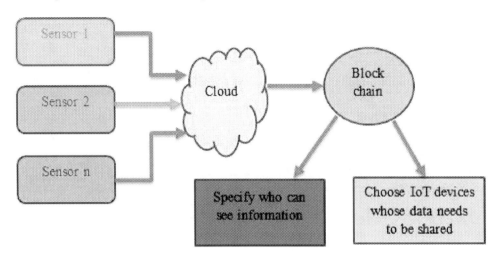

heart rate, quality of sleep, blood pressure, blood glucose level, and so on. The user's daily data from their smart health gadgets may be combined with their EHR thanks to blockchain technology (EHRs) (Techskill Brew, 2021; Jaiswal, & Anand, 2021). Doctors now have access to real-time information on their patients because to the data collected by these wearable as shown in Figure 6 (Techskill Brew, 2021). Since data is continually collected, not all tests need to be done during in-person patient visits. This will save the doctor and patient time and money on routine diagnostic procedures (Chowdhary, 2020; Chowdhary & Acharjya, 2018; Somayaji et al., 2020). As a result, in emergency situations where providing timely care is the largest problem, combining IoT devices with EHRs might be a benefit.

Figure 6. Better IoT-EHR integration on Blockchain (Techskill Brew, 2021)

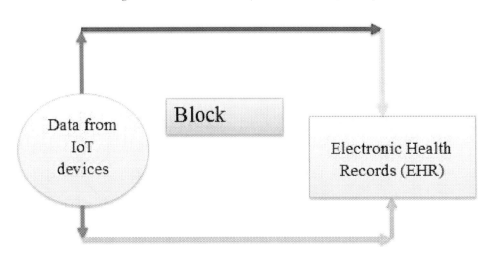

FUTURE DIRECTION AND INTEGRATION OF OTHER TECHNOLOGIES

Blockchain may be used with cloud computing, mobile edge computing, fog computing, big data analytics, AI, and cyber physical systems. Cloud computing may be used to process more data. As the cloud grows well, many backend applications may be installed there. When Blockchain becomes too large for other nodes to manage, cloud storage may be utilised. Cyberphysical systems allow machines to interact and function autonomously. Blockchain-based smart contracts may amplify autonomous behaviour by predefining actions. These principles may be simply implemented for many advantages.

Big data analytics has given individuals business information by examining enormous data volumes. Side-by-side study of cloud-stored data may assist users boost efficiency and revenues. Fog computing deploys nodes between the cloud and end user to bring processing closer. The architecture's management nodes also assist process data before sending it. Fog computing decreases latency, which is important in certain circumstances. Software defined networking is one of several new industry ideas. Most of these can improve service quality and user/device data processing.

RESULTS AND DISCUSSION

IoT in healthcare will give numerous benefits, from monitoring patient health to devising disease-prevention measures. IoT promises to transform healthcare and is delivering. In the next years, remote monitoring, wearables, and disease testing will be common. After the COVID-19 epidemic, health-conscious patients have grown more aware of their lifestyle and diet. Tech-savvy equipment like smartwatches, blood pressure and sugar monitoring devices offer daily results of diet and a sedentary lifestyle. IoT in wearable devices is now commercial. Using healthcare mobility technologies, IoT can automate patient care. Data transfer, machine-to-machine connectivity, and interoperability have improved healthcare. IoT saves time for patients and doctors. Smartphones allow people consult physicians at any time, saving them money by eliminating hospital visits. IoT will be incorporated into almost all healthcare industries. IoT will disrupt everything from medical prescriptions to ambulance services. IoT will improve the present healthcare system by making it quicker and more efficient.

CONCLUSION

This article looked into the healthcare industry's problems with blockchain-enabled applications and the Internet of Things. We also highlighted some of the difficulties that have been encountered when doing research on blockchain's potential applications in the healthcare industry. Additional difficulties arise for blockchain-based healthcare systems because to the complexity of healthcare involvement and rules. These difficulties include system evolution, privacy leakage, energy consumption, and scalability in communication. The current works on healthcare blockchain are summarized in this overview. In order to accelerate healthcare technology, future studies may want to think about including additional technical features to enhance feasibility evaluation and shorten the time between concepts and implementations. However, miners must use more power to keep nodes in sync after each ledger update, which is necessary because the ledger is updated based on the most recent transaction. Also, without a Distributed Computing strategy, applications built on the blockchain rely on individual nodes for proper operation.

REFERENCES

Abu-Elezz, I., Hassan, A., Nazeemudeen, A., Househ, M., & Abd-Alrazaq, A. (2020). The benefits and threats of blockchain technology in healthcare: A scoping review. *International Journal of Medical Informatics, 142*, 104246. doi:10.1016/j.ijmedinf.2020.104246 PMID:32828033

Aktas, F., Ceken, C., & Erdemli, Y. E. (2018). IoT-based healthcare framework for biomedical applications. *Journal of Medical and Biological Engineering, 38*(6), 966–979. doi:10.100740846-017-0349-7

Angraal, S., Krumholz, H. M., & Schulz, W. L. (2017). Blockchain technology: Applications in health care. *Circulation: Cardiovascular Quality and Outcomes, 10*(9), e003800. doi:10.1161/CIRCOUTCOMES.117.003800 PMID:28912202

Anjum, H. F., Rasid, S. Z. A., Khalid, H., Alam, M. M., Daud, S. M., Abas, H., Sam, S. M., & Yusof, M. F. (2020). Mapping research trends of blockchain technology in healthcare. ieee. *Access, 8*, 174244–174254. doi:10.1109/ACCESS.2020.3025011

Aujla, G. S., & Jindal, A. (2020). A decoupled blockchain approach for edge-envisioned IoT-based healthcare monitoring. *IEEE Journal on Selected Areas in Communications, 39*(2), 491–499. doi:10.1109/JSAC.2020.3020655

Azogu, I., Norta, A., Papper, I., Longo, J., & Draheim, D. (2019, April). A framework for the adoption of blockchain technology in healthcare information management systems: A case study of Nigeria. In *Proceedings of the 12th International Conference on Theory and Practice of Electronic Governance* (pp. 310-316). ACM. 10.1145/3326365.3326405

Azzawi, M. A., Hassan, R., & Bakar, K. A. A. (2016). A review on Internet of Things (IoT) in healthcare. *International Journal of Applied Engineering Research, 11*(20), 10216–10221.

Ben Fekih, R., & Lahami, M. (2020, June). Application of blockchain technology in healthcare: a comprehensive study. In *International Conference on Smart Homes and Health Telematics* (pp. 268-276). Springer, Cham. 10.1007/978-3-030-51517-1_23

Bharadwaj, H. K., Agarwal, A., Chamola, V., Lakkaniga, N. R., Hassija, V., Guizani, M., & Sikdar, B. (2021). A review on the role of machine learning in enabling IoT based healthcare applications. *IEEE Access : Practical Innovations, Open Solutions, 9*, 38859–38890. doi:10.1109/ACCESS.2021.3059858

Chakraborty, S., Aich, S., & Kim, H. C. (2019, February). A secure healthcare system design framework using blockchain technology. In *2019 21st International Conference on Advanced Communication Technology (ICACT)* (pp. 260-264). IEEE. 10.23919/ICACT.2019.8701983

Chanchaichujit, J., Tan, A., Meng, F., & Eaimkhong, S. (2019). Blockchain technology in healthcare. In *Healthcare 4.0* (pp. 37-62). Palgrave Pivot, Singapore. doi:10.1007/978-981-13-8114-0_3

Chowdhary, C. L. (2020). Growth of financial transaction toward bitcoin and blockchain technology. In *Bitcoin and blockchain* (pp. 79–97). CRC Press. doi:10.1201/9781003032588-6

Chowdhary, C. L., & Acharjya, D. P. (2018). Singular Value Decomposition–Principal Component Analysis-Based Object Recognition Approach. In Bio-Inspired Computing for Image and Video Processing (pp. 323-341). Chapman and Hall/CRC. doi:10.1201/9781315153797-12

Clauson, K. A., Breeden, E. A., Davidson, C., & Mackey, T. K. (2018). Leveraging Blockchain Technology to Enhance Supply Chain Management in Healthcare:: An exploration of challenges and opportunities in the health supply chain. *Blockchain in healthcare today*.

Daniel, J., Sargolzaei, A., Abdelghani, M., Sargolzaei, S., & Amaba, B. (2017). Blockchain technology, cognitive computing, and healthcare innovations. *J. Adv. Inf. Technol*, *8*(3), 194–198. doi:10.12720/jait.8.3.194-198

Darshan, K. R., & Anandakumar, K. R. (2015, December). A comprehensive review on usage of Internet of Things (IoT) in healthcare system. In *2015 International Conference on Emerging Research in Electronics, Computer Science and Technology (ICERECT)* (pp. 132-136). IEEE. 10.1109/ERECT.2015.7499001

Dash, S., Gantayat, P. K., & Das, R. K. (2021). Blockchain technology in healthcare: opportunities and challenges. *Blockchain Technology: Applications and Challenges*, 97-111.

de Morais Barroca Filho, I., Aquino, G., Malaquias, R. S., Girão, G., & Melo, S. R. M. (2021). An IoT-based healthcare platform for patients in ICU beds during the COVID-19 outbreak. *IEEE Access : Practical Innovations, Open Solutions*, *9*, 27262–27277. doi:10.1109/ACCESS.2021.3058448 PMID:34786307

Djenna, A., & Saïdouni, D. E. (2018, October). Cyber-attacks classification in IoT-based-healthcare infrastructure. In *2018 2nd Cyber Security in Networking Conference (CSNet)* (pp. 1-4). IEEE. 10.1109/CSNET.2018.8602974

Dutta, P., Choi, T. M., Somani, S., & Butala, R. (2020). Blockchain technology in supply chain operations: Applications, challenges and research opportunities. *Transportation research part e: Logistics and transportation review*, *142*, 102067.

Elhoseny, M., Ramírez-González, G., Abu-Elnasr, O. M., Shawkat, S. A., Arunkumar, N., & Farouk, A. (2018). Secure medical data transmission model for IoT-based healthcare systems. *IEEE Access : Practical Innovations, Open Solutions*, *6*, 20596–20608. doi:10.1109/ACCESS.2018.2817615

Engelhardt, M. A. (2017). Hitching healthcare to the chain: An introduction to blockchain technology in the healthcare sector. *Technology Innovation Management Review*, *7*(10).

Faisal, M., Sadia, H., Ahmed, T., & Javed, N. (2022). Blockchain Technology for Healthcare Record Management. In *Pervasive Healthcare* (pp. 255–286). Springer. doi:10.1007/978-3-030-77746-3_17

Govindan, K., Nasr, A. K., Saeed Heidary, M., Nosrati-Abarghooee, S., & Mina, H. (2022). Prioritizing adoption barriers of platforms based on blockchain technology from balanced scorecard perspectives in healthcare industry: A structural approach. *International Journal of Production Research*, 1–15. doi:10.1080/00207543.2021.2013560

Haleem, A., Javaid, M., Singh, R. P., Suman, R., & Rab, S. (2021). Blockchain technology applications in healthcare: An overview. *International Journal of Intelligent Networks*, *2*, 130–139. doi:10.1016/j.ijin.2021.09.005

Hamza, R., Yan, Z., Muhammad, K., Bellavista, P., & Titouna, F. (2020). A privacy-preserving cryptosystem for IoT E-healthcare. *Information Sciences*, *527*, 493–510. doi:10.1016/j.ins.2019.01.070

Hasselgren, A., Kralevska, K., Gligoroski, D., Pedersen, S. A., & Faxvaag, A. (2020). Blockchain in healthcare and health sciences—A scoping review. *International Journal of Medical Informatics*, *134*, 104040. doi:10.1016/j.ijmedinf.2019.104040 PMID:31865055

Hou, J. L., & Yeh, K. H. (2015). Novel authentication schemes for IoT based healthcare systems. *International Journal of Distributed Sensor Networks*, *11*(11), 183659. doi:10.1155/2015/183659

Hussien, H. M., Yasin, S. M., Udzir, N. I., Ninggal, M. I. H., & Salman, S. (2021). Blockchain technology in the healthcare industry: Trends and opportunities. *Journal of Industrial Information Integration*, *22*, 100217. doi:10.1016/j.jii.2021.100217

Hussien, H. M., Yasin, S. M., Udzir, S. N. I., Zaidan, A. A., & Zaidan, B. B. (2019). A systematic review for enabling of develop a blockchain technology in healthcare application: Taxonomy, substantially analysis, motivations, challenges, recommendations and future direction. *Journal of Medical Systems*, *43*(10), 1–35. doi:10.100710916-019-1445-8 PMID:31522262

Idrees, S. M., Nowostawski, M., Jameel, R., & Mourya, A. K. (2021). Security aspects of blockchain technology intended for industrial applications. *Electronics (Basel)*, *10*(8), 951. doi:10.3390/electronics10080951

Jaiswal, K., & Anand, V. (2021). A survey on IoT-based healthcare system: potential applications, issues, and challenges. In *Advances in Biomedical Engineering and Technology* (pp. 459–471). Springer. doi:10.1007/978-981-15-6329-4_38

Javaid, M., & Khan, I. H. (2021). Internet of Things (IoT) enabled healthcare helps to take the challenges of COVID-19 Pandemic. *Journal of Oral Biology and Craniofacial Research*, *11*(2), 209–214. doi:10.1016/j.jobcr.2021.01.015 PMID:33665069

Kashani, M. H., Madanipour, M., Nikravan, M., Asghari, P., & Mahdipour, E. (2021). A systematic review of IoT in healthcare: Applications, techniques, and trends. *Journal of Network and Computer Applications*, *192*, 103164. doi:10.1016/j.jnca.2021.103164

Kodali, R. K., Swamy, G., & Lakshmi, B. (2015, December). An implementation of IoT for healthcare. In *2015 IEEE Recent Advances in Intelligent Computational Systems* (RAICS) (pp. 411-416). IEEE.

Krishnamoorthy, S., Dua, A., & Gupta, S. (2021). Role of emerging technologies in future IoT-driven Healthcare 4.0 technologies: A survey, current challenges and future directions. *Journal of Ambient Intelligence and Humanized Computing*, 1–47.

Kumar, T., Ramani, V., Ahmad, I., Braeken, A., Harjula, E., & Ylianttila, M. (2018, September). Blockchain utilization in healthcare: Key requirements and challenges. In *2018 IEEE 20th International conference on e-health networking, applications and services (Healthcom)* (pp. 1-7). IEEE.

Laroiya, C., Saxena, D., & Komalavalli, C. (2020). Applications of blockchain technology. In *Handbook of research on blockchain technology* (pp. 213–243). Academic Press. doi:10.1016/B978-0-12-819816-2.00009-5

Leeming, G., Cunningham, J., & Ainsworth, J. (2019). A ledger of me: Personalizing healthcare using blockchain technology. *Frontiers in medicine*, *6*, 171. doi:10.3389/fmed.2019.00171 PMID:31396516

Li, J., Cai, J., Khan, F., Rehman, A. U., Balasubramaniam, V., Sun, J., & Venu, P. (2020). A secured framework for sdn-based edge computing in IOT-enabled healthcare system. *IEEE Access : Practical Innovations, Open Solutions*, *8*, 135479–135490. doi:10.1109/ACCESS.2020.3011503

Li, Y., Shan, B., Li, B., Liu, X., & Pu, Y. (2021). Literature review on the applications of machine learning and blockchain technology in smart healthcare industry: A bibliometric analysis. *Journal of Healthcare Engineering*, *2021*, 2021. doi:10.1155/2021/9739219 PMID:34426765

Mallikarjuna, B., Shrivastava, G., & Sharma, M. (2022). Blockchain technology: A DNN token-based approach in healthcare and COVID-19 to generate extracted data. *Expert Systems: International Journal of Knowledge Engineering and Neural Networks*, *39*(3), e12778. doi:10.1111/exsy.12778 PMID:34511692

Mathew, P. S., Pillai, A. S., & Palade, V. (2018). Applications of IoT in healthcare. In *Cognitive Computing for Big Data Systems Over IoT* (pp. 263–288). Springer.

McBee, M. P., & Wilcox, C. (2020). Blockchain technology: Principles and applications in medical imaging. *Journal of Digital Imaging*, *33*(3), 726–734. doi:10.100710278-019-00310-3 PMID:31898037

Radanović, I., & Likić, R. (2018). Opportunities for use of blockchain technology in medicine. *Applied Health Economics and Health Policy*, *16*(5), 583–590. doi:10.100740258-018-0412-8 PMID:30022440

Ray, P. P., Dash, D., Salah, K., & Kumar, N. (2020). Blockchain for IoT-based healthcare: Background, consensus, platforms, and use cases. *IEEE Systems Journal*, *15*(1), 85–94. doi:10.1109/JSYST.2020.2963840

Saeed, H., Malik, H., Bashir, U., Ahmad, A., Riaz, S., Ilyas, M., Bukhari, W. A., & Khan, M. I. A. (2022). Blockchain technology in healthcare: A systematic review. *PLoS One*, *17*(4), e0266462. doi:10.1371/journal.pone.0266462 PMID:35404955

Saha, A., Amin, R., Kunal, S., Vollala, S., & Dwivedi, S. K. (2019). Review on "Blockchain technology based medical healthcare system with privacy issues". *Security and Privacy*, *2*(5), e83. doi:10.1002py2.83

Said, O., & Tolba, A. (2021). Design and evaluation of large-scale IoT-enabled healthcare architecture. *Applied Sciences (Basel, Switzerland)*, *11*(8), 3623. doi:10.3390/app11083623

Selvaraj, S., & Sundaravaradhan, S. (2020). Challenges and opportunities in IoT healthcare systems: A systematic review. *SN Applied Sciences*, *2*(1), 1–8. doi:10.100742452-019-1925-y

Shahnaz, A., Qamar, U., & Khalid, A. (2019). Using blockchain for electronic health records. *IEEE Access : Practical Innovations, Open Solutions*, *7*, 147782–147795. doi:10.1109/ACCESS.2019.2946373

Sharma, A., Kaur, S., & Singh, M. (2021). A comprehensive review on blockchain and Internet of Things in healthcare. *Transactions on Emerging Telecommunications Technologies*, *32*(10), e4333. doi:10.1002/ett.4333

Sharma, N., Bhushan, B., Kaushik, I., & Debnath, N. C. (2021). Applicability of Blockchain Technology in Healthcare Industry: Applications, Challenges, and Solutions. In *Efficient Data Handling for Massive Internet of Medical Things* (pp. 339–370). Springer. doi:10.1007/978-3-030-66633-0_15

Sharma, N., Bhushan, B., Kaushik, I., & Debnath, N. C. (2021). Applicability of Blockchain Technology in Healthcare Industry: Applications, Challenges, and Solutions. In *Efficient Data Handling for Massive Internet of Medical Things* (pp. 339–370). Springer. doi:10.1007/978-3-030-66633-0_15

Siyal, A. A., Junejo, A. Z., Zawish, M., Ahmed, K., Khalil, A., & Soursou, G. (2019). Applications of blockchain technology in medicine and healthcare: Challenges and future perspectives. *Cryptography*, *3*(1), 3. doi:10.3390/cryptography3010003

Somayaji, S. R. K., Alazab, M., Manoj, M. K., Bucchiarone, A., Chowdhary, C. L., & Gadekallu, T. R. (2020, December). A framework for prediction and storage of battery life in iot devices using dnn and blockchain. In *2020 IEEE Globecom Workshops* (pp. 1-6). IEEE.

Srivastava, A., Jain, P., Hazela, B., Asthana, P., & Rizvi, S. W. A. (2021). Application of Fog Computing, Internet of Things, and Blockchain Technology in Healthcare Industry. In *Fog Computing for Healthcare 4.0 Environments* (pp. 563–591). Springer. doi:10.1007/978-3-030-46197-3_22

Srivastava, G., Parizi, R. M., & Dehghantanha, A. (2020). The future of blockchain technology in healthcare internet of things security. *Blockchain cybersecurity, trust and privacy*, 161-184.

Subramaniyaswamy, V., Manogaran, G., Logesh, R., Vijayakumar, V., Chilamkurti, N., Malathi, D., & Senthilselvan, N. (2019). An ontology-driven personalized food recommendation in IoT-based healthcare system. *The Journal of Supercomputing*, *75*(6), 3184–3216. doi:10.100711227-018-2331-8

Tanwar, S., Parekh, K., & Evans, R. (2020). Blockchain-based electronic healthcare record system for healthcare 4.0 applications. *Journal of Information Security and Applications*, *50*, 102407. doi:10.1016/j.jisa.2019.102407

Tao, H., Bhuiyan, M. Z. A., Abdalla, A. N., Hassan, M. M., Zain, J. M., & Hayajneh, T. (2018). Secured data collection with hardware-based ciphers for IoT-based healthcare. *IEEE Internet of Things Journal*, *6*(1), 410–420. doi:10.1109/JIOT.2018.2854714

Techskill Brew. (2021). Blockchain and IoT in smart healthcare. *Medium*. https://medium.com/techskill-brew/blockchain-and-iot-in-smart-healthcare-814287551300

Thakar, A. T., & Pandya, S. (2017, July). Survey of IoT enables healthcare devices. In *2017 International Conference on Computing Methodologies and Communication (ICCMC)* (pp. 1087-1090). IEEE. 10.1109/ICCMC.2017.8282640

Trayush, T., Bathla, R., Saini, S., & Shukla, V. K. (2021, March). IoT in Healthcare: Challenges, Benefits, Applications, and Opportunities. In *2021 International Conference on Advance Computing and Innovative Technologies in Engineering (ICACITE)* (pp. 107-111). IEEE. 10.1109/ICACITE51222.2021.9404583

Tyagi, S., Agarwal, A., & Maheshwari, P. (2016, January). A conceptual framework for IoT-based healthcare system using cloud computing. In *2016 6th International Conference-Cloud System and Big Data Engineering (Confluence)* (pp. 503-507). IEEE.

Verdejo Espinosa, Á., López, J. L., Mata Mata, F., & Estevez, M. E. (2021). Application of IoT in healthcare: Keys to implementation of the sustainable development goals. *Sensors (Basel)*, *21*(7), 2330. doi:10.339021072330 PMID:33810606

Wang, H. (2020). IoT based clinical sensor data management and transfer using blockchain technology. *Journal of ISMAC, 2*(03), 154–159. doi:10.36548/jismac.2020.3.003

Wong, M. C., Yee, K. C., & Nøhr, C. (2018). Socio-technical considerations for the use of blockchain technology in healthcare. In *Building Continents of Knowledge in Oceans of Data: The Future of Co-Created eHealth* (pp. 636–640). IOS Press.

Wu, T., Wu, F., Redoute, J. M., & Yuce, M. R. (2017). An autonomous wireless body area network implementation towards IoT connected healthcare applications. *IEEE Access : Practical Innovations, Open Solutions, 5*, 11413–11422. doi:10.1109/ACCESS.2017.2716344

Yaeger, K., Martini, M., Rasouli, J., & Costa, A. (2019). Emerging blockchain technology solutions for modern healthcare infrastructure. *Journal of Scientific Innovation in Medicine, 2*(1), 1. doi:10.29024/jsim.7

Yang, Y., Zheng, X., Guo, W., Liu, X., & Chang, V. (2019). Privacy-preserving smart IoT-based healthcare big data storage and self-adaptive access control system. *Information Sciences, 479*, 567–592. doi:10.1016/j.ins.2018.02.005

Yeole, A. S., & Kalbande, D. R. (2016, March). Use of Internet of Things (IoT) in healthcare: A survey. In *Proceedings of the ACM Symposium on Women in Research 2016* (pp. 71-76). ACM. 10.1145/2909067.2909079

Yoon, H. J. (2019). Blockchain technology and healthcare. *Healthcare Informatics Research, 25*(2), 59–60. doi:10.4258/hir.2019.25.2.59 PMID:31131139

Zubaydi, H. D., Chong, Y. W., Ko, K., Hanshi, S. M., & Karuppayah, S. (2019). A review on the role of blockchain technology in the healthcare domain. *Electronics (Basel), 8*(6), 679. doi:10.3390/electronics8060679

Chapter 5
Mapping Research Trends in Eco-Innovation:
A Citespace-Based Scientometric Analysis

Waleska Yone Yamakawa Zavatti Campos
https://orcid.org/0000-0001-5050-1557
Pontifical Catholic University of Rio de Janeiro, Brazil

Fábio de Oliveira Paula
https://orcid.org/0000-0002-1926-2241
Pontifical Catholic University of Rio de Janeiro, Brazil

Maria J. Sousa
https://orcid.org/0000-0001-8633-4199
University Institute of Lisbon, Portugal

Luciana Aparecida Barbieri da Barbieri da Rosa
https://orcid.org/0000-0001-9240-0236
Pontifical Catholic University of Rio de Janeiro, Brazil

Marcos Cohen
https://orcid.org/0000-0003-3248-7776
Pontifical Catholic University of Rio de Janeiro, Brazil

Maria Carolina Martins Martins Rodrigues
https://orcid.org/0000-0003-2575-8611
Universidade do Algarve, Portugal

ABSTRACT

Eco-innovation contributes to a more sustainable environment through the development of green technologies. In this context, eco-innovations are capable of improving the performance of organizations in order to reduce environmental and social impacts. Knowing the behavior of studies on eco-innovation can help researchers in understanding the scientific and intellectual structure of the field. Therefore, the general objective of this research is to evaluate the literature on eco-innovation on the world stage from bibliometric and Scientometric analysis, with the help of CiteSpace software. The results show a highly fruitful, dynamic, and rapidly expanding field of studies, marked by a high degree of interdisciplinarity and multidisciplinarity. The conclusions of the research reveal the trends for studies in the area, from the understanding of the behavior of the field of studies in eco-innovation.

DOI: 10.4018/978-1-6684-6123-5.ch005

INTRODUCTION

As a result of the Rio-92 Conference on environment and development, the signatory countries committed themselves to the adoption of sustainable development principles (Rennings et al., 2006). Given this, it became evident that sustainability would imply significant changes in the long-term technological framework so that the mere adaptation of existing technologies would not be sufficient, and in this context, innovations focused on sustainable development were defined as environmental technical innovations or eco-innovations (Rennings, 2000; Rennings et al., 2006).

Eco-innovation contributes to a more sustainable environment through the development of ecological improvements and technologies (Horbach, Rammer, & Rennings, 2012; Xavier, Naveiro, Aoussat, & Reyes, 2017). Such technologies impact the way companies operate and business objectives for sustainability (Bocken et al., 2014).

Knowing the behavior of the field of studies on eco-innovation can help researchers understand the different nuances on the subject. This is because, eco-innovation can be expressed in different ways, however, with the same central meaning. Innovation that generates a reduced impact on the environment or, which favors the proper use of natural resources, can be identified by different expressions, used in several systematic reviews of literature and theoretical-empirical works, such as sustainability-oriented innovation (Klewitz & Hansen, 2014), or sustainable innovation, environmental innovation, green innovation, clean innovation, ecology innovation (Pacheco et al., 2017), environmentally sustainable innovation (Hellström, 2007), eco-friendly innovation (Jeong & Ko, 2016) and finally eco-innovation (Bossle, Dutra de Barcellos, Vieira, & Sauvée, 2016). The nine distinct forms found in the literature to address eco-innovation were the object of the bibliometric survey and the scientometric analysis performed in this study, and from now on, they will be related to eco-innovation.

This work seeks to complement the bibliometric study conducted by Yin, Gong, & Wang (2018) whose work addressed only the keyword 'green innovation' for conducting the survey. In this sense, and while the studies on eco-innovation have received extensive attention from researchers, few studies have sought to answer the following question to use the expressions related to eco-innovation: How do the field of research on eco-innovation behave on the world stage? For this, the general objective is: To evaluate the literature on eco-innovation on the world stage from bibliometric and scientific analysis. To achieve the proposed object, the following specific objectives were defined: (1) to characterize the literature on eco-innovation through the number of articles and citations, research categories, and journals; (2) identify the level of scope and scope of research concerning countries, institutions, and main authors; (3) to know the themes of interest within the field of eco-innovation studies, longitudinally. To achieve these objectives, documents were collected in the Science Citation Index Expanded (SCI-EXPANDED), a database of the Web of Science (WoS), published since the first occurrence, with the subsequent realization of bibliometric and scientometric analysis to address the research objectives.

Bibliometric analysis was chosen to answer the research question because it is a quantitative and statistical method that, if applied with methodological rigor, can bring important information to researchers about the performance of various fields of knowledge, being an instrument for directing research from the knowledge of the most important themes (Araújo, 2006; Chen et al., 2016; Ho, 2019a). In addition, visual maps and networks elaborated through scientometric and mapping analyses can assist in data mining for the exploration of different relationships (Lazar & Chithra, 2021; Zárate-Rueda et al., 2021), to determine gaps and future directions for research when seeking the identification of intellectual structures in scientific fields (Huang et al., 2020; Kumari & Kumar, 2020; Zhou, Chen, & Huang,

2019). In this perspective, one of the most influential scientometric software is CiteSpace, developed by Chen (2006) and used in this work.

Due to the peculiarities of the research questions and the objectives that underlie the bibliometric analyses, in general, recent bibliometric studies do not present a theoretical reference design. Therefore, studies from different fields (from those related to critical or environmental studies, for example) do not address theoretical reference as the very characteristic of the survey does not justify its elaboration (Aboelmaged & Mouakket, 2020; Deng et al., 2020; Duan et al., 2020; Guo et al., 2020; Mallawaarachchi et al., 2020).

Therefore, the work is organized into four sections: after this introduction, methodological choices are discussed; the third section discusses bibliometric and scientometric results, and the fourth section highlights the final considerations.

METHOD

The flowchart used in the methodology is shown in Figure 1. The process of survey and bibliometric analysis was carried out in three phases: definition of the research strategy, eligibility criteria for data collection, and analysis of the results. In the definition of the research strategy, the keywords were searched primarily through systematic reviews of the literature and were listed to include the greatest possible variety of occurrences. This methodological stage is important because the choice of keywords incongruous with the theme or incomplete can compromise the quality of the survey, and therefore the validity of the study data (Ho, 2019a). To expand the bibliometric analysis (Yin et al., 2018)[6]elaborated by Yin, Gong, & Wang (2018), whose survey was based on only one keyword, namely, 'green innovation, the present study was based on the premise that other keywords could have been employed, as sustainability-oriented innovation (Klewitz & Hansen, 2014), or sustainable innovation, environmental innovation, clean innovation, ecology innovation (Pacheco et al., 2017), environmentally sustainable innovation (Hellström, 2007), eco-friendly innovation (Jeong & Ko, 2016) and finally, eco-innovation (Bossle et al., 2016). Therefore, the bibliometric approach performed in this study used nine different keywords, using stemming in the database, through the use of an asterisk (*), from which, variations of the searched keywords are included.

This work observed the propositions of several studies for the best adequacy of the bibliometric method using the WoS database (Ho, 2019a, 2019c). This was necessary since the main collection of the Web of Science was designed for literature consultation and not for the elaboration of bibliometric studies, which imposes on researchers the need to manually perform a bibliometric treatment when using this database (Ho, 2018).

In the eligibility criteria, and the study, following the recommendations of Ho (2019a), to avoid the occurrence of duplicate results, used only the SCI-Expanded of the WoS, and in addition, used the method of reading the "first-page" of the works as a filter for refinement of the bibliometric method as proposed by Fu, Wang, & Ho (2012), to verify the presence of the keywords in the topic (title, abstract or keywords of the author), so that if the keyword used for the survey was asked only in Keywords Plus, the article should be removed from the sample. Such bibliometric treatment was necessary because the existence of Keywords Plus may represent a bias of the Web of Science platform, including as a result of the survey's unrelated works (Ho, 2019b). After reading the first page and analyzing keywords plus, there was the exclusion of and unrelated papers, at which time the sample fell from 2,885 to 2,495, that

Figure 1. Research method

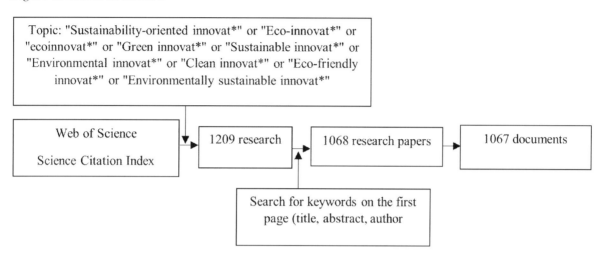

is, 88.34% of the initial sample. Before 1991, only one paper was published in 1978, and therefore, it was decided to remove this document, resulting in a final sample of 2,494 papers, related to the 1991-2022 timespan. Data collection was carried out on November 23, 2022, and therefore, data for the year 2022 may differ from data collected after the end of the year.

For the analysis of the results, bibliometry and scientometrics were used. Bibliometry is a quantitative and statistical technique whose focus is on the measurement of scientific indicators (Araújo, 2006). The bibliometric indicators used in this work were proposed by Zupic & Čater (2015), among which stand out the analysis of citation, co-citation, co-authorship, and co-words. The analysis of bibliographic copying was not performed in this study. Scientometrics' objective involves understanding science from its social perspective to attribute meaning to data, goes beyond bibliometric quantification, because it presupposes the use of findings by researchers and public policies (Chen & Song, 2019; Santos & Kobashi, 2009). The scientific mappings undertaken were based on the scientific perspective. The data collected from the WoS database were organized and visualized according to Figure 2.

Figure 2. Tools for data visualization

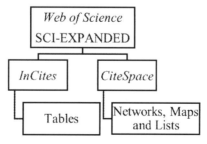

The InCites (Clarivate Analytics) tool provides productivity, impact, and collaboration indicators from researchers, articles, institutions, countries, journals, research areas, and funding agencies (Bornmann

& Leydesdorff, 2013). The results of InCites were integrated into the tables of this study and supported the bibliometric analysis.

About scientometric analyses, the mapping of the field was performed through CiteSpace software – version 5.7. R1, which is considered one of the most influential instruments of analysis in the literature (Chen, 2006). CiteSpace was used to develop networks of subject categories, countries, authors, and keywords. The parameters defined in CiteSpace were: (1) Time 1991-2022; (2) Node type: subject categories, countries, authors, keywords, references; (3) Selection criteria: first 50 articles per year. It is important to note that the size of the nodes is equivalent to the participation of the actor in the network.

The networks were measured according to the centrality of degree and/or intermediation. In this sense, degree centrality measures the number of contacts caught directly by each actor in the network, or, in other words, the number of edges that a node has (Masquietto et al., 2011), while the centrality of intermediation measures the level at which an act as a means to reach others (Masquietto et al., 2011). The nodes surrounded by purple rings point to the presence of high centrality of intermediation for that actor, or, still, the actors are important in mediating the relationships between different nodes. The edges between the nodes reflect the relationships between the actors, and bind pairs of nodes (Bródka, Skibicki, Kazienko, & Musiał, 2011; Chen & Song, 2019).

The results were delineated in terms of literature characteristics (number of articles and citations, research categories, analysis of international journals), the foreignness of research (countries and institutions, analysis of authorship), and research points of interest (main keywords, clusters, explosion of citations).

RESULTS

Characteristics of the Literature

Number of Articles and Citations

The number of articles and citations over time is an important tool for analyzing scientific production in a given field (Huang et al., 2020; Zupic & Čater, 2015). Figure 3 shows the trend in the publication of articles and citations over time.

Regarding the number of articles and publications, it is evident that the papers published between 1991 and 2022 have increasingly developed, especially in the last decade. There was a significant increase in the number of publications in the 2011-2013 interstitium, when the annual publication exceeded 100 documents for the first time, totaling 123 papers, an increase of 123,6% compared to the previous interstice (2008-2010). The annual average of published articles is 227, at an average annual growth rate of 33%.

Concerning the number of citations, the trend is also increasing. The total number of citations from the first year of study to the november of 2022 is 68,315 citations, where the interstice 2005-2007 is a decisive milestone, in which the number of citations grows by 1,277.8% compared to the previous period. The average annual citation rate is 6,210 citations each year, with an average annual growth rate of 33.28%.

The oldest work in the sample is that of Edwards (1991), which is made efforts to raise the status of eco-innovation in the international context. The most recent work is that of Imran et al. (2022), in which the authors discussed how to minimize the impacts of agro-industrial waste during its disposal in the environment. with this, the authors discuss the application of eco-innovative technologies in waste treatment.

Figure 3. Trends in the number of articles and citations

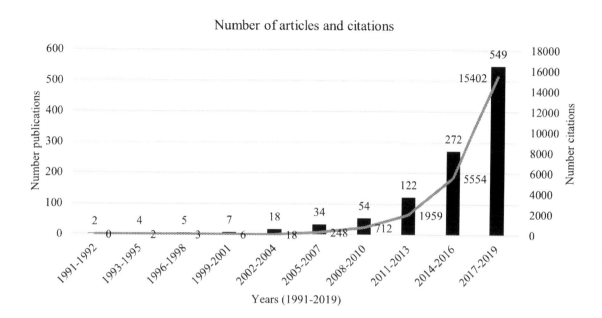

Subject Categories

Subject categories are related to the disciplines to which the articles are linked (Peng et al., 2018). Table 1 presents indicators related to the ten main subject categories in eco-innovation studies.

In the environmental science's subject category, 699 papers were published, corresponding to 17,219 citations, the two largest amounts among the categories. Regarding the percentage of documents cited once or more, in percentage terms, the *Economics* category has the highest value (96.39%). This category is also the one with the highest percentage of studies located among the 10% most cited in the database, but in absolute terms, the *Environmental Sciences* category has the highest number of studies in this condition (147 documents or 21.03% of 699).

From the point of view of international collaboration, the areas generally have a percentage above 30% of authors and co-authors from different countries. Because it is a field of global interest, it would be interesting for more international networks to be formed among the authors. Figure 4 shows the interrelationship between subject categories so that it is possible to visualize the contribution of each area to the field of eco-innovation studies.

The most prominent subject category in the network is *Environmental Sciences,* which also presents the highest centrality of intermediation, that is, it is a discipline that plays a mediating role among other fields of knowledge. Another important category is *Green & Sustainable Science & Technology,* followed by *Engineering, Environmental* and *Environmental Studies*. The network analysis allows us to conclude that the field of study in eco-innovation is marked by interdisciplinary relationships (due to the existence of edges between the areas) and because it is multidisciplinary because it presents numerous nodes from different subject categories.

Table 1. Subject categories productivity indicators

Subject Category	Ps	TC	% DC	Top10%	% IC
Environmental Sciences	699	17219	88,13%	21,03%	30,47%
Green & Sustainable Science & Technology	513	11236	87,91%	17,35%	32,16%
Engineering, Environmental	354	10672	96,05%	17,23%	34,75%
Environmental Studies	265	6540	81,13%	26,79%	27,17%
Economics	83	5314	96,39%	56,63%	33,73%
Ecology	49	3577	91,84%	38,78%	34,69%
Energy & Fuels	68	1950	95,59%	11,76%	36,76%
Operations Research & Management Science	34	1504	94,12%	41,18%	32,35%
Engineering, Industrial	46	1400	86,96%	26,09%	41,3%
Management	37	1038	89,19%	21,62%	48,65%

Note: Ps: number of papers published. Timespan: 1991-2019. TC: total citations. %DC: percentage of published documents cited one or more times. Top 10% is a high-performance research indicator, evidences the percentage and studies that are among the 10% most cited in the world. %IC: percentage of international co-authors.

The heterogeneity of the various categories demonstrates the importance of the theme for engineering, sustainability, economics, energy and fuels, operations, management, public environment, construction, and food science.

Analysis of International Journals

The papers on eco-innovation were published in 285 journals, containing articles, which in turn cited 17.047 out the magazines. Table 2 lists the top ten international journals in the field. In general, it is evident that the fifteen journals come from developed countries, members of the OECD, and take a turn between the USA (2), The Netherlands (3), England (4), and Switzerland (1). The half-life of citations is the main indicator adopted by the Journal Citation Reports (JCR) on the obsolescence of journal litera-

Figure 4. Subject categories

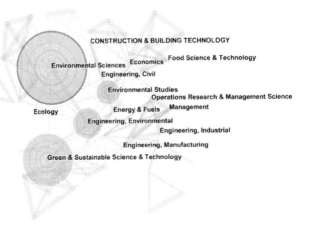

Table 2. Main international journals

Journal	Country	Ps	TC	%DC	HL	JIF
Journal of Cleaner Production	United States	304	9,501	96.71%	2.9	6.4
Ecological Economics	Netherlands	41	3,405	97.56%	9.4	4.28
Energy Policy	England	29	1,25	100%	7.1	4.88
Sustainability	Switzerland	162	975	70.99%	2.2	2.59
Technovation	Netherlands	7	528	100%	10	5.25
Building And Environment	England	2	423	100%	6.8	4.82
Journal of The Royal Society Interface	England	1	344	100%	6	3.22
Global Environmental Change-Human and Policy Dimensions	England	4	314	100%	6.8	10.43
Environmental Innovation and Societal Transitions	Netherlands	13	299	92.31%	3.7	7.51
Journal of Industrial Ecology	United States	11	262	90.91%	6.6	4.83

Note: Ps: number of papers published per journal. Timespan: 1991-2019. TC: total citations of the articles of a journal. %DC: percentage of published documents cited one or more times. HL: half-life of citations is a metric in years, in which half of the citations obtained are for items published below the said value and the other half of the citations for items published longer than that. JIF: an impact factor of the journal, according to Journal Citation *Reports*.

ture (Diniz, 2013; Strehl, 2005). The concept represents an eloquent metric in the analysis of journals to evidence whether older or more recent studies are getting more attention by researchers (Strehl, 2005). Half-life, as well as the impact factor, acts as a measure of the influence of journals, since important publications, in addition to being highly cited, are cited for a longer period. Another important consideration is that the half-life of citations in the areas of social sciences tends to be greater than the half-life of 'hard *sciences*, since the former has a half-life in general greater than eight years, while the latter is less than five years (Diniz, 2013).

The impact factor of journals is a metric adopted by the Journal Citation Reports (JCR) based on citations to measure the performance of scientific journals, with wide use in bibliometric studies, due to their informative and stable character (Glänzel & Moed, 2002). The most productive journal on eco-innovation is the *Journal of Cleaner Production,* whose impact factor is 6.4, and published 304 papers, which were cited 9,501 times so that 96,71% of the papers were cited at least once. The half-life of citations for this journal is 2.9, which means that half of the citations have been directed to articles published approximately 2.9 years ago, the second most recent metric among all journals. The most recent half-life belongs to the journal *Sustainability,* with 2.2 years.

On the other hand, the half-life of the most mature citations belongs to the journal *Technovation*, that is, the age of the most cited articles in this journal is 10 years. From the point of view of the percentage of documents cited (%DC), except for one, all journals have a percentage above 90% of their articles cited one or more times. The biggest impact factors are from the journals *Global Environmental Change-Human and Policy Dimensions* (10.43) and *Environmental Innovation and Societal Transitions* (7.51).

The dual map overlay of Figure 5, whose labels come from the journal titles, shows that the themes of the journals are diverse. The 285 journals that published the articles (left) are especially from the areas of physics, chemistry, materials, ecology, land, marine, mathematics, and systems.

The articles cited 17,047 other journals in their references (right side of the map), from the areas of chemistry, materials, ecology, earth, plants, zoology, environment, nutrition, veterinary, systems, biology, and economics, which reaffirms the interdisciplinary and multidisciplinary character of the

Figure 5. Dual map overlay of journals

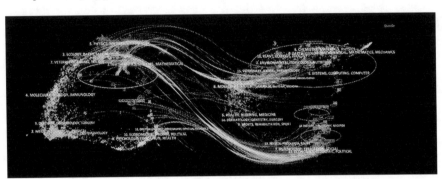

field. In general, it is evident that the fifteen main journals (5.3% of 285 journals) published 55.2% of the papers (589) and received 79.3% of the total citations (18,965 out of 23,904 citations) in a highly interdisciplinary and multifaceted field.

Scope of Research

Countries and Institutions

The scope of a research field can be discussed by analyzing the number of publications or citations by countries and institutions (Huang et al., 2020). Table 3 shows the productivity indicators and countries with publications on the subject. The works come from countries located mainly in Europe and North America. The continents of Asia and South America also published on the subject in the period analyzed. Only China, Taiwan, and Brazil are not OECD member countries. China was the country that published the most on eco-innovation, accounting for 158 jobs, followed by Italy (128 jobs) and the United Kingdom (126).

The countries whose works obtained the highest number of citations are Germany (4377 citations) and the United Kingdom (4090 citations). Concerning the percentage of papers with at least one citation, all countries have values that can be considered high, above 74%. A continuous act, the percentage of studies of the countries present in the 10% most cited in the world varies between 21.05% and 31.18%, showing that research on eco-innovation has potential in the field.

Finally, about the percentage of international collaboration of authors and co-authors, the highest percentage is obtained by the USA, with 75.25%. In general, the authors of the most productive countries establish international research networks for the publication of the papers. Figure 6 shows the network of collaboration between countries, between 1991-2019.

The network consists of 48 nodes, 247 edges, and has a density of 0.219, that is, 21.9% of the possible relations between the countries were established. The countries with different and higher degrees of the centrality of intermediation are the USA, Italy, Germany, The Netherlands, England, Australia, and China, as they are surrounded by purple rings, indicating that these countries, therefore, play a role of connection between the countries of the network. Brazil appears well-connected in the network, besides presenting a prominent position worldwide when it is among the 15 countries in the world with the highest number of citations.

Table 3. Main countries in studies on eco-innovation

Country	TC	Ps	%DC	Top 10%	%IC
Germany	4,377	94	81.91%	28.72%	42.55%
United Kingdom	4.090	126	91.27%	32.54%	60.32%
Netherlands	3,597	93	95.70%	31.18%	69.89%
Spain	3,227	121	92.56%	21.49%	39.67%
United States	2,434	101	82.18%	24.75%	75.25%
China Mainland	2,241	158	74.68%	26.58%	24.68%
Italy	2.240	128	89.06%	26.56%	38.28%
Australia	1,301	29	93.10%	27.59%	58.62%
Taiwan	1,168	49	89.8%	22.45%	26.53%
France	1,062	50	88%	22%	62%
Sweden	1,020	40	95%	32.5%	55%
Canada	932	38	84.21%	21.05%	50%
Brazil	707	46	95.65%	26.09%	41.3%
Switzerland	640	26	84.62%	26.92%	61.54%

Note: TC: total citations of articles from a country. Timespan: 1991-2019. Ps: number of papers published by the author(s) in the country. %DC: percentage of published documents cited one or more times. Top 10% is a high-performance research indicator, indicates the percentage of jobs that are among the 10%most cited in the world. % IC: international collaborations are the percentage of the country's ability to attract partnerships.

Figure 6. Countries most prominent in research on eco-innovation

Authors

The results of Table 4 show that the author with the highest total number of citations in the references of the sample papers is Klauss Rennings, with 1890 citations, where the 8, or 100% research papers of this author are situated among the 10% most cited in the world, and finally, 38% of his works were carried out in collaboration with international co-authors.

Nicholas Owen is the author with the highest 'h index' of the sample, 97, followed by Adrian Bauman (91) and James F. Sallis (87), which means that the authors have at least h citations in h number of articles (Huang et al., 2020). About the Top 10 indicator, all fifteen authors have at least one study located among the 10% most cited in the world. Most authors whose articles have a high number of citations have some level of collaboration with international co-authors. The authors are most co-cited in the references of the studies culminated in the co-citation network of authors in Figure 7.

Table 4. Authors with the highest number of citations

Authors	Institution	Ps	TC	H-index	Top 10%	%IC
Rennings, Klaus	Erasmus University Rotterdam	8	1890	15	100%	38%
Boons, Frank	University of Manchester	2	786	18	100%	100%
Bocken, Nancy M. P.	Lund University	5	760	25	60%	40%
Owen, Nicholas	Swansea University	1	729	97	100%	100%
Leslie, Eva	Flinders University South Australia	1	729	40	100%	100%
Bauman, Adrian	University of Sydney	1	729	91	100%	100%
Sallis, James F.	University of California San Diego	1	729	87	100%	100%
Humpel, N	University of Wollongong	1	729	14	100%	100%
Evans, Steve	University of Cambridge	2	654	4	50%	50%
Rana, Padmakshi	University of Cambridge	1	626	6	100%	0%

Note: Ps: number of papers. Timespan: 1991-2019. TC: total citations of the articles. %DC: percentage of published documents cited one or more times. Top 10% is an indicator of high-performance research, indicates the percentage of papers that are among the 10% most cited in the world. %IC: is the percentage of the author's ability to attract international partnerships.

The network consists of 399 nodes (authors) and 1194 edges with a density of 0.015, considered low because only 1.5% of the possible relationships between the authors are locked. Because of the analysis of Figure 7, it is evident that the network of authors co-cited in the references of the papers on eco-innovation involves recognized authors in strategy and innovation studies, such as David J. Teece, Michael Eugene Porter, Jay Barney, Kathleen Marie Eisenhardt, Wesley M. Cohen, Claes Fornell, Henry Chesbrough, among others and organisms such as OECD.

In addition, it is perceived that the basic studies for research in eco-innovation involve authors of the area of sustainability such as John Elkington, and Charbel Jose Chiappetta Jabbour. The relationship between the authors co-mentioned in the studies surveyed allows us to unveil the established cooperations and also to deepen the knowledge about the author's importance to the field (Zupic & Čater, 2015).

Main Themes and Keywords of Interest to Surveys

The co-words are characterized by the occurrence of two or more keywords in the same document, from which it can be seen that they may have some kind of relationship, so that the greater the number of occurrences, the closer the relationship (Chen et al., 2016; Courtial, 1994). The analysis of co-words can

Figure 7. Co-citation network of authors

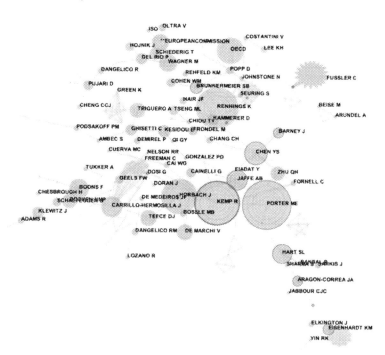

be performed by several methods, among which we highlight factor analysis, multivariate, cluster, and social networks and has as main benefit the knowledge of the general field (Chen et al., 2016). Therefore, the network of co-occurrence of keywords in Figure 8 pieces of evidence, through cluster analysis, the themes most discussed in studies on eco-innovation and correlates.

Figure 8. Clusters on eco-innovation

Seven major clusters of co-citation of keywords were identified: *environmental innovation, sustainable business model archetype, technological change, industrial symbiosis, induced eco-innovation, green innovation practice, and eco-innovation determinant*. The two largest clusters are summarized in the Table 5.

Table 5. Thematic clusters in eco-innovation studies

#	Silhouette	Main themes	Examples of *cluster jobs*
0	0,733	**Label: Environmental innovation** open innovation ecosystem; environmental strategic orientation; creating sustainable innovation; green public procurement; technology; services sector; environmental sustainability; sustainable supply chain.	**Social and legal level:** 1. As political pressures, corporative, legal affect the adoption of green innovation (Lin, Zeng, Ma, Qi, & Tam, 2014). 2. The patenting and legislation of countries and organizations (Albino et al., 2014).
1	0,689	**Label: Sustainable business model archetype** consumer; environmental strategic orientation; key environmental indicator; progressive product innovation; technology push; safe operating range; resource constraint; low-carbon economy; risk finance.	**Institutional level:** 1. Sustainable business projects, relating to eco-innovations, eco-efficiency, and corporate social responsibility (Bocken et al., 2014). 2. Development of a tool to support brainstorming changes for *eco-innovation* (Bocken et al., 2012).

According to Chen (2014), the silhouette of a cluster reflects its configuration quality, whose value varies between -1 and 1 so that a high value indicates a cluster consistent from the point of view of its members. According to Table 5, cluster 0, whose label is *environmental innovation* has a high silhouette of 0.733, where the main topics addressed are open innovation ecosystem, environmental strategic orientation, sustainable innovation, green public procurement, persuasive technology, service sector, environmental sustainability, and sustainable supply chain. Cluster 1, on the other hand, has a silhouette of 0.689, and a *sustainable business model archetype* label, whose main themes are: consumer organization, environmental strategic guidance, key environmental indicator, progressive product innovation, technological boost, safe operating range, resource restriction, low carbon economy, and risk financing.

Regarding the network of most used keywords, Figure 9 shows the interrelationship between them. The network consists of 188 nodes (keywords), 1289 edges, with a density of 0.0733, indicating that 7.33% of the relationships were established. The keyword network shows that the most prominent node (fact indicated by its largest size in the network) is *eco-innovation,* followed by *performance, management, sustainability, innovation, and environmental innovation.* Table 6 shows the metrics on frequency, degree centrality, and centrality of intermediation of the ten main keywords of the network.

It is not surprising that the word *eco-innovation* is one of the most prominent on the network since it was one of the words used to survey the works. *Eco-innovation* contains the highest frequency of occurrences (207), higher centrality of grade 36, and high centrality of intermediation 0.08, which means to say that *eco-innovation* occurred concomitantly in the same study in 207 studies, being a node that relates to 36 other nodes, with the ability to mediate relationships in 8% of the time.

The highest degree of centrality is obtained by *management*, which relates to 51 other keywords, and holds the highest centrality of intermediation (10%, as well as *innovation).* The main keywords show that the themes of eco-innovation gravitate around performance, management, sustainability, technology, sustainability, and policies.

Concerning the most cited subjects longitudinally in the research on eco-innovation, Figure 10 shows the explosion of citations of the five main keywords with greater use in studies on eco-innovation between 2018 and 2019, present in the titles, abstracts, and keywords of the author of the works.

In the first column, the keywords are presented, the following column shows the force of the explosion, and finally, the beginning and end years of greater use of these words are presented. It is noticed

Figure 9. Network of keyword co-occurrence

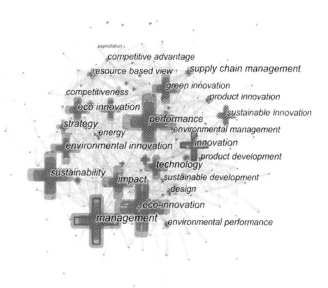

that the most recent keywords are *environmental regulation, network, agriculture, circular economy,* and *dynamic capability*.

From the point of view of environmental regulation, the impacts of the regulatory environment on the intensity of environmental innovation are of particular importance (Liao & Tsai, 2019). The importance of partnerships and alliances between companies, universities, and research agencies to foster eco-innovation stand out (Triguero et al., 2015).

In turn, eco-innovation has as its main contributions to agriculture, whose focus is on reducing spending and increasing productivity through increased environmental technology (Viaggi, 2015). Concerning the circular economy, its relationship with eco-innovation is notorious and much discussed in the

Table 6. Keyword co-occurrence network metrics

Keyword	Frequency	Degree Centrality	Intermediation Centrality
Eco-innovation	207	36	0.08
performance	189	43	0.07
Management	170	51	0.10
Sustainability	163	41	0.08
Innovation	135	38	0.10
Environmental innovation	126	36	0.07
Green innovation	106	33	0.03
Technology	103	41	0.06
Policy	92	24	0.04
Sustainable innovation	82	28	0.06

Figure 10. Keyword citation burst

Keywords	Strength	Begin	End	1991 - 2019
environmental regulation	3.4367	2018	2019	
network	5.021	2018	2019	
agriculture	3.5109	2018	2019	
circular economy	6.2637	2018	2019	
dynamic capability	5.5251	2018	2019	

scientific field, since eco-innovations are implemented and necessary for the realization of the circular economy (Prieto-Sandoval et al., 2018). Finally, as far as dynamic capacities are concerned, studies analyze, proprietarily, the role of dynamic capacities and their micro-foundations for the development of sustainable innovations (Mousavi et al., 2019).

CONCLUSION

The general objective of this work was to evaluate the literature on eco-innovation in the international scenario from the bibliometric and scientometric analysis. Through the SCI-Expanded database and the applied bibliometric procedure for reading the first page of all papers, it was possible to collect the documents related to the theme of eco-innovation (sustainability-oriented innovation, green innovation, sustainable innovation, environmental innovation, clean innovation, ecology, eco-friendly innovation, environmentally sustainable innovation), for the relevant capture of the field of knowledge, relatively not observed in previous studies, which allowed to meet the proposed objective and adequately respond to the research question.

When evaluating the production of studies on eco-innovation, it is verified that the field is in the process of frank expansion, especially in the last decade. The most prominent subject categories are *Environmental Sciences, Green & Sustainable Science & Technology and Engineering, Environmental*, in a field marked by high interdisciplinarity and multidisciplinarity. The fifteen main journals (5.3% of 285 journals) published 55.2% of the papers and received 79.3% of the total citations (18,965 out of 23,904 citations), a scenario in which the most productive journal on eco-innovation is *the Journal of Cleaner Production*. The papers were published by authors from 83 countries, located mainly in Europe and North America. Countries such as China, Taiwan, and Brazil are not OECD member countries but have presented relevant production on the subject. The author with the highest number of citations and publications is Klauss Rennings. Concerning the most used keywords are *eco-innovation, performance, management, sustainability, innovation, and* environmental *innovation*. The three words with citation burst are *environmental management, circular economy*, and *dynamic capability*.

This study contributes to the theoretical field as it performs a mapping of the trends of research on eco-innovation, performing comprehensive and relatively unobserved capture in previous studies, due to the amplitude of the keywords used, extracted from systematic reviews of the literature and theoretical-empirical studies.

Finally, as a gap, it was identified that the numerous forms of registration of the term make the field diffuse. The limitations involve the use of only one international database, so that future changes can broaden the methodological strategy about the inclusion of more databases, or also perform theoretical analyses on the reasons why there is the dispersion of concepts in different themes and expressions, an essential measure for understanding the area of studies on eco-innovation, whose theoretical question is not properly pacified in the field.

REFERENCES

Aboelmaged, M., & Mouakket, S. (2020). Influencing models and determinants in big data analytics research: A bibliometric analysis. *Information Processing & Management*, *57*(4), 102234. doi:10.1016/j.ipm.2020.102234

Albino, V., Ardito, L., Dangelico, R. M., & Messeni Petruzzelli, A. (2014). Understanding the development trends of low-carbon energy technologies: A patent analysis. *Applied Energy*, *135*, 836–854. doi:10.1016/j.apenergy.2014.08.012

Araújo, C. A. (2006). Bibliometria: Evolução histórica e questões atuais [Bibliometrics: Historical evolution and current issues]. *Questao*, *12*(1), 11–32.

Bocken, N. M. P., Allwood, J. M., Willey, A. R., & King, J. M. H. (2012). Development of a tool for rapidly assessing the implementation difficulty and emissions benefits of innovations. *Technovation*, *32*(1), 19–31. doi:10.1016/j.technovation.2011.09.005

Bocken, N. M. P., Short, S. W., Rana, P., & Evans, S. (2014). A literature and practice review to develop sustainable business model archetypes. *Journal of Cleaner Production*, *65*, 42–56. doi:10.1016/j.jclepro.2013.11.039

Bornmann, L., & Leydesdorff, L. (2013). Macro-Indicators of Citation Impacts of Six Prolific Countries: InCites Data and the Statistical Significance of Trends. *PLoS One*, *8*(2), 1–5. doi:10.1371/journal.pone.0056768 PMID:23418600

Bossle, M. B., Dutra De Barcellos, M., Vieira, L. M., & Sauvée, L. (2016). The drivers for adoption of eco-innovation. *Journal of Cleaner Production*, *113*, 861–872. doi:10.1016/j.jclepro.2015.11.033

Bródka, P., Skibicki, K., Kazienko, P., & Musiał, K. (2011). A degree centrality in multi-layered social network. *Proceedings of the 2011 International Conference on Computational Aspects of Social Networks*, (pp. 237–242). IEEE. 10.1109/CASON.2011.6085951

Chen, C. (2006). CiteSpace II: Detecting and Visualizing Emerging Trends and Transient Patterns in Scientific Literature. *Journal of the American Society for Information Science and Technology*, *57*(3), 359–377. doi:10.1002/asi.20317

Chen, C. (2014). *The CiteSpace Manual*, 94. College of Computing and Informatics Drexel -- Drexel University. http://cluster.cis.drexel.edu/~cchen/citespace/

Chen, C., & Song, M. (2019). Visualizing a field of research: A methodology of systematic scientometric reviews. *PLoS One*, *14*(10), e0223994. doi:10.1371/journal.pone.0223994 PMID:31671124

Chen, X., Chen, J., Wu, D., Xie, Y., & Li, J. (2016). Mapping the Research Trends by Co-word Analysis Based on Keywords from Funded Project. *Procedia Computer Science, 91*(Itqm), 547–555. doi:10.1016/j.procs.2016.07.140

Courtial, J. P. (1994). A coword analysis of scientometrics. *Scientometrics, 31*(3), 251–260. doi:10.1007/BF02016875

Deng, W., Liang, Q., Li, J., & Wang, W. (2020). Science mapping: a bibliometric analysis of female entrepreneurship studies. Gender in Management. doi:10.1108/GM-12-2019-0240

Diniz, E. (2013). Editorial. *RAE - Revista de Administração de Empresas, 53*(3), 223.

Dos Santos, R., & Kobashi, N. (2009). Bibliometria, cientometria, infometria: Conceitos e aplicações [Bibliometrics, scientometrics, infometrics: concepts and applications]. *Pesquisa Brasileira Em Ciência Da Informação [Brazilian research in information science], 2*(1), 155–172.

Duan, G., Bai, Y., Ye, D., Lin, T., Peng, P., Liu, M., & Bai, S. (2020). Bibliometric evaluation of the status of Picea research and research hotspots: Comparison of China to other countries. *Journal of Forestry Research, 31*(4), 1103–1114. doi:10.100711676-018-0861-9

Edwards, F. (1991). The Banff Centre for Management's Recent or Impending Initiatives in Environmental Innovation. *Environmental Conservation, 18*(4), 369–370. doi:10.1017/S0376892900022736

Fu, H. Z., Wang, M. H., & Ho, Y. S. (2012). The most frequently cited adsorption research articles in the Science Citation Index (Expanded). *Journal of Colloid and Interface Science, 379*(1), 148–156. doi:10.1016/j.jcis.2012.04.051 PMID:22608849

Glänzel, W., & Moed, H. F. (2002). Journal impact measures in bibliometric research. *Scientometrics, 53*(2), 171–193. doi:10.1023/A:1014848323806

Guo, P., Tian, W., Li, H., Zhang, G., & Li, J. (2020). Global characteristics and trends of research on construction dust: Based on bibliometric and visualized analysis. *Environmental Science and Pollution Research International, 27*(30), 37773–37789. doi:10.100711356-020-09723-y PMID:32613507

Hellström, T. (2007). Dimensions of environmentally sustainable Innovation: The structure of eco-innovation concepts. *Sustainable Development, 15*(3), 148–159. doi:10.1002d.309

Ho, Y. S. (2018). Comment on: "A bibliometric analysis and visualization of medical big data research" Sustainability 2018, 10, 166. *Sustainability (Switzerland), 10*(12), 2017–2018. doi:10.3390u10124851

Ho, Y. S. (2019a). Comments on "A Bibliometric Analysis of Research on Intangible Cultural Heritage Using CiteSpace" by Su et al. (2019). *SAGE Open, 9*(4), 0–1. doi:10.1177/2158244019894291

Ho, Y. S. (2019b). Comments on Research trends of macrophage polarization: A bibliometric analysis. *Chinese Medical Journal, 132*(22), 2772. doi:10.1097/CM9.0000000000000499 PMID:31765362

Ho, Y. S. (2019c). Rebuttal to: Su et al. "The neurotoxicity of nanoparticles: A bibliometric analysis," Vol. 34, pp. 922–929. *Toxicology and Industrial Health, 35*(6), 399–402. doi:10.1177/0748233719850657 PMID:31244406

Horbach, J., Rammer, C., & Rennings, K. (2012). Determinants of eco-innovations by type of environmental impact - The role of regulatory push/pull, technology push and market pull. *Ecological Economics*, *78*, 112–122. doi:10.1016/j.ecolecon.2012.04.005

Huang, L., Zhou, M., Lv, J., & Chen, K. (2020). Trends in global research in forest carbon sequestration: A bibliometric analysis. *Journal of Cleaner Production*, *252*, 1–17. doi:10.1016/j.jclepro.2019.119908

Imran, A., Humiyion, M., Arshad, M. U., Saeed, F., Arshad, M. S., Afzaal, M., Imran, M., Usman, I., Ikram, A., Naeem, U., Hussain, M., & Al Jbawi, E. (2022). Extraction, amino acid estimation, and characterization of bioactive constituents from peanut shell through eco-innovative techniques for food application. *International Journal of Food Properties*, *25*(1), 2055–2065. doi:10.1080/10942912.2022.2119999

Jeong, H. J., & Ko, Y. (2016). Configuring an alliance portfolio for eco-friendly innovation in the car industry: Hyundai and Toyota. *Journal of Open Innovation*, *2*(4), 24. doi:10.118640852-016-0050-z

Klewitz, J., & Hansen, E. G. (2014). Sustainability-oriented innovation of SMEs: A systematic review. *Journal of Cleaner Production*, *65*, 57–75. doi:10.1016/j.jclepro.2013.07.017

Kumari, P., & Kumar, R. (2020). Scientometric Analysis of Computer Science Publications in Journals and Conferences with Publication Patterns. *Journal of Scientometric Research*, *9*(1), 54–62. doi:10.5530/jscires.9.1.6

Lazar, N., & Chithra, K. (2021). Comprehensive bibliometric mapping of publication trends in the development of Building Sustainability Assessment Systems. *Environment, Development and Sustainability*, *23*(4), 4899–4923. doi:10.100710668-020-00796-w

Liao, Y. C., & Tsai, K. H. (2019). Innovation intensity, creativity enhancement, and eco-innovation strategy: The roles of customer demand and environmental regulation. *Business Strategy and the Environment*, *28*(2), 316–326. doi:10.1002/bse.2232

Lin, H., Zeng, S. X., Ma, H. Y., Qi, G. Y., & Tam, V. W. Y. (2014). Can political capital drive corporate green innovation? Lessons from China. *Journal of Cleaner Production*, *64*, 63–72. doi:10.1016/j.jclepro.2013.07.046

Mallawaarachchi, H., Sandanayake, Y., Karunasena, G., & Liu, C. (2020). Unveiling the conceptual development of industrial symbiosis: Bibliometric analysis. *Journal of Cleaner Production*, *258*, 120618. doi:10.1016/j.jclepro.2020.120618

Masquietto, C. D., Sacomano Neto, M., & Giuliani, A. C. (2011). Centrality and Density in Interfirm Networks: a Study of an Ethanol Local Productive Arrangement. *Review of Administration and Innovation - RAI*, *8*(1), 122–147. doi:10.5773/rai.v8i1.456

Mousavi, S., Bossink, B., & van Vliet, M. (2019). Microfoundations of companies' dynamic capabilities for environmentally sustainable innovation: Case study insights from high-tech innovation in science-based companies. *Business Strategy and the Environment*, *28*(2), 366–387. doi:10.1002/bse.2255

Pacheco, D. A. de J., ten Caten, C. S., Jung, C. F., Ribeiro, J. L. D., Navas, H. V. G., & Cruz-Machado, V. A. (2017). Eco-innovation determinants in manufacturing SMEs: Systematic review and research directions. *Journal of Cleaner Production*, *142*, 2277–2287. doi:10.1016/j.jclepro.2016.11.049

Peng, B., Guo, D., Qiao, H., Yang, Q., Zhang, B., Hayat, T., Alsaedi, A., & Ahmad, B. (2018). Bibliometric and visualized analysis of China's coal research 2000–2015. *Journal of Cleaner Production*, *197*, 1177–1189. doi:10.1016/j.jclepro.2018.06.283

Prieto-Sandoval, V., Jaca, C., & Ormazabal, M. (2018). Towards a consensus on the circular economy. *Journal of Cleaner Production*, *179*, 605–615. doi:10.1016/j.jclepro.2017.12.224

Rennings, K. (2000). Redefining innovation— Eco-innovation research and the contribution from ecological economics. *Ecological Economics*, *32*(2), 319–332. doi:10.1016/S0921-8009(99)00112-3

Rennings, K., Ziegler, A., Ankele, K., & Hoffmann, E. (2006). The influence of different characteristics of the EU environmental management and auditing scheme on technical environmental innovations and economic performance. *Ecological Economics*, *57*(1), 45–59. doi:10.1016/j.ecolecon.2005.03.013

Strehl, L. (2005). O fator de impacto do ISI e a avaliação da produção científica: Aspectos conceituais e metodológicos [The impact factor of the ISI and the evaluation of scientific production: Conceptual and methodological aspects]. *Ci. Inf.*, *34*(1), 19–27. doi:10.1590/S0100-19652005000100003

Triguero, A., Moreno-Mondéjar, L., & Davia, M. A. (2015). Eco-innovation by small and medium-sized firms in Europe: From end-of-pipe to cleaner technologies. *Innovation (North Sydney, N.S.W.)*, *17*(1), 24–40. doi:10.1080/14479338.2015.1011059

Viaggi, D. (2015). Research and innovation in agriculture: Beyond productivity? *Bio-Based and Applied Economics*, *4*(3), 279–300. doi:10.13128/BAE-17555

Xavier, A. F., Naveiro, R. M., Aoussat, A., & Reyes, T. (2017). Systematic literature review of eco-innovation models: Opportunities and recommendations for future research. *Journal of Cleaner Production*, *149*, 1278–1302. doi:10.1016/j.jclepro.2017.02.145

Yin, J., Gong, L., & Wang, S. (2018). Large-scale assessment of global green innovation research trends from 1981 to 2016: A bibliometric study. *Journal of Cleaner Production*, *197*, 827–841. doi:10.1016/j.jclepro.2018.06.169

Zárate-Rueda, R., Beltrán-Villamizar, Y. I., & Murallas-Sánchez, D. (2021). Social representations of socioenvironmental dynamics in extractive ecosystems and conservation practices with sustainable development: A bibliometric analysis. *Environment, Development and Sustainability*, *23*(11), 16428–16453. Advance online publication. doi:10.100710668-021-01358-4

Zhou, W., Chen, J., & Huang, Y. (2019). Co-Citation Analysis and Burst Detection on Financial Bubbles with Scientometrics Approach. *Economic Research Journal*, *32*(1), 2310–2328. doi:10.1080/1331677X.2019.1645716

Zupic, I., & Čater, T. (2015). Bibliometric Methods in Management and Organization. *Organizational Research Methods*, *18*(3), 429–472. doi:10.1177/1094428114562629

Chapter 6
Education for Sustainability:
Promoting the Sustainable Development Goals in the Development of Mobile Applications

Clara Silveira
https://orcid.org/0000-0003-2809-4208
Polytechnic of Guarda, Portugal

Cristiano Teixeira
Polytechnic of Guarda, Portugal

Leonilde Reis
https://orcid.org/0000-0002-4398-8384
Instituto Politecnico de Setubal, Portugal

ABSTRACT

Information and Communication Technologies enhance human progress, bringing value to people and society. The role of software in society requires a paradigm shift for software development. The Karlskrona Manifesto reflects this change by establishing a focus on sustainability education. The objective of the chapter is to present the development of an Android mobile application inspired by the Sustainable Development Goals to promote sustainability. The methodology adopted used agile development integrated with the Software Engineering Method and Theory - SEMAT approach. SEMAT and agile development are two complementary initiatives, and perfectly aligned, both are structured and non-prescriptive that help to think and improve software development capability. The developed mobile application, Android, thus allows learning more about sustainability by answering questionnaires, thus contributing for the target audience to apply knowledge in environmental and social domains enhancing human progress, bringing value to people and society.

DOI: 10.4018/978-1-6684-6123-5.ch006

INTRODUCTION

The involvement of higher education institutions in the promotion of sustainable development goals is considered central. Higher Education, in particular in Portugal, can play a decisive role among an age group of students strongly motivated and sensitive to the problem of the SDGs. In this way, it is essential to enhance the appeal to responsible life among students, in view of the principles of Sustainable Development. In this sense, it is considered that the various digital technologies can interact, reason, perceive, learn and act in a given environment.

Some of these technologies enhance the development of paradigms that could play a key role in driving the Sustainable Development Goals (SDGs) in the current contexts of digital transformation. It is considered that, given the potential of Information and Communication Technologies (ICT), these can contribute to achieving the 17 Sustainable Development Goals. Namely exploiting the Information Systems (IS) to make information more reliable and enhancing decision making, strategy definition and implementation of policies based on data analysis. In this sense, the conditions for optimization in terms of resource allocation may be created.

One of the objectives of this research is to promote the use of ICT in favor of people, namely in the search for sustainable behaviors and attitudes, creating better habits of life. In this project, it was also defined as an objective the inclusion of the five dimensions of sustainability from the perspective of the Karlskrona Manifesto, thus connecting the concerns underlying software development in the social, human, economic, technical and environmental dimensions.

This chapter is organized into six sections. The first is the introduction in which the need identified in the organizational context is presented, specifying the objective of the chapter. In the second section, the theoretical framework is presented with regard to the various themes that are addressed in the chapter. The development of the Mobile Application is presented in the third section. The fourth section describes the solutions and recommendations. The fifth section, Future Research Directions, describes the importance of including the five dimensions of sustainability from the perspective of the Karlskrona Manifesto. Finally, in section six, the main conclusions of the work are presented.

BACKGROUND

This chapter discusses the study of some currently available options, in this case applications, and analyzes them to identify their strengths and weaknesses. This study may benefit the development of this project's application. Thus, an explanation of what the Sustainable Development Goals are is presented, followed by an analysis of the applications: "SDGs in Action", "Educa 2030", and "ODS Research and Action".

Sustainable Development Goals

The SDGs define global sustainable development priorities and aspirations for 2030 and seek to mobilize global efforts around a set of common goals and targets (UNDP, 2020). There are 17 SDGs, in areas that affect the quality of life of all the world's citizens and those yet to come. Figure 1 shows the 17 Sustainable Development Goals, and they will be listed below.

The SDGs have implicitly 17 global goals (Figure 1), set by the United Nations General Assembly, (United Nations, 2022) in which it mentions the importance of transforming our world according to

Education for Sustainability

Figure 1. The 17 Sustainable Development Goals
Source: (United Nations, 2022).

the 2030 Agenda for Sustainable Development. The goals are broad and interdependent, but each has a separate list of targets to be achieved. Achieving all 169 goals would indicate the achievement of all 17 goals. Figure 2 emphasizes SDG 4 as it highlights the role of education that aims to ensure access to inclusive, quality and equitable education and promote lifelong learning opportunities for all. One of the most prominent goals of the SDG is to provide learners with high-quality education (Saini, Sengupta, Singh, Singh, & Singh, 2022).

In fact, when considering the increasing daily use of mobile technologies in education, it is believed that these technologies hold important potential in raising awareness for sustainable development goals (Çimşir & Uzunboylu, 2019). Studies reveal positive effects (Çimşir & Uzunboylu, 2019), in using mobile applications (Alrabaiah & Medina-Medina, 2021) to raise awareness among university students (Almazroa, Alotaibi, & Alrwaythi, 2022) about sustainable development goals.

The mobile application presented, in the next section of this chapter, emphasizes the importance of the SDG introduced in Figure 2. The promotion of Education and Learning Opportunities brought about by ICT, currently allows educators to teach knowledge to students more conveniently using the Internet and mobile apps, compared to conventional classroom-based approaches conducted face-to-face (Wu, Guo, Huang, Liu, & Xiang, 2018). This is thus considered to be a contribution to SDG 4.

It is also considered relevant to address sustainability requirements in the development of software systems. It is intended to motivate software engineers to apply lessons learned and identify possible sustainability impacts. According to (Penzenstadler, Raturi, Richardson, & Tomlinson, 2014), software engineers have the potential to considerably improve the sustainability of civilization.

SDG in Action

The "SDGs in Action" application contains all the lists of tasks to end poverty, reduce inequalities and combat climate change, i.e., all the information about the 17 SDGs, information about partnerships

Figure 2. SDG 4 – Quality Education (Ensure inclusive and equitable quality education and promote lifelong learning opportunities for all)
Source: (United Nations, 2022).

that are part of the SDGs and information about news and events taking place. Figure 3 shows some screenshots of the application.

Source: https://sdgsinaction.com/

From the analysis of the "SDGs in Action" application came the idea of integrating all the information about the SDGs in the Grow+ application, since this is also an application focused mainly on the SDGs.

Educa 2030

EDUCA 2030 is a tool for sensitization, communication and development of learning Objectives of Sustainable Development through the game. It is based on the EdTech methodology, where they combine teaching and learning options, instruction, evaluation, gamification, in an attractive and developed environment with didactic methodologies. EDUCA 2030 is an educational game to improve social awareness processes in an innovative way, through game methodologies. The "Educa 2030 ODS" tool presented in Figure 4 shows some screenshots of the application.

This app is similar to the application described in this chapter as it presents a game supported by general questions on the themes inherent to the 17 SDGs.

Education for Sustainability

Figure 3. App "SDGs in Action"

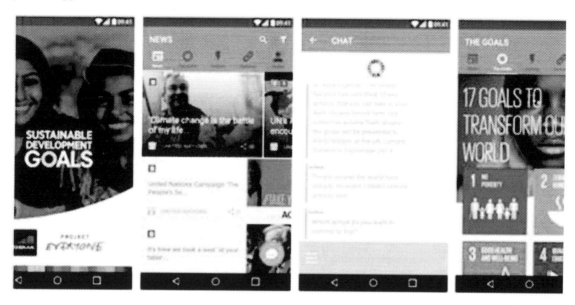

Figure 4. "Educa 2030 ODS"
Source: https://m.apkpure.com/br/educa-2030-ods/com.cricket.educa2030.

Figure 5. App ODS Research and Action
Source: https://odsresearch.com/.

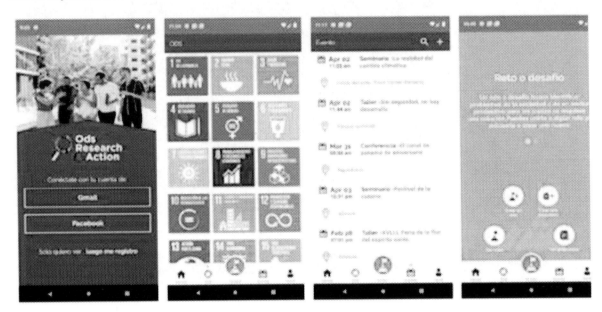

SDG Research and Action

Figure 5 illustrates some screenshots of the "ODS Research and Action" application. It is an app that allows to identify critical problems (straight or challenges) linked to sustainable development and generate proposals in different areas, which serve as a basis for the development of research that can be translated into programs and public policies or contribute to the development of new business opportunities, through an accessible, open and collaborative research model (FUDIS Desarrollo Sostenible, 2020).

Figure 5 shows information about the 17 SDGs, documents, and research areas about them. It also informs the user about upcoming events/conferences. It also contains some questionnaires on current issues.

MOBILE APPLICATION DEVELOPMENT

Based on the analysis of some existing applications on the market and with the reference that the main goal is to develop a mobile application of educational quizzes based on the SDGs, the development process is presented.

Methodology

The methodology of this research followed the principles of the Agile Manifesto (Beck, et al., 2001) for software development (Rubin, 2013), integrating the Software Engineering Method and Theory (SEMAT) approach (Object Management Group, 2022). From the perspective of (Jacobson, Spence, & Ng, 2013) SEMAT and agile development are two complementary initiatives, both are structured and non-prescriptive that help to think and improve software development capability.

Education for Sustainability

Figure 6. Architecture Schema SEMAT
Source: (Object Management Group, 2018).

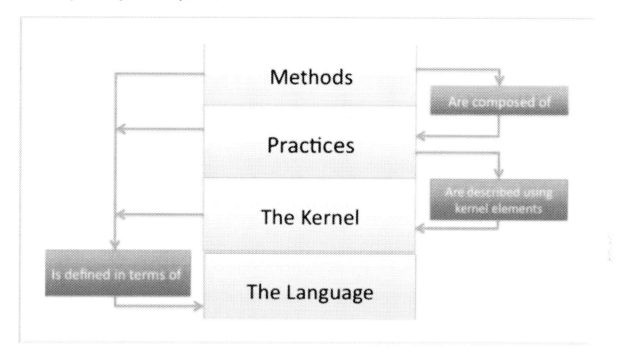

It should be noted that the main characteristic of agile development is the constant validation of functionality by the user as the code is developed, thus avoiding discrepancy between the intended and the final project. In this sense, the first principle of the Agile Manifesto states: "Our highest priority is, from the earliest stages of the project, to satisfy the customer through the rapid and continuous delivery of valuable software" (Beck, et al., 2001). It is also ensured that at any point in the project development it will be possible to accommodate changes simply, quickly and effectively.

SEMAT (Jacobson, Lawson, Ng, McMahon, & Goedicke, 2019), (Object Management Group, 2018) was created as a new approach to software engineering built on the experience of software experts that captures their understanding to educate and support a new generation of practitioners. It is an initiative that focuses on both supporting the art (methods) and building fundamental understanding (theory). The result of this initiative was the creation of the Essence language, which is the core (Kernel) of a software engineering theory, with methods, practices, and a language to describe the theory shown in Figure 6.

Figure 6 shows the software engineering kernel includes the set of elements that are found in all types of software system designs. Essence also defines the language that can be used to represent the kernel and describe practices and methods. Importantly, the language is intended to be used by practitioners and methodologists. The Essence language was considered, by the Object Management Group, to be a standard as of 2015. It is currently at version 1.2 (Object Management Group, 2018).

The practices developed based on the Kernel allow the use of agile methods. Scrum for example, can be seen as a composition of three smaller practices: Daily Stand-up, Backlog-Driven Development and Retrospective. A practice can be expressed by, (Jacobson, Lawson, Ng, McMahon, & Goedicke, 2019):

Figure 7. Key Elements Language Essence
Souce: (Jacobson, Lawson, Ng, McMahon, & Goedicke, 2019)

Element Type	Syntax	Meaning of Element Type
Alpha	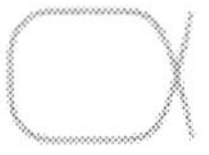	An essential element of the development endeavor that is relevant to an assessment of the progress and health of the endeavor.
Work Product	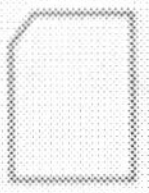	A tangible thing that practitioners produce when conducting software engineering activities.
Activity	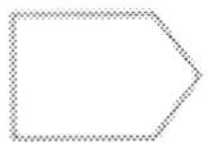	A thing that practitioners do.
Competency		An ability, capability, attainment, knowledge, or skill necessary to do a certain kind of work.

- Identify the areas in which the project is advancing;
- Describe the activities used to achieve this progress and the work products produced;
- Describe the specific skills required to carry out these activities.

The kernel thus provides a common framework for describing all practices and allowing them to be combined into methods. Inserting a set of practices into this system allows gaps and overlaps to be identified more easily. Figure 7. Presents some of the most important elements of the Essence language.

Figure 8 shows the most important elements of the Essence language. The core defines seven dimensions for measuring progress, known as alphas. The seven dimensions are: opportunity, stakeholders (interested parties), requirements, software system, work, team, and way of working (Jacobson, Lawson, Ng, McMahon, & Goedicke, 2019).

Figure 8 presents three areas: Customer, Solution, and Effort that are worth highlighting, (Jacobson, Lawson, Ng, McMahon, & Goedicke, 2019).

In the Customer area, we highlight:

- Opportunity: an opportunity is a possibility to do something of value to customers, including fixing an existing problem through the software system;
- Stakeholders: stakeholders are individuals, organizations or groups who have some interest or concern either in the system to be developed or in its development.

With regard to the Solution area, this includes:

Figure 8. Seven Dimensions SEMAT
Source: (Jacobson, Lawson, Ng, McMahon, & Goedicke, 2019)

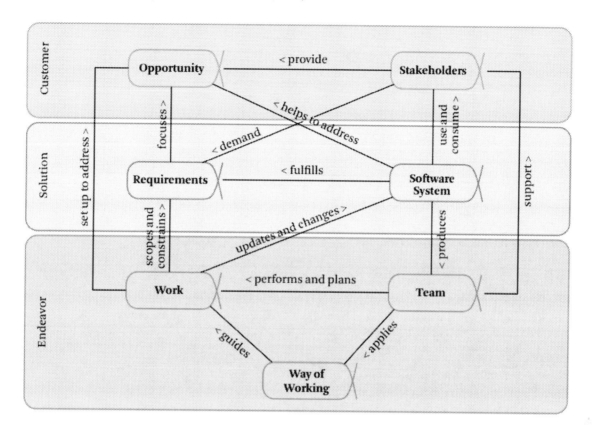

- The requirements that provide stakeholders with the vision of what they expect the software system to do;
- Software system: this refers to the functionality that is developed.

The area Endeavor, features:

- The team that must have enough people (not too many), with an appropriate mix of skills, must work collaboratively and adapt to changing environments, as good teamwork is essential;
- The work of bringing the opportunity to reality. Effort and time are the most important measures of work;
- Way of working: team members must agree on their way of working.

The agile methodology was used as it is considered the most suitable for software development projects (Dhir, Kumar, & Singh, 2019), with constant concern for validating functionality (Manisha, Khurana, & Kaur, 2021) as code is developed. In fact, the practice of delivering products continuously and frequently is very important in agile methods (Telemaco, Oliveira, Alencar, & Cowan, 2020). The process involved several workflows represented by dimensions (alphas). The alpha states provided a useful way to discuss

Table 1. Actors and their goals

Actors	Objectives
User (in the role of learning)	Create account. Answer an educational quiz. Search quiz highscores. View information about the 17 SDGs. Delete highscores. See information about the planet and pollution. Edit account.
App Manager	Create Quiz.

responsibilities and involvement (Jacobson, Spence, & Ng, 2013). Progress through the requirements, in terms of use cases/user stories, represented the main focus in developing the application.

Requirements Specification

Requirements are characteristics that the software or system to be developed should meet (Sommerville, 2016). They clearly define what the software will have to accomplish to meet the needs of the organization/users/stakeholders. In specifying requirements use cases will be used, as they provide a vehicle to: capture requirements about the system; communicate with end users and subject matter experts; test the system.

Table 1 presents the actors that integrate the "Grow+" app as well as their objectives. The goals presented correspond to the use cases.

As shown in Table 1, the use cases are represented in the "objectives" column of the two main actors "User, in the role of learning" and "Application Manager", where:

- User (in the role of learning) - any end user of the application who, after valid login, can: create and edit their account, answer educational/cultural quizzes, query scores of answered quizzes, query information about the 17 SDGs, delete scores, query information on how to help the planet be a better place;
- App Manager - user with access privileges to create quizzes and keep the App updated.

The use cases, representing the goals of the main actors, will be described with the flow of events from the actors' point of view. It should detail what the system must provide to the actor when the use case is executed. Typically, it will show how the use case starts and ends. As a key to understanding this concept, it should be noted that a use case is not just a software module - rather it is something that provides value to the actor.

The description of a use case shows why the system is needed. They are very useful to assist in requirements analysis in communicating with stakeholders. What should the description of a use case have? On the one hand there is the normal flow of events (Main Path), where everything runs smoothly, showing the dialog between the actor and the system. On the other hand, there is the flow of alternative events, where things do not go normally (Alternative Paths), showing the alternatives to the "Main Path" steps.

The description of each use case will have the following fields (template):

Education for Sustainability

Table 2. Use case description "Answer an educational quiz"

Name	Answer an educational quiz.
Description	This use case aims to describe the interaction of the actor "User in the role of learning" with the App when he is answering a quiz. The educational quiz can be general culture or specific to the SDG.
Precondition	Valid login.
Main Path	1. The actor selects a quiz that he/she intends to answer; 2. The system displays an initial page with the highscore value; 3. The actor selects the option "Start"; 4. The system presents the questions to answer, with the option to confirm the answer; advance to the next question; 5. The actor at the final question selects the option to end the quiz; 6. After finishing the system presents the new highscore resulting from the answers.
Alternative Paths	1.a) Empty list of quizzes, a quiz must be inserted; 3.a) The actor selects "Exit"; 5.a) The actor selects "Cancel".
Postcondition	--
Adornments	Test whether a highscore was saved after a quiz was answered. Test if all highscores are reset to zero when they are deleted.

- Name: name of the use case that is going to be described;
- Description: short and succinct description of the use case in question, indicating the purpose of the main actor;
- Precondition: initial condition necessary for the use case to start;
- Main Path: description of how the user must proceed in order for everything to run successfully;
- Alternative Paths: description of what could go wrong in a particular step of the main path;
- Postcondition: condition the system is in after this use case is finished;
- Adornments: description of tests to be performed or non-functional requirements.

Table 2 presents the template with the description of the use case "Answer an educational quiz".

Table 3 describes the use case "Search quiz highscores." Note that there are fields in the template that do not need to be completed, such as the "Adornments" field.

Table 3. Use Case description "Search quiz highscores"

Name	Search quiz highscores.
Description	The use case aims to search the highscores of each answered quiz.
Precondition	Valid login.
Main Path	1. The actor selects the "Highscores" search option. 2. The system displays all the highscores for each quiz. 3. The actor selects the exit option.
Alternative Paths	3.a) The actor can also select "Delete". 3.b) The system erases all the results obtained, that is, it sets all the highscores to zero.
Postcondition	The system sends a toast that the score points have been saved in the actor's account.
Adornments	--

Figure 9. Mobile Application Architecture

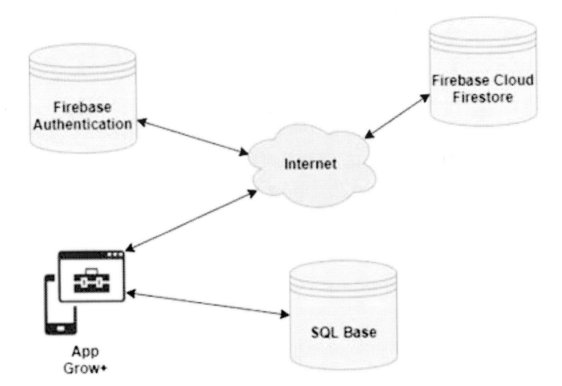

The template used in the description of the use case "Search quiz highscores", shows the sequence of steps when everything goes well - without errors - in the "Main Path"; it shows the possible flaws in the "Alternative Paths"; in the "Postcondition" shows a notification of the sent message.

Architecture and Technologies

The Grow+ mobile app architecture, in Figure 9, shows the infrastructures used: contains the Firebase Authentication component that authenticates users, Firebase Cloud Storage that stores app grow+ data, and SQL Base that stores questions and answers from surveys.

The Grow+ app, as we can see in Figure 9, can be accessed by users via a smartphone, obtaining from SQL Base all questions and answers from the questionnaires and the authentication and user data part at Firebase. The operation of the App with internet connection allows the user never to miss the results obtained in the quizzes, thus being able to see all their highscores on another smartphone with the installation of the App.

Thus, the technologies/tools used and that contributed to the development of the application were: the integrated development platform dedicated to Android programming - Android Studio (Android Studio, 2022); the Android mobile operating system (Android, 2022); the Java programming language based on the object-oriented paradigm; the application development platform for mobile devices and the web - Firebase - uses an authentication service, a real-time database and cloud storage to store all

Education for Sustainability

Figure 10. Java Code Extract of class "MostraPontuacaoQuizErradicarPobreza"

```
package grow.plus.app.PaginasIniciaisDosQuizzes;

import ...

public class MostraPontuacaoQuizErradicarPobreza extends AppCompatActivity {

    private static final int REQUEST_CODE_QUIZ_ERRADICAR_POBREZA = 1;

    private TextView textViewHighscoreErradicarPobreza;

    private int highscoreErradicarPobreza;


    private Button Start;
    private ImageView botaoretroceder;

    FirebaseFirestore fStore;
    String userID;
    FirebaseAuth fAuth;
    @Override
    protected void onCreate(Bundle savedInstanceState) {
        super.onCreate(savedInstanceState);
        setContentView(R.layout.activity_mostra_pontuacao_quiz_erradicar_pobreza);
        getWindow().setFlags(WindowManager.LayoutParams.FLAG_FULLSCREEN, WindowManager.

        textViewHighscoreErradicarPobreza = findViewById(R.id.textViewHighScoreQuizErra
        Start = findViewById(R.id.buttonComecarQuizErradicarPobreza);
        botaoretroceder = (ImageView) findViewById(R.id.imageViewRetrocederQuizErradica

        fAuth = FirebaseAuth.getInstance();
        fStore = FirebaseFirestore.getInstance();

        userID = fAuth.getCurrentUser().getUid();
```

data related to the Grow+ application; the SQL language (Structured Query Language), as a standard declarative search language for relational databases; the GitHub repository (GitHub, 2022), as a version control platform. It is important to highlight the importance of the GitHub repository, because all the project files that are stored in the repository, it is possible to add changes (commits) and therefore create different versions throughout the iterations of the process.

Figure 10 shows a code extract, in Java, used in the App development project. The "MostraPontuacaoQuizErradicarPobreza" class, is the class that displays the quiz home page, in this case the quiz related to the Sustainable Development Goal, Eradicate Poverty. First the variables used are declared. Within the onCreate class all the functional code is implemented. In this code extract the variables are assigned the ID of each feature displayed on the screen and also the search of the user id to Firebase to be able to display all the user data on the screen.

Modules of the Mobile Application

Projects in Android Studio contain everything that defines the workspace for an app, from source code and resources to test code and build setup. Each time a new project is created, Android Studio creates the necessary structure for all the surrounding files. In each Android app module, the files are grouped into three main folders:

Figure 11. Modules of the application structure

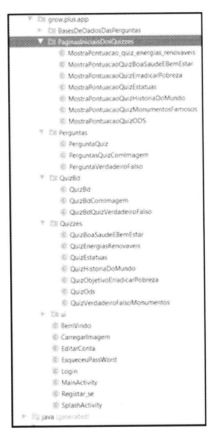

- The folder that contains all the app settings and configurations;
- The folder with the Java source code files, separated by name, activities, fragments, and classes;
- A folder with non-code resources such as XML layouts, UI strings and bitmap images divided into corresponding subfolders.

Figure 11 indicates all modules of the Android Grow+ app, namely the folder with the quizzes (questions and answers), the User Interface (UI) folder and the Java folder.

Figure 12 displays the code used to register the user in Firebase. The user registration module is responsible for creating the user and saving all the access data in Firebase. In this class some validations were also used so that the user has a character limit and that nothing is filled in blank.

After the access data are all created, the user will move to a new activity where it is possible to select a profile picture (Figure 13).

Figure 14 presents the interfaces developed in the application to: register the user and select the profile picture.

The quiz module starts with the class "ShowQuizODSscore" and is responsible for presenting the user with the score made in the SDG quiz and for storing that score in the FireBase Cloud. Figure 15 shows a small code snippet of this class, responsible for the function of the "Start" button that starts the

Education for Sustainability

Figure 12. Code to register the user in Firebase

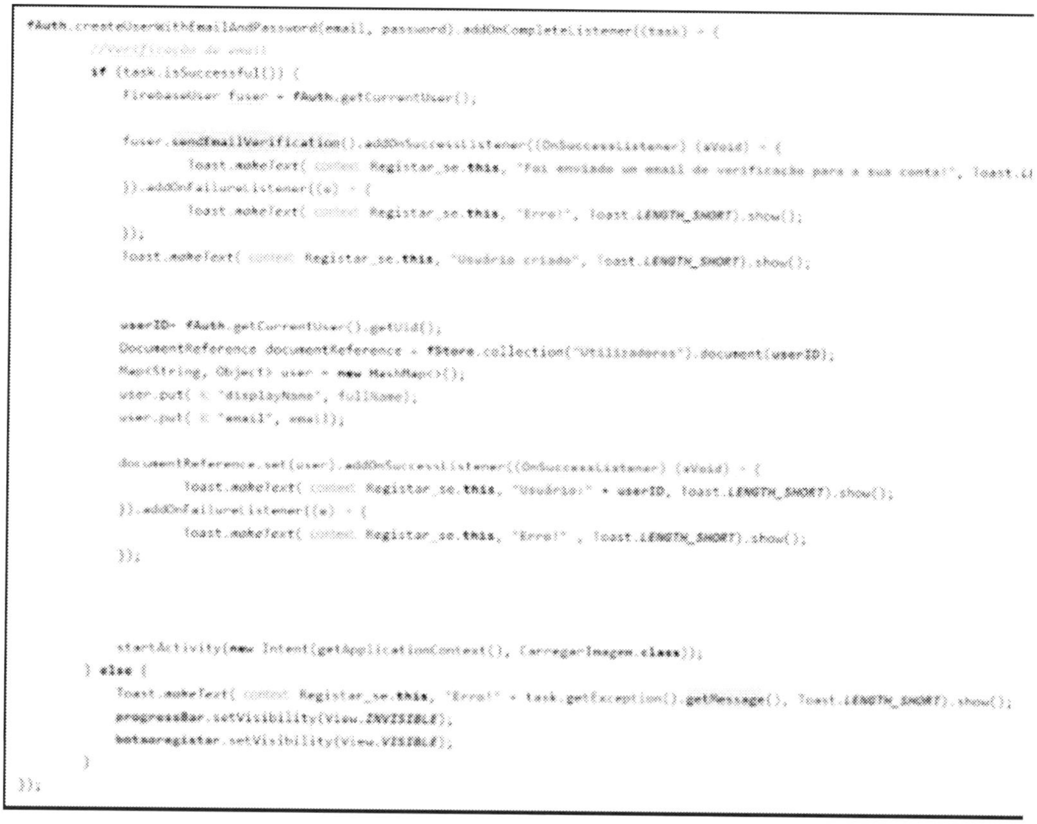

quiz. It also serves to compare the final score obtained in the Quiz with the previous one and save it in Firebase if it is higher than the previous one.

The QuizODS class is responsible for presenting the questions and answers (Figure 16), score and number of questions in the quiz, in this case the quiz on the Sustainable Development Goals.

To save the quiz score in Firebase, the updateHighscore() function is used, with the code in Figure 17.

To illustrate the Grow+ app, Figure 18 shows some screenshots: quiz selection menu, application menu, information on how to help the planet be a better place.

The mobile app Grow+ enables learning through quizzes and pedagogically promotes the increase of the level of knowledge in the fields of SDGs, monuments, and world history, in order to contribute to awareness actions on how to help the Planet become a more sustainable world.

SOLUTIONS AND RECOMMENDATIONS

One of the objectives of this research is to use technology in favor of people, namely in the search for sustainable behaviors and attitudes, creating habits of life. It is considered that this goal was achieved with the inclusion of the five dimensions of sustainability from the perspective of the Karlskrona Manifesto (Becker, Chitchyan, Duboc, & Easterbrook, 2015).

Figure 13. Code for uploading the profile picture into Firebase

```
@Override
protected void onCreate(Bundle savedInstanceState) {
    super.onCreate(savedInstanceState);
    setContentView(R.layout.activity_carregar_imagem);
    getWindow().setFlags(WindowManager.LayoutParams.FLAG_FULLSCREEN, WindowManager.LayoutParams.FLAG_FULLSCREEN);

    imagemDePerfil=(ImageView) findViewById(R.id.imageDePerfilcarregar);
    concluir=(Button) findViewById(R.id.buttonConfirmarCarregarImagem);
    carregarImagem=(Button) findViewById(R.id.buttonCarregarImagem);
    retroceder = (ImageView) findViewById(R.id.imageViewRetrocederCarregarImagem);

    fStore = FirebaseFirestore.getInstance();
    fAuth = FirebaseAuth.getInstance();
    userID = fAuth.getCurrentUser().getUid();

    retroceder.setOnClickListener((view) -> {
            startActivity(new Intent(getApplicationContext(), MainActivity.class));
    });
    concluir.setOnClickListener((view) -> {
            startActivity(new Intent(getApplicationContext(), MainActivity.class));
    });

    carregarImagem.setOnClickListener((view) -> {
            //abrir galeria
            Intent galleryIntent = new Intent(Intent.ACTION_PICK, MediaStore.Images.Media.EXTERNAL_CONTENT_URI);
            startActivityForResult(galleryIntent,REQUESCODE);
    });

}
```

The application developed impacts the various dimensions of sustainability, showing several examples for each: individual, environmental, social, technical and economic. Applying the dimensions of sustainability to the Grow+ project, we obtained:

- Social dimension: A more educated, self-confident and responsible population is built up in terms of the SDGs;
- Economic dimension: Reduces paper and learning costs;
- Individual or human dimension: Familiarization with the various cultures that exist in the world, also contributing to a more autonomous and confident person;
- Technical dimension: Facilitates access to cultural and educational goods (information/quizzes);
- Environmental dimension: It contributes to an environmental and sustainable education.

In this sense, Figure 19 presents the SDGs in which the Grow+ app may have more direct impacts.

It is considered that the chapter presents innovative and distinctive features that encourage the user to act on some of the goals related to the SDGs, since it promotes their practical application. Because it is an informative application, encouraging users to search and learn more, improving the knowledge of the SDGs, to contribute to the change of behaviors of the target audience.

Education for Sustainability

Figure 14. User register interface

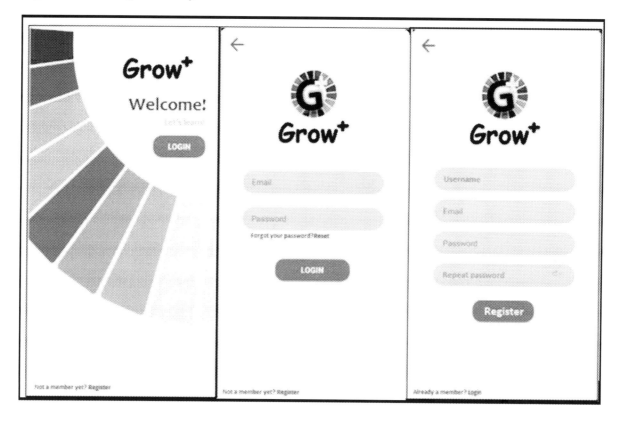

When designing software for sustainability, it is critical that all stakeholders are involved to maximize understanding of the sustainability impacts of a software application. Importantly, education for sustainability allows students to visualize the sustainability challenges they will face as future professionals in sustainable software development. This will be a mission to implement in higher education institutions, following examples such as (Liswani & Scarioni, 2021), or (Seyff, et al., 2021).

FUTURE RESEARCH DIRECTIONS

It is considered that future work is urgent for the application to include a quiz automatically from a smartphone with administrator login, or even for users themselves to be able to create and pass an evaluation to see if it is good enough to publish in the app. Another of the valences to be developed will be for the application to have links to social networks as a feature to be explored given the current dissemination of the use of social networks, thus enhancing the dissemination of the application with the goal of acquiring new users.

Given the goal of massifying the use of the application as a pedagogical tool at the service of the SDGs, it could be to share the highscores, or to create a ranking table for users, all competing among themselves to dispute the first place in the table.

Figure 15. Code used to compare the highscore

```java
@Override
protected void onActivityResult(final int requestCode, final int resultCode, @Nullable final Intent data) {
    super.onActivityResult(requestCode, resultCode, data);

    userID = fAuth.getCurrentUser().getUid();

    DocumentReference documentReference = fStore.collection( collectionPath: "HighScoreQuizOds").document(userID);
    documentReference.addSnapshotListener( activity: this, (documentSnapshot, e) -> {
        if (requestCode == REQUEST_CODE_QUIZ) {
            if (resultCode == RESULT_OK) {
                if (documentSnapshot.exists()) {
                    String highscorestr = documentSnapshot.getString( field: "highScoreQuizOds");
                    highscore = Integer.parseInt(highscorestr);
                    int score = data.getIntExtra(QuizOds.EXTRA_SCORE, defaultValue: 0);
                    if (score > highscore) {
                        updateHighscore(score);
                        loadHighscore();
                    }
                }else if(documentSnapshot.getString( field: "highScoreQuizOds")==null){
                    highscore = 0;
                    int score = data.getIntExtra(QuizOds.EXTRA_SCORE, defaultValue: 0);
                    if (score > highscore) {
                        updateHighscore(score);
                        loadHighscore();
                    }
                }
            }
        }
    });
```

Figure 16. Code to check the quiz answer

```java
private void checkResposta(){

    countDownTimer.cancel();//para o tempo
    respostas = true;
    RadioButton rbSelected = findViewById(rbGroup.getCheckedRadioButtonId());
    int perguntaNr = rbGroup.indexOfChild(rbSelected) + 1;
    if (perguntaNr == currentPergunta.getRespostaNr()) {
        score++;
        textViewScore.setText("Score: " + score);
    }
    showSolution();
}
```

Education for Sustainability

Figure 17. Code of the updateHighscore() function to store the quiz score

```
private void updateHighscore(int highscoreNew) {
    highscore = highscoreNew;

    userID= fAuth.getCurrentUser().getUid();

    String HighScoreSTR = String.valueOf(highscoreNew);
    DocumentReference documentReference = fStore.collection( collectionPath: "HighScoreQuizOds").document(userID);
    Map<String, Object> user = new HashMap<>();
    user.put( k: "highScoreQuizOds", HighScoreSTR);
    documentReference.set(user);

}
```

CONCLUSION

The basic motivation for this project arises from the perception of the planet's emergency, given the increase in global warming due to people's pollution, the differences that still exist in the world between people/cultures, and to encourage people to learn more and evolve in a more sustainable way, (Reis, Cagica Carvalho, Silveira, Marques, & Russo, 2021).

In this sense, and with the goal of promoting the SDGs in order to contribute to a more cultured, educated, and environmentally friendly society, an App was developed with pedagogical intent, fostering learning more and helping the planet, by conducting quizzes so that users can learn and test their knowledge.

The chapter presents a project in which the implementation of an Android Grow+ mobile application was outlined that allows the user to learn more about the 17 SDGs in a more dynamic way. The Project's underlying methodology was supported on agile development integrated with SEMAT, being

Figure 18. Some screenshots from the Grow+ app

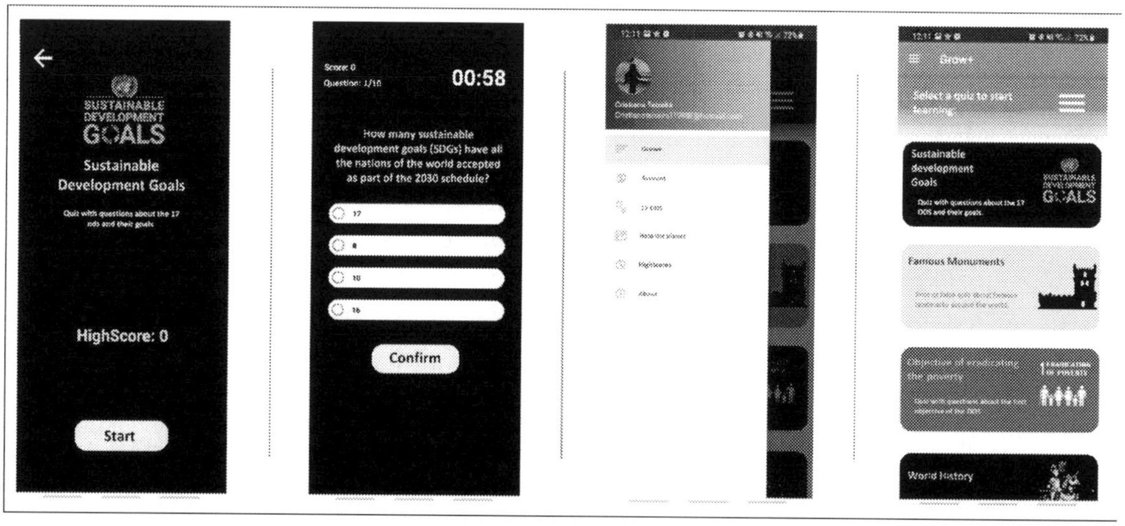

Education for Sustainability

Figure 19. Impacts of the Grow+ app on the SDGs
Source: Adapted from (United Nations, 2022).

methodologies structured to improve software development capacity and that proved adequate to the specificity of the Project.

Validation tests were performed on the app, and it was implemented on Google Play and tested on several devices with different android versions and different screen sizes. The Grow+ app will soon be available for free and can be downloaded directly to an Android device.

Finally, the mobile application developed in Android, allows learning more about sustainability by answering questionnaires, promoting SDG 4 "Ensure inclusive and equitable quality education and promote lifelong learning opportunities for all", 10 " Reduce inequality within and among countries", and 13 " Take urgent action to combat climate change and its impacts". It is considered a contribution for the target audience to apply knowledge in environmental and social domains that improve human progress, bringing value to people and society.

REFERENCES

Almazroa, H., Alotaibi, W., & Alrwaythi, E. (2022). Sustainable Development Goals and Future-Oriented Teacher Education Programs. *IEEE Transactions on Engineering Management*, 1–14. doi:10.1109/TEM.2022.3165686

Alrabaiah, H. A., & Medina-Medina, N. (2021). Agile Beeswax: Mobile app development process and empirical study in real environment. *Sustainability*, *13*(4), 1909. doi:10.3390u13041909

Android. (2022). *Android*. https://www.android.com/intl/pt-BR_br/

Beck, K., Beedle, M., Bennekum, A. v., Cockburn, A., Cunningham, W., & Fowler, M. (2001). Princípios do Manifesto Ágil. *Agile Manifesto*. https://agilemanifesto.org/iso/ptpt/principles.html

Becker, C., Chitchyan, C., Duboc, R., & Easterbrook, L. (2015). Sustainability Design and Software: The Karlskrona Manifesto. *37th International Conference on Software Engineering (ICSE 15)*. IEEE. 10.1109/ICSE.2015.179

Çimşir, B. T., & Uzunboylu, H. (2019). Awareness Training for Sustainable Development: Development, Implementation and Evaluation of a Mobile Application. *Sustainability, 11*(3), 1–17. doi:10.3390u11030611

Desarrollo Sostenible, F. U. D. I. S. (2020). *Ods Research & Action*. https://odsresearch.com/

Dhir, S., Kumar, D., & Singh, V. (2019). Success and failure factors that impact on project implementation using agile software development methodology. *Software Engineering, 731*, 647–654. doi:10.1007/978-981-10-8848-3_62

GitHub. (2022). *GitHub*. Obtido de https://github.com

Jacobson, I., Lawson, H. "., Ng, P.-W., McMahon, P. E., & Goedicke, M. (2019). *The Essentials of Modern Software Engineering: Free the Practices from the Method Prisons*. ACM Books.

Jacobson, I., Spence, I., & Ng, P. (2013). Agile and SEMAT: Perfect partners. *Communications of the ACM, 11*(9), 1–12.

Liswani, H., & Scarioni, B. (2021). Tech4Dev - Operational Report. *EPFL*. https://www.epfl.ch/innovation/domains/wp-content/uploads/2022/03/Tech4Dev-2021-Long.pdf

Manisha, K., M., & Kaur, K. (2021). Impact of Agile Scrum Methodology on Team's Productivity and Client Satisfaction – A Case Study. *3rd International Conference on Advances in Computing, Communication Control and Networking (ICAC3N)*, (pp. 1686-1691). 10.1109/ICAC3N53548.2021.9725505

Object Management Group. (2018). *Essence—Kernel and Language for Software Engineering Methods*. OMG. www.omg.org/spec/Essence/

Penzenstadler, B., Raturi, A., Richardson, D., & Tomlinson, B. (2014). Safety, Security, Now Sustainability: The Nonfunctional Requirement for the 21st Century. *IEEE Software, 31*(3), 40–47. doi:10.1109/MS.2014.22

Reis, L., Cagica Carvalho, L., Silveira, C., Marques, A., & Russo, N. (2021). *Inovação e Sustentabilidade em TIC*. Silabo.

Rubin, K. S. (2013). *Essential Scrum: A Practical Guide to the Most Popular Agile Process*. Adisson-Wesley.

Saini, M., Sengupta, E., Singh, M., Singh, H., & Singh, J. (2022). Sustainable Development Goal for Quality Education (SDG 4): A study on SDG 4 to extract the pattern of association among the indicators of SDG 4 employing a genetic algorithm. *Education and Information Technologies*, 1–39. doi:10.100710639-022-11265-4 PMID:35975216

Seyff, N., Penzenstadler, B., Betz, S., Brooks, I., Oyedeji, S., Porras, J., & Venters, C. (2021). The Elephant in the Room-Educating Practitioners on Software Development for Sustainability. *IEEE/ACM International Workshop on Body of Knowledge for Software Sustainability (BoKSS)* (pp. 25-26). IEEE. 10.1109/BoKSS52540.2021.00017

Sommerville, I. (2016). *Software Engineering* (10th ed.). Pearson.

Studio, A. (2022). *Developer Android*. Obtido de https://developer.android.com/studio?gclid=CjwKCAjwqvyFBhB7E iwAER786XUrp2bo8yfAcorObMLYazNeRtNEzEXk63p-qsAe7DvgPaf8HECUA RoC2PkQAvD_BwE&gclsrc=aw.ds

Telemaco, U., Oliveira, T., Alencar, P., & Cowan, D. (2020). A Catalogue of Agile Smells for Agility Assessment. *IEEE Access: Practical Innovations, Open Solutions, 8*, 79239–79259. doi:10.1109/ACCESS.2020.2989106

UNDP. (2020). *Integrated Solutions for Sustainable Development*. United Nations Development Programme: //sdgintegration.undp.org/

United Nations. (2022). *Objetivos de Desenvolvimento Sustentável*. UNRIC. https://unric.org/pt/objetivos-de-desenvolvimento-sustentavel/

Wu, J., Guo, S., Huang, H., Liu, W., & Xiang, Y. (2018). Information and communications technologies for sustainable development goals: State-of-the-art, needs and perspectives. *IEEE Communications Surveys and Tutorials, 20*(3), 2389–2406. doi:10.1109/COMST.2018.2812301

KEY TERMS AND DEFINITIONS

Agile Software Development: Software development process that favors direct communication between all stakeholders and simplifies documentation.

Information and Communication Technologies: A technological resource set used to process information and ensure communication. When used in an integrated way it enhances information transmission and communication processes.

Information Systems: This is the organized set of components such as people, processes of collection and transmission of data and material resources, automated or manual. The interaction of components enhances the processing and dissemination of information.

Karlskrona Manifesto: Establishes the principles and dimensions for the design of sustainable software systems.

Requirements Analysis: Iterative process to identify features and restrictions with a view to developing or changing a software product. Usually use cases are used.

Software Systems Development: Set of activities involved in the production of software. These activities are related to each other in an iterative and incremental process.

Sustainability: Ability to sustain life on the planet, considering the five dimensions: individual, social, economic, technical, and environmental.

Section 2
Governance for Sustainability

Chapter 7
Entrepreneurship and Local Government:
A Study of the Information Available on Web Pages and Its Evolution Over Time

Teresa Nevado Gil
https://orcid.org/0000-0002-4924-0908
University of Extremadura, Spain

María Pache Durán
https://orcid.org/0000-0002-6670-5818
University of Extremadura, Spain

Luisa Cagica Carvalho
https://orcid.org/0000-0002-9804-7813
Instituto Politécnico de Setúbal, Portugal & CEFAGE, Universidade de Évora, Portugal

Boguslawa M. B. Sardinha
Instituto Politécnico de Setúbal, Portugal

ABSTRACT

Entrepreneurship is one of the main drivers of social development, innovation, global competitiveness, and the economy growth. Because of that, local governments around the world, carry out initiatives focused on the promotion of entrepreneurship in order to support an economic growth and social development. This paper has two objectives. In the first place, the authors analyse the degree and type of information that local governments of the Alentejo region offer to entrepreneurs through their web pages, and calculate the index of information disclosure. Then the authors analyse the evolution of this index between 2015 and 2019. The results show a generalized increase in the information offered using the content analysis technique.

DOI: 10.4018/978-1-6684-6123-5.ch007

INTRODUCTION

In recent years, increasing attention has been paid concerning a role of the entrepreneur as a guide for economic growth and evolution. For decades, many authors (Schumpeter, 1934; Baumol, 1996; González & Ballesta, 2018) considered the importance of entrepreneurship on economic growth. Based on those finding, governments around the world carried out initiatives focused on promoting entrepreneurship and economic growth (Minniti, 2012). Mechanisms, which promote innovative entrepreneurship alongside enterprise policy instruments, are still not well understood (Audretsch 2022). The social context thereby decisively affects latent and emergent entrepreneurship (Hahn et al., 2021), but is also shaped by the entrepreneur's power relations with other stakeholders (Finn, 2021). One of the ways to promote this social context is to assemble the important information on available places. It might be a role of the local governments.

The importance of national, even local institutions for entrepreneurship development has been discussed in the literature in recent years, from the theoretical (see, for example, Baumol, 1990; Rodríguez-Pose, 2013; Williamson, 2000) and empirical perspectives (Fuentelsaz et al., 2015; Holmes et al., 2013). The European Commission (E.C.) as well as the European Parliament also indicate an important of the institutional support for entrepreneurship development and economic growth for the harmonious development of all European Union (see, for example, the Entrepreneurship 2020 Action Plan).

The objective of the research is to study the information that Alentejo local governments (minicipalites) offer their entrepreneurs through the web pages, as well as observe their evolution over time. This issue has not been treated so far despite its importance, for the economic growth.

Using the content analysis technique, based on the indicators proposed by Carvalho, Gallardo, and Nevado (2018), the websites of the municipalities were analyzed to deepen the study of the information offered to entrepreneurs. The importance of the study lies in its contribution to encourage local municipalities to promote the economic development through the advancement of entrepreneurship. In addition, the results of this study can be a base for the formulation of local public policies that contribute to job creation. With this, it is intended to contribute to the knowledge in this area in Portugal.

The work is structured as follows. Following this introduction, a review of the literature on entrepreneurship and economic growth is carried out. Next, the methodology used, and the results achieved are analyzed, to conclude with some final conclusions of the study and future lines of research.

LITERATURE REVIEW

Despite the absence of an officially accepted definition of the term "entrepreneur" (Galindo & Méndez, 2014), there are several authors who have tried to define this concept. Entrepreneurship is one of the main engines of social development, innovation, global competitiveness, and the growth of the economy. Thus, the green paper of the European Commission emphasizes these qualifications defining the entrepreneurial spirit as "the attitude and the process of creating an economic activity combining the assumption of risks, creativity and innovation with a solid management, in a new organization or in an existing one" (European Commission, 2003).

Regarding the empirical evidence, the results obtained do not keep a uniform aspect, but vary depending on the variables studied, the geographical scope, the used indicators, even if the character of the venture is by necessity or by opportunity (Acs et al., 2004; Wong, Ho & Autio, 2005; Valliere &

Peterson, 2009). Despite this, most of the research has shown a positive and solid relationship between both variables (Audretsch & Keilbach, 2005; Audretsch, 2007; Galindo, Ribeiro, and Méndez, 2012; Glaeser, Kerr, & Kerr, 2015). Although they develop different proposals, they all agree on the necessity to create a climate which favors the entrepreneur, as well as the economic perspectives of the local country and support policies of the entity studied, as essential for the development of the activity (Carrillo, Bergamini, & Navarro, 2014).

Regarding the role that local governments can play in supporting and promoting entrepreneurship, it should be noted that, thanks to the new Information and Communication Technologies (ICTs), governments can carry out an important job in supporting business behavior. In recent years, ICTs have become a fundamental tool in organizations (Puron & Rodríguez, 2016; Garde, Rodríguez, & López, 2015), and the dissemination of information through the Internet (e-government) is one of the main mechanisms of communication with citizens (Chaín, Muñoz, & Más, 2008; Ayuso & Martínez, 2005). One of the main strands in e-government research focuses on evaluating the government website. Previous studies have developed several models of government website evaluation (Irawan & Hidayat, 2022). Therefore, public administrations struggle to progress, to innovate and to transform the delivery of public services through websites (Camarero, 2003; Rodrigues, Sarabdeen & Balasubramanian, 2016; Pache & Nevado, 2021), resulting in a more efficient, responsible, and transparent government (Bannister & Connolly, 2011). Then it would be interesting to analyze whether administrations use the full potential of ICTs to promote entrepreneurship, offering information to entrepreneurs and thus promoting economic growtn, since studies on website evaluation have become an integral part of the so-called "information society", whose infrastructure is based on information technologies, computers, and electronic communication systems (Mehrad, Eftekhar, & Goltaji, 2020).

The creation of a new business can be associated with an opportunity, innovation or simply the need to create your job. However, with or without innovation, creation increases competition, boosting reorganizations and adjustments in the market and new market structures, resulting in greater efficiency and economic dynamism, which have a direct reflection on GDP levels and levels. of employment.

Figure 1 presents a synthesis on the relationships between entrepreneurship and economic development. As a result of the entrepreneurial dynamic, new businesses emerge in the market, introducing an innovation, as defended by the Schumpeterian vision, which can be translated into a new product, a new market, a new manufacturing or distribution method, a new source of raw materials or a new source of raw materials new organizational form. In this way, the presence of a new company in the market increases competition and can cause changes in the universe of its competitors, which react through the exit, mergers or significant changes in the companies that remain in the market. Consequently, a new market structure emerges in which greater efficiency and greater dynamism are achieved, resulting in positive changes in the GDP and employment indicators.

In enabling environments "entrepreneurship can be developed and lead to greater competitiveness and innovation causing development effects" (Gezer & Cardoso, 2015, 45). In a broad way we can indicate the combination of three distinctive factors in the entrepreneurial spirit:

- The risk, given that the results are uncertain.
- The innovation or creativity that characterizes the new business.
- The detection of an opportunity whose perspective is to obtain benefits.

Entrepreneurship and Local Government

Figure 1. Entrepreneur and economic performance
Source: Adapted from Barros e Pereira (2008)

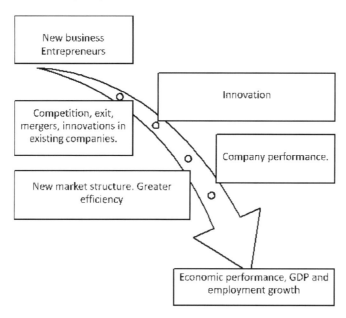

An important contribution to the analysis and understanding of the phenomenon of entrepreneurship is the consideration of a determining variable that refers to the context or environment in which the entrepreneur is implanted, a factor that will influence his behavior and actions (Sarasvathy, 2003).

Entrepreneurship is intrinsically linked to economic growth. Economic growth creates the conditions and frees up the resources for the development of regions and municipalities, which constitutes the main desire of local entities, and the main objective of intervention of all municipalities. Although development is a broad concept that implies the increase of human capacities (Sen, 1999), it needs the increase in the life patterns of individuals and the growth of the economy. This is one of the goals of the current economy and, although it is a narrow goal, it is probably one of the most important goals for development policies.

Economic growth can be seen at the macro level - nations, regions and even sectoral, but to connect growth to entrepreneurship we must go to the micro - individual level - of the entrepreneur and his ability to identify and take advantage of market opportunities (Wennekers and Thurik, 1999) and to to transform latent entrepreneurship into effective entrepreneurship (Audretsch 2021). In this field, municipalities can intervene both at the micro level, encouraging the development of attitudes and skills of individuals, and at a higher level, creating favorable conditions for local businesses and promoting the region nationally and internationally. Carree and Thurik (2002) argue that different types of ventures have differentiated impacts on economic growth. For this purpose, it identifies three types of entrepreneurs: Schumpeterian entrepreneurs, intra-entrepreneurs, and proprietary managers. The former generally represent small businesses that create new market solutions and may or may not eventually become proprietary managers. Intra-entrepreneurs or business managers make decisions on behalf of the owners of medium and large companies, risky their time and work and finally the owners managers manage most of the small businesses belonging to the so-called "central body of the economy" (Kirchhoff, 1994).

The greatest contribution of Schumpeterian and intra-entrepreneurial entrepreneurs to economic growth is related to the ability to introduce "novelties" in the economy from start-up to "transformation of ideals into economically viable entities regardless of whether there is a new company or not (intra-entrepreneurism)" (Baumol, 1993). New ideas introduced in the market increase the competitiveness of companies, or by increasing their number, or by fighting for the market share of existing companies. The novelty introduced through start-up is a stimulus for the increase in competitiveness that connects entrepreneurship to economic growth. The proprietary managers, on the other hand, have a very important role of organization, coordination of production and distribution, but they will not be the engines for innovation and introduction of novelties (Carree and Thurik, 2002).

At the macro level, countries, Global Entrepreneurship Monitor (GEM) studies seem to show the positive link between levels of entrepreneurship and economic growth indicators (Reynolds et al., 2005). Descending at the regional level, Acs and Armington (2010) established the link between entrepreneurship and economic growth, based on regional data from the United States for the period 1980-1992. They concluded that, correcting for the specific factors of each region such as: agglomeration effect next to the cities and size of the establishments, higher levels of entrepreneurship led to the highest economic growth rates. Similarly, Audretsch and Fritsch (2002) studied this relationship in the regions of Germany and reached the same conclusions that entrepreneurship is the engine of economic growth. All these studies suggest a positive relationship between entrepreneurial spirit levels and economic growth.

Although entrepreneurial action occurs at the company level, entrepreneurs need a vehicle that transforms their personal qualities and ambitions into actions. Small businesses, where the entrepreneur has control of participation, provide this opportunity. By creating small subsidiaries, the largest companies manage to create this dimension of smallness to facilitate the entrepreneurial and intra-entrepreneurial process (Carree & Thurik, 2010). At the macro level, the entrepreneurial actions of companies introduce the novelties in the market, stimulating competitiveness and promoting the selection of the best ideas and the most robust companies. The variety, competition, selection and the imitation, expand and transform the productive potential of a region or national economy (by substitution or displacement of obsolete companies, by greater productivity and by the expansion of new niches and industries) (Wennekers & Thurik, 1999). They also suggest a set of favorable conditions for the process to occur, identified in Figure 2, such as the national (or regional) cultural environment, and the internal culture of corporations, the institutional structure, both nationally and within companies, which define incentives for individuals to transform their ambitions into actions.

The existence of institutions for the development of entrepreneurship is paramount in this process. Municipalities, through the creation of various support structures for entrepreneurs and businesses, through national and international promotion of territory, play a fundamental role in stimulating entrepreneurship and economic growth by creating bases for the development of the regions. The need for development to be sustainable poses new challenges to local institutions for promoting entrepreneurship for the emergence of a new type of entrepreneur who can recognize, develop and exploit opportunities to innovate and create future goods and services with social and ecological economic gains.

METHODOLOGY

In the study an analysis of the information offered to entrepreneurs was carried out in a sample of 58 Alentejo municipalities (Portugal) in 2015 as proposed by Nevado, Gallardo, & Carvalho (2019). In order

Figure 2. Framed structure of entrepreneurship union to economic growth
Source: Wennekers and Thurik (1999)

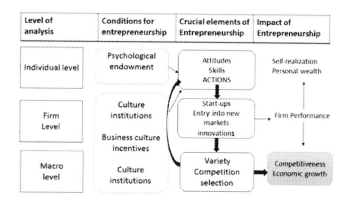

Table 1. Indicators by dimensions

Dimensions	N° of indicators
1. General information	10
2. Information on resources and entrepreneur support	12
3. Information on active entrepreneurship	10
4. Information on digital entrepreneurship	6
5. Information on disclosure and communication with the entrepreneur	11
Total	49

Source: Adapted from Carvalho *et al.* (2018)

to compare the dissemination practices during the time and verify if local governments are involved in an evolution process oriented to the satisfaction of the informative needs of entrepreneurs, a temporary study has been carried out on the same sample, collecting data from the web pages in 2019.

In order to achieve the objectives of this research, the present study starts from the compilation of data related to information addressed to entrepreneurs through the websites of Alentejo municipalities, using the content analysis technique. According to López (2002, p. 173) "this technique is an instrument of response to that natural curiosity of man to discover the internal structure of information, either in its composition, in its form of organization or structure, or in its dynamics". In addition, "it should be clarified immediately that, in many cases, the content analysis is not limited to the content, but takes into account the continent. The content analysis can be an analysis of the meanings, but it can also be an analysis of the signifiers" (Bardin, 1996, p. 25).

Data collection was carried out through access to the search engine of the main page of each municipality between the months of January and February 2019. Each indicator was assign a dichotomous scalethat is, the value 1 if the item is disclosed by the municipality and 0, otherwise (Escamilla, Plaza and Flores 2016). The instrument used to collect information was a questionnaire proposed by Carvalho et al. (2018), which has a total of 49 indicators divided into 5 dimensions (Table 1).

Subsequently, the disclosure indices were prepared (Table 2), with the objective of analyzing the degree and type of information offered by the municipalities of the Alentejo region to entrepreneurs,

Table 2. Disclosure indices

Índices	Concept	Expression
Disclosure index by local government and dimension (IDGD)	Measure the percentage of total disclosure of each local government in each of the dimensions	$IDGD_j = \left(\dfrac{\sum_{i=1}^{M}(Aij)}{M} * weighing \right) * 100$
Disclosure index by local government (IDG)	Measure the total disclosure of each local government	$IDA_j = \sum_{i=1}^{D}(IDGD_j)$
Disclosure index by item (IDI)	Measure the percentage of local governments that report each item	$IDI_i = \dfrac{\sum_{i=1}^{N}(Aij)}{N} * 100$
Disclosure index by dimension (IDD)	Measure the total disclosure of each dimension	$IDD_i = \left(\dfrac{\sum_{i=1}^{d}(IDI_i)}{M} * weighing \right) * 100$
Total Disclosure Index (IDT)	Measure the total disclosure of the sample	$IDT = \sum_{i=1}^{D}(IDD_i)$

Source: Adapted from Carvalho et al. (2018)

as well as their evolution over time. The use of indexes to measure the level of information has been carried out in previous studies such as those of Navarro et al. (2011) and Navarro et al. (2015), Moneva and Martín (2012), Gandía and Archidona (2008), and Beuren and Angonese (2015), among others.

Where, M = number of items that make up each dimension; Aij = takes the value of 1 if the characteristic that defines the indicator (i) is present in the local government (j), and 0 in the opposite case; D = number of dimensions; N = number of local governments; Having no empirical evidence on the importance of the different partial indices that make up the total index, the same specific weight has been assigned for each of the dimensions (weighting = 20%).

RESULTS

The results achieved by the disclosure rates by local government (IDG) are included in Annex I. It is observed that, in 2015, the municipality that offers the most information was Santarém, with a disclosure rate of 57.39%, followed from Ourique (55.91%), Odemira (55.39%) and Aljustrel (55.06%). However, in 2019, the municipalities that disseminate the most information were Odemira (75.51%), Vendas Novas (71.43%), Aljustrel (69.39%) and Salvaterra de Magos (65.31%). As it's possibe to see, Oldemira and Ajustrel are still among the local governments that offer more information to the entrepreneur.

At the opposite extreme, in 2015 the municipalities of Portel and Vila Viçosa are located, with almost zero disclosure rates of 2%, and Reguengos de Monsaraz and Mourão, with disclosure rates of 5.15%

Entrepreneurship and Local Government

Figure 3. Evolution of the Disclosure Index of each indicator in dimension 1 (IDI)
Source: Own elaboration

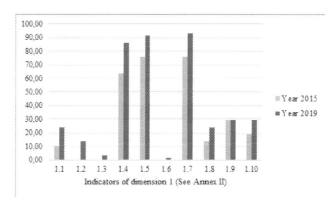

and 5, 33%, respectively. In 2019, Portel remains the city that least informed, with an index of 2.04%. It was followed by the municipalities of Avis, Barrancos and Estremoz, with disclosure rates of 8.16%. It can also be seen that in 2015 there were six municipalities that have rates that exceed 50%, while in 2019, there are fourteen local governments that exceed this percentage.

Likewise, dissemination indexes by dimensions (IDD) (Annex II) have been calculated, which measure the dissemination of information from the local governments of the sample in each of the five dimensions and represent the contribution of each of them to the index of total disclosure (RTD). There is a general increase in all dimensions with respect to 2015. Specifically, in both years, general information is the dimension that offers more information, with indices of 7.93% and 5.76%, respectively, while the dimension three, on active entrepreneurship, is the least disclosed dimension (5.59% and 3.38%). In addition, the dimension that has experienced a greater increase in the index corresponds to the fifth dimension, dissemination and communication with the entrepreneur.

On the other hand, the results achieved in the information disclosure indexes by items (IDI) for the years 2015 and 2019 are shown in Annex II. Regarding the first dimension, general information, a significant increase in the items analyzed in the evolution of the years can be observed in Figure 3. We observe, in 2015, that 75.86% of the websites of the analyzed municipalities contain links for social networks, while, in 2019, most of the websites own it (93.10%). Also, in 2015, 75.86% of the websites have an internal search engine that facilitates their navigation while, in 2019, it exists in 91.38% of the cases. In addition, in 2019, 13.79% of the websites analyzed include priorities and strategies to be achieved, 3.45% includes events, achievements and failures recorded by the entity, and 1.72%, offers the possibility to listen the page. In contrast, in 2015, these percentages were zero. With regard to the publication of key economic information, such as GDP or the unemployment rate, the figure reached in 2015 (29.31%) has been maintained in 2019.

On average, it's possible to observe a bigger investment of all municipalties to provide important information for the entrepreneur concering the priorities and strategies, economic data, development strategic issues or reasons to invest. This seems to indicate a bigger preoccupation of the municipalities about the importance of entrepreneurship as means of achieving economic growth.

Regarding the analysis of the second dimension, resources and support to the entrepreneur, it can be seen in Figure 4 that, in 2015, 58.62% of the websites analyzed report on the contacts of interest, increasing significantly in 2019 (87, 93%). The same happens when talking about the existence of a

Figure 4. Evolution of the Disclosure Index of each indicator in dimension 2 (IDI)
Source: Own elaboration

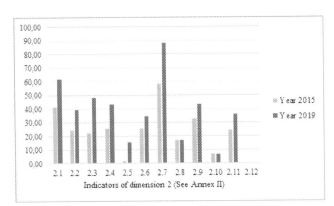

cabinet or support body for the entrepreneur. Specifically, in 2015, 41.38% of the local governments analyzed have an entrepreneur support body, while in 2019, the figure increases considerably (62.07%). On the other hand, the percentage regarding the recognition of specific needs of young people, women and other target groups, is maintained for both years (17.24%), as is the case with regard to information on protection of knowledge, brand, etc.., which offer the websites analyzed (6.90%) In the two periods analyzed, no city council offers financing solutions or reports on the cost of starting a business.

What seems to be important for the local authorities on this dimension is to provide useful contacts and contact to cabinet or organ in support of the entrepreneur, much less important protection of knowledge, brand, or information about population with special needs or costs to start business.

In relation to active entrepreneurship, Figure 5 shows an increase in most of the items studied. It follows that, in 2015, in 34.48% of the websites of the analyzed municipalities there is a nest of companies or municipal technology park, while, in 2019, the figure increases, being its disclosure rate of 55.17%. As can be seen, in 2015, 46.55% of municipalities offer services to help start businesses, while in 2019, 53.45% of municipalities do. However, initiatives to encourage Corporate Social Responsibility have not grown in the last year analyzed (1.72%).

Concerning the active entrepreneurship that is still a lot to do by local authorities, despair some advances concerning support on business creation in compliance with regulatory requirements and incubators, its necessary to reinforce courses and tutorials to develop the business skills.

Regarding the fourth dimension, digital entrepreneurship, it can be seen in Figure 6, how, in 2015, 67.24% of the municipalities have a mailbox for the citizen or a section for complaints and suggestions, reaching this figure, in 2019, an index of 79.31%. Also, in 2015 13.79% of the websites contain a platform for business exchange while, in 2019, the figure increases to 25.86%. But the biggest increase can be seen in the ease of support by the municipality for the improvement of technology in small businesses, obtaining, in 2015, an index of 10.34% and growing, in 2019, to 39.66%. However, there are few municipalities in which monitoring is carried out in the processing status and incidents in the proposed procedures, with an index of 3.45%, in 2015, and increasing only one point in 2019 (5,17%).

It is possible to see the bet for the simple and cheap solutions leaving the background solutions which need more human and technology investment, to be treated probably in the future.

With respect to the fifth dimension (Figure 7), corresponding to disclosure and communication with the entrepreneur, we observe that the greatest increases occur in the existence of a specific area for

Figure 5. Evolution of the Disclosure Index of each indicator in dimension 3 (IDI)
Source: Own elaboration

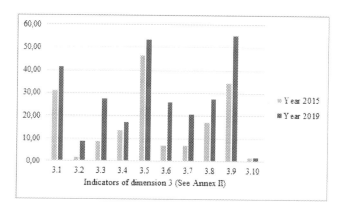

the entrepreneur, going from 32.76%, in 2015, to 65.52%, in 2019, and accessibility on the web, from 25.86%, in 2015, to 58.62%, in 2019. We can also observe that the high percentage of governments is maintained premises that publish a municipal newsletter with news related to business activity, around 55%, in 2015, and 62.07%, in 2019. However, the percentage of local governments that advertise newly created companies is maintained at the same level for both years (1.72%).

For local authority seem to gain a lot of importance is a disclosure and communication with the entrepreneur. The publication of municipal newsletter allows to publicize campaigns to promote entrepreneurship and developments as new approved projects but non publicity is given to new companies recently created.

Finally, the total disclosure index (RTD) has been analyzed, which indicates that all municipalities in the Alentejo region disclose on average, in 2019, 34.42% of the total information on entrepreneurship (Annex III), observing a noticeable increase with respect to the information they offered in 2015 (22.25%).

Figure 6. Evolution of the Disclosure Index of each indicator in dimension 4 (IDI)
Source: Own elaboration

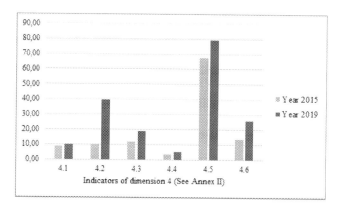

Figure 7. Evolution of the Disclosure Index of each indicator in dimension 5 (ID
Source: Own elaboration

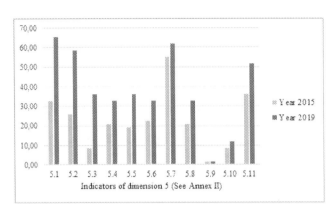

CONCLUSION

In the present work, a study of the degree and type of information that local governments in the Alentejo region offer to entrepreneurs through their web pages and, secondly, an analysis of their work is carried out evolution over time, in the period between 2015 and 2019 and indicate the involvement of the local governments into the process oriented to the satisfaction of the informative needs of entrepreneurs,

Based on the results obtained, we can say that local governments in the Alentejo region offer information to entrepreneurs through their websites. At the same time, we have been able to verify that the information they offer is scarce, reaching a degree of disclosure of 34.42%, as shown by the IDT.

Regarding the nature of the information disclosed, of the five analysis axes analyzed, the dimension that provides more information is the general information, while the dimension on active entrepreneurship is the least disclosed. In relation to the second of our objectives, the results show a general increase in the information offered in 2019, compared to 2015. Likewise, as in 2015, the information that is most widely disclosed is that of in general, while the disclosure on active entrepreneurship is the one that presents the greatest deficiencies.

These gaps show the need to propose information guides to local governments in general that will help them improve their information systems for their dissemination through their websites. Therefore, this work is intended to contribute to the generation of knowledge in this area in the Portuguese field, given that it is a topic that has not been treated so far despite its importance due to its impact on economic growth. We are aware that our research is not exempt from aspects that limit the results obtained. On the one hand, the effects of the region and the country must be considered. In this case, all the municipalities analyzed belong to the Alentejo region (Portugal), however, the consideration of entities belonging to other regions and even to other countries would conclude with deeper results. Consequently, and in terms of future research, it would be convenient to approach the study with a larger sample that allows comparative analyzes at the country level. On the other hand, we can point out that it is a purely descriptive, cross-sectional study, so it is not possible to make inferences.

National, but mainly local, public authorities face the scarcity of resources and many of them budget problems, therefore, they can no longer afford to allocate resources in a way that does not generate economic and social results from local development activities. Local actors can promote different forms of innovation it does not need to be a radical transformation, taking full advantage of business opportunities through publicity by local authorities of the region's economic development objectives (ex. strategic development plans, municipal master plan) as well as creating opportunities for collaboration and sharing of ideas between entrepreneurs and the general population. Collaboration brings with it competitive advantages. At the very least, it brings new ideas and can challenge existing practices while mobilizing new resources. Innovative ideas must be disseminated by local authorities and partner organizations through the means of their power (ex. municipal bulletin or municipality page dedicated to business activity).

Thus, the communication and availability of information and activities related to the effective promotion of entrepreneurship to transform latent entrepreneurship (Audretsch 2021) into effective entrepreneurship can form the basis of local transformation policy.

This reason leads us to propose a new line of research aimed at analyzing the explanatory factors of the development of these dissemination practices. The next step will be a comparison of the entrepreneurial economic activities in different municipalities with each disclosure index (RTD) to evaluate the impact of it on local entrepreneurship.

REFERENCES

Acs, Z. J., Audretsch, D. B., Braunerhjelm, P., & Carlsson, B. (2004). *The missing link: The knowledge filter and entrepreneurship in endogenous growth.* (Working Paper 4783). London Center for Economic Policy Research.

Acs, Z., & Armington, C. (2010). The Determinants of Regional Variation in New Firm Formation. *Regional Studies*, *36*(1), 33–45. doi:10.1080/00343400120099843

Audretsch, D. B., Belitski, M., Caiazza, R., Günther, C., & Menter, M. (2021). From latent to emergent entrepreneurship: The importance of context. *Technological Forecasting and Social Change*, 121356.

Audretsch, D. B. (2007). Entrepreneurship capital and economic growth. *Oxford Review of Economic Policy*, *23*(1), 63–78. doi:10.1093/oxrep/grm001

Audretsch, D., & Fritsch, M. (2002). Growth Regimes over Time and Space. *Regional Studies*, *36*(2), 113–124. doi:10.1080/00343400220121909

Audretsch, D. B., & Keilbach, M. (2005). Entrepreneurship capital and regional growth. *The Annals of Regional Science*, *39*(3), 457–469. doi:10.100700168-005-0246-9

Ayuso, M. D., & Martínez, V. (2005). Gobierno electrónico. Contenidos y organización de las sedes webs de los parlamentos autonómicos. [E-government. Contents and organization of the headquarters websites of the autonomous parliaments.]. *Revista Espanola la de Documentacion Cientifica*, *28*(4), 462–478.

Bardin, L. (1996). *El análisis de contenido [Content analysis.].* Ediciones Akal Universitaria.

Bannister, F., & Connolly, R. (2011). The trouble with transparency: A critical review of openness in e-government. *Policy and Internet*, *3*(1), 1–30. doi:10.2202/1944-2866.1076

Baumol, W. J. (1993). *Entrepreneurship, management, and the structure of payoffs.* New York University.

Baumol, W. J. (1996). Entrepreneurship: Productive, Unproductive and Destructive. *Journal of Business Venturing*, *11*(1), 3–22. doi:10.1016/0883-9026(94)00014-X

Beuren, I. M., & Angonese, R. (2015). Instruments for determining the disclosure index of accounting information. *Revista Eletrônica de Estratégia e Negócios – REEN, 8*(1), pp. 120-144.

Camarero, C. G. (2003). Las nuevas formas de comunicación de la administración con el ciudadano. [The new forms of communication of the administration with the citizen.] *Anales de documentación, 6*, 109-119.

Carree, M. A., & Thurik, A. R. (2010). The Impact of Entrepreneurship on Economic Growth. In Z. J. Acs & D. B. Audretsch (eds.) Handbook of Entrepreneurship Research, 557-594. Springer. doi:10.1007/978-1-4419-1191-9_20

Carrillo, L. L., Bergamini, T. P., & Navarro, C. L. C. (2014). El emprendimiento como motor del crecimiento económico. [Entrepreneurship as an engine of economic growth.] *Boletín económico de ICE. Información Comercial Española, 3048*, 55–63.

Carvalho, L., Gallardo, D., & Nevado, M. T. (2018). Local municipalities' involvement in promoting entrepreneurship: An analysis of web page orientation to the entrepreneurs in Portuguese municipalities. In L. Carvalho (ed.) Handbook of Research on Entrepreneurial Ecosystems and Social Dynamics in a Globalized World, 1-19. Évora, Portugal: IGI Global. doi:10.4018/978-1-5225-3525-6.ch001

Hahn, D., Spitzley, D. I., Brumana, M., Ruzzene, A., Bechthold, L., Prügl, R., & Minola, T. (2021). Founding or succeeding? Exploring how family embeddedness shapes the entrepreneurial intentions of the next generation. *Technological Forecasting and Social Change, 173*, 121182. doi:10.1016/j.techfore.2021.121182

Chaín, C., Muñoz, A., & Más, A. (2008). La gestión de información en las sedes webs de los ayuntamientos españoles. [The management of information in the headquarters websites of the Spanish municipalities.]. *Revista Espanola la de Documentacion Cientifica, 31*(4), 612–638. doi:10.3989/redc.2008.4.662

Comisión Europea. (2003). *El libro verde. El espíritu empresarial en Europa.* [*The Green Paper. Entrepreneurship in Europe.*] Comisión Europea, Bruselas. 0027-final. https://eur-lex.europa.eu/legal-content/ES/TXT/?uri=celex:52003DC0027

European Commission. (2013). Entrepreneurship 2020 Action Plan. Reigniting the entrepreneurial spirit in Europe. EC. https://eur-lex.europa.eu/legal-content/EN/TXT/PDF/?uri=CELEX:52012DC0795&from=EN

Escamilla, S., Plaza, P., & Flores, S. (2016). Análisis de la divulgación de la información sobre la responsabilidad social corporativa en las empresas de transporte público urbano en España. [Analysis of the disclosure of information on corporate social responsibility in urban public transport companies in Spain.]. *Revista de Contabilidad, 19*(2), 195–203. doi:10.1016/j.rcsar.2015.05.002

Finn, P. (2021) Organising for entrepreneurship: How individuals negotiate power relations to make themselves entrepreneurial. *Technological Forecasting and Social Change, 166*, 120610. doi:10.1016/j.techfore.2021.120610

Fuentelsaz, L., González, C., Maícas, J. P., & Montero, J. (2015). How different formal institutions affect opportunity and necessity entrepreneurship. *BRQ Business Research Quarterly*, *18*(4), 246–258. doi:10.1016/j.brq.2015.02.001

Holmes, R. M. Jr, Miller, T. J., Hitt, M. A., & Salmador, M. P. (2013). The interrelationships among informal institutions, formal institutions and inward foreign direct investment. *Journal of Management*, *39*(2), 531–566. doi:10.1177/0149206310393503

Galindo, M. A., & Méndez, M. T. (2014). Entrepreneurship, economic growth, and innovation: Are feedback effects at work? *Journal of Business Research*, *67*(5), 825–829. doi:10.1016/j.jbusres.2013.11.052

Galindo, M. Á., Ribeiro, D., & Méndez, M. T. (2012). Innovación y crecimiento económico: Factores que estimulan la innovación. [Innovation and economic growth: Factors that stimulate innovation.]. *Cuadernos de Gestión*, *12*(Esp), 51–58. doi:10.5295/cdg.110309mg

Gandía, J. L., & Archidona, M. (2008). Determinants of web site information by Spanish city councils. *Online Information Review*, *32*(1), 35–57. doi:10.1108/14684520810865976

Garde, R., Rodríguez, M. P., & López, A. M. (2015). Are Australian universities making good use of ICT for CSR reporting? *Sustainability*, *7*(11), 14895–14916. doi:10.3390u71114895

Gezer, I., & Cardoso, S. P. (2015). Entrepreneurship and its impact on innovation and development. A multivariate analysis with socioeconomic indicators. Globalization, *Competitiveness & Governability*, *9*(2), 43–60. doi:10.3232/GCG.2015.V9.N2.02

Glaeser, E. L., Kerr, S. P., & Kerr, W. R. (2015). Entrepreneurship and urban growth: An empirical assessment with historical mines. *The Review of Economics and Statistics*, *97*(2), 498–520. doi:10.1162/REST_a_00456

González, B., & Ballesta, J. A. (2018). Caracterización del emprendimiento femenino en España: Una visión de conjunto. [Characterization of female entrepreneurship in Spain: An overview.] *REVESCO: Revista de estudios cooperativos,* (129), pp. 39-65.

Irawan, B., & Hidayat, M. N. (2022). Evaluating Local Government Website Using a Synthetic Website Evaluation Model. [IJISM]. *International Journal of Information Science and Management*, *20*(1).

Kirchhoff, B. A. (1994). *Entrepreneurship and Dynamic Capitalism*. Praeger.

López, F. (2002). El análisis de contenido como método de investigación. [Content analysis as a research method.]. *XXI Revista de Educación*, *4*, 167–179.

Mehrad, J., Eftekhar, Z., & Goltaji, M. (2020). Vaccinating users against the hypodermic needle theory of social media: Libraries and improving media literacy. [IJISM]. *International Journal of Information Science and Management*, *18*(1), 17–24.

Moneva, J. M., & Martín, E. (2012). Universidad y Desarrollo sostenible: Análisis de la rendición de cuentas de las universidades públicas desde un enfoque de responsabilidad social. [University and Sustainable Development: Analysis of the accountability of public universities from a social responsibility approach.]. *Revista Iberoamericana de Contabilidad de Gestión*, *10*(19), 1–18.

Minniti, M. (2012). El emprendimiento y el crecimiento económico de las naciones. [Entrepreneurship and the economic growth of nations.]. *Economía Industrial, 383*, 23–30.

Navarro, A., Alcaraz, F. J., & Ortiz, D. (2010). La divulgación de información sobre responsabilidad corporativa en administraciones públicas: Un estudio empírico en gobiernos locales. [The disclosure of information on corporate responsibility in public administrations: an empirical study in local governments.]. *Revista de Contabilidad, 13*(2), 285–314. doi:10.1016/S1138-4891(10)70019-4

Navarro, A., Ruiz, M., De los Ríos, A., & Tirado, P. (2011). Responsabilidad social y administración pública local: un análisis del grado de divulgación de información en Reino Unido e Irlanda. [Social responsibility and local public administration: an analysis of the degree of disclosure of information in the United Kingdom and Ireland.] *In Actas del XVI Congreso AECA*, Granada: Asociación Española de Contabilidad y Administración de Empresas.

Navarro, A., Tirado, P., Ruiz, M., & De los Ríos, A. (2015). Divulgación de información sobre responsabilidad social de los gobiernos locales europeos: El caso de los países nórdicos. [Dissemination of information on social responsibility of European local governments: The case of the Nordic countries.]. *Gestión y Política Pública, 24*(1), 229–269.

Nevado, M. T., Gallardo, D., & Carvalho, L. (2019). Entrepreneurship in a local government: An empirical study of information in the websites of Alentejo region municipalities (Portugal). *Innovar (Universidad Nacional de Colombia), 29*(71), 97–112. doi:10.15446/innovar.v29n71.76398

Pache, M., & Nevado, M. T. (2021). Compromiso de los Ayuntamientos Malagueños con la divulgación de información responsable. [Commitment of Malaga City Councils to the dissemination of responsible information.]. *Transinformação, 33*.

PURON., G., & Rodríguez, M. P. (2016). Financial transparency in Mexican municipalities: An empirical research. En *Proceedings of the 17th International Digital Government Research Conference on Digital Government Research*. ACM.

Reynolds, P., Bosma, N., Autio, E., Hunt, S., De Bono, N., Servais, I., Lopez-Garcia, P., & Chin, N. (2005). Global entrepreneurship monitor: Data collection design and implementation 1998–2003. *Small Business Economics, 24*(3), 205–231. doi:10.100711187-005-1980-1

Rodríguez-Pose, A. (2013). Do Institutions Matter for Regional Development?. Regional Studies, *47*(7), 1034–1047. [Taylor & Francis Online], doi:10.1080/00343404.2012.748978

Rodrigues, G., Sarabdeen, J., & Balasubramanian, S. (2016). Factors that influence consumer adoption of e-government services in the UAE: A UTAUT model perspective. *Journal of Internet Commerce, 15*(1), 18–39. doi:10.1080/15332861.2015.1121460

Sarasvathy, S. D. (2003). *Effectuation: Elements of entrepreneurial expertise. The Darden School*. University of Virginia.

Schumpeter, J. A. (1934). *The Theory of Economic Development*. Harvard University Press.

Sen, A. (1999). *Development as freedom*. Oxford University Press.

Valliere, D., & Peterson, R. (2009). Entrepreneurship and economic growth: Evidence from emerging and developed countries. *Entrepreneurship and Regional Development*, *21*(5-6), 459–480. doi:10.1080/08985620802332723

Wennekers, S., & Thurik, R. (1999). Linking Entrepreneurship and Economic Growth. *Small Business Economics*, *13*(1), 27–55. doi:10.1023/A:1008063200484

Williamson, O. E. (2000). The new institutional economics: Taking stock, looking ahead. Journal of Economic Literature, *38*(3), 595–613. [Crossref], [Web of Science ®], . doi:10.1257/jel.38.3.595

Wong, P. K., Ho, Y. P., & Autio, E. (2005). Entrepreneurship, innovation, and economic growth: Evidence from GEM data. *Small Business Economics*, *24*(3), 335–350.

APPENDIX A

Disclosure Rates of Local Governments

	IDG 2015	IDG 2019
Alandroal	9.33	10.20
Alcácer do Sal	16.82	59.18
Aljustrel	55.06	69.39
Almeirim	30.09	44.90
Almodóvar	7.82	48.98
Alpiarça	7.33	8.16
Alter do chao	14.48	12.24
Alvito	7.82	32.65
Arraiolos	9.48	22.45
Arronches	8.00	8.16
Avis	9.33	8.16
Azambuja	37.73	8.60
Barrancos	11.00	8.16
Beja	48.91	55.10
Benavente	32.24	38.78
Borba	11.15	44.90
Campo maior	14.97	18.37
Cartaxo	22.30	34.69
Castelo de Vide	16.82	18.37
Castro verde	24.45	28.57
Chamusca	13.00	32.65
Coruche	35.12	40.82
Crato	9.15	10.20
Cuba	11.48	14.29
Elvas	28.97	32.65
Estremoz	5.82	8.16
Évora	26.76	59.18
Ferreira do Alentejo	20.97	26.53
Froteira	34.42	40.82
Gaviao	30.42	40.82
Golega	25.24	36.73
Grândola	**36,12**	**61,22**
Marvao	45.58	46.94
Mértola	23.48	22.45
Monforte	13.15	14.29
Montemor-o-Novo	16.64	57.14
Mora	9.15	26.53
Moura	37.45	46.94
Mourao	5.33	16.33
Nisa	10.97	34.69
Odemira	55.39	75.51
Ourique	55.91	57.14
Ponte de Sor	7.82	16.33
Portalegre	25.09	28.57
Portel	2.00	2.04
Redondo	17.48	44.90
Reguengos de Monsarat	5.15	28.57
Rio Maior	52.39	55.10
Salvaterra de Magos	54.39	65.31
Santarem	57.39	57.14
Santiago do Cacém	22.48	46.94
Serpa	16.82	22.45
Sines	32.61	51.02
Sousel	23.82	32.65
Vendas Novas	9.15	71.43
Viana do Alentejo	6.82	20.41
Vidigueira	9.33	51.02
Vila Viçosa	2.00	44.90

Entrepreneurship and Local Government

APPENDIX B

Disclosure Rates by Items and Dimensions

DIMENSION 1: GENERAL INFORMATION	IDI 2015	IDI 2019
There is a statement of the maximum authority about the importance of entrepreneurship and economic development	10,34	24,14
Priorities and strategies to achieve are included in this statement	0,00	13,79
Events, achievements and failures by the entity are included	0,00	3,45
1.4. There is a map of the web itself	63,79	86,21
There is an internal search engine for easy navigation	75,86	91,38
There is a possibility to hear the page	0,00	1,72
There are links to social networks	75,86	93,10
There is a list of companies in the territory	13,79	24,14
Key economic data is published, as GDP or the unemployment rate	29,31	29,31
1.10. It reports on strategic issues or reasons to invest in that municipality	18,97	29,31
DISCLOSURE INDEX DIMENSION 1 (IDD)	**5,76**	**7,93**
DIMENSION 2: RESOURCES AND SUPPORT FOR ENTREPRENEURS	**IDI 2015**	**IDI 2019**
2.1. There is a cabinet or organ in support of the entrepreneur	41,38	62,07
2.2. It is available the rules for creating a business	24,14	39,66
2.3. It is available a reference to a physical space for the start of the activity	22,41	48,28
2.4. It is published information about procedures to be followed for creating a business	25,86	43,10
2.5. There are specific objectives and measurable targets for increasing business activity	1,72	15,52
2.6. Information about entrepreneurship is disseminated, including social entrepreneurship and its impact on the economy	25,86	34,48
2.7. There is reference to useful contacts	58,62	87,93
2.8. Specific needs for young people, women and other target groups are recognized	17,24	17,24
2.9. It reports on tax incentives for investment (financing solutions venture capital)	32,76	43,10
2.10. There is information on the protection of knowledge, brand etc.	6,90	6,90
2.11. It reports on formalities for creating enterprises	24,14	36,21
2.12. Reference is made to the time and cost of starting a business	0,00	0,00
DISCLOSURE INDEX DIMENSION 2 (IDD)	**4,68**	**7,24**
DIMENSION 3: ACTIVE ENTREPRENEURSHIP	**IDI 2015**	**IDI 2019**
3.1. Mechanisms such as multi-stakeholder forums to promote dialogue on entrepreneurship	31,03	41,38
3.2. There are tutorials available to help start a business	1,72	8,62
3.3. There are courses available to develop the skills of entrepreneurs	8,62	27,59
3.4. There are contests, prizes or similar events to publicly recognize entrepreneurs	13,79	17,24
3.5. The city offers services to help business creation in compliance with regulatory requirements	46,55	53,45
3.6. There is clarity on priorities and the type of project that the municipality wants to encourage	6,90	25,86
3.7. Days of exchange of experiences and best practices are held	6,90	20,69
3.8. There are educational programs for entrepreneurship in younger schools	17,24	27,59
3.9. There is an incubator companies nest or municipal technology park	34,48	55,17
3.10. There are initiatives to encourage SR in companies	1,72	1,72
DISCLOSURE INDEX DIMENSION 3 (IDD)	**3,38**	**5,59**
DIMENSION 4: DIGITAL ENTREPRENEURSHIP	**IDI 2015**	**IDI 2019**
4.1. There are awareness and capacity development campaigns in the use of ICT and digital economy	8,62	10,34
4.2. The municipality facilitates support to improvement of technology in small businesses	10,34	39,66
4.3. Ability to make administrative procedures, permits, online licenses	12,07	18,97
4.4. Online monitoring of processing status and incidents of the posed proceedings	3,45	5,17
4.5. There is a citizen mailbox or a section for complaints, suggestions	67,24	79,31
4.6. There is a platform for business exchange, enterprise portals, trade fairs, business associations and clubs	13,79	25,86
DISCLOSURE INDEX DIMENSION 4 (IDD)	**3,85**	**5,98**
DIMENSION 5: DISCLOSURE AND COMMUNICATION WITH THE ENTREPRENEUR	**IDI 2015**	**IDI 2019**
5.1. There is a specific area for entrepreneurs	32,76	65,52
5.2. It is accessible on the web and easy to identify	25,86	58,62
5.3. Investment opportunities are disclosed	8,62	36,21
5.4. It reports on developments as new approved projects, news highlights …	20,69	32,76
5.5. There is a space for user satisfaction to improve support cabinets	18,97	36,21
5.6. There is a space for users to express your reviews	22,41	32,76
5.7. A Municipal Bulletin is published related to business news	55,17	62,07
5.8. It reports on what it takes to start a business	20,69	32,76
5.9. It spreads or publicity is given to new companies recently created	1,72	1,72
5.10. There is a relationship to the university or community college	8,62	12,07
5.11. campaigns are conducted to promote entrepreneurship	36,21	51,72
DISCLOSURE INDEX DIMENSION 5 (IDD)	**4,58**	**7,68**

APPENDIX C

Full Disclosure Index

	2015	2019
FULL DISCLOSURE INDEX (IDT)	**22,25**	**34,42**

Chapter 8
The Challenges Cities Face on Their Way Towards Creativity:
Indicative Case Studies of Creative Cities

Antonia Stefanidou
Technical University of Crete, Greece

ABSTRACT

The chapter highlights how important it is for modern cities to invest in the field of creative economy, which has dominated the international scene in recent decades. Promoting innovation, with emphasis on competitiveness and cultural diversity, creative economy's sectors are developing dynamically. The chapter aims to define the concept of the city, to present cities' categorization nationally (in Greece), as well as in European and global level. It highlights the relationship between culture and development, creative cities' characteristics, as well as examining the creative sectors, with emphasis to the challenges that cities face in their effort to benefit from the creative economy. For the empirical analysis of the theoretical part, indicative case studies of cities, such as Nantes and Medellin, models of urban revival at European and world level, are used. The challenges faced by these cities and the policies they have implemented, aimed at development, are presented.

Culture: "A whole complex of mental, material, spiritual, emotional elements that characterize a society / social group. It includes not only the arts, but also beliefs, lifestyles, value systems, fundamental human rights" (World Conference on Cultural Policy, Mexico, 1982).

Development: "A process that extends beyond simple economic growth, incorporating all aspects of life, the activities of communities, whose members are called upon to contribute and are expected to benefit" (World Conference on Cultural Policy, Mexico, 1982).

DOI: 10.4018/978-1-6684-6123-5.ch008

INTRODUCTION

Orville (2019) argues that creativity, a special kind of renewable resource and human talent, is at the heart of sustainability, as culture, creativity as well as artistic innovation are drivers of development. Cultural expression may have an inner value; however it also provides energy and empowerment, building better ways of living together in different societies.

Moreover, creativity plays a vital role in strengthening communities. The expansion of cities, due to globalization, transformed culture and creativity into new keywords in understanding the urban transformations. Nowadays, cities are at the heart of development and innovation, facing unprecedented challenges. According to UNESCO, cities shelter half the world's population, consume 60% of global energy, release 75% of greenhouse gas emissions and they produce 70% of global waste. Mass tourism and uncontrolled development put cultural heritage sites and living heritage at risk.

At the same time, as centers of creativity, cities combine culture and technology, promoting economic growth through Creative and Cultural Industries (CCIs). Urban areas transform themselves into eco-cities; new technologies give access to digital content, increasing interaction and exposure, reducing production costs, offering a range of digital solutions.

In the 2030 Agenda of the UN for Sustainable Development it is recognized that culture, including the world cultural heritage and Creative Industries, can play a vital role in achieving sustainable development. Therefore, culture is promoted by many cultural actors as the fourth pillar of sustainable development, along with the other three dimensions (economic, social and environmental) and a precondition for the achievement of the Sustainable Development Goals (SDGs).

The power of culture should not be underestimated, as cultural activity through its different sectors is a highly effective tool and also a vector for the implementation of sustainable development. Culture contributes to the increase of businesses' competitiveness, promoting job creation and competitiveness both inside and outside the EU. Moreover, it promotes international cooperation among cities, as numerous city alliances have been created.

The Chapter approaches the ways cities are classified based on specific criteria as well as creative cities' characteristics. The difficulties / challenges faced by cities globally during their transition to a creative city status are presented, focusing on the ways these difficulties were addressed.

Among the Chapter's expected results is the proof that culture operates as a *"social glue"*, a powerful engine, spreading awareness in favor of the SDGs, the urgent call for action by all countries. Moreover, the proof that the development of creative economy and its sectors is a basic condition in the way towards cities' creative transition, is attempted, as well as the classification of the difficulties that cities face in their quest to develop creatively.

Firstly, we will present the relationship between city, culture and development and the concept of *"city"*. The classification of cities follows, based on specific characteristics, at national and international level. The relationship between culture and development with a presentation of the economic and cultural-centered approach is emphasized.

Moreover, we will present the characteristics of creative cities and creative sectors, while we examine in detail the challenges and obstacles that cities have to face in their effort to develop creatively. An indicative presentation of the international environment regarding the connection between culture and development is undertaken, with a brief presentation of the UNESCO Network of Creative Cities, as well as the evaluation mechanisms of cities.

Secondly, we will examine the cases of the cities of Medellin in Colombia and Nantes in France. The cities were selected due to the obstacles they had to face during their *"creative"* transition. Creative economy's benefits are highlighted, as well as the image both cities present in their effort to develop.

Secondary sources, literature, articles and internet sources are used in the Chapter. Chapter's originality results from the fact that it focuses on the challenges that cities face in their effort to benefit from creative economy, highlighting at the same time the relationship between culture and development.

The theoretical framework shows that changes that took place in the second half of the 20th century in cities led to a series of issues at urban, economic and social level (unemployment, economic misery, decline of urban parts, etc.). The global trend of investing in the cultural and creative sector provides an answer to this problem, as its utilization helps to increase cohesion, rejuvenate *"tired"* and abandoned areas, generate income and economic benefits, end social exclusion and in general improve the quality of citizens' life.

Empirical data show that cities such as Medellin and Nantes, which have faced significant problems in the past (crime, unemployment, economic downturn, etc.), have succeeded to reverse problematic situations, implementing a specific urban revitalization strategy, based on Cultural and Creative Industries.

BACKGROUND

From antiquity the city was the place where political institutions were developed, social morals and innovations were connected with culture. It is no coincidence that the term *"culture"* derives from the terms *"city"* and *"citizen"* (Babiniotis, 2002, p. 1441). Industrialism placed the city at the center of modern life, transforming it into a multidimensional body with advantages and disadvantages.

With half of the world's population concentrated in cities, cities obtain characteristics such as: innovation, pluralism, dynamism, but also social and economic inequalities, pollution, alienation, gigantism. Cities reflect the trends of modern society as places where different cultures coexist, new ideas are developed; social, artistic, technological, etc. changes are promoted, while at the same time issues arise, such as unemployment, abandonment / overdevelopment of areas.

According to Hall (1998), since cities developed in structure and size, they have become more complex, developing urban management problems, transformed into receptors for problem-solving processes that arose from their development.

Moreover, the economy of culture has dominated the international economic scene in recent decades connected with the development strategies of both developed and emerging economies. At international level, Creative Industries are developing dynamically, being the means for innovative developments, enhancing competitiveness and cultural diversity. In this aspect characteristic is the case of the European financing program, Horizon 2020[1], focusing on research and innovation.

Social change and technological developments, mainly in the field of information and communication, have helped significantly to this direction, with Internet and its applications offering citizens new forms of entertainment and recreation, interactive participation and easier access to cultural products. According to Lazaretou (2014), willing consumers of cultural goods, human resources and funds, as well as the availability of diverse cultural products for the citizens resulted in the connection of economy with creation and culture.

According to the Institute of Statistics of UNESCO report (2016), global trade in creative products more than doubled from 2004 to 2013, while culture is a key factor in the creative economy, based on creativity, innovation and access to knowledge. In 2016, Creative Industries accounted for almost 3% of global GDP, offering 30 million jobs. At European level, these industries provided more than 7 million jobs, while, as the report points out, in developing countries, creative sectors promote sustainable, inclusive growth.

The first two decades of 2000 were particularly effective at promoting the subject of culture and sustainable development in the academic debate. Several Master's Theses findings used in the Chapter are of interest regarding creative economy and creative cities, such as: a) Sterioti, Th. (2016). *Creative Economy: culture and creativity, levers of development of modern cities.* Hellenic Open University. Athens, b) Bersi, D. (2019). *Orange Economy and Open Innovation Systems: The Case of Hubs in the Culture and Creativity Sectors.* Harokopio University. Athens and c) Papadopoulou, E. (2019*). Creative city and policies: Theoretical approaches and the case of Thessaloniki.* Aristotle University of Thessaloniki. The findings prove that culture, creativity and innovation are enablers of development, presenting creativity as a new means that leads to sustainable development.

CITY, CULTURE AND DEVELOPMENT

Cities' Classification

In Greece, Municipalities constitute the first degree of Local Government. They are divided into the following categories, based on population, special geomorphologic characteristics, economic activity, urbanization, etc.:

1. Municipalities of Metropolitan Centers,
2. Large Continental Municipalities and Municipalities of Capitals of Prefectures (population over 25,000 inhabitants),
3. Medium Mainland Municipalities (population over 10,000 to 25,000 inhabitants),
4. Small Continental and Small Mountain Municipalities (population less than 10,000 inhabitants),
5. Large and Medium Island Municipalities (population over 3,500 inhabitants),
6. Small Island Municipalities (population up to 3,500 inhabitants).

Internationally, according to EUROSTAT, a city is a Local Government Unit, whose 50% of population at least lives in one or more urban centers. At European level cities are ranked according to functional classification criteria, such as the intensity of urban problems, urban characteristics, their operation as decision-making centers, seats of international organizations, etc. The following figure is indicative:

Furthermore, based on various approaches, different terms have been formulated for big cities globally, such as: capitalist centers, world cities, mega - cities, ecumenical cities, etc. These approaches are classified, according to Beaverstock, Smith & Taylor (1999), mainly into two categories: a) the demographic approach, where cities are classified according to their population, and b) the functional approach, where cities are classified according to global economic system, based on their economic activities. The relationship between culture and development as well as the economic and cultural-centered model of development are examined below.

Figure 1. Top European cities (Source: Petrakos & Oikonomou 1999)

Ranking by:	Intensity of problems	Urban features	Attracting High-tech Activities	Function as Decision-making Centers and Headquarters of international organizations	Centers of multinational companies - advanced productive services
Studies	Cheshire et.al.(1986) Cheshire and Hay (1989) Cheshire (1990)	Reclus/Datar (1989)	Conti and Spriano (1990)	Palomaki (1991)	Taylor & Hoyler (2000)
Top European Cities	Frankfurt Brussels Venice Munich Amsterdam	London Paris Milan Madrid Munich Frankfurt Rome Brussels Barcelona Amsterdam	London Paris Brussels Amsterdam Rome Frankfurt Milan	London Paris Frankfurt Randstad	London Paris Frankfurt Milan

Economic Versus Cultural-Centered Approach

According to Moren (1998, p. 534) the *"myth of development"* prevailed in the industrialized countries in the 1950s and 1960s, as a result of post-war economic development. As Paschalidis (2002) points out, in the economic approach it is argued that social progress is ensured by the progress of technology, science and industrial-economic development, with culture sharing a secondary role. In fact, development is treated as economic and not as social / cultural factor, focusing exclusively on economic indicators.

In the field of arts, only the economic benefits are examined, with emphasis on: a) the direct effects of artistic activities on employment and income growth, b) the spin-off consequences of activities' implementation and audience presence and c) the indirect consequences, focusing on the economic results. Hansen (1995) describes this approach as *"instrumental"*, presenting the arts as a means of achieving economic goals, without appreciating their multidimensional role.

During the 1970s and especially from 1980 onwards the economic-centered model was in crisis, while the cultural-centered model of development was gradually implemented in various countries. In this approach, according to Paschalidis (2002), development is not linked exclusively to economic indicators, but it is human-based, connected with political, socio-economic freedoms, equal opportunities for living, education, care, equal human rights regardless of class, gender / race, cultural self-expression, benefits to the groups with special needs.

Therefore, the concept of culture includes everything that gives meaning to life and it is a goal of development. In this model culture is treated as a whole way of life, with emphasis on humans, including all the creations of society.

International institutions, such as UN and UNESCO, contributed to make culture a key development goal. In fact, as Paschalidis (2002, p. 230) points out, in World Conference on Cultural Policy in Mexico, in 1982, the concepts of culture and development were redefined. Culture was defined as *"a whole complex of mental, material, spiritual, emotional elements that characterize a society / social group. It includes not only the arts, but also beliefs, lifestyles, value systems, fundamental human rights"* (World Conference on Cultural Policy, Mexico, 1982).

At the same time, development was defined as a *"process that extends beyond simple economic growth, incorporating all aspects of life, the activities of communities, whose members are called upon to contribute and are expected to benefit"* (World Conference on Cultural Policy, Mexico, 1982).

The UN Decade of Action for Cultural Development (1988-1997) followed, while in 1992 UNESCO established World Commission on Culture & Development, adopting proposals such as the protection of cultural rights as human rights, the convening of a World Summit on Culture and Development, etc. The Commission in 1996 published the Report *"Our Creative Diversity"*, an important framework of principles on the 21st century's cultural development.

Creativity and Creative Cities

As Babiniotis (2002, p. 471) argues, *"the person who has the ability to create, to produce new and original forms"* is characterized as creative. According to Kunzmann (2004), the concept of creativity was introduced in 1985 by the Swedish, Ake Andersson.

A process characterized as *"creative"* must also have added value, as it is not enough to combine existing knowledge or be innovative (Higgins, 1999). In the European Commission's Communication (2018) creativity is defined *as "a complex process of innovation, combining some / all of the following dimensions: ideas, skills, technology, management, production processes and culture"*.

The Creative City Movement, according to Landry (2005), began with studies in the 1980s that used concepts such as: cultural design, cultural industries and cultural resources. Florida (2002) attributes the emergence of the term *"creative city"* to the changes that took place in society, production and the economy, comparing them with the changes during the transition from the rural to the industrial age.

According to UNESCO (2004), a city is characterized as creative when it considers creativity as a strategic factor of sustainable development in the economic, social, cultural, environmental field. Landry (2005) argues that global change is linked to the emergence of new technologies, Internet-based economy, the shift from manual to brain work, the value given to innovation, etc.

Creative city, a tool of urban development, consists of its material and intangible infrastructure, promoting residents' creativity, urban problems' treatment through creative solutions and the utilization of cultural resources (architecture, artistic heritage, different cultures, built environment, etc.). At the same time, it relies on the *"creative industry"*[2], without specifying policies or results.

Buitrago and Duque (2013) conducted a research on the creative economy in Latin America and the Caribbean and replaced the term with *"orange economy"*[3]. As they noted, the development of orange / creative economy depends on the existence of *"7Is"* (Sterioti, 2016).

- Information (mapping of the Creative Industries, cultural consumption research, exchange of best practices, etc.).
- Institutions (wide participation of public and private sector, NGOs and community for a common development strategy).
- Industry (Cultural and Creative Industries).
- Infrastructures (roads, parks, shopping malls, fiber optics, antennas, which facilitate communication and interaction between the audience, artists, entrepreneurs, etc.).
- Integration (training and support for artists, cooperation of countries in the field of production, distribution and consumption of creative goods).
- Inclusion (empowering all kinds of minorities, creating new jobs, bringing in beneficiaries around a common agenda to bridge social gaps).
- Inspiration (an adequate educational system is needed, which will promote the creative forces of curiosity and imagination).

Characteristics of Creative Cities

According to Hospers (2003), the following four factors increase the chances of developing urban creativity:

- Concentration of a significant number of inhabitants in an urban area leads to urban creativity, mainly due to dense interactions.
- Diversity, which includes, in addition to the diversity of the inhabitants, the variety of the urban image, which also results from the built environment.
- Instability that characterizes the external environment of the city. A city can develop creatively under changing conditions, during rapid socio-economic changes (Hall 2000).
- Reputation - recognition of a city as creative by the rest of the world, in order to attract residents, tourists, etc. Marketing policies are used, promoting the image of the city.

According to Florida (2002), cities with high concentration of creative people find it easier to attract new businesses, making them more competitive. The *"4Ts"* of economic development figure prominently in the approach, which correspond to Technology, Talent, Tolerance and Territory assets of a region. The city is considered creative when the following factors coexist:

- Technology: *"High technology"* areas attract creative workforce, which contributes to the production of new technologies.
- Talent: Creative people are mainly concentrated in places that are centers of creativity.
- Tolerance: The *"creative class"* chooses cities with an international character and tolerance of national, cultural, racial, etc., differences.
- Spatial advantages: The quality of the environment (authenticity of the place), the unique characteristics of a location.

The following figure is also indicative:
A short description of the creative sectors, closely related to the concept of the creative city, follows.

Figure 2. Structural components of intelligent cities (Source: Siokas 2018, own processing)

Creative Sectors

The term *"creative sectors[4]"*, as defined by the European Commission in Creative Europe Program, includes all areas whose activities are based on cultural values or other artistic, individual / collective, creative expressions. According to the United Nations Conference on Trade and Development report (UNCTAD, 2010) the Creative Industries are classified into the following models, as shown in the figure:

Indicative of the exports of cultural goods and the activity in the creative sectors are the following detailed figures:

Challenges / Obstacles of Creative Cities

Creative cities, as shown in the previous analysis, focus on infrastructure, productive sectors, technology, culture, education and art. According to Higgins & Morgan (2000), the creative thinking of individuals and groups is a key skill in solving problems, while for the transition of the city from *"normal"* to *"creative"* its priorities change.

Culture has an important role in Sustainable Development Goal 11 *"Make cities and human settlements inclusive, safe, resilient and sustainable"*, which is one of the few goals mentioning culture under Target 11.4 *"Strengthen efforts to protect and safeguard the world's cultural and natural heritage"*. However, nowadays the goal for sustainable communities seems to be more difficult than ever, due to administrative, political and policy challenges. According to the report results from brainstorming sessions organized by the EU (2021) as a Voices of Culture Structured Dialogue polarization and social crisis in cities are growing. At local / national level, there is a lack of trust for political prioritizations and culture is excluded from strategic planning. There is a need for collaboration among public - private actors, decision-makers, politicians, cultural actors, different fields and sectors. Moreover, there are a few platforms and forums on the meaning and purpose of the SDGs, in order for the public to agree on actions.

Figure 3. Classification of Creative Industries (Source: UNCTAD 2010, Papadopoulou 2019)

1. UK DCMS model	2. Symbolic texts model	3. Concentric circles model	4. WIPO copyright model
Advertising Architecture Art and antiques market Crafts Design Fashion Film and video Music Performing arts Publishing Software Television and radio Video and computer games	**Core cultural industries** Advertising Film Internet Music Publishing Television and radio Video and computer games **Peripheral cultural industries** Creative arts **Borderline cultural industries** Consumer electronics Fashion Software Sport	**Core creative arts** Literature Music Performing arts Visual arts **Other core cultural industries** Film Museums and libraries **Wider cultural industries** Heritage services Publishing Sound recording Television and radio Video and computer games **Related industries** Advertising Architecture Design Fashion	**Core copyright industries** Advertising Collecting societies Film and video Music Performing arts Publishing Software Television and radio Visual and graphic art **Interdependent copyright industries** Blank recording material Consumer electronics Musical instruments Paper Photocopiers, photographic equipment **Partial copyright industries** Architecture Clothing, footwear Design Fashion Household goods Toys

Figure 4. Exports of cultural goods worldwide, 2000 – 2015 (Source: UNCTADStat database 2022)

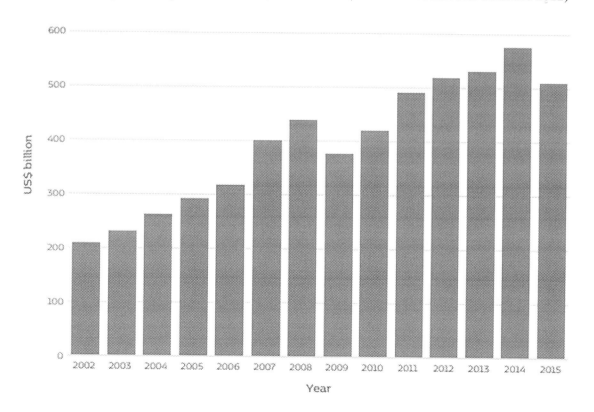

Figure 5. Percentage of creative professions in selected cities, 2008 – 2019 (Source: UNESCO 2021)

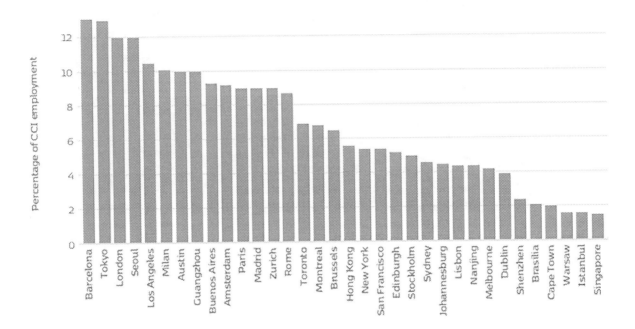

Another challenge is also funding, legislation and regulation as there is a large disparity as far as where funding is directed in the cultural sector. According to the report, the approach from funders and policymakers is a top-down one, while the cultural sector doesn't participate when decisions are made on development programs. Communities also lack the resources to provide a vision.

Another point is that the funding instruments for culture focus on goal-oriented and results-driven approaches, without recognizing the creative interventions as a process just as important as the end goal. Moreover, culture can add value to other sectors and it should be seen as an investment and industry that sustains national economies. Local authorities and cities should work as enablers to strength the collaboration between cultural actors and communities.

The challenges that European cities face, according to Van den Berg & Braun (2001), are the following:

- Cities, due to the competitive market conditions, should attract investment opportunities, by promoting their comparative advantages.
- Identification of new sub-markets, aiming at the development of new activities, the encouragement of investments, the creation of income, jobs and the increase of cities' financial capacity.
- Balancing economic growth and quality of life.
- Balancing economic development and spatial organization.
- Balancing economic and social development.

Globally, the factors that negatively affect the development of a city to creative, creating even sustainability issues, as Kelaidi (2020) notes, are related to the main challenges, such as climate change, lack of resources and energy. According to International Organizations' reports the demand in these sectors

The Challenges Cities Face on Their Way Towards Creativity

is constantly increasing, as a result of global population growth, urbanization, changing eating habits and economic growth.

Inadequate transport systems, traffic jams make everyday life difficult; noise pollution affects the quality of life, while the poorly maintained infrastructure in urban centers also acts as a deterrent. At the same time, obstacles to creativity and sustainability are being raised by concerns about human health, with coronavirus disease as a typical example in recent years (COVID-19)[5]. The pandemic has presented how vulnerable the cultural field can be as it relies on physical venues, participation, public funding and interaction among people. The disease significantly affected the cities, due to their urban planning, turning them into over-transmission areas.

The following figure presents the percentage change in creative jobs, in selected cities:

Figure 6. Percentage change in creative jobs, in selected cities February - May 2020 (Source: unesco.org 2021)

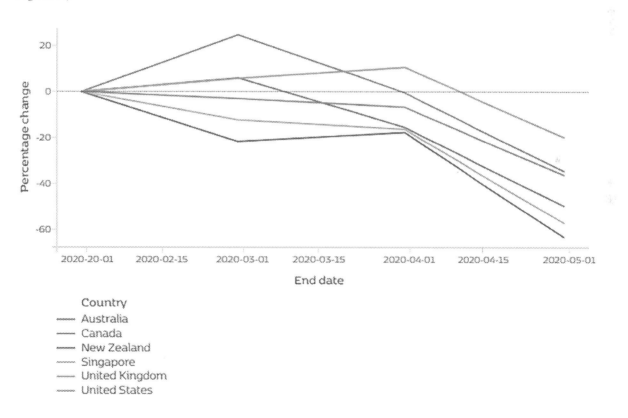

Networks and Platforms

The United Nations Educational, Scientific and Cultural Organization -UNESCO having prioritized the cooperation of cities and local authorities and promoting creativity worldwide, established in 2004 the Creative Cities Network. The Network aims at sustainable urban development and cultural diversity, while, through pilot programs, meetings and research, contributes to the development of local Creative Industries. The network has 246 cities[6], distinguished in one of the following areas of creativity:

Figure 7. Geographic distribution of Creative Cities in each creative field (as of July 2019) (Source: official website of UCCN)

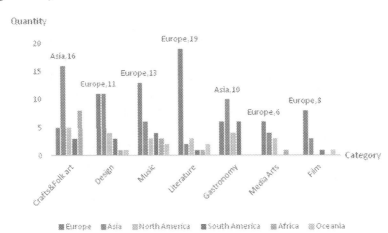

- Crafts & Folk Art
- Gastronomy
- Literature
- Film
- Music
- Media Arts
- Design

According to the Observatory on Creative Cities (2019), the Creative Cities Network has developed steadily with global impact and the number of Creative Cities has exhibited a growing trend since the program's launch. According to the level of national economic development developed countries have an advantage in the fields of Media Arts, Literature, Film and Music and developing countries are dominant in Crafts, Folk Arts and Gastronomy. Therefore, Creative Cities in the core fields of cultural industry are located in developed European countries and Creative Cities that have advantages in Crafts, Folk Arts and Gastronomy are located in developing countries of Asia and Africa with important cultural background.

In November 2021 Thessaloniki joined the Network, being the first Greek participation since its establishment. The sector in which the city is distinguished worldwide is that of gastronomy, while the goals of the local authorities are to prove that actions of sustainability are implemented in the city, in addition to its gastronomic value.

As an evaluation tool for European cities in the field of creativity and culture, EU has created the *"Cultural and Creative Cities Monitor"* platform, a *"fount"* of qualitative and quantitative data, in order to assist national, regional and municipal authorities in their effort to utilize culture and creative economy. A list of creative cities from the international arena follows:

As Bersi (2019) notes, at European level the ideal creative city could result from the combination of the best performance of cities in each category. Medellin and Nantes are examined below as case studies of cities at global and European level, which managed to face important issues and obstacles and become creative cities.

Figure 8. Examples of creative cities from the international arena (Source: Gatzis 2020, own processing)

Creative City	Creative Fields
Santa Fe – Argentina	Crafts & Folk Art
Sasayama – Japan	Crafts & Folk Art
Bamiyan – Afghanistan	Crafts & Folk Art
Leeds – England	Media Arts
Lyon – France	Media Arts
York – England	Media Arts
Bogota – Colombia	Music
Kingston – Jamaica	Music
Seville – Spain	Music
Sydney – Australia	Film
Bradford – England	Film
Santos – Brazil	Film
Bergen – Norway	Gastronomy
Florianapolis – Brazil	Gastronomy
Saint Francis – USA	Performing Arts
New York – USA	Fashion, Arts, Film
London – England	Museums

CASE STUDY OF CREATIVE CITIES

Medellin

Brief Background

Medellin is the second largest city in Colombia, the capital of the Antioquia region, with an area of 382 sq.km. The city is built at an altitude of 1,459 m in the Andes Mountains and its population counts 2.569[7] million inhabitants.

The economy was based on coffee production, textiles and furniture trade. In the 1980s, however, the city was dominated by a drug cartel, led by Pablo Escobar. Medellin was one of the most violent areas in the world in 1991, with the homicide rate approaching 380 per 100,000 inhabitants (Sterioti, 2016).

The Transition to a Creative City

According to Clemons (2016), Medellin was gradually transformed from a murder capital into a miracle city, due to the efforts that began in 1990, when the first cultural policy plan was formulated, in order for culture to be a *"bulwark"* in the social crisis. From 1991 onwards, the Colombian government, in order to deal with violence, decentralized resources and responsibilities, providing greater autonomy to the cities. Escobar was assassinated by the authorities in 1993, as well as other gang leaders were arrested.

Figure 9. The city of Medellin (Source: wikipedia.org)

In 1999 the municipal authorities adopted a program for the international promotion of Medellin as a cultural and artistic city, while in the period 2003-2007, during the mayoral tenure of S. Fajardo, a great effort was made in this field. According to Eveland (2016), with the motto *"the most beautiful for the most humble"*, a policy of holistic revival was promoted, with the best projects being addressed at the most violent and poorest neighborhoods. Emphasis was placed on culture and education, while the policy of social urbanization was promoted, aiming at enhancing inhabitants' dignity, revitalizing decadent neighborhoods, combating violence, promoting social cohesion and improving quality of life.

The construction of ten libraries - parks followed (Parque Bibliotequa Espana, etc.) in degraded neighborhoods, functioning as places of study, allowing access to information, as well as places of recreation and social contact, promoting social integration. Also, cultural centers were created; music festivals and music schools were established, with the city participating in the UNESCO Network of Creative Cities, due to its distinction in music.

The creation of the Medellinnovation District was also important, which offers digitally equipped venues that attract businesses and creative people globally. The reduction of bureaucracy and tax credits for companies, which are established in the specific creative district, lead to the creation and development of start-ups (Bahl, 2012).

The Challenges Cities Face on Their Way Towards Creativity

Figure 10. Homicide rate (Source: Bahl 2012)

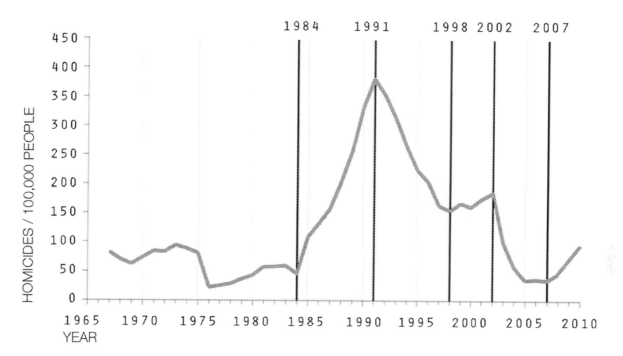

According to Sterioti (2016), Medellin's citizens were by far the greatest source of assistance to the effort, participating in the design and implementation of the projects, showing confidence in local leaderships, and protecting the projects. In the reform of the marginalized neighborhoods, transport networks were also important, connecting inaccessible areas with the shopping center.

At this stage, the city is an example of urban recovery with an attractive environment (public-private partnerships, infrastructure, upgraded urban environment, city - branding, etc.) to the creative class. According to Saldarriaga (2022), the National Statistical Office shows that the unemployment rate is 11.7%[8], below the average of the 13 major cities in the region, which is up to 13%. Moreover, the total number of employees from January 2021 to February 2022 has moved up from 1,603,000 to 1,828,000.

However, despite the development, in order to implement projects (transport network, green spaces, etc.) residents were relocated to different areas away from their neighborhoods. In addition, the gap between strong and economically weaker groups increased (Sterioti, 2016). The case of Nantes, a creative city that utilizes innovation, improving citizens' quality of life, is examined below.

Brief Background

Nantes, the birthplace of Jules Verne, is located in western France and it is one of the largest port cities in the country. As the capital of the Loire Region it has a population of 656,275[9]. The island (Île de Nantes) originated from the union of smaller islands in the Loire River, has a length of 4.9 km, while thirteen bridges connect it with the mainland.

Figure 11. The city of Nantes (Source: nantes.maville.com)

Nantes used to be a commercial channel and later became an industrial area. It was an important shipyard in the 1960s, with facilities on the Île de Nantes, but after a long period of recession, the area's shipyards stopped operating in 1986, resulting in 1,500 unemployed people and countless abandoned areas.

The Transition to a Creative City

Urban revitalization projects begun in 2000, aiming to transform the island (île de Nantes) from a symbol of decline to a symbol of creativity. In the first phase of the cycle of planning and programming (2000 - 2010) the goal was to create an attractive environment on the island and its connection with the city. Connection paths and bridges were built, shipyards were converted into urban parks, warehouses were renovated, housing cultural activities, as well as 4,400 housing units were created.

During the second phase (2010 - 2030) the goal is the development of the areas located in the southwest of the island and their transformation into a development center. Emphasis is placed on developing dialogue with professional associations, private stakeholders and citizens, while, as Perrin & Delvainquière (2015) point out the French Ministry of Culture mobilizes Municipalities to implement their own policies in the cultural field. Also, in the period 2010 - 2012 the Region of Nantes implemented the initiative *"Ma ville demain"*, during which organizations and citizens prepared a plan regarding the image of Nantes by 2030 (Durand, 2012).

Emphasis was placed on the development of the Quartier de la Creation (QdlC), an ecosystem of Cultural and Creative Industries, offering services such as: housing to businesses, supporting their projects, promoting networking, collaborating with universities, public institutions and private investors. The goal of the QdlC is to become a European center of excellence, strengthening Nantes image internationally, with emphasis on culture, creativity and innovation.

In 2019, the European Commission awarded Nantes the title of *"European Innovation Capital"*, in recognition of its ability to leverage innovation. The title was accompanied by a cash prize of € 1 million, funded by the EU's research and innovation program, *"Horizon 2020"*. The Mayor of Nantes, Johanna Rolland, after the distinction, in her statement stressed, that *"innovation by all and for all"* is at the heart of the policies implemented in the city.

The policies and plans the city was distinguished for, according to the European Commission, included:

- The *"Big Debates"* initiative, in which more than 55,000 citizens participated. A specific map with commitments on issues such as the energy transition emerged.
- The *"Creative factory"* initiative, a system for supporting Creative and Cultural Industries, etc.

However, criticism has been levelled mainly at the population composition of the island. On the west side the inhabitants are more educated, mostly young, with artistic and spiritual interests, while in the eastern quarters, families live for years. According to Vileou et all (2013), there are concerns that the island may *"lose"* its roots in the working class, exploited by creative groups.

FUTURE RESEARCH DIRECTIONS

As aforementioned, despite the multiple benefits of Creative Industries for modern cities' sustainable development, there are also elements that raise concerns. Specifically, the investments in the urban area may lead to the change of the physiognomy of a place and changes in population composition. Therefore, equal access to cultural goods for all must be ensured, with emphasis on vulnerable groups.

Moreover, as Scott (2006) has pointed out cities may have unprecedented creative capabilities, but they are also places where social, cultural, and economic inequalities prevail. As a result there can be no truly final achievement of the creative city where these problems remain.

Basic issues of citizenship and democracy have to be reinforced into the active life of the city, so as for creativity and diversity to be empowered. Therefore, *"creativity is not something that can be simply imported into the city on the backs of peripatetic computer hackers, skateboarders, gays, and assorted bohemians but must be organically developed through the complex interweaving of relations of production, work, and social life in specific urban contexts"* (Scott, 2006, p. 15).

CONCLUSION

Culture can be an important and powerful driver for development, according to statistics and data throughout the past decades. With social, environmental, and economic impact, Cultural and Creative Industries and cultural infrastructure have proven to be excellent means in the effort of revitalizing the economy of cities. Therefore, many cities use cultural events and institutions to stimulate urban development, attract investments and visitors as well as improve their image.

Cities, in their quest to utilize creative economy, have to correspond to administrative, political and policy challenges as well as to funding, legislation and regulation issues connected with cultural sector. The competitive market, economic growth and spatial planning are among the factors that affect the transition to a creative city.

Key challenges such as climate change, lack of resources and energy, inadequate transport systems, poor infrastructure in urban centers negatively affect the transition of a city into a creative one. Significant obstacles are also caused by concerns for human health, with the typical example of coronavirus disease (COVID-19).

Medellin case study presented the effort of national, local authorities and residents, to move to creativity, as in the 1980s the city was among the most violent areas. In the decades that followed, specific cultural policy plans were formulated, with an emphasis on Medellin's international promotion as a cultural and artistic city. As a result of the joint effort, the city today is a model of urban recovery, with an attractive environment for the creative class.

Nantes, a major shipbuilding center in the 1960s, faced issues such as the decline of entire areas and unemployment of thousands of workers. Urban revitalization plans for the city began in 2000, aiming to island's transformation (île de Nantes) from a symbol of decline to a symbol of creativity. The creation of an attractive environment on the island and its connection with the rest of the city, the transformation of entire areas into development centers, resulted in the title of *"European Capital of Innovation"* for the city in 2019.

The cities' case studies prove that the development of Cultural and Creative Industries contributes to the revitalization of *"degenerate"* areas, revenue generation and economic benefits, the end of social exclusion and in general the improvement of citizens' quality of life. The support of the advantages that cities have, which make them unique, in combination with the utilization of the creative economy, are key factors on the way to development.

Moreover, the allocation of funds for sharing best practices and cooperation, investments in action research and pilot projects for sustainable cities and culture's promotion as a driver of resilient communities are among some of the measures that have to be implemented towards creativity. Additionally, giving more responsibilities to communities and enabling more dialogue between authorities, policy makers and the cultural sector may contribute significantly to the overall effort.

Culture should be viewed as the central pillar of sustainable development and must be placed at the core of future thinking and planning. In the face of global events such as COVID-19 pandemic, it is of great importance to value our culture so as to secure a more prosperous future. In the years to come, cultural initiatives will need to include issues such as the health crisis, longstanding inequalities and climate emergency, with emphasis on how culture can respond to the challenges which we all face at local, regional, national and international level.

REFERENCES

Avdikos, V. (2014). Οι Πολιτιστικές και Δημιουργικές Βιομηχανίες στην Ελλάδα [The Cultural and Creative Industries in Greece]. Epikentro Ed. Athens.

Babiniotis, G. (2002). Λεξικό της Νέας Ελληνικής Γλώσσας [Dictionary of the Modern Greek Language]. Lexicology Center. 2nd ed. Athens

Bahl, V. (2012), *Murder Capital to Modern Miracle? The Progression of Governance in Medellin.* Colombia. Development Planning Unit. The Bartlett, University College London. https://opendocs.ids.ac.uk/opendocs/bitstream/handle/20.500.12413/11792/Murder_capital.pdf?sequence=1&isAllowed=y (5/4/2022)

Beaverstock, J. V., Smith, R. G., & Taylor, P. J. (1999). A roster of World Cities. In *Cities*, *16* (6), 445-458. https://www.sciencedirect.com/science/article/pii/S0264275199000426?via%3Dihub (2/4/2022)

Bersi, D. (2019). *Πορτοκαλί Οικονομία και Ανοιχτά Συστήματα Καινοτομίας: Η Περίπτωση των Hubs στους Κλάδους Πολιτισμού και [Orange Economy and Open Innovation Systems: The Case of Hubs in the Culture and Creativity Sectors]* [Unpublished master's thesis, Harokopio University. Athens]. https://tinyurl.com/yc3c7m75

Buitrago, R., & Duque, M. (2013). *The Orange Economy: An Infinite Opportunity*. (Online). Available at: https://publications.iadb.org/el/orange-economy-infinite-opportunity (5/4/2022)

Clemons, A. (2016). From World's Murder Capital to the Medellin Miracle. *The culture trip*. https://theculturetrip.com/south-america/colombia/articles/from-world-s-murder-capital-to-the-medellin-miracle/ (2/4/2022)

Commission of the European Communities. (2007). *Ανακοίνωση σχετικά με μια ευρωπαϊκή ατζέντα για τον πολιτισμό σ' έναν κόσμο παγκοσμιοποίησης [Communication on a European Agenda for Culture in a Globalizing World]*. Brussels. https://eur-lex.europa.eu/legal-content/EL/TXT/PDF/?uri=CELEX:52007DC0242&from=DE

Durand, C. (2012). Trois visions pour Nantes Métropole en 2030. [Three visions for Nantes Metropolis in 2030.] *Le courrier du pays de Retz journal*.

European Commission. (2018). *Creative Europe program*. Brussels. https://eur-lex.europa.eu/legal-content/EN/TXT/?uri=COM%3A2018%3A366%3AFIN

European Commission. (2020). *Horizon 2020, Details of the EU funding program*. EC. https://ec.europa.eu/info/research-and-innovation/funding/funding-opportunities/funding-programmes-and-open-calls/horizon-2020_en#latest (3/4/2022)

European Commission. (2022). *Cultural and Creative Cities Monitor*. EC. https://composite-indicators.jrc.ec.europa.eu/cultural-creative-cities-monitor

Eurostat. (2022). *Statistics explained, Glossary*. EC. https://ec.europa.eu/eurostat/statistics-explained/index.php?title=Glossary:City (1/4/2022)

Eveland, J. (2014). *Medellin transformed: from murder capital to modern city*. Lee Kuan Yew World City Prize. https://www.leekuanyewworldcityprize.gov.sg/resources/features/medellin-transformed/ (29/3/2022)

Flew, T. (2002). Beyond ad hocery: Defining Creative Industries. In *Cultural Sites, Cultural Theory, Cultural Policy. The Second International Conference on Cultural Policy Research*. Wellington. https://eprints.qut.edu.au/256/1/Flew_beyond.pdf (29/3/2022)

Florida, R. (2002). The rise of the creative class. Basic Books Ed., New York

Gkatzis, Th. (2020). *Η συμβολή της δημιουργικής οικονομίας στην αστική αναγέννηση και βιώσιμη ανάπτυξη των πόλεων: Θεωρητική προσέγγιση* [The contribution of the creative economy to urban regeneration and sustainable urban development: A theoretical approach], Το Βήμα των Κοινωνικών Επιστημών [Lexical characteristics of Greek language] [The Step of the Social Sciences]. V. 72. University of Thessaly. https://tinyurl.com/nxfzxxr2 (2/5/2022)

Hall, P. (1998). Cities in Civilization: Culture, Innovation and Urban Order (Weidenfeld & Nicolson Ed). Pantheon Book.

Hall, P. (2000). Creative Cities and Economic Development. *Urban Studies, 37*(4), 639 – 649. https://journals.sagepub.com/doi/pdf/10.1080/00420980050003946

Hansen, T. B. (1995). *Measuring the value of culture* (Vol. 1). Cultural Policy.

Hellenic Statistical Authority. (2022), *Ελλάς με αριθμούς, Ιανουάριος – Μάρτιος 2022* [Lexical characteristics of Greek language] [Greece in numbers, January - March 2022]. https://www.statistics.gr/documents/20181/17831637/GreeceInFigures_2022Q1_EN.pdf/82613203-ee3c-eb7b-600b-503211d9dc91 (29/3/2022)

Higgins, L. (1999). Applying principles of creativity management to marketing research efforts in high-technology markets. *Industrial Marketing Management, 28*(3), 305–317. https://tinyurl.com/yms62zxb

Higgins, M., & Morgan, J. (2000). The Role of Creativity in Planning: The "Creative Practitioner". *Planning Practice and Research, 15*(1–2), 117–127. https://www.tandfonline.com/doi/abs/10.1080/713691881

Hospers, G. J. (2003). Creative Cities in Europe: Urban Competitiveness in the Knowledge Economy. *Intereconomics, 38*(5), pp. 260–269. https://www.researchgate.net/publication/47872007_Creative_Cities_in_Europe_Urban_Competitiveness_in_the_Knowledge_Economy

Hospers, G. J. (2003). Creative cities: Breeding places in the knowledge economy. *Knowledge, Technology & Policy, 16*(3), 143–162. https://www.researchgate.net/publication/240357359_Creative_cities_Breeding_places_in_the_knowledge_economy (3/4/2022)

International Center for Creativity and Sustainable Development (ICCSD). UNESCO. (2019). *Observatory on Creative Cities. The Development of the UNESCO Creative Cities Network (2004-2019)*. UNESCO. https://f2.cri.cn/M00/21/8A/rBABC2BF0MqAWMGXAMBztfzOTqQ662.pdf (01/04/2022)

Kelaidi, E. (2020), *Έξυπνες Πόλεις, Ψηφιακές Συνεργασίες και Επιχειρηματικότητα* [Smart Cities, Digital Collaborations and Entrepreneurship]. [Unpublished master's thesis, National Technical University of Athens, Athens]. https://dspace.lib.ntua.gr/xmlui/bitstream/handle/123456789/51797/diplomatiki_ekelaidi.pdf?sequence=2 (1/4/2022)

Kunzmann, K. (2004). Culture, Creativity and Spatial Planning. *The Town Planning Review, 75*(4), 383–404. http://www.scholars-on-bilbao.info/fichas/KUNZMANN%20CultureCreativitySpatialPlanningTPR2004.pdf

Landry, C. (2005). Lineages of the Creative City. In *Creativity and the City: How the creative economy is changing the city*, pp. 42–55. Charles Landry. http://charleslandry.com/panel/wp-content/uploads/downloads/2013/03/Lineages-of-the-Creative-City.pdf (28/4/2022)

Lazaretou, S. (2014). *Η έξυπνη οικονομία: πολιτιστικές και δημιουργικές βιομηχανίες στην Ελλάδα. Μπορούν να αποτελέσουν προοπτική διεξόδου από την κρίση*. [The smart economy: cultural and creative industries in Greece. Can they be a way out of the crisis?]. Bank of Greece. https://www.bankofgreece.gr/Publications/Paper2014175.pdf

Medellin. Official city portal. (2022). Medellin. https://www.medellin.gov.co/irj/portal/medellin (1/4/)

Moren, E. (1998). Η πολιτική πολιτισμού. [The politics of culture]. In E. Moren & S. Nair, *Μια πολιτική πολιτισμού [Lexical characteristics of Greek language] [A policy of culture]*. Athens Nantes, Official city portal. https://metropole.nantes.fr/

Νέα Αρχιτεκτονική της Αυτοδιοίκησης και της Αποκεντρωμένης Διοίκησης – Πρόγραμμα Καλλικράτης. [New Architecture of Local Government and Decentralized Administration - Kallikratis Program]. http://www.et.gr (22/4/2022)

Orville, H. (2019). The Relationship between Sustainability and Creativity. *Cadmus, Promoting Leadership in Thought that Leads to Action, 4*(1). https://tinyurl.com/2s3hsame

Papadopoulou, E. (2019*). Δημιουργική πόλη και πολιτικές: Θεωρητικές προσεγγίσεις και η περίπτωση της Θεσσαλονίκης [Creative city and policies: Theoretical approaches and the case of Thessaloniki]* [Unpublished master's thesis, Aristotle University of Thessaloniki. Thessaloniki.] http://ikee.lib.auth.gr/record/308004/files/PAPADOPOULOUERMIONH_DE.pdf (1/4/2022)

Paschalidis, Gr. (2002). Η συμβολή του πολιτισμού στην κοινωνική και οικονομική ανάπτυξη [The contribution of culture to social and economic development]. In Paschalidis, Gr. & Hambouri-Ioannidou Aik., Οι Διαστάσεις των Πολιτιστικών Φαινομένων: Εισαγωγή στον Πολιτισμό [Lexical characteristics of Greek language] [The Dimensions of Cultural Phenomena: Introduction to Culture]. Hellenic Open University. Patras

Perrin, T., & Delvainquière, J. C. (2015). *France/ 1. Historical perspective: cultural policies and instruments*. Compedium: *Cultural Policies and Trends in Europe*. https://tinyurl.com/33j7nkn8 (29/03/2022)

Petrakos, G., & Oikonomou, D. (1999). Διεθνοποίηση και διαρθρωτικές αλλαγές στο Ευρωπαϊκό σύστημα αστικών κέντρων. [Internationalization and structural changes in the European system of urban centers] In D. Economou & G. Petrakos (Eds.), *Η ανάπτυξη των Ελληνικών πόλεων. Διεπιστημονικές προσεγγίσεις αστικής ανάλυσης και πολιτικής* [The development of Greek cities. Interdisciplinary approaches to urban analysis and politics]. (pp. 13–44). University Publications of Thessaly – Gutenberg.

Saldarriaga, G. (2022). *Medellín continues to advance as one of the cities with the lowest unemployment rate in the country.* News, Economic Development, Portal de Medellin. https://www.medellin.gov.co/irj/portal/medellin?NavigationTarget=contenido/12142-Medellin-sigue-avanzando-como-una-de-las-ciudades-con-menor-tasa-de-desempleo-del-pais#google_translate_element (2/4/2022)

Scott, A. (2006). Creative cities: Conceptual issues and policy questions. *Journal of Urban Affairs, 28*(1). University of California, Los Angeles. https://escholarship.org/content/qt77m9g2g6/qt77m9g2g6.pdf?t & (3/4/2022)

Siokas, G. (2018). Οι Ευφυείς Πόλεις και ο ρόλος της Τοπικής Αυτοδιοίκησης: Θεωρητικό πλαίσιο και ελληνικά [The Intelligent Cities and the Role of Local Government: Theoretical Framework and Greek Examples]. *Ermoupolis Seminar on the Information Society and the Knowledge Economy.* Technical University of Athens. https://www.researchgate.net/publication/349916627_Prosdioristikoi_paragontes_kai_strategikes_sto_schediasmo_mias_Euphyous_Poles (1/5/2022)

Sterioti, Th. (2016). *Δημιουργική Οικονομία: ο πολιτισμός και η δημιουργικότητα, μοχλοί ανάπτυξης των σύγχρονων πόλεων* [Creative Economy: culture and creativity, levers of development of modern cities] [Unpublished Master Thesis, Hellenic Open University. Athens]. https://apothesis.eap.gr/handle/repo/32067

UNCTAD. (2010). *Creative Economy Report 2010.* UNCTAD. https://unctad.org/system/files/official-document/ditctab2010 03_en.pdf

UNESCO. (2004). *Creative Cities Network.* UNESCO. https://en.unesco.org/creative-cities/home (5/4/2022)

UNESCO. (2021). *Cities, culture, creativity: leveraging culture and creativity for sustainable urban development and inclusive growth.* World Bank and UNESCO. Washington. (Online). Available at: https://unesdoc.unesco.org/ark:/48223/pf0000377427 (10/4/2022)

UNESCO. Institute for Statistics. (2016). *The globalization of cultural trade: A shift in consumption. International flows of cultural goods and services 2004-2013.* Canada. http://uis.unesco.org/sites/default/files/documents/the-globalisation-of-cultural-trade-a-shift-in-consumption-international-flows-of-cultural-goods-services-2004-2013-en_0.pdf (23/4/2022)

United Nations. (2015). *Transforming our world: The 2030 agenda for sustainable development.* UN. https://sustainabledevelopment.un.org/content/documents/2125 2030%20Agenda%20for%20Sustainable%20Development%20web.pdf (4/4/2022)

United Nations. (2022). *Sustainable Development Goals.* UN. https://www.un.org/sustainabledevelopment/cities/ (1/4/2022)

Van den Berg, L., & Braun, E. (2001). Growth clusters in European cities: An integral approach. *Urban Studies, 38*(1), 185–205. Sage. https://journals.sagepub.com/doi/pdf/10.1080/00420980124001 (29/4/2022)

Vileou, G. (2015). *Portrait-robot des habitants de l'île de Nantes*. [*Portrait-robot of the inhabitants of the island of Nantes*.] https://datajournalisme2013.hyblab.fr/projets/population/ (28/04/2022)

Voices of Culture. (2021). Culture and the United Nations Sustainable Development Goals: Challenges and Opportunities. *Brainstorming Report*. European Union. https://voicesofculture.eu/wp-content/uploads/2021/02/VoC-Brainstorming-Report-Culture-and-SDGs.pdf (12/10/2022)

ADDITIONAL READING

Eastman, S. (2015). Medellín, en la red creativa musical de la Unesco. Ecolombiano [Online]. Available at: https://www.elcolombiano.com/cultura/medellin-en-la-red-creativa-musical-EC3271655 (06/03/201622

Florida, R. (2003). Cities and the Creative Class. New York: Routledge. (01/04/2022) doi:10.1177/0739456X9901900202

Florida, R. and Cooper, R. N. (2005). *The Flight of the Creative Class: The New Global Competition for Talent*, Foreign Affairs, (84), p. 170. (01/04/2022) doi:10.2307/20031721

Glickhouse, R. (2014). LatAm Minute: Ruta N and Transforming Medellin into an Innovation Center. AC/COA. (Online). Available at: https://www.as-coa.org/articles/latam-minute-ruta-n-and-transforming-medellin-innovation-center (26/04/2022)

Nolen, S. (2014). *'Social urbanism'* experiment breathes new life into Colombia's Medellin. The Globe and Mail. (Online). Available at: https://www.theglobeandmail.com/news/world/social-urbanism-experiment-breathes-new-life-into-colombias-medellin/article22185134/ (01/03/2022)

Οικονόμου, Δ. (2004). *Αστική αναγέννηση και πολεοδομικές αναπλάσεις* [Lexical characteristics of Greek language] [Urban renaissance and urban renewal], Τεχνικά Χρονικά [Lexical characteristics of Greek language] Technical *chronicles* 5(3): 1-10

Turok, I. (2014). Medellin's 'social urbanism' a model for city transformation. Mail & Guardian. https://mg.co.za/article/2014-05-15-citys-social-urbanism-offers-a-model/ (28/04/2022)

KEY TERMS AND DEFINITIONS

CCIs: This stands for Creative and Cultural Industries
EUT: his stands for European Union

GDP: This stands for Gross Domestic Product
OECD: This stands for Organization for Economic Co-operation and Development
QdlC: This stands for Quartier de la Creation
SDGs: This stands for Sustainable Development Goals
UN: This stands for United Nations
UNCTAD: This stands for United Nations Conference on Trade and Development
UNESCO: This stands for United Nations Educational, Scientific and Cultural Organization

ENDNOTES

1. The program lasted from 2014 to 2020, with a budget of about 80 billion euros.
2. *"Creative industry",* according to Avdikos (2014), includes professions related to photography, advertising, publishing, architecture, design, media, such as writers, artists, etc.
3. The choice of orange color was made, as it has been associated several times with culture, creativity, identity.
4. According to Flew (2002), the term first appeared in England in the 1980s. It was officially presented in 1997 when an attempt was made to map the activity in the specific sectors by the Ministry of Culture, Media and Sport of England (DCMS).
5. Coronavirus disease (SARS-CoV-2, COVID-19) was classified as a pandemic by the WHO on 11/3/2020.
6. Data collection 2022.
7. According to 2020 data.
8. Data collection May 2022.
9. Data collection May 2022.

Chapter 9
The Contribution of Urban Domestic Waste Management to the Circular Economy:
The Perspective of Six European Countries

Maria de Fátima Nunes Serralha
Instituto Politécnico de Setúbal, Portugal

Alexandra Anderluh
https://orcid.org/0000-0001-7337-0799
St. Pölten University of Applied Sciences, Austria

Beatriz Sara Santos
Instituto Politécnico de of Setúbal, Portugal

Dalma Radványi
Hungarian University of Agriculture and Life Sciences, Hungary

Maira Leščevica
https://orcid.org/0000-0001-9172-3607
Vidzeme University of Applied Sciences, Latvia

Zahra Mesbahi
St. Pölten University of Applied Sciences, Austria

Nelson Carriço
https://orcid.org/0000-0002-2474-7665
Instituto Politécnico de of Setúbal, Portugal

Pamela Nolz
St. Pölten University of Applied Sciences, Austria

Sarah De Coninck
UC Leuven-Limburg, Belgium

Sergiu Valentin Galatanu
Poitehnica University Timisoara, Romania

ABSTRACT

In line with the European community's goal, each EU Member State should recycle at least 60% of municipal waste or prepare them for reuse. In this chapter, the authors intend to show the waste management strategies implemented in six European countries, namely, Austria, Belgium, Hungary, Latvia, Portugal, and Romania. The methodology used was to analyse reports and publications on the management of urban waste and dialogue with some technicians of the municipalities. This knowledge of what is done in each country allows others to learn from the best and most innovative solutions and reflect on the various waste management forms implemented, according to environmental, economic, and social

DOI: 10.4018/978-1-6684-6123-5.ch009

perspectives. The analysis identifies several challenges to bring up in further research and projects, with the contribution of the different countries and the synergies that might be obtained. The authors intend to promote a decrease in consumption and an increase in reuse, separately collected waste and recycling, contributing to circular economic growth and the sustainability of the planet.

INTRODUCTION

The increase in the world population has caused a growing demand for raw materials, many of them scarce. The extraction and processing of natural resources have negative environmental impacts (e.g., loss of biodiversity and increase in water stress) but also contribute to total greenhouse gas emissions (GHG). In the current economy, we take materials from the earth, make products, and finally throw them away as waste in a linear process. As the urbanisation and industrialization of modern societies increase, so does the amount of municipal waste. If waste is not managed correctly, it negatively impacts our health and the environment and causes economic losses.

The European Union (EU) has adopted strategies to develop a sustainable, low carbon, resource-efficient, and competitive economy. Decoupling economic growth from environmental harm is a critical component of the European Green Deal (European Commission, 2020). Circular economy (CE) is a concept of decoupling economic growth from resource consumption (The Ellen MacArthur Foundation, 2022). CE policies aim to improve waste management and encourage a responsible production and consumption culture. CE has been a recurring theme on the international, European and national agendas, which aims to extend the life cycle of products so that they fit into the R's of circular economy: rethink, refuse, reduce, regift, recycle, repair and recover. Also reducing waste to a minimum is a key element to promote the decoupling of economic growth and the increase in resource consumption (Kaza et al., 2018).

In addition, one of the main building blocks of the European Green Deal is the new Circular Economy Action Plan (CEAP) (European Commission, 2020). The European Commission adopted the CEAP in March 2020. Some of the measures introduced under the new action plan include ensuring less waste and making the circular economy work for people, regions, and cities.

According to the World Bank, the world is expected to generate 2.59 billion tonnes of waste per year by 2030 (Kaza et al., 2018). However, global waste generation is expected to reach 3.40 billion tonnes by 2050 (Kaza et al., 2018). Therefore, waste management is a global issue. It also affects the three dimensions of sustainable development: social, environmental, and economic.

Almost every human activity generates waste. The amount of municipal waste generated per person in the EU is around 500 kg annually. About 48% of this amount is recycled yearly (European Environment Agency, 2022b). Although the percentage of waste recycled has increased, the amount of waste generated is also growing.

Considering waste as a resource is the first step toward sustainable waste management. Waste not only can be a health or environmental problem but also an economic loss. However, whether the waste is a problem, or a resource depends on how we manage it. Therefore, the EU Waste Framework Directive has created a five-step waste hierarchy for managing waste. According to this hierarchy, waste prevention and reuse are the preferred options, followed by recycling and energy recovery, and landfill disposal is the least preferable option and should be limited to the minimum (European Commission, 2022).

A key principle of EU waste policy is to advance waste management further and follow circular economy principles. Recycling is one of the most important ways to reduce the consumption of primary resources. Therefore, the EU has set two targets to achieve by 2030. First, each EU Member State should

The Contribution of Urban Domestic Waste Management to the Circular Economy

Figure 1. The six European countries analysed in this chapter (own source)

recycle at least 60% of municipal waste or prepare them for reuse. Second, to halve residual municipal waste that is landfilled or incinerated (European Environment Agency, 2022a). However, it is unlikely to achieve this goal by 2030 without reducing waste generation. Therefore, waste prevention is of the highest importance.

Waste prevention is actions taken before a substance, material, or product has become waste. The goal is to reduce the amount of waste, the waste-related harmful substances, and the undesirable effects on the environment and human health (European Environment Agency, 2022a). Waste prevention is challenging because it depends on behavioural changes and requires a whole new infrastructure to enable those changes.

The problems of waste generation and management increase as societies continue to develop. How much waste we generate is closely linked to our consumption and production patterns. Municipal waste accounts for about 10% of the total waste generated in Europe (European Environment Agency, 2021b). Although it is not high in amount, its prevention can reduce the environmental impact. Citizens play an essential role in waste management. Therefore, education and raising awareness in this regard play an indispensable role. For example, a large proportion of municipal solid waste is kitchen and garden waste. Such waste can be converted into a source of energy or fertiliser if it is collected separately (European Environment Agency, 2021c).

Improving waste management and achieving a circular economy in Europe requires joint efforts, starting with small steps. It is essential to intensify the exchange of knowledge, insights, and practical experiences to make the necessary changes.

In this chapter, the authors intend to show the waste management programs implemented in six European countries (Figure 1), showing the strategies implemented and the way they are contributing to sustainability.

WASTE MANAGEMENT PRACTICES IN SIX EU COUNTRIES

Austria

In Austria, starting at the end of the 19th century, the first attempts at waste recycling were considered. From 1950 onwards, the Austrian provinces began to include the issues of the waste collection into legal texts. At the end of the 1960s aesthetic and environmental concerns were incorporated into the waste laws. This was enforced due to many environmental catastrophes in the 1970s, which gave rise to an increased concern about environmental aspects, and due to the impact of the report "Limits of Growth" by the "Club of Rome" (European Environment Agency, 2022a).

Today, individual regional magistrate departments take care of waste management in specific regions. There are usually several municipalities in a province assigned to one of the Austrian waste associations. These are responsible for collecting the waste. At the state level, there are laws, such as the Waste Management Act (Abfallwirtschaftsgesetz, AWG), and from the individual provinces, the collection of municipal waste is regulated by directives, such as the provincial waste laws, whereby the organisation of the collection of municipal waste is subject to the municipalities (BMK, 2022a).

Companies have to take care of the disposal of commercial waste themselves, and for this purpose they commission companies to dispose of the waste. All households, on the other hand, are obliged to participate in the public waste disposal system of their municipality. The waste disposal fees are regulated by the individual municipalities and are included in the municipal taxes (Austrian Government, 2022).

The costs incurred in Austria for recycling packaging, the so-called licence fees, are paid by the companies that put the packaging on the market and are usually included in the purchase price of the packaged product (WKO, 2022a). On the other hand, there is a deposit on returnable glass bottles, which can be returned to retailers to be refilled (Austrian Government, 2022). Used clothes, i.e. clean and wearable clothing and underwear, can be handed in at used clothes collections by charitable organisations; there are also official used clothes containers for this purpose (Austria Glasrecycling, 2022). Since 2005 electrical appliances and, since 2008, used batteries can be disposed of free of charge. There is even a 1:1 take-back in retail, where the dealers are obliged to take back old devices, if an equivalent, new device is bought at the same time. And for batteries, there is nowadays a collection point in almost every supermarket (Austria Glasrecycling, 2022).

In some waste material collection centres, fees have to be paid for the disposal of special waste, such as construction waste, bulky waste or hazardous waste. Such waste collection centres exist several times in a waste association and, depending on the area, in all larger municipalities. In these collection centres, citizens can bring their self-collected and separated waste. Depending on the municipality, certain types of waste are collected from households by the waste collection services of different companies. For garbage collection, garbage is collected and separated either in garbage cans or in large plastic bags, such as the yellow bag for plastic garbage. Glass is collected by Austria Glas Recycling (AGR) in certain communities or cities, and the cost of this is covered by the companies for "licensed" glass packaging (Austria Glasrecycling, 2022). In addition, there are other collection and recycling systems of companies such as Altstoff Recycling Austria AG, AGR Austria Glas Recycling GmbH, Bonus Holsystem für Verpackungen, GmbH & Co KG, European Recycling Platform (ERP) Austria GmbH, Interseroh Austria GmbH, Reclay UFH GmbH or the GUT GmbH (WKO, 2022b).

In 2019, the total amount of waste generated in Austria was approximately 71.26 million tons. The increase, which can be seen when looking back at past years, can be explained by a larger amount of excavation materials and waste from the construction industry (BMK, 2021). There is no data yet for the year 2020 for the whole of Austria. However, based on the values of individual federal states, it is possible to compare 2019 and the pandemic year 2020. For example, in the province of Upper Austria, there was a 3% increase in the total amount of waste from 2019 to 2020, the amount of residual waste increased by 2%, and there was a 7% increase in the amount of biogenic waste collected separately and treated mainly in composting and biogas plants; there was a slight decrease in the amount of used/valuable materials collected (Land Oberösterreich, 2022). At least at first glance, there are no outliers in the amount of waste that can be attributed to home offices and quarantine of the COVID 19 pandemic.

In terms of waste treatment, the figures for 2019 are as follows: 41% of waste was materially recovered (recycled and backfilled), 7% was thermally treated in plants subject to the Waste Incineration Ordinance, 46% was landfilled, and 6% of waste was treated in some other way, for example by chemical-physical or mechanical-biological waste treatment plants. However, if we exclude the 42.02 million tons of excavated materials from the total waste generated in Austria in 2019 (about 71.26 million tons), 64% of the waste was recycled and 2% backfilled, 17% was thermally treated, 11% was landfilled, and 6% of the waste was treated in other ways (BMK, 2021).

The rate of plastic recycling in 2019 was 28%. A total of 0.92 million tons of plastic waste was generated, of which 20% was production waste and 80% was post-consumer waste (WKO, 2022a). In 2020, 1.3 million tons of wastepaper were collected in Austria. As the paper industry processed 2.6 million tons of recovered paper in Austria in 2020, recovered paper is also imported to Austria, as the recovered paper is an important source of raw material in paper production. The recycling rate for wastepaper in Austria in 2019 is 77.6% (Austro Papier, 2022). The recycling rate for glass was even expected to exceed 80%. To achieve this, more than 270,000 tons of waste glass were collected in 2020, or around 29.4 kilograms per person (Riedmann et al., 2021).

Approximately 4.5 million tons of the total waste generated in 2019 comes from municipal waste from households and similar establishments. Thus, an average of 507 kg of municipal waste was generated per capita. Of this, 1.7 million tons of mixed municipal waste were sent for treatment as residual or bulky waste via public waste collection, 648,200 tons were wastepaper, and 165,400 tons were accounted for by used plastics and composite materials. Of this around 52% - more than half of the approximately 4.5 million tons of municipal waste from households and similar establishments - was sent for material recycling in 2019. Around 43% was treated thermally and less than 5% was treated mechanically-biologically (BMK, 2021). Untreated municipal waste has not been landfilled since 2008, and biogenic material has not been allowed to be landfilled since the landfill ban came into force in 2004. This landfill ban is still being extended and by 2026 there will also be a gradual landfill ban for most mineral building materials (VOEB, 2022).

In general, the Austrian federal government is also making efforts to advance the circular economy, and in addition to the European Green Deal, there are also federal guidelines to avoid single-use plastic products (BMK, 2022b). For example, the ban on single-use plastic carrier bags, also known colloquially as the "plastic bag ban," was put into effect in 2020 and a general ban on single-use plastic products was introduced in 2021. There will also be reusable beverage packaging quotas from 2024 and deposits on single-use beverage bottles in 2025 (BMK, 2022c). Generally, waste is also collected separately in public buildings, such as schools or train stations. This could be a motivation to also separate one's own household waste, which is why public institutions have a role model function both for children and adults when it comes to waste separation.

Belgium

In Belgium, household waste and industrial waste are collected differently. In addition, there are specific rules for waste from construction. Citizens can choose to bring their waste to the collection centres or have it collected from their homes. Waste collection is organised at the regional level (Vlaanderen, 2022). The Openbare Vlaamse Afvalstoffenmaatschappij (OVAM) is responsible for waste management and soil remediation in Flanders.

The categories of household waste being collected are glass, paper and cardboard, PMD (a combination of certain plastics, metal packaging and beverage packaging), food and gardening waste, electric and electronic appliances, small and hazardous waste and undifferentiated waste. All packaging from the PMD category is recycled within Belgium or neighbouring countries. Plastic waste is washed and ground up into pellets or granulates. The pellets and granulates are then melted or moulded into new products (Fost Plus, 2022a). With the growth of the number of plastics being recycled, it is important to have a large enough sales market. For this reason, an exploratory study on implementing the use of plastic recyclates in construction is being carried out by Centexbel-VKC (OVAM, 2022a).

White and coloured glass are collected separately at a collection point, not at home. Glass waste streams are purified and melted into new products (*Glas*, 2022b). Metal from the PMD category mostly contains steel and aluminium and is collected at home. Afterwards, steel and aluminium are magnetically separated, ground up, purified, melted, and made into new products (Fost Plus, 2022c). Paper and cardboard are collected at home or brought to collection points. Afterwards, it is sorted by quality, made into a pulp, purified, and removed of ink. The pulp is then crushed and dried. Depending on the desired end product, the pulp undergoes different processes (Fost Plus, 2022d). There are separate collection points for solar panels. Concerning car tyres, there are several possibilities to recycle these in an environmentally friendly way through garages, farmers or tyre centrals (Recytyre, 2022).

Municipalities can choose the prices for each waste stream being collected, within certain guidelines provided by the government (Vlaanderen, 2022). Municipalities have the choice to pay for waste collection by their own means, by asking for a flat rate tax or by asking for a variable cost price. When municipalities choose to ask for a variable cost price, the Flemish government provides a certain minimum and maximum rates for different waste streams, although some household waste streams, such as asbestos and small and hazardous waste, need to be collected for free. The government also recommends collecting glass, paper and cardboard for free, but this is not obligatory. Metal and textile waste are often collected for free as well, given the value of these waste streams.

Throughout the years there is a small decline in the amount of waste collected per citizen. However, in 2020 the amount of undifferentiated household waste for each Flemish citizen increased (from 143,5kg to 147kg) for the first time in 25 years (OVAM, 2022a). Given that the amount of industrial waste decreased in 2020 by 60,000 tons, the restrictions due to the COVID-19 pandemic could be to blame for the increase of undifferentiated household waste (OVAM, 2022a). A similar evolution can be seen in the amount of waste being littered and burned. In total there was a 4% decline in littering and burning waste in 2020, while the amount of household waste that was burned increased to 5% (OVAM, 2022a).

To get a more in-depth understanding of the recycling behaviour of households, OVAM organised a quantitative and qualitative research study concerning the recycling of household waste (OVAM, 2022b). More than 80% of respondents claim to consequently recycle waste into the different selective waste categories. 6% of respondents admit to burning waste (mainly garden and food waste or paper and cardboard) every once in a while. Categories that are recycled the least (16% to 36%) are food and garden waste, aluminium, small hard plastics and plastic foil or bags. Reasons respondents gave for sometimes not recycling properly were 1) being able to throw out food waste sooner, 2) not wanting to separate food waste from the packaging or 3) doubts about which category waste belongs to.

75% of respondents state that they try their best to minimise household waste by 1) recycling (80%), 2) selling things second hand or giving them away (70%), 3) buying second hand (40%) or 4) buying new goods with a long lifespan (40%). Other strategies mentioned in the qualitative research are consuming less, reusing, repairing, buying less packaging, and composting.

Respondents indicate that the most important motivating factors for recycling are doing the right thing, being part of the daily routine, contributing to the environment, having clear recycling guidelines, the reduced cost of selective waste streams and social or legal obligations. Barriers to recycling are thinking that most waste will become part of undifferentiated waste in the end, unclear rules concerning recycling, laziness, not having the space for different bins for each waste stream and thinking that they already recycle enough so it's okay to not recycle every once in a while, and not having enough waste of a specific category to fill an entire bag or bin.

More than 80% of respondents indicate that there are possibilities to be stimulated to improve their recycling behaviour. They propose decreasing the cost of selective waste streams, clear recycling guidelines, awareness, reward, or penalty systems and clearly indicated on the packaging how it should be recycled as possible stimulating factors. When it comes to waste prevention, offering composting bins, providing reusable grocery bags, organising second-hand markets, and providing a platform for the sharing economy as possible motivators that municipalities can provide.

In 2021, a new charter against littering in Flanders was launched (OVAM, 2022a). This led to several proposed actions such as:

1. **Working with littering reinforcers.** Littering reinforcers can address litterers and write out fines when necessary. During the World Championship cycling in September, these littering reinforcers were applied for the first time.
2. **Click-app.** With the Click app, people can scan and take pictures of the waste that they throw away. When they do this correctly, they earn circular coins that can be spent at local stores.

In addition to the charter on littering, there is the 'mooimakers' ('prettymakers') initiative organised by OVAM, the Union of Flemish Cities and Municipalities and Fost Plus. Mooimakers invests in communication and awareness, encouraging vendors to find alternatives to single-use packaging, reinforcing initiatives against littering and connecting with citizen projects (OVAM, 2022a).

Since 2020 single-use packaging has been prohibited in large-scale events in Flanders (OVAM, 2022a). The "Green Deal Differently Packaged" was created where several Flemish organisations and knowledge institutions engage themselves to look for alternatives to single-use packaging in the distribution sector. OVAM launched a campaign to raise awareness about single-use food packaging in November 2021. There was a special chip shop where citizens could get free french fries if they brought their own packaging. In addition, the shopcakes campaign was implemented at the same time to raise awareness concerning food waste. Shopcakes are small cakes made specifically to shop without hunger. If you shop when hungry, you often buy food that you don't need and ultimately goes to waste. Shopcakes are made from food that often goes to waste but makes you feel very saturated.

Hungary

Hungary has a well-developed policy and legal framework for waste management. It is mainly driven by EU requirements and supported by quantitative targets and economic instruments. There are a lot of positive waste management trends: decoupling of waste generation from economic growth, increased recycling, and recovery rates, and decreased use of landfills. However, Hungary remains an average performer in some cases, like glass recycling. A good monitoring system was developed for waste management, more precisely waste generation, and treatment. At the national level, the Ministry of Agriculture is the main authority for waste management policies. It is also the lead ministry for the transition to a circular economy, however, this transition is perceived as an extension of waste management policies. There are ongoing efforts to include resource efficiency and a circular economy (OECD, 2018).

In Hungary landfilling was the dominant treatment of municipal solid waste for decades. When Hungary joined the European Union (EU) in 2004 the national waste management policy priorities have been primarily driven by EU waste legislation (Herczeg, 2013). The waste management strategy of Hungary is defined in National Waste Management Program (NWMP). National programmes and regional and local waste management plans were prepared according to various waste types.

According to EU legislation, Hungary needed to improve its waste management program and develop some aspects. The National Waste Prevention Programme identified needs related to recycling and recovery in several aspects, namely municipal waste, non-hazardous production waste, non-hazardous agricultural and industrial food waste, sewage sludge, hazardous waste, particularly high priority hazardous waste stream, packaging waste, biodegradable waste, waste tyres and construction waste (Herczeg, 2013). Considerable results were achieved in the field of supply, level of service, modernisation, re-cultivation, and selective collection, which are the basis for the developments of the coming period. The municipal waste management public service is now available in almost 100% of the settlements. Hungary collected 3,203,367 t residual municipal solid waste and separately 236,673.619 tons of municipal solid waste in 2019. The total waste generation decreased by 17%, while GDP increased by 3%, which is a good achievement. Municipal solid waste generation decreased by around 19% (OECD, 2018). The separately collected waste, i.e., packaging materials, undergoes the sorting plant technology to recycle materials. Biowaste is also separately collected and further treated in composting- or biogas plants. The residual fraction is also treated in mechanical-biological plants for the sake of energy- and material recycling.

The main results for municipal solid waste management:

- The number of dwellings involved in the regular collection reached 93% rate, which means full coverage, and it is carried out by modern, closed, dust-free technologies.
- The implementation of the recultivation programme for old landfills is ongoing.
- The mixed municipal waste incinerator was upgraded, and its energy efficiency improved. In addition, the co-burning of the combustible components of municipal waste combined with energy recovery started for example in the Matra Power Plant and in some cement factories.
- Today, the selective collection system is available at more than 1 200 municipalities nationwide, for 55% of the population.
- Selective waste collection rates increased to 12% of the total volume of municipal solid waste. Taking into account the amount of organic waste collected separately, the rate exceeds 15%. Together with energy recovery in the capital, the overall recovery rate is 23-24%, which means more than 1 million tonnes of municipal waste.

Figure 2. Amount of recycled and reused municipal solid waste (in tonnes) in Hungary over time (Hungarian Central Statistical Office, 2022)

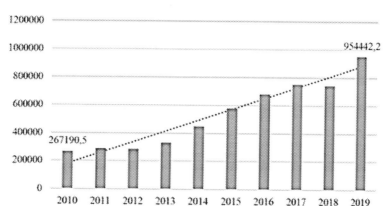

- In the case of priority waste streams, the EU standards have been achieved. Separate waste collection systems were established for wastes belonging to the responsibility of the producers (batteries, electronic devices, fluorescent tubes, pharmaceuticals, tyres), thereby reducing the risk and quantity of mixed waste (European Environment Agency, 2013).
- The amount of municipal solid waste recycled has been raised over a decade. The amount of recycling and reuse of solid municipal waste in 2009 was 267,190 tonnes and increased to 954,442 tonnes in 2019 or by 350% in ten years (Figure 2.)

Hungary has recently implemented significant waste sector reforms. Still, the new plan must be adapted to the different social and economic environments, taking into account changes in production, the transformation of technologies, improved service levels, and the consequences of the economic crisis. Municipalities are still in charge of public waste management services such as waste collection, transportation, and treatment, as well as the operation and maintenance of waste management facilities. There were approximately 130 public service operators (European Commission, 2019). The emphasis is placed on waste prevention and stimulating selective waste collection and recycling in order to increase resource savings. In addition, the safety of treatment, recultivation, and the elimination of illegal waste disposal continue to play an important role.

Significant progress is most likely for residential household waste. The goal is waste reduction, collection network development, improvement of recycling rates and the further reduction of the share of disposal through incentives. A number of comprehensive measures are required for waste prevention – and for recycling, recovery and disposal – in the area of product, technology and infrastructure development. To improve waste management at the local level, environmentally appropriate behaviour has to be acquired from an early age, and waste management skills are taught in all educational institutions as part of the National Curriculum (European Environment Agency, 2013).

To sum up, Hungary was successful in developing proper municipal solid waste and wastewater management due to clear regulations, substantial financial support, integrated separative collection, the building of sorting plants, and mechanical-biological and biological (composting, biogas facilities) ones for the treatment of municipal solid waste. It established an advanced wastewater network and treatment (Janetasari & Bokányi, 2022).

Latvia

A waste management system (WMS) is a complex system involving numerous waste streams, collection schemes, treatment processes and various actors, and WMS in Latvia is recently experiencing more sustainable development trends like circular economy and increased material recycling.

In 2017, household and similar waste generators produced around 850,000 tons of waste. A large part of the generated domestic waste ended up in landfills. In 2018, 462,358 tons of domestic waste were produced representing 58.9% of the total amount of generated waste. It is the value locked in the unused materials, which is not only not returned to the national economy but causes economic losses as an unused opportunity in promoting employment, developing innovations and building the local economy. With the introduction of measures such as a natural resource tax on landfills as well as the gradual introduction of separate collection, recycling rates have gradually risen from 5% in 2004, 9% by 2010 to finally 41% by 2019 (European Environment Agency, 2021a). Landfill still accounted for 45% of municipal waste in Latvia in 2016 though that number is now steadily decreasing. Where the country used to have over 500 unregulated landfills in the 1990s since 2000 most have been closed or re-cultivated according to EU standards. The existing recycling infrastructure is well maintained due to EU funding (VARAM, 2021).

The waste management companies have established separate waste collection points (which include both – door-to-door collection and containers in publicly available places, as in Latvia they are not observed separately). Also, civic amenities were established throughout the country, which along with paper, metal, plastic and glass provide the possibility to turn over hazardous household waste, waste of electric and electronic equipment (WEEE) and other household waste types. Separate collection of specific waste types (e.g., WEEE, batteries, end-of-life vehicles, packaging) has been facilitated by the producer responsibility system implemented through nature resource tax. Several waste management regions facilitate separate waste collection by providing eco-bags and eco-boxes intended for a particular type of waste (paper, plastic, metal) to private houses which are emptied by the waste management company free of charge. Yet the situation differs from region to region and even from municipality to municipality. The State Waste Management Plan of Latvia for 2021-2028 prepared by the Ministry of Environmental Protection and Regional Development, aims to expand the system for the separate collection of waste, develop the institutional framework for waste management, create stronger waste management regions and implement the principles of circular economy in order to: substantially increase waste recycling and reduce waste disposable. In order to reduce waste generation and ensure more efficient management, achieve ambitious targets for waste recycling and reduce waste storage, the necessary resources should be planned, and a strong waste management repository developed. Developing action programmes and measures to achieve the environmental policy objectives for the reduction of waste generation, the separate collection and recycling of waste, and the reduction of waste stored in landfills. The proposed action lines are in line with the European Union's (EU) growth strategy, the European Green Course, as well as the New Circular Economy Action Plan "Towards a cleaner and more competitive Europe" (VARAM, 2021).

In the late 20th century, the terms "waste recovery" and "recycling networks" began to appear in the scientific literature to describe the management of consumer recyclable waste materials through chain activity and involved stakeholders (Brouwer et al., 2018; Wilson et al., 2006). First of all, the topic of waste processing has become particularly relevant. It is increasingly common to hear that it is necessary to produce less and less waste and recycle it. The concept of zero waste has recently emerged in Latvia.

There is a need for the implementation of an improved waste management system that requires a significant and thorough planning stage, the results of which will significantly depend on the availability of detailed information on the possible waste flows and waste composition. The experimental results indicated that the unsorted municipal waste stream comprises up to 32.9% of recyclable materials and 29.2% of biodegradable wastes. Almost 60% of the waste currently being subjected to unsorted waste management systems in Latvia could potentially be source separated, ensuring higher quality of the recovered materials, and promoting circular value chains. The results from the recent research indicate a slight difference between waste composition in different waste management regions, noting that, in addition to the number of inhabitants and their habits, the local system in each waste management region may influence the composition of the collected wastes (Kubule et al., 2019).

The amount of unsorted waste in Latvia is higher than in other Baltic countries (2018), the amount of sorted waste was 25.2% of the total amount of waste (in Lithuania and Estonia, 52.5% and 28.0%, respectively (Eurostat, 2022)). Unsorted waste contains a significant amount of economically valuable materials - ferrous and non-ferrous metals, glass, plastic, and paper - and their reuse in the production of products would reduce the costs of both product production and raw material transportation since raw materials can be obtained locally due to waste sorting and recycling. Better separate collection and recycling of waste would reduce the country's dependence on imported resources and return the value of materials to the domestic economy. However, it needs better infrastructure. According to the Cabinet of Ministers Regulation No. 328 of June 13, 2017, on average, in the country, there is one separate collection point for 598 inhabitants, and one separate collection area for 14,816 inhabitants, which is not sufficient.

Used packaging constitutes a significant part of household waste. In 2017, more than 237,000 tons of used packaging were created in Latvia (Eurostat, 2022). More than 144,000 tons of material (60.88%) were processed (including regenerated and prepared for processing). By type of packaging, 38% of paper and cardboard, 5% of metal, 11% of plastic, 27% of glass and 19% of used wooden packaging were recycled.

On February 1, 2022, finally, there was a deposit system for beverage packaging in Latvia. As part of this, residents are able to hand over certain categories of beverage packages for recycling at sales points, container points and sorted waste collection points – plastic or glass bottles or cans of different volumes of beer, non-alcoholic and low-alcohol drinks. It will promote the development of circular economy solutions to replace these types of packaging and products with more environmentally friendly ones, as well as new behavioural/consumption practices in society.

Portugal

In Portugal, the treatment of waste arising from professional activities is governed by general waste legislation; the responsibility for waste management, including waste costs, lies with the original producer of the waste. The original producer of the waste or the holder must ensure that the waste is treated and to do so may use 1) a dealer; 2) a licensed entity carrying out waste collection or treatment operations; or 3) a licensed entity responsible for systems managing specific waste streams. Entrepreneurs who collect or transport waste must deliver it to licensed waste treatment operators (ePortugal, 2022).

Urban domestic waste is an exception, whose collection and treatment constitute a public service provided by municipal or multi-municipal systems. This exception includes small retail, service and catering establishments, school establishments, health care units, tourist developments, or other sources whose waste is similar in terms of nature and composition to that of households, and comes from a single establishment that produces less than 1,100 litres of waste per day (APA, 2022).

Therefore, waste management is based on a hierarchical organisation that promotes prevention in the first place, followed by reuse, recycling, recovery and, finally, disposal. The management of waste collection, transport and treatment is regulated at the level of municipal waste and non-urban waste (namely hospital and industrial waste, including hazardous waste). Some types of waste have a differentiated approach, in order to optimize their management - such as the following specific waste streams: construction and demolition waste, waste from batteries and accumulators, used oils, packaging and packaging waste, waste from electrical and electronic equipment, used tires and end-of-life vehicles.

The entity responsible for the implementation of environmental and waste management policies in Portugal is the Portuguese Environment Agency - APA, which aims to ensure the planning and management of waste, to prevent or reduce its production, the character of harmful and possible adverse impacts. It also seeks to promote efficiency in the use of resources, based on the principles of the waste hierarchy and CE, as well as to protect and enhance the environment.

The APA, in addition to the aforementioned objectives, ensures and monitors the execution of the national strategy for waste, also carries out the operational and administrative control of waste transfers in national territory, and ensures the collection and treatment of waste information, guaranteeing the validation of the information necessary for the application of the economic and financial regime of waste management, the so-called Waste Management Fee (TGR). This fee is applied by each municipality, in the collection and disposal of undifferentiated waste in landfills, that is, those residues that are not separated for recovery. In 2022, has been applied the value of 22€ per ton.

Regarding urban waste, the entities responsible for the municipal, inter-municipal or multi-municipal management system ensure the selective collection of the following fractions of waste:

- islands with paper/cardboard (packaging and non-packaging) and metal, plastic and glass packaging;
- used cooking oils.

The selective collection of bio-waste is being planned (with a target of implementation from the beginning of 2024) and of textiles, furniture waste and other bulky waste and hazardous waste (from 2025).

Based on these future targets, users of municipal waste management services are subject to the waste tariff, which is charged by the service provider, to cover the respective costs. The waste tariff should encourage the increase in adequate waste separation, and from 2026, it should no longer be adjusted to the water consumption of each user, as it is nowadays, and should be applied to the amount of waste collected, measured in units of weight, or estimated by its volume.

APA has recorded since the beginning of its activity the production of municipal waste per capita in Portugal. From 2019 to 2020 there was a decrease of 0.05%, from 5,281 million tons to 5,279 million tons, respectively. Thus, for 2020 the daily production of municipal waste was 1.40 kg per inhabitant, corresponding to an annual capitation of 512 kg per inhabitant (APA, 2022).

Regarding the collection of municipal waste, undifferentiated collection, unfortunately, continued to be, in 2020, the one with the highest percentage (79%).

Concerning the amount landfilled, a decrease was observed, as in 2018 and 2019 the percentages were, respectively, 58.3% and 57.7%. However, in 2020, this positive trend suffered an inversion, with an 8% increase compared to the previous year. The justification for this could be the lockdowns decreed by the government during the Covid-19 pandemic situation.

With the pandemic situation that was suffered, we also analyzed the results provided by APA regarding the values of the destination of waste for recycling and composting, obtained by the Eurostat methodology, in which these final destinations, both directly and indirectly, had the same repercussion as the destination to landfill, retreating in 2020 in relation to the years 2018 and 2020.

In this way, there is an increasing need to implement measures to improve the management of urban waste, not only in Portugal but all over the world. Even though there has been an increase in the amount of infrastructure for selective collection since 2020, we must continue to develop campaigns to raise awareness of the importance of waste separation. Currently, in some municipalities in Portugal, such as the regions of Setúbal and Porto, a form of awareness-raising is being carried out by companies, namely, the Amarsul and Suldouro Groups, which are investing in a door-to-door collection of packaging (paper and plastic) and organic waste of domestic origin, the latter of which will become a mandatory measure from 2023.

In other places, for example, in the municipality of Maia, the Lipor company developed a pilot project of a PAYT (pay-as-you-throw) system where citizens only pay for the undifferentiated garbage they produce/do not separate, instead of paying based on water consumption.

A system for depositing disposable plastic bottles and aluminium cans is also in place, which allows consumers to receive a discount voucher for each plastic bottle and aluminium can that is placed in a machine, ensuring its forwarding for recycling. With this initiative, the objective is to recycle more than 90% of the bottles and avoid their release into the environment. Several companies and environmental associations have developed apps to help citizens know where they should put their waste and how to sort it out.

Therefore, new challenges will continue to appear, mainly due to the obligations established by the European Community, which require new investments in infrastructure, in particular for waste recovery. In what concerns recycling, there should occur an efficiency increase in waste collection and sorting. As for organic recovery, efforts will be intensified to avoid sending biodegradable waste to landfills, which will require a strong investment in biological treatment technologies.

Another important aspect is the identification of improvement opportunities in the way production of urban solid waste is managed at production at all stages of the economic system, intending to guarantee, in the long term, a stabilization of the demand for natural resources and of the final volume of waste to be landfilled. There is a need to continue to raise awareness and emphasize the important role of the citizens, which is decisive for the success of these present and future goals, assuming increasingly responsible conduct, so that all these fundamental factors collaborate in the pursuit of sustainable development of a CE.

Romania

Romania is at the bottom of Europe in terms of recycling, with a rate four times lower than the European average. By 2020, Romania should have reached a 50% recycling rate. This target has been missed, as only 14% of municipal waste was recycled, making Romania the second-lowest recycling rate country in the EU, above Malta with only 6.5%. A shameful percentage, considering that the European average is 47%, the highest rate being in Germany, 67%. The next threshold must be reached by 2025 when Romania must have a 55% recycling rate, a hard-to-reach target for Romania, which is still struggling with the closure of illegal landfills and illegal waste imports. Another challenge for Romania is to reduce the amount of municipal waste stored by up to 10% by 2035. At present, Romanians deposit 70% of the waste they produce, while the European average is 24% and the rest is recycled (Friedrich Ebert Stiftung, 2022).

Waste management activities in Romania are based on Law 211/2011 on waste, republished, which implements a series of directives of the Council of Europe. The amendments to the waste management process is closely related to the European Union legislation in the field of waste that has as its main legislative instrument the Waste Framework Directive. It starts with prevention, followed by preparation for reuse, recycling, and recovery and ends with the disposal. The main goal of waste management is to prevent as much as possible the generation of waste, to use the generated waste as a resource, and minimise the amount of waste that ends up in landfills.

Romania relies on agencies at three levels to manage waste: the Ministry of Environment (MoE) and Ministry of Administration and Interior (MAI), the County Councils, and municipalities, which must ensure that all non-complying landfills and illegal dumps are closed, existing municipal landfills rehabilitated or extended, and new landfills constructed where needed. Moreover, according to Law 211/2011, local authorities are obliged to ensure separate collection of at least paper, metal, plastic and glass and to achieve, by 2020, the 50% preparation for reuse and recycling target (European Commission, 2011).

The municipality of Timisoara is located 571 km from the country's capital. Timisoara is the capital of Timiş County, it is the largest city in the western part of Romania, with a population of 323734 according to the Timis Regional Directorate of Statistics on 01.07.2020. According to the 2011 census, the city's residents were grouped into more than 21 ethnic groups and 18 religions, reflecting two major features of the population, namely interculturality and a high degree of tolerance.

SC RETIM Ecologic Service SA is a company established in 1997 following the association of the Timişoara Local Council with the German company RWE. The company holds an ANRSC Class I License for the public sanitation service of municipalities and is TÜV certified according to the quality, environment, occupational health and safety integrated management system. The society places special emphasis on civic and ecological education actions, especially among students (RETIM, 2022)

Through RETIM Ecological Service SA, waste stored/abandoned in the public domain of the municipality of Timişoara, in the amount of 12,764.45 tons, was collected and transported to the Ghizela Non-hazardous Waste Repository (RETIM, 2022).

TransClean S.R.L provides the services of collection, transport, and storage/utilization of plant waste and construction waste from demolitions, redevelopment, and rehabilitation activities, abandoned in the public domain of the municipality of Timişoara (Transclean, 2022). Likewise, 4,774.24 tons of vegetable waste and 677.16 tons of abandoned construction waste were collected, transported, and stored in the public domain.

In order to prevent the abandonment of waste and the sanctioning of citizens, the Timişoara City Hall together with RETIM Ecologic SA has opened four free collection points for vegetable waste, bulky household waste and waste resulting from construction and demolition, where citizens of Timişoara can voluntarily deposit their waste in order to avoid the fines for abandoning waste on public domain. Starting from January 2022, polystyrene and cardboard waste from the packaging of bulky items or household appliances can be handed in free of charge at RETIM centres. In accordance with the provisions of HCL 405 of 19.07.2019 approving the Prevention Plan and measures regarding the avoidance and reduction of the quantities of waste abandoned on the public domain within the radius of the Municipality of Timişoara, the quantities collected free of charge in these four centres are:

The Contribution of Urban Domestic Waste Management to the Circular Economy

1. Vegetable waste, in maximum quantities of 3 cubic meters per month;
2. Bulky and household waste, in maximum quantities of 5 m^3/month;
3. Waste from homes, generated by their interior/exterior remodelling activities, in maximum quantities of 1 m^3/month (RETIM, 2022).

The waste sorting station of Timișoara municipality operates within the framework of the "Integrated Waste Management System in Timiș County", which serves the municipality of Timișoara and 9 peri-urban municipalities (Dumbrăvița, Ghiroda, Giarmata, Giroc, Moșnița Nouă, Ortisoara, Remetea Mare, Șag, Sânmihaiu Romanian), starting from November 1, 2018 (Primăria Municipiului Timișoara, 2022).

Through the Waste Sorting Station SSDM, the following categories of waste are managed: residual waste from household waste from the population (wet fraction) taken from the urban and rural environment and recyclable waste from the population (dry fraction) taken from the urban and rural environment: paper/cardboard, glass, PET, other types of plastic, metal.

At the Waste Sorting Station in 2020, a quantity of 79,017.00 tons of household waste (wet fraction) was processed from the population in the urban environment, and 2,925.58 tons were used as energetic waste (Colterm, 2022).

The amount of recyclable waste (dry fraction) from the urban population, processed at the sorting station, is 16,384.192 tons, of which a quantity of 2,211.96 tons was recovered through recycling, as follows:

- Paper/cardboard in the amount of 1674.32 tons;
- Glass in the amount of 74.33 tons;
- PET in the amount of 399.38 tons;
- Metal in the amount of 18.96 tons;
- Other types of plastic in the amount of 44.98 tons

For the collection of glass waste from the population, a number of 350 bell-type containers (green) are placed in the public domain, in several locations, including all the neighbourhoods in the municipality of Timișoara. Through the Partnership Agreement concluded between the Municipality of Timișoara and the Caritas Federation of the Diocese of Timișoara, 30 special containers are placed for the collection of used clothing and other used textile products from the population. The locations of the containers are established by Caritas in partnership with the representatives of Timișoara Municipality to avoid obstructing pedestrian visibility. The collection action has both a charitable purpose, to support people in need, and an ecological purpose.

CHALLENGES RELATED TO WASTE MANAGEMENT IN EUROPE

There are several societal challenges related to waste management in Europe and worldwide. The amount of waste is increasing in many countries as their population increases in purchasing power. The waste management authorities are continuously requesting to separate different types of waste, for example, coffee capsules, medicines, electronic devices, batteries, used cooking oil, and textiles. One great challenge is how to do this separation in cities where the apartments have very little space to store all the separated waste until the collection is made, or the streets are narrow to have public bins or containers for all types of waste. It is urgent to reduce the amount of waste produced by changing the consumers' behaviour for buying according to their real needs.

Another challenge is the bulky waste generated every year in each country. The main goal of the households is the cheapest and most practical way to get rid of the unneeded bulky waste. How might we reuse vast quantities of second-hand, good-quality bulky waste with reuse potential (ex. doors, windows, pieces of wood)? Repair services are essential to the process of reusing bulky waste, and local authorities may be able to provide support. The "Repair Network Vienna" is a meaningful example for the prevention of waste in a sustainable and resource-efficient economy through reuse and repair support services. The Repair Network Vienna, founded in 1999, consists of about 80 member companies and provides a hotline for inhabitants to quickly identify appropriate repair services for their needs. According to the network, the members carry out more than 50.000 repairs per year, which corresponds to about 750 tons of waste prevented every year (Stadt Wien, 2018).

The European pattern of textile consumption has increased textile waste since the clothes and fabric shops sell them very cheaply, and people's consumption habits have changed over time. In the current fast fashion economy, clothes are bought to be used a few times and then thrown away. In some countries, there are already containers for collecting textiles. These textiles are made of different materials which are not yet recyclable and most of them are disposed of in landfills or incinerated. So, textile waste prevention and innovative recycling methods are needed. A big challenge is how to change people's habits of consumption and closet replacement and elimination. The SDG Watch Europe campaign "Wardrobe Change", for example, asks EU leaders to take urgent action in the textile industry sector to change how clothes are made, sold, worn – and re-worn (SDG Watch Europe, 2022).

Electrical and electronic equipment is one of the growing rapid waste streams. These types of equipment comprise a complex mixture of different materials, some of which are hazardous. Some types of equipment have rare and expensive resources that can be recycled if the waste is properly managed. Some countries need to improve the collection, treatment, and recycling of electrical and electronic equipment. For electronic waste, an initiative called StEP (Solving the E-waste Problem) was founded in 2004 as an independent, multi-stakeholder platform. The StEP aims at designing strategies that address the management and development of environmentally, economically, and ethically sound e-waste resource recovery, reuse, and prevention (StEP, 2022).

When it comes to municipal waste, the six countries analysed in this chapter, present different recycling rates as shown in Figure 3.

One of the main reasons for the low rate of municipal waste may be the lack of people's literacy concerning waste management in their countries. The influence of individual and socio-demographic characteristics (age, gender, education level, and occupational status), contextual variables (rural or urban, country of origin), and the awareness about environmental importance have an impact on literacy concerning waste management. The challenge here is how to increase people's literacy and communicate better and increase the rate of reuse and recycling for a sustainable and resource-efficient economy.

CONCLUSION

The EU waste policy states to advance waste management further and to follow more circular economy principles – to recycle at least 60% of municipal waste or prepare them for reuse, and to decrease by halve residual municipal waste landfilled or incinerated, by 2030. This illuminates need for waste reduc-

The Contribution of Urban Domestic Waste Management to the Circular Economy

Figure 3. The recycling rates of municipal waste per EU country (European Environment Agency, 2021)

tion and waste prevention. Waste prevention can be challenging due to the consumer behaviour, waste management systems and a completely new infrastructure to enable those changes.

This chapter analyses the waste management strategies and approaches in six European countries, namely, Austria, Belgium, Hungary, Latvia, Portugal, and Romania. Research performed reflects the various waste management forms implemented, according to each country environmental, economic, and social perspectives, also showing some very innovative solutions. The authors intend to promote a decrease in consumption and waste generation, and an increase in reuse, separately collected waste and recycling, contributing to circular economic growth and the sustainability of the planet. Knowing what is done in different countries allows us to learn from the best examples and reflect on the various waste management forms implemented according to environmental, economic, and social perspectives. The identified and described examples can be adapted and implemented in countries with less efficient waste management systems.

It can be seen that all countries analysed align or are trying to align themselves with European norms on waste management. The main categories of household waste that are collected by the six countries are glass, paper and cardboard, PMD (a combination of certain plastics, metal packaging and beverage packaging), food and gardening waste, electric and electronic appliances, small and hazardous waste, and undifferentiated waste. However, each country is in different degrees of implementation, due to greater or increased involvement of the population and incentive policies implemented by local authorities and governments, in some cases with the use of support from the European Union to improve waste system collection and recycling.

Analysing the separation for recycling and making an analogy with the number of years that each country has already been concerned about this subject and the policies applied, it is perceived that Austria is the country that has earlier demonstrated concern and applied policies for collection of municipal waste such as the provincial waste laws, whereby the organisation of municipal waste collection is subject to the municipality's costs. Entry into the European community has helped countries to increase the separation % of waste through the implementation of European policies, as happened in Hungary where landfilling was the dominant treatment of municipal solid waste for decades, but in 2004 the national waste management policy priorities became driven by EU waste legislation.

Then, for circular economy growth compliance and implementation of waste management improvements must be promoted, for example, the application of tax revenues, recycling fees for waste produced by the municipality, private investments, promoting more employability, gross value added related to CE, and patents related to recycling.

The analysis identifies several challenges to bring up in next research and projects. One of the main challenges is future changes in consumer's behaviour from generation of different types of waste to reduction by raising citizen's awareness of the problem in general and in details. These research and developments should primarily include food, medicine, and textiles, due to the large volumes produced and the possible scarcity of these products in the near future.

The future research will contribute to building a sustainable future, including reflection on consumption and waste recycling and reuse processes, and raising citizens' awareness of the problem to develop transformative social action.

ACKNOWLEDGMENT

The authors want to acknowledge the E^3UDRES^2 alliance for supporting this research. This research was supported by the ERASMUS+ Programme [101004069].

REFERENCES

APA. (2022). *Waste.* Relatório do Estado do Ambiente. https://rea.apambiente.pt/environment_area/waste?language=en

Austria Glasrecycling. (2022). *Glas entsorgen.* [*Dispose of glass*]. AGR.. https://www.agr.at/glasrecycling/glas-entsorgen

Austrian Government. (2022). *Abfallentsorgung/Müllabfuhr.* [*Waste disposal/garbage collection*] Österreichs digitales Amt. oesterreich.gv.at https://www.oesterreich.gv.at/themen/bauen_wohnen_und_umwelt/umzug/5/Seite.180301.html

Austro Papier. (2022). *Positionen—Altpapier.* Austropapier. https://austropapier.at/positionen-altpapier/

BMK. (2021). *Die Bestandsaufnahme der Abfallwirtschaft in Österreich—Statusbericht 2021.* [*Taking Stock of Waste Management in Austria—Status Report 2021*] Bundesministerium für Klimaschutz, Umwelt, Energie, Mobilität, Innovation und Technologie. https://www.bmk.gv.at/dam/jcr:04ca87f4-fd7f-4f16-81ec-57fca79354a0/BAWP_Statusbericht2021.pdf

BMK. (2022a). *Allgemeines zur Abfallwirtschaft.* [*General information on waste management.*]. Österreichs digitales Amt. oesterreich.gv.at - https://www.oesterreich.gv.at/themen/bauen_wohnen_und_umwelt/abfall/1/Seite.3790060.html

BMK. (2022b). *Europäischer Green Deal.* BMK. https://www.bmk.gv.at/themen/klima_umwelt/eu_international/euop_greendeal.html

BMK. (2022c). *Kunststoffabfälle in Österreich.* [*Plastic waste in Austria.*]. BMK. https://www.bmk.gv.at/themen/klima_umwelt/kunststoffe/kunststoffabfaelle.html

Brouwer, M. T., Thoden van Velzen, E. U., Augustinus, A., Soethoudt, H., De Meester, S., & Ragaert, K. (2018). Predictive model for the Dutch post-consumer plastic packaging recycling system and implications for the circular economy. *Waste Management (New York, N.Y.), 71,* 62–85. doi:10.1016/j.wasman.2017.10.034 PMID:29107509

Colterm. (2022). *Deseuri—Colterm S.A. [Waste—Colterm S.A.].* Colterm. https://www.colterm.ro/anunturi/protectia-mediului/1078-deseur

ePortugal. (2022). *Gestão de resíduos—EPortugal.gov.pt. [Waste management—EPortugal.gov.].* ePortugal. *t*https://eportugal.gov.pt/cidadaos-europeus-viajar-viver-e-fazer-negocios-em-portugal/bens-e-mercadorias-em-portugal/gestao-de-residuos

European Comission. (2022). *Waste Framework Directive.* EC. https://environment.ec.europa.eu/topics/waste-and-recycling/waste-framework-directive_en

European Commission. (2011). *Country factsheet Romania* [File]. EC. https://www.eea.europa.eu/themes/waste/waste-prevention/countries/romania-waste-prevention-country-profile-2021/view

European Commission. (2019). *The EU Environmental Implementation Review 2019 Country Report—Hungary.* EC. https://ec.europa.eu/environment/eir/country-reports/index_en.htm

European Commission. (2020). *A new Circular Economy Action Plan.* EUR-Lex. https://eur-lex.europa.eu/legal-content/EN/TXT/HTML/?uri=CELEX:52020DC0098&from=EN

European Environment Agency. (2013). *Waste—National Responses (Hungary)* [SOER 2010 Common environmental theme (Deprecated)]. EEA. https://www.eea.europa.eu/soer/2010/countries/hu/waste-national-responses-hungary

European Environment Agency. (2021a). *Waste recycling in Europe.* EEA. https://www.eea.europa.eu/ims/waste-recycling-in-europe

European Environment Agency. (2021b). *Municipal waste management across European countries.* European Environment Agency. https://www.eea.europa.eu/publications/municipal-waste-management-across-european-countries

European Environment Agency. (2021c). *Waste: A problem or a resource?* European Environment Agency. https://www.eea.europa.eu/publications/signals-2014/articles/waste-a-problem-or-a-resource

European Environment Agency. (2022a). *Reaching 2030's residual municipal waste target—Why recycling is not enough—European Environment Agency* [Briefing]. EEA. https://www.eea.europa.eu/publications/reaching-2030s-residual-municipal-waste/reaching-2030s-residual-municipal-waste/

European Environment Agency. (2022b). *Europe is not on track to halve non-recycled municipal waste by 2030—European Environment Agency* [News]. EEA Europa. https://www.eea.europa.eu/highlights/europe-is-not-on-track

European Parliament & Council of the European Union. (2008). *Directive 2008/98/EC of the European Parliament and of the Council of 19 November 2008 on waste and repealing certain Directives (Text with EEA relevance)*. EUR-Lex. https://eur-lex.europa.eu/eli/dir/2008/98/oj/eng

European Parliament, & Council of the European Union. (2018). *Directive (EU) 2018/851 of the European Parliament and of the Council of 30 May 2018 amending Directive 2008/98/EC on waste*. https://eur-lex.europa.eu/legal-content/EN/TXT/?uri=celex%3A32018L0851

Eurostat. (2022). *Statistics*. Eurostat. https://ec.europa.eu/eurostat/databrowser/view/sdg_11_60/default/table?lang=en

Fost Plus. (2022a). *Plastic verpakkingen*. Fost Plus. https://www.fostplus.be/nl/recycleren/plastic-verpakkingen

Fost Plus. (2022c). *Metalen verpakkingen*. [*Metal packaging*.] Fost Plus. https://www.fostplus.be/nl/recycleren/metalen-verpakkingen

Fost Plus. (2022d). *Papier—Karton*. Fost Plus. https://www.fostplus.be/nl/recycleren/papier-karton

Friedrich Ebert Stiftung. (2022). *Monitor Social*. Monitor Social. https://monitorsocial.ro/

Glas. (2022b). Fost Plus. https://www.fostplus.be/nl/recycleren/glas

Herczeg, M. (2013). *Municipal waste management in Hungary*. European Environment Agency. https://www.eea.europa.eu/publications/managing-municipal-solid-waste/hungary-municipal-waste-management/view

Hungarian Central Statistical Office. (2022). *Központi Statisztikai Hivatal 2020*. [*Central Statistical Office 2020*.]. KSH. https://www.ksh.hu/?lang=en

Janetasari, S. A., & Bokányi, L. (2022). Challenges on creation of sustainable municipal waste and wastewater management in Indonesia using experience of Hungary. *IOP Conference Series: Earth and Environmental Science, 1017*(1), 012028. doi:10.1088/1755-1315/1017/1/012028

Kaza, S., Yao, L. C., Bhada-Tata, P., & Van Woerden, F. (2018). *What a Waste 2.0: A Global Snapshot of Solid Waste Management to 2050*. World Bank. doi:10.1596/978-1-4648-1329-0

Kirchherr, J., Reike, D., & Hekkert, M. (2017). Conceptualizing the circular economy: An analysis of 114 definitions. *Resources, Conservation and Recycling, 127*, 221–232. https://doi.org/10.1016/j.resconrec.2017.09.005

Kubule, A., Klavenieks, K., Vesere, R., & Blumberga, D. (2019). Towards Efficient Waste Management in Latvia: An Empirical Assessment of Waste Composition. *Environmental and Climate Technologies, 23*(2), 114–130. doi:10.2478/rtuect-2019-0059

Land Oberösterreich. (2022). *Land Oberösterreich—Entwicklung der Abfallmengen 1990 bis 2020. [Province of Upper Austria—Development of waste volumes 1990 to 2020.]* Land Oberösterreich. https://www.land-oberoesterreich.gv.at

OECD. (2018). *OECD Environmental Performance Reviews: Hungary 2018*. Organisation for Economic Co-operation and Development. https://www.oecd-ilibrary.org/environment/hungary-2018_9789264298613-en

OVAM. (2022a). *Jaarverslag 2021*. OVAM. https://jaarverslag.ovam.be/sites/default/files/2022-05/OVAM_jaarverslag_2021-JB.pdf

OVAM. (2022b). Preventie- en sorteergedrag van de Vlaamse Bevolking. Samenvatting.... [Prevention and sorting behaviour of the Flemish Population. Summary]. OVAM. www.vlaanderen.be. https://www.vlaanderen.be/publicaties/preventie-en-sorteergedrag-van-de-vlaamse-bevolking-samenvatting-kwantitatieve-en-kwalitatieve-bevraging-2021

Primăria Municipiului Timișoara. (2022). *Acasa—Primăria Municipiului Timișoara. [Home—Timișoara City Hall.]* PMT. https://www.primariatm.ro/

Recytyre. (2022). *Waar naartoe? [Where to go?]* Recytyre. https://www.recytyre.be/nl/waar-naartoe

RETIM. (2022). *Acasă—RETIM SA*. RETIM. https://retim.ro/

Riedmann, C., Schwarzenhofe, H., Huemer, C., & Reidelshöfer, K. (2021). *Nachhaltig und sicher: Glasrecycling in Österreich. [Sustainable and safe: glass recycling in Austria.]*. OIT. https://www.oesterreich-isst-informiert.at/verantwortung/nachhaltig-und-sicher-glasrecycling-in-oesterreich/

Stadt Wien. (2018). *Repair Network Vienna*. Reparatur Netzwerk. https://www.reparaturnetzwerk.at/repair-network-vienna

StEP. (2022). *Organisation*. StEP Initiative. https://www.step-initiative.org/organisation-rev.html

The Ellen MacArthur Foundation. (2022). *What is a circular economy?* Ellen MacArthur Foundation. https://ellenmacarthurfoundation.org/topics/circular-economy-introduction/overview

Transclean. (2022). *Transclean*. http://transclean.ro/

VARAM. (2021). *Minister Plešs: State waste management plan will ensure the development of the sector [Pinister Pless: State waste management plan will ensure the development of the sector.]*. Vides aizsardzības un reģionālās attīstības ministrija [Ministry of Environmental Protection and Regional Development]. https://www.varam.gov.lv/en/article/minister-pless-state-waste-management-plan-will-ensure-development-sector

Vlaanderen. (2022). *Afvalinzameling en sorteren [Waste collection and sorting]*. Vlaanderen. www.vlaanderen.be. https://www.vlaanderen.be/afvalinzameling-en-sorteren

VOEB. (2022). *Erreichung der EU-Klimaziele: Recyceln statt Deponieren*. [*Achieving EU climate targets: recycling instead of landfilling.*] VOEB. https://www.voeb.at/service/voeb-blog/detail/show-article/erreichung-der-eu-klimaziele-recyceln-statt-deponieren/

S. D. G. Watch Europe. (2022). *Wardrobe Change*. SDG Watch Europe. https://www.sdgwatcheurope.org/wardrobe-change/

Wilson, D. C., Velis, C., & Cheeseman, C. (2006). Role of informal sector recycling in waste management in developing countries. *Habitat International, 30*(4), 797–808. https://doi.org/10.1016/j.habitatint.2005.09.005

WKO. (2022a). *Abfallwirtschaft im Betrieb*. [*Waste management in the company.*] WKO. https://www.wko.at/service/umwelt-energie/Abfallwirtschaft_im_Betrieb.html

WKO. (2022b). *Information zur Verpackungsverordnung*. [*Information on the Packaging Ordinance.*] WKO. https://www.wko.at/service/umwelt-energie/information-verpackungsverordnung.html

KEY TERMS AND DEFINITIONS

Composting: Process for transforming organic waste into fertilizers and/or energy. The compound that results from this process is a highly nutritious material and can be used in gardens, vegetable gardens and orchards.

Municipal Waste: mixed waste and separately collected waste from households, including paper and cardboard, glass, metals, plastics, biowaste, wood, textiles, packaging, waste electrical and electronic equipment, waste batteries and accumulators, and bulky waste, including mattresses and furniture.

Recover: One of the points of the 7R policy proposed by the circular economy, where food scraps and other organic materials can be reinstated into nature, through organic composting; this is the best process for transforming organic waste into fertilizers.

Recover: One of the points of the 7R policy proposed by the circular economy, where food scraps and other organic materials can be reinstated into nature, through organic composting; this is the best process for transforming organic waste into fertilizers.

Recycle: One of the points of the 7R policy proposed by the circular economy, where each material must be conditioned on a specific collector to be recycled, according to its nature; people can separate materials anywhere and take them directly to recycling centres or look for collecting services that pass-through neighbourhoods.

Reduce: One of the points of the 7R policy proposed by the circular economy, that suggests people to consume less amounts of goods, by using products with greater durability and packaging in the right measure.

Refuse: One of the points of the 7R policy proposed by the circular economy, that suggests people not to opt for products from companies that do not respect nature or harm the environment, giving preference to those who benefit society and produce with low impact on the environment.

Regift: One of the points of the 7R policy proposed by the circular economy, which can also be named reusing, that proposes to people to give someone a product that is no longer useful for them; in alternative, that given object can gain totally different functions from the original one and remain very useful.

Repair: One of the points of the 7R policy proposed by the circular economy, that suggests people to fix products in poor condition, instead of just buying a new one.

Rethink: One of the points of the 7R policy proposed by the circular economy, that refers to the need of consumers to think about their habits, choosing to buy only what they really need.

Separately Collected Waste: Waste that is collected for a specific purpose separately from mixed waste, sorting it according to its nature (plastics, paper and cardboard, glass, etc).

Section 3
Entrepreneurship and Corporate Innovations

Chapter 10
Teleworking:
New Challenges and Trends

Andreia Raquel Pinto
Institute Polytechnic of Setúbal, Portugal

Dulce Matos Coelho
Polytechnic Institute of Setúbal, Portugal

Raquel Pereira
Institute Polytechnic of Setúbal, Portugal

ABSTRACT

In light of new trends and realities, today's society promotes and values the importance of innovation and sustainability as a solution to the challenges ahead, where companies contribute to the attraction and retention of young talent through Telework. This study addresses Telework as a solution to new challenges. The focus of the study is five small and medium-sized companies. It was concluded that companies opted for Telework at home, with advantages of cost reduction, conciliation of personal and professional life, and reduction of pollutant gases; and as disadvantages, the investment in equipment, social isolation, and precariousness of the labor market. The impacts of telework were verified at the level of productivity, commitment, and performance, where it was concluded that telework can be partially maintained. The contribution of this work focuses on the importance that virtual experiences have in Generation Z and the flexibility of companies in adopting pre-existing work models.

INTRODUCTION

Society is currently facing new trends and realities, where the existence of mechanisms that promote and value the importance of innovation and sustainability is imperative, as solution for the great challenges ahead.

In labour market, those challenges are huge and require that organizations have the capability to adapt to the society restructuration, through the improvement of economic models, where companies play an

DOI: 10.4018/978-1-6684-6123-5.ch010

important role in attracting and retaining young talents. Young people entering the labour market bring new skills, new characteristics, and new challenges for enterprises.

As a way of retaining talents, companies are developing new working methodologies, where teleworking, as long as structured and thought out, emerges as an innovative reality that benefits both parts. Companies see their productivity rates increase and young people, on the other hand, seek a dynamic and flexible environment in the company, promoting economic and social development with a win-win partnership.

This chapter approaches the teleworking modality as a practice of human resource management to be adopted by companies as a possible solution for the new societal challenges through sustainable practices. Considering today's society and the way it is currently constituted, the contribution of this study focuses on the importance that virtual experiences have in Generation Z and how companies must be flexible to adapt existing working models, developed and improved in line with the characteristics, skills and requirements of this new generation.

Teleworking, in its various modalities, appears as a reality in the labour market, which brings advantages and disadvantages for companies, workers and, in a global way, the whole society. The objectives of the present study are to identify the main impacts associated with teleworking. Moreover, it equates the possibility of systematic adoption of telework in the future, which may have impacts within companies and on the evolution of society, especially focusing on the new generations of workers and the adoption of new working models.

BACKGROUND

Literature review discusses teleworking, starting with the concept and exploring types and modalities. The impacts of teleworking on the activity of small and medium-sized enterprises (SMEs) will be clarified. Generation Z will be described, regarding its origin and definition, as well as its characteristics and behaviours. Finally, the relationship between Generation Z and their entrance into an increasingly exigent labour market is analysed.

Teleworking: Concept, Types, and Modalities

Over time, working organization methods have been suffering changes due to constant technological developments, economic and financial crises, and other dynamics of the environment in which companies are placed.

The traditional manual work has taken new forms. The use of Information and Communication Technologies (ICT) has facilitated the achievement of fundamental competitive advantages for the survival and improvement of the competitiveness of business enterprises.

Teleworking appears, this way, as a method of working at a distance, although the concept of teleworking is more complete, as there is an exchange of information and data between the worker and the employer (Araújo & Bento, 2002). Tasks usually performed with the physical presence of the worker, are replaced by tasks that involve the use of ICT, giving rise to a new way of performing those functions (López & Rodriguez, 2020). An example of this change resides in the way of communicating, in which communication via the Internet is used, instead of using physical paper and other traditional means.

Over time, several authors have added new definitions to the concept of teleworking, where the common feature is the use of ICT by workers to perform their tasks (Herrera et al., 2022; Belzunegui & Erro-Garcés, 2020; López & Rodriguez, 2020; ILO, 2020; Chiru, 2017; Goulart, 2009; Kobal et al., 2009; Sakuda & Vasconcelos, 2005; Rebelo, 2004; Araújo & Bento, 2002).

Given the different concepts presented by several authors, we propose to define teleworking as a form of working relationship carried out at a distance, with the use of ICT, which should provide the teleworker with flexibility in the working organization, allowing an improvement in their work, both in terms of their economic and productive conditions.

The attempt to define teleworking raises doubts about who should provide the ICT indispensable for working. In general, it is argued that companies have to supply the technologies, as well as adequate training actions and even the possibility of contributing to increasing costs that workers have to support (for example, electricity, gas, consumables, Internet).

Teleworking is based, in most cases, on work carried out at the worker's home, but several studies point out that teleworking can also be developed in other modalities: teleworking in satellite centres, teleworking in tele-centres or mobile teleworking (Belzunegui & Erro-Garcés, 2020; ILO, 2020; Fiolhais, 2007; Pérez-Pérez et al., 2003; Bailey & Kurland, 2002; Tremblay, 2002).

Teleworking at home is the most common form of teleworking, in which the worker develops his activity at home, using ICT that allows him to establish a relationship with the employer - online, direct contact or indirect contact (Belzunegui & Erro-Garcés, 2020; ILO, 2020; Fiolhais, 2007).

Teleworking in satellite centres is defined as the activity carried out in branches of the company, strategically relocated on a different geographic location with ICT connections to the main company. This form of teleworking is adopted, for instance, by companies located in large cities where traffic congestion is present (Belzunegui & Erro-Garcés, 2020; ILO, 2020; Fiolhais, 2007; Pérez-Pérez et al., 2003; Bailey & Kurland, 2002; Tremblay, 2002).

The sharing of spaces properly equipped with ICT by workers from different companies, implemented close to the workers' homes, is called teleworking in tele-centres (Belzunegui & Erro-Garcés, 2020; ILO, 2020; Fiolhais, 2007; Pérez-Pérez et al., 2003; Tremblay, 2002).

Mobile teleworking is any activity that is carried out remotely by the worker, in various places, such as at the customers' companies, almost always on moving and with constant connection to the employer via ICT (Belzunegui & Erro-Garcés, 2020; ILO, 2020; Martínez-Sánchez et al., 2008; Fiolhais, 2007, 2002; Pérez-Pérez et al., 2003; Bailey & Kurland, 2002; Tremblay, 2002).

Some authors argue that one can characterize teleworking modalities by the proportion of work that is performed in telework, that is, full-time, part-time, and *ad hoc* or casual teleworking (Belzunegui & Erro-Garcés, 2020; ILO, 2020; Gajendran & Harrison, 2007; Tremblay, 2002).

Full-time teleworking occurs when the worker performs all his activity at home, while in part-time teleworking the worker carries out his activity partially at home, partially at the office or partially at the customer company. *Ad hoc* or casual teleworking occurs when work is occasionally performed at home due to a temporary impediment (worker illness or child caring, for instance).

However, teleworking modalities may also vary according to the professional bond that the worker has with the employer, deriving from teleworking at home totally carried out for a single employer, to freelance teleworking where the work is also done at home or in a place other than the company, but related to several employers (ILO, 2020; Martínez-Sánchez et al., 2008).

Duxbury, Higgins & Thomas (2006) add a different modality of teleworking observing the time when teleworking occurs, as it can be done during the normal working period or it can be carried out outside the working hours (evenings or weekends) in order to compensate for work not carried out during normal working hours.

Impacts of Teleworking

Teleworking can be seen as an alternative to traditional labour relations and, according to most authors, it brings several benefits to companies, such as increased organizational flexibility, cost reduction and diversity in recruitment. Globally, teleworking facilitates flexibility at work and a better work-family balance, while reducing the environmental impacts of mobility (Lodovici et al., 2021; Belzunegui & Erro-Garcés, 2020; Lopez & Rodriguez, 2020).

For workers, it may provide a better balance of professional and family life, increase disposable income and increase productivity and levels of satisfaction (Herrera et al., 2022; Lodovici et al., 2021; Gajendran & Harrison, 2007).

For society, in a macro level, there are benefits in the reduction of fuel consumption and the consequent reduction of polluting gases, as well as possible improvements in the policies of labour market integration (Belzunegui & Erro-Garcés, 2020; Lodovici et al., 2021).

Productivity is fundamental for competitiveness and for the survival of companies. However, what will be the impact of teleworking on productivity? Most authors are persuasive in stating that teleworking increases productivity through effective cost reductions.

The efficient use of ICT combined with training geared to the needs of the company and the teleworker (knowledge and skills) is crucial for increasing productivity. The work performed during the time that the worker spent before on their journeys to the workplace and the experience of teleworkers are also pointed out as reasons for increased productivity (Lodovici et al., 2021; Torten et al., 2016; Aboelmaged & Subbaugh, 2012; Lupton & Haynes, 2000).

For teleworkers, telework is seen as something positive. However, the perception of an increase in productivity is based on assumptions related to a positive reaction to teleworking and an improvement in the productivity of these workers. Positive reactions are motivated by work-life balance (Herrera et al., 2022; Lodovici et al., 2021; Gajendran & Harrison, 2007; Rasmussen & Corbett, 2008).

Also seen as contributing factors to productivity increases are the following: work flexibility for teleworkers; hiring staff with more skills and exceptional qualifications who reside far from the workplace; the reduction of costs with staff; efficient use of energy; reduction of environmental impact and creation of job opportunities for people with disabilities (Lodovici et al., 2021; Chiru, 2017; Perincherry, 2009; Mills et al., 2001).

According to Martin and MacDonnell (2012), productivity can be measured by increasing or reducing the effective production of an economy, using the same available resources. However, some argue that it is necessary to obtain information of a more comprehensive scope, which is generally beyond the access of researchers and is difficult to measure, through broadband analyses, ways of traveling by teleworkers, fuel consumption, earnings per hour and even contribution to Gross Domestic Product (Perincherry, 2009).

One of the obstacles to the real measurement of productivity is related to the execution of knowledge-based tasks, where knowledge, in addition to being intangible and difficult to quantify, has different ways of being performed. The same happens with non-fixed, non-routine tasks, without standard times, that can be performed in diversified ways by teleworkers (Lodovici et al., 2021; Bosch-Sijtsema et al., 2009; Ramírez & Nembhard, 2004).

Teleworking

Besides productivity, talent retention is also assumed as differentiating and as such a source of competitive advantage for companies. Teleworking can improve talent retention, allowing improvements in human resources indicators such as turnover and abstention rates (Martin and MacDonnell, 2012). In fact, the perception of talent retention intentions and turnover from the perspective of employers and teleworkers is one way to measure talent retention (Iscan & Naktiyok, 2005; Mills et al., 2001).

Teleworking can be also assumed as a relevant factor that influences the talent retention, as the teleworking options can eliminate the need for a worker to change residence of their household or other related factors, thus allowing companies to hire the best candidate without having to assume the relocation expenses of that individual. Thus, teleworking allows overcoming the obstacle of geographic location, as well as a better balance of family and professional life (Timsal & Awais, 2016; Mello, 2007).

However, teleworking may lead to the departure of talent from the company and, in some cases, it may increase turnover rates (Marx et al., 2021; Caillier, 2017), since the job opportunities for these talents are greater, as the geographic issue is less important.

Telework can also be a way of making work more flexible, which provides autonomy in working hours for teleworkers, allowing them to increase their commitment to the company (Barros & Silva, 2010; Golden, 2006). Organizational commitment is positively related to teleworking and negatively related to turnover intentions (Marx et al., 2021; Barros & Silva, 2010; Golden, 2006).

Teleworkers have higher rates of overall job satisfaction and organizational commitment compared to those who do not adopt telework. Thus, companies should invest in a strategic management of human resources based on duties and tasks to be assigned in telework, avoiding dissatisfaction and even the departure of workers from the company (Taborosi et al., 2020; Kelliher & Anderson, 2010).

Teleworkers can create their own organizational commitment, which, according to Jacobs (2006), occurs when values and attitudes are perceived by the organization through the special effort of teleworkers, leading to a reciprocal commitment for both parts, fulfilling the psychological contract and emphasizing the definitions of organizational commitment.

However, teleworking can lead to the isolation and independence of teleworkers, threatening and defragmenting the company and the organizational culture. To combat this possible occurrence, the company should make efforts to affiliate teleworkers by providing them with social support, based on the work they perform. The perception of social support will moderate the relationship between teleworkers' need for affiliation and their strength of organizational commitment (Wiesenfeld et al., 2001).

In addition to productivity, talent retention and organizational commitment, performance may be another benefit of teleworking. It is important to note that performance can be seen as the perception of how teleworkers and company are positively performing their tasks (Martin & MacDonnell, 2012).

Teleworking gives autonomy to the teleworker and the fact that they are evaluated by their performance leads them to be proactive, to dedicate more time to planning their tasks and do their activities more efficiently, to work longer hours compared to normal work and to be more creative (Lodovici et al., 2021; Aderaldo et al., 2017; Noonan & Glass, 2012; Gajendran & Harrison 2007; Raghuram et al., 2003; Amabile et al., 2002).

There is a positive relationship between flexibility in the workplace, performance and telework, which can be explained through the practice of flexible schedule and the possible demand on the teleworker of more working hours in order to be able to deal with the overload activities and tasks (Giovanis, 2018; Martin & MacDonnell, 2012).

A better performance by teleworkers leads to better company performance, which provides increasing profits and growth, allowing them to award salary increases, thus improving the quality of labour relations (Lodovici et al., 2021; Vega et al., 2015; Sánchez et al., 2007).

Thus, one can conclude that teleworking causes positive impacts for companies, which should determine whether to include teleworking in their short-term growth strategies.

This possible inclusion of telework could make companies more productive, strengthening organizational commitment, ensuring talent retention, improving the performance of teleworkers, allowing a competitive advantage that contributes to the growth of the company and the economy in general (Lodovici et al., 2021; Martin & MacDonnell, 2012).

Generation Z: Origin and Definition

According to Latkovikj et al. (2016), it was in the 1950s that greater emphasis was placed on generational differences. All individuals within a society belong to a certain generation, which is different from its previous as well as from the next.

A generation can be defined as a group of individuals who have in common a set of characteristics such as age, geographic location, physical condition, social status, or others (Parry and Urwin, 2011). The generation is the period that individuals live illustrating the way they define their lifestyle (Glass, 2007; Salleh et al., 2017).

Six distinct generations were identified in the literature, as described in figure 1:

Figure 1. Generations Timeline
Source: Adapted from Bencsik et al. (2016)

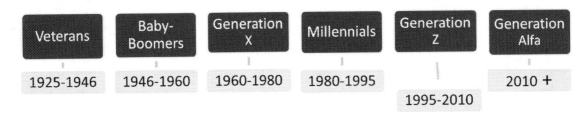

Generation Z is made up of young people born between 1995 and 2010, who may be known as Digital Natives, Me Generation, Generation N, Gen Z, Zs, Gen Z-ers, iGeneration, Gen Tech, Gen Wii, Net Gen, Gen Next, Post Gen, Plurals and Centennials (Feiertag & Berge, 2008; Prensky, 2001; Turner, 2015), being the successors of the Millennials generation.

These young people were born in an era of new challenges, in terms of terrorism, world political instability, strong concerns about environment and sustainability, being the first generation that grew up under the direct influence of information and communication technologies (ICT), remaining among the offline and online state, absorbing overloads of information from the Internet (Prensky 2001; Turner, 2015).

Characteristics and Behaviors of Generation Z

The young people who make up Generation Z are characterized by a technological dependence, which is introduced by their parents at an early age of their life. This makes them avid for new knowledge, using new languages acquired through the assimilation of different information when using the Internet, knowing its verbal and visual world, i.e., knowing in depth how to interact with the different approaches to the Internet (Feiertage & Berge, 2008; PRENSKY, 2001; Salleh et al., 2017; Torocsik et al., 2014).

They are eager, restless, always connected and learn through direct observation, being known as multitaskers, which differs from previous generations (Erwin and Shatto, 2016; Levickaite, 2010). In fact, Generation Z faces the same problems as previous generations, however through their technological opportunities they can positively circumvent or even subtract these problems from their lives (Torocsik et al., 2014).

They are also individualistic, egocentric and less team-oriented than previous generations. They seek flexibility in the work environment, giving extreme importance to novelties, but showing a reduced tolerance towards individuals who do not master technologies in the same way as they do (Dolot, 2018; Erwin & Shatto, 2016; Etter et al., 2017; Levickaite, 2010; Mohr & Mohr, 2017). According to Torocsik et al. (2014), young people of Generation Z have a little attention span, so the messages that are transmitted to them should be short, direct, and assertive, prevailing the "less is more".

However, Levickaite (2010) draws attention to the fact that these characteristics cannot be generalized. They may vary according to the culture, region, and social environment. Generation Z behaviour, as in previous generations, is influenced and adjusted by the environment and current circumstances with new challenges and trends.

Young people of Generation Z, growing up with a background in technological environments, become Internet savvy, but this obsession results in a lack of physical interaction and a weak use of logic in situations that require cognitive thinking (Salleh et al., 2017).

The lack of work experience, the social justice movements and the dependence on smartphones and social networks can also be pointed out as a pattern of behaviour, although mitigated by growth in a culture of safety (Schroth, 2019).

Relationship of Generation Z and the Labor Market

Social and economic globalization leads to constant changes in the business environment. The Internet and technological evolution are central on these changes and, if well explored, could make companies good harbours for young people at the beginning of their professional careers.

Human resources have a key impact on the creation and implementation of modern solutions and concepts, namely the optimization of human resources in the context of their location and requirements for working conditions can solve societal challenges related to young generations (Graczyk et al., 2022).

Latkovikj and Popovska (2020) and Marginean (2021) argue that the entry of young people into the workforce is having a strong impact on the way companies communicate and organize themselves and how they thrive in the business world, developing working methods in a more efficient way, to respond to demands and needs of this new workforce, including Generation Z.

Gabrielova and Buchko (2021) refer that at the same time Generation Z is beginning to enter the workforce and the workplace continues to change and adapt to those new professionals, millennials generation are now progressing in their careers and will challenge with managing these incoming workers and their characteristics, behaviours and requirements.

Companies must consider the characteristics and behaviours of Generation Z so that they can understand how technologies and new means of communication in working environments will respond to the challenges imposed by this new workforce (Gabrielova & Buchko, 2021; Bezerra et al., 2019; Zivnuska et al., 2019; Fusi & Zhang, 2018; Etter et al., 2017).

Companies should establish sustainable business goals and strategies focusing on the skills developed by Generation Z, focusing on improving or even remodelling their current framework of skills (Hecklau et al., 2016).

In a study developed by Marginean (2021) is concluded that older generations try to attract and retain younger generations. On the one hand, through mentoring, they help them to adapt to the new work reality, but on the other hand, they shape them to respond to what companies want and need.

Barhate and Dirani (2022) argue that Generation Z has well-defined career expectations and career development plans. Generation Z values job security, have no hurry to assume leadership positions and give importance to mastering new skills, where the opportunity to learn from mentors is relevant. They prefer large or multinational companies to small and medium-sized companies and want to have their own workspace, which in most cases is not the reality due to the open space design, but they refer that teleworking can be a positive solution (Marginean, 2021).

Companies must consider that the workplace must be designed to accommodate Generation Z, but they cannot in any way neglect the demands of the older generations that are still in the company, in a multigenerational coexistence prolonged by increasing the duration of working life (Gabrielova & Buchko, 2021; Bezerra et al., 2019; Latkovikj & Popovska, 2020).

Generation Z is persuaded more by visual rather than verbal information, so companies can use digital technologies and social media to recruit, motivate and retain talents, as well as use new strategies, with digital and technological tools, to increase organizational commitment (Smith & Strawser, 2022).

According to Graczyk et al. (2022), the general characteristics and demands of Generation Z can help companies to adapt themselves and contribute to resolve societal challenges, namely because Generation Z representatives desire to work in innovative companies and have higher level of commitment to innovation and sustainability. This new workforce will force companies to innovate and can represent a qualitative leap towards a more innovative and sustainable society.

As the most digitally generation entering the labour market, Generation Z should be able to adapt quickly and efficiently to new technologies (Smith & Strawser, 2022). So that, this new workforce can be a strong value, a competitive issue, for companies to become more innovative, more flexible, more adaptable, more able to cope with the growing demands of an increasingly changing global economy.

A CASE STUDY ABOUT TELEWORKING

Methodological Considerations

The focus of the chapter will be the study carried out by the authors to five small and medium-sized enterprises located in the municipality of residence of one of the authors, in Barreiro, district of Setúbal, Portugal.

The start of activity of the chosen companies varies between the years 1986 and 2016, which in itself was an important fact for the selection of companies due to the diversity of opinions of the representatives of the companies, which vary according to the solidity of the organizational culture of each company allied to permanence in the market.

They are, for the most part, companies with the main activity in service sector, with sizes ranging from 4 to 130 workers, where the diversity of the number of workers can be considered an asset for the study, since this way it was possible to assess different opinions of those responsible for human resources, taking into account the company size.

Taking into account that the study was developed during the COVID-19 pandemic, there were companies in which 100% of the workers were in telework, while at the opposite extremes there was an adoption to telework for only two weeks. The latter was due to the incompatibility of the adoption of telework with the activity developed by the company.

The main subject of study was the perception of those responsible for the companies about the impact of telework on the activity of SMEs and the analysis of whether companies would intend to maintain the practice of telework in the near future, assessing what would be the modality and what the strategy would be.

The research work had an exploratory character, using a predominantly qualitative methodology, based on case study, more specifically multi-case. Information collected through semi-directive interviews was analysed, using previously prepared interview guides, to representatives of the companies and workers of the five companies. The interviews were transcribed, accurately preserving the maximum amount of information.

A content analysis was performed, based on the assumption that the themes that emerged from the interviews were significant and full of meaning. More than analysing how frequent certain message elements appears, the significance was privileged in the analysis, through the themes that emerged, those that did not emerge, and the sequences in which they appear (Bardin, 2011).

The category system has been provided by existing theories and enunciation analysis – commonly applied to non-directive interview data – was performed, assuming that each interview, as the base unit, is studied in itself, as an organized and singular totality. Following Bardin (2011), the authors tried to find the intrinsic logic that structures each interview.

Impacts of Telework on the Companies' Activity: The Perspective of Managers and Workers

After analysing the information collected in the interviews, it was possible to conclude that the impacts of telework on the activity of companies were productivity, talent retention and the reduction of the absenteeism rate, organizational commitment, and the performance of teleworkers.

About productivity, the opinions of the representatives of the companies varied, perhaps due to the sector of activity of each company, where they stated that the productivity rates with the adoption of telework in their companies remained or even decreased in some cases.

However, company representatives reported productivity considering the performed work without taking into account the time spent to perform it, which contradicts the concept of productivity, since it concerns the ability to do as much work as possible with the least number of resources needed.

When analysing the opinions of company workers, it appears that they do not use the concept of productivity correctly either. For workers, the fact that they worked longer hours was crucial for them to feel more productive, which is a perception that is not in line with the concept under analysis.

With regard to the fact that telework contributes in some way to the retention of talent, company representatives mentioned that they do not perceive a direct relationship between telework and the retention of talent, but a relationship between telework and the reduction of absenteeism rates. Most said they had not laid off workers yet. In fact, reductions in absenteeism rates were noted, as with the adoption of telework, workers had more time to deal with their personal affairs without having to miss work. The majority of the workers shared this opinion, since they had more time to solve personal problems by remaining working at home, so they did not need to be away from work to deal with these issues.

The indices of commitment to telework, in the opinion of the representatives of the companies in the study, did not suffer any kind of influence, as in some cases it was maintained and in others, it was found that when objectives were established, they were met within the established deadlines. The workers stated that they felt more committed to the company, but once again, it can be concluded that there is a confusion in the real meaning of commitment, since it is not by working longer hours that commitment increases. Commitment to the company should not be confused with this feeling of commitment.

The performance, according to the analysis of the representatives of the companies in the study, was evaluated through the work and objectives completed and not through the way that work was carried out.

In this impact, it was not possible to establish a relationship between the responses of the company representatives and their workers, as the latter were questioned about autonomy. It can be concluded that there were two different points of view to talk about performance. Companies evaluated the work and objectives completed, workers evaluated themselves by the autonomy they had to perform their work.

Through the study, it can be concluded that there is not a perception of a direct relationship between telework and talent retention but telework can contribute to making the work method attractive to Generation Z, since according to the theoretical framework it is possible to verify that this generation develops hand in hand with technology.

Telework, when well-structured and thought out, allows companies to expand in terms of technologies, thus becoming more attractive to new workforces. The use of ICT does not create any obstacles for Generation Z; in fact, it makes the work experience much more attractive and stimulating. For Generation Z, the fact of working remotely does not bring any kind of inconvenience, a characteristic that should be well explored by companies.

The set of new skills that Generation Z brings to companies is in itself already a decisive factor in retaining these talents, as the introduction of "new blood" is welcome for the competitiveness of companies, as it cannot be copied.

In short, despite the fact that telework, in the study carried out by the authors, does not appear to directly relate with the retention of talents, when incorporated into working methods, may allow companies to attract and retain young people from Generation Z, who are eager for development opportunities using ICTs that they know so well.

Are companies prepared to adapt to new generations of workers? As described earlier, Generation Z is the new workforce. However, are companies prepared for Generation Z? In the study carried out by the authors, the existence of interviews with workers made it possible to open the door to conclusions that could help to answer this very pertinent question.

On the one hand, workers over 40 years of age were interviewed, these belonging to Generation X and Millennials, who showed difficulties in adapting to telework, just because they changed their daily

Teleworking

routine and had to learn new forms of communication. This occurred despite workers being in favour of telework as a way of reconciling professional and personal life, where the care of children and direct relatives was the most given justification.

On the other hand, despite being a minority, young people belonging to Generation Z were interviewed, who revealed that teleworking was a way for them to be at home managing their time as they wished, although the productivity rates were not what was expected, because they worked longer hours at home than if they were at the company.

They also recognized that they had no difficulty adapting to this method of work, the training they had was just what they needed to know how to work remotely, and the lack of contact with colleagues was something that did not affect them psychologically. They felt "freer" in time management, but with the deadlines and goals set, they had a sense of daily control.

Both the more mature generation and Generation Z were peremptory in stating that the adoption of telework brought benefits. However, over time telework should be adopted in a more hybrid way – a few days at home, a few days at the company – in order to escape the routines and lessen the feeling of isolation that many said they felt at some point in time.

Companies increasingly feel that they lack human resources and find it difficult to attract and retain new talent from new generations. Talent retention is associated with motivation, which is the crucial key for workers to feel that they are important and that they make a difference in the workplace.

The young, qualified people, who make up Generation Z, in addition to their salary, seek stability, security and mentoring to have the opportunity to demonstrate their new skills and to accept or keep the job. But they also seek their well-being, conciliation between professional and personal life, favour forms of digital communication, work flexibility and do not fail to include in their requirements a good range of benefits.

Analysing what Generation Z is looking for and what companies have to offer, telework emerges as a working method that responds positively to the requirements of this generation.

It can be concluded that telework can contribute to the retention of talents of the new generation of workforce, Generation Z, insofar as it can allow keeping the best professionals in the company. Organizations will be able to obtain competitive advantages that set them apart from the others, and motivate young people of Generation Z to perform their tasks in the best way possible.

SOLUTIONS AND RECOMMENDATIONS

One of the solutions pointed out in the study carried out by the authors to the five SMEs was the adoption of telework in a hybrid form, allowing increasing motivation and unlocking the feelings of discomfort caused by staying at home for a long period.

As can be seen, Generation Z is driven by new technologies and is eager for digital communication and the efficient use of ICT. Companies must present, in addition to the use of technologies that so attract young people, other sets of benefits (gym memberships, health insurance, among others) that in themselves convey a positive image of the company as a good place to work.

A talent is something that is constantly changing, so there is nothing better than creating constant challenges, such as new role(s) where young people feel valued, given that Generation Z have no hurry to assume leadership positions. Training combined with the use of telework can be a tool that allows young talents to develop new knowledge and skills on an ongoing basis.

Work flexibility is one of the advantages of adopting telework, as it allows young people of Generation Z to work in a flexible schedule that allows them to work from home, in order to reconcile their professional and personal lives. This will allow, in some way, to reduce stress situations, increasing their well-being and motivating them for their work.

FUTURE RESEARCH DIRECTIONS

Very little is yet known about the entry of Generation Z into the job market, as they are young people who are mostly just starting their professional careers.

Paid professional internships are well regarded in the eyes of Generation Z, but they do not allow them the stability they are looking for because, in most cases, when internships end they are not replaced by employment contracts with the company.

The study of Generation Z under different scenarios and in the face of new challenges is crucial as future research directions, since what is a reality today is not tomorrow.

Generation Z lives and develops allied to technological development, so it is pertinent to investigate the impacts of this alliance on the labour market, taking into account the different employment relationships, the different types of companies (namely, size, the sector of activity), and the willingness of young people to create their own business.

CONCLUSION

Today's society brings new challenges to the labour market, for not only the older generations, especially because of technological and social changes, but also for young people who enter the labour market for the first time, bringing with them new skills, new features and new challenges.

As a way of adapting to this new reality, companies are investing in new work methodologies, including telework, which, if well-structured and thought out, emerges as an innovative reality that allows benefits to both workers and companies, as well as to society in general.

This study allowed us to analyse how the very characteristics of Generation Z can influence the job market and, on the other hand, how companies are flexible to adapt existing work models, thus bringing advantages, but also disadvantages for companies and workers. One of the solutions pointed out was the adoption of telework in a hybrid form – a few days at home, a few days at the company -, allowing increasing motivation and simultaneously avoiding feelings of discomfort caused by staying at home for a long period.

REFERENCES

Aboelmaged, M., & Subbaugh, S. (2012). Factors influencing perceived productivity of Egyptian teleworkers: An empirical study. *Measuring Business Excellence*, *16*(2), 3–22. doi:10.1108/13683041211230285

Aderaldo, I., Aderaldo, C., & Lima, A. (2017). Aspectos críticos do teletrabalho mu ma companhia multinacional [Critical aspects of teleworking in a multinational company]. *Cadernos EBAPE.BR, 15*(8), 511–533. doi:10.1590/1679-395160287

Amabile, T., Hadley, C., & Kramer, S. (2002). Creativity under the gun. *Harvard Business Review, 80*(8), 52–61. PMID:12195920

Araújo, E. R., & Bento, S. C. (2002). *Teletrabalho e Aprendizagem: Contributos para uma problematização [Telework and Learning: Contributions to a problematization]*. Fundação Calouste Gulbenkian.

Bailey, D., & Kurland, N. (2002). A Review of Telework Research: Findings, New Directions, and Lessons for the Study of Modern Work. *Journal of Organizational Behavior, 23*(4), 383–400. doi:10.1002/job.144

Bardin, L. (2011). *Content analysis* (5th ed.). Edições.

Barros, A. & Silva, J. (2010). Percepções dos indivíduos sobre as consequências do teletrabalho na configuração home-office: estudo de caso na Shell Brasil [Perceptions of individuals about the consequences of telework in the home-office configuration: A case study at Shell, Brazil]. *Cadernos EBAPE. BR, 8*(1) – artigo 5, 71-91.

Bencsik, A., Horváth-Csikós, G., & Juhász, T. (2016). Y and Z Generations at Workplaces. *Journal of Competitiveness, 6*(3), 90–106. doi:10.7441/joc.2016.03.06

Barhate, B., & Dirani, K. M. (2022). Career aspirations of generation Z: A systematic literature review. *European Journal of Training and Development, 46*(1/2), 139–157. doi:10.1108/EJTD-07-2020-0124

Belzunegui, A., & Erro-Garcés, A. (2020). Teleworking in the Context of the Covid-19 Crisis. *Sustainability, 12*(9), 3662. doi:10.3390u12093662

Bezerra, M. M., Lima, E. C., Brito, F. W., & Santos, A. C. (2019). Geração Z: Relações de uma Geração Hipertecnológica e o Mundo do Trabalho [Generation Z: Relationships of a hypertechnological generation and the world of work]. *Revista Gestão em Análise [Management in Analysis Magazine], 8* (1), 136-149.

Bosch-Sijtsema, P., Ruohomaki, V., & Vartiainen, M. (2009). Knowledge Work Productivity In Distributed Teams. *Journal of Knowledge Management, 13*(6), 533–546. doi:10.1108/13673270910997178

Caillier, J. (2017). Do Work-Life Benefits Enhance the Work Attitudes of Employees? Findings from a Panel Study. *Public Organization Review, 7*(3), 393–408. doi:10.100711115-016-0344-4

Chiru, C. (2017). Teleworking: Evolution and trends in USA, UE and Romania. Economics. *Management and Financial Markets, 12*(2), 222–229.

Dolot, A. (2018). The characteristics of Generation Z. The Characteristics of Generation Z. *E-mentor, 74*(2), 44–50. doi:10.15219/em74.1351

Erwin, K., & Shatto, B. (2016). Moving on From Millenials: Preparing for Generation Z. *Journal of Continuing Education in Nursing, 47*(6), 253–254. doi:10.3928/00220124-20160518-05 PMID:27232222

Etter, M., Ravasi, D., & Colleoni, E. (2017). Social Media and the Formation of Organizational Reputation. *Academy of Management Review, 44*(1), 28–52. doi:10.5465/amr.2014.0280

Feiertag, J., & Berge, Z. L. (2008). Training Generation N: How Educators Should Approach the Net Generation. *Education + Training*, *50*(6), 457–464. doi:10.1108/00400910810901782

Fiolhais, R. (2007). Teletrabalho e Gestão dos Recursos Humanos [Telework and Human Resources management]. In A. Caetano, & J. Vala (Org.), Gestão de Recursos Humanos: Contextos, processos e técnicas [Human resource management: contexts, processes, and techniques] (3ª ed., pp. 235-262). Editora RH.

Fusi, F., & Zhang, F. (2018). Social Media Communication in the Workplace: Evidence From Public Employees Network. *Review of Public Personnel Administration*, *0*(0), 1–27. doi:10.1177/0734371X18804016

Gabrielova, K., & Buchko, A. (2021). Here comes Generation Z: Millennials as managers. *Business Horizons*, *64*(4), 489–499. doi:10.1016/j.bushor.2021.02.013

Gajendran, R., & Harrison, D. (2007). The Good, the Bad, and the Unknown About Telecommuting: Meta-Analysis of Psychological Mediators and Individual Consequences. *The Journal of Applied Psychology*, *92*(6), 1524–1541. doi:10.1037/0021-9010.92.6.1524 PMID:18020794

Giovanis, E. (2018). The Relationship Between Flexible Employment Arrangements and Workplace Performance in Great Britain. *International Journal of Manpower*, *39*(1), 51–70. doi:10.1108/IJM-04-2016-0083

Glass, A. (2007). Understanding Generational Differences for Competitive Success. *Industrial and Commercial Training*, *39*(2), 98–103. doi:10.1108/00197850710732424

Golden, T. (2006). Avoiding depletion in virtual work: Telework and the intervening impact of work exhaustion on commitment and turnover intentions. *Journal of Vocational Behavior*, *69*(1), 176–187. doi:10.1016/j.jvb.2006.02.003

Goulart, J. (2009). *Teletrabalho: Alternativa de trabalho flexível [Telework: flexible work alternative]*. Editora Senac.

Graczyk, M., Olszewski, R., Golinski, M., Spychala, M., Szafranski, M., Weber, G. W., & Miadowicz, M. (2022). Human resources optimization with MARS and ANN: Innovation geolocation model for generation Z. *Journal of Industrial and Management Optimization*, *18*(6), 4093–4110. doi:10.3934/jimo.2021149

Hecklau, F., Galeitzke, M., Bourgeois, S., & Kohl, H. (2016). Holistic Approach for Human Resource Management In Industry 4.0. *Procedia CIRP*, *54*, 1–6. doi:10.1016/j.procir.2016.05.102

Herrera, J., De las Heras-Rosas, C., Rodríguez-Fernández, M., & Ciruela-Lorenzo, A. M. (2022). Teleworking: The Link between Worker, Family and Company. *Systems*, *10*(5), 134. doi:10.3390ystems10050134

ILO. (2020). *Teleworking during the COVID-19 pandemic and beyond: a practical guide*. International Labour Office. Geneva, Switzerland. https://www.ilo.org/travail/whatwedo/publications/WCMS_751232/lang--n/index.htm

Iscan, O., & Naktiyok, A. (2005). Attitudes towards telecommuting: The Turkish case. *Journal of Information Technology*, *20*(1), 52–63. doi:10.1057/palgrave.jit.2000023

Jacobs, G. (2006). Communication for commitment in remote technical workforces. *Journal of Communication Management (London)*, *10*(4), 353–370. doi:10.1108/13632540610714809

Kelliher, C., & Anderson, D. (2010). Doing more with less? Flexible working practices and the intensification of work. *Human Relations*, *63*(1), 83–106. doi:10.1177/0018726709349199

Kobal, F., Agner, T., & Oliveira, A. (2009). Vantagens e desvantagens do teletrabalho: Uma pesquisa de campo em uma multinacional [Advantages and disadvantages of telecommuting: a field survey in a multinational]. XXIX Encontro Nacional de Engenharia de Produção - A Engenharia de Produção e o Desenvolvimento Sustentável: Integrando Tecnologia e Gestão [XXIX National Meetinf of Production Engineering—Production engineering and sustainable Development: Integrating technology and management]. Brasil.

Latkovikj, M. T., &, Popovskab, M. B. (2020). *How Millennials, Gen Z, and Technology are Changing the Workplace Design?* Conference: STPIS 2020 Socio-Technical Perspective in IS Development 2020. Grenoble, France.

Latkovikj, M. T., Popovska, M. B., & Popovski, V. (2016). Work Values and Preferences of the New Workforce: HRM Implications for Macedonian Millennial Generation. *Journal of Advanced Management Science*, *4*(4), 312–319. doi:10.12720/joams.4.4.312-319

Levickaite, R. (2010). Generations X, Y, Z: How Social Networks Form the Concept of the World Without Borders (the case of Lithuania). *LIMES: Cultural Regionalistics*, *3*(2), 170–183. doi:10.3846/limes.2010.17

Lodovici, M. S. (2021). *The impact of teleworking and digital work on workers and society*. Study Requested by the EMPL Committee.

López, P., & Rodríguez, P. (2020). Who is Teleworking and Where from? Exploring the Main Determinants of Telework in Europe. *Sustainability*, *12*(21), 87–97. doi:10.3390u12218797

Lupton, P., & Haynes, B. (2000). Teleworking – the perception-reality gap. *Facilities*, *18*(7/8), 323–328. doi:10.1108/02632770010340726

Martin, B., & MacDonnell, R. (2012). Is telework effective for organizations? A meta-analysis of empirical research on perceptions of telework and organizational outcomes. *Management Research Review*, *35*(7), 602–616. doi:10.1108/01409171211238820

Martínez-Sánchez, A., Pérez-Pérez, M., Vela-Jiménez, M., & Carnicer, P. (2008). Telework adoption, change management and firm performance. *Journal of Organizational Change Management*, *21*(1), 7–31. doi:10.1108/09534810810847011

Marx, C., Reimann, M., & Diewald, M. (2021). Do Work-Life Measures Really Matter? The Impact of Flexible Working Hours and Home-Based Teleworking in Preventing Voluntary Employee Exits. *Social Sciences*, *10*(1), 1–22. doi:10.3390ocsci10010009

Mello, J. (2007). Managing Telework Programs Effectively. *Employee Responsibilities and Rights Journal*, *19*(4), 247–261. doi:10.100710672-007-9051-1

Mills, J., Wong-Ellison, C., Werner, W., & Clay, J. (2001). Employer liability for telecommuting employees. *The Cornell Hotel and Restaurant Administration Quarterly, 42*(5), 48–59. doi:10.1016/S0010-8804(01)80057-4

Mohr, K. A., & Mohr, E. (2017). Understanding Generation Z Students to Promote a Contemporary Learning Environment. *Journal on Empowering Teaching Excellence, 1*(1), 84–94. doi:10.15142/T3M05T

Noonan, M., & Glass, J. (2012). The Hard Truth About Telecommuting. *Monthly Labor Review*, U. S. Department of Labor. *Bureau of Labor Statistics., 135*, 38–45.

Parry, E., & Urwin, P. (2011). Generational Differences in Work Values: A Review of Theory and Evidence. *International Journal of Management Reviews, 73*(1), 79–96. doi:10.1111/j.1468-2370.2010.00285.x

Pérez-Pérez, M., Sánchez, A., & Carnicer, M. (2003). The organizational implications of human resources managers' perception of teleworking. *Personnel Review, 32*(6), 733–755. doi:10.1108/00483480310498693

Perincherry V. (2009). A Framework for Evaluating Regional Impacts of Broadband Internet Access: Application to Telecommuting Behavior. doi:10.2139/ssrn.1489377

Prensky, M. (2001). Digital Natives, Digital Immigrants. *On the Horizon, 9*(5), 1–6. doi:10.1108/10748120110424816

Raghuram, S., Wiesenfield, B., & Garud, R. (2003). Technology enabled work: The role of self-efficacy in determining telecommuter adjustment and structuring behavior. *Journal of Vocational Behavior, 63*(2), 180–198. doi:10.1016/S0001-8791(03)00040-X

Ramírez, Y., & Nembhard, D. (2004). Measuring Knowledge Worker Productivity – a taxonomy. *Journal of Intellectual Capital, 5*(4), 602–628. doi:10.1108/14691930410567040

Rasmussen, E., & Corbett, G. (2008). Why Isn't Teleworking Working? *New Zealand Journal of Employment Relations, 33*(2), 20–32.

Rebelo, G. (2004). *Trabalho e privacidade: contributos e desafios para o direito do trabalho*. Editora RH.

Sakuda, L. & Vasconcelos, F. (2005). Teletrabalho: Desafios e Perspectivas. *O&S., 12* (33), 39-49.

Salleh, M. S., Mahbob, N., & Baharudin, N. S. (2017). Overview of "Generation Z" Behavioural Characteristic and Its Effect Towards Hostel Facility. *International Journal of Real Estate Studies, 11*(2), 59–74.

Sánchez, A., Pérez-Pérez, M., Carnicer, P., & Jiménez, P. (2007). Teleworking and workplace flexibility: A study of impact on firm performance. *Personnel Review, 36*(1), 42–64. doi:10.1108/00483480710716713

Schroth, H. (2019). Are You Ready for Gen Z in the Workplace? *California Management Review, 61*(3), 5–18. doi:10.1177/0008125619841006

Smith, S., & Strawser, M. (2022). Welcome Gen Z to the Workforce. In Atay, A. and Ashlock, M. Z. (eds) Social Media, Technology and New Generations. Lexington Books.

Taborosi, S., Strukan, E., Postin, J., Konjikusic, M., & Nikolic, M. (2020). Organizational Commitment and Trust at Work by Remote Employees. *Journal of Engineering Management and Competitiveness, 10* (1), 48-60.

Timsal, A., & Awais, M. (2016). Flexibility or ethical dilemma: An overview of the work from home policies in modern organizations around the world. *Human Resource Management International Digest*, *14*(7), 12–15. doi:10.1108/HRMID-03-2016-0027

Torten, R., Reaiche, C., & Caraballo, E. (2016). Teleworking in the New Millennium. *The Journal of Developing Areas. Special Issue on Kuala Lumpur Conference Held in N*, *50*(5), 317–326. doi:10.1353/jda.2016.0060

Tremblay, D. G. (2002). Organização e satisfação no contexto do teletrabalho [Organization and satisfaction in the context of telework]. *ERA – Revista de Administração de Empresas [ERA- Business Administration Magazine]*, *42* (3), 54-65. doi:10.1590/S0034-75902002000300006

Turner, A. (2015). Generation Z: Technology and Social Interest. *Journal of Individual Psychology*, *71*(2), 103–113. doi:10.1353/jip.2015.0021

Vega, R., Anderson, A., & Kaplan, S. (2015). A Within-Person Examination of the Effects of Telework. *Journal of Business and Psychology*, *30*(2), 313–323. doi:10.100710869-014-9359-4

Wiesenfeld, B., Raghuram, S., & Garud, R. (2001). Organizational identification among virtual workers: The role of need for affiliation and perceived work-based social support. *Journal of Management*, *27*(2), 213–229. doi:10.1177/014920630102700205

Zivnuska, S., Carlson, J., Carlson, D., Harris, R., & Harris, K. (2019). Social Media Addiction and Social Media Reactions: The Implications for Job Performance. *The Journal of Social Psychology*, *159*(6), 745–760. doi:10.1080/00224545.2019.1578725 PMID:30821647

ADDITIONAL READING

Kerman, K., Korunka, C., & Tement, S. (2022). Work and home boundary violations during the COVID-19 pandemic: The role of segmentation preferences and unfinished tasks. *Applied Psychology*, *71*(3), 784–806. doi:10.1111/apps.12335 PMID:34548734

Mărginean, A. E. (2021). Gen Z Perceptions and Expectations upon Entering the Workforce. *European Review of Applied Sociology*, *14*(22), 20–34. doi:10.1515/eras-2021-0003

Moens, E., Lippens, L., Sterkens, P., Weytjens, J., & Baert, S. (2022). The COVID-19 crisis and telework: A research survey on experiences, expectations and hopes. *The European Journal of Health Economics. HEPAC*, *23*(4), 729–753. doi:10.100710198-021-01392-z PMID:34761337

Mullins, L. B., Scutelnicu, G., & Charbonneau, É. (2022). A Qualitative Study of Pandemic-Induced Telework: Federal Workers Thrive, Working Parents Struggle. *Public Administration Quarterly*, *46*(3), 258–281. doi:10.37808/paq.46.3.4

Pérez-Pérez, M., Sánchez, A., & Carnicer, M. (2002). Benefits and barriers of telework: Perception differences of human resources managers according to company's operations strategy. *Technovation*, *22*(12), 775–783. doi:10.1016/S0166-4972(01)00069-4

Restrepo, B. J., & Zeballos, E. (2022). Work from home and daily time allocations: Evidence from the coronavirus pandemic. *Review of Economics of the Household, 20*(3), 735–758. doi:10.100711150-022-09614-w PMID:35729933

Törőcsik, M., Szűcs, K., & Kehl, D. (2014). How Generations Think: Research on Generation Z. *Acta Universitatis Sapientiae. Communicatio, 1*, 23–45.

Towers, I., Duxbury, L., Higgins, C., & Thomas, J. (2006). Time thieves and space invaders: Technology, work and the organization. *Journal of Organizational Change Management, 19*(5), 593–618. doi:10.1108/09534810610686076

Tweng, J. M. (2010). A Review of the Empirical Evidence on Generational Differences in Work Attitudes. *Journal of Business and Psychology, 25*(2), 201–210. doi:10.100710869-010-9165-6

KEY TERMS AND DEFINITIONS

Generation Z: Young people born between 1995 and 2010, usually associated with technological dependence, avid for new knowledge, multitaskers, individualistic, having a little attention span, less team-oriented than previous generations, seeking flexibility and novelty in the work environment.

Organizational Commitment: The individual's identification with, and involvement in, a particular organization. According to Allen and Meyer's (1990) three-component model, commitment has three different forms, namely affective commitment (the affective attachment of the individual to the organization), continuance commitment (the costs that would take place if the individual left the organization), and normative commitment (the feelings of the individual to remain a member of the organization).

Organizational Performance: The degree to which an organization achieves its objectives, taking into account its resources and means, as well as the dynamic capabilities of the organization in adapting to change.

Productivity: Ratio between the output volume and the volume of inputs or, in other words, effective production of a unit of analysis during a period, in relation with the resources used for that production. There are several different measures of productivity.

Small and Medium-sized Enterprises: Enterprises which employ fewer than 250 persons and which have an annual turnover not exceeding EUR 50 million, and/or an annual balance sheet total not exceeding EUR 43 million. Defined in the EU recommendation 2003/361, they represent 99% of all businesses in the European Union.

Talent Retention: The ability of an organization, through policies and practices, to retain the best employees.

Teleworking: A form of working relationship carried out at a distance, with the use of Information and Communications Technologies, which should provide the teleworker with flexibility in the working organization, allowing an improvement in their work, both in terms of their economic and productive conditions.

Chapter 11
Sustainability Through Innovation:
The Case of Indian Startup Thaely

Mrudula Risbud
Vishwakarma University, Pune, India

Rahul Baburao Waghmare
https://orcid.org/0000-0001-7245-0869
Vishwakarma University, Pune, India

ABSTRACT

The highlighting aspect of the Indian startup ecosystem is the inclusion of startups and SMEs in the various sectors. The Indian startups vary from traditional business sectors to technology-based businesses, and from traditional to social entrepreneurs. A significant number of Indian startups have focused on solving social and environmental problems through creativity and innovation. This chapter discusses one of the social entrepreneurs, Mr. Ashay Bhave, who recently got international recognition for his innovative startup, Thaely. Thaely is involved in manufacturing sneakers from recycling plastic bags with an innovative process. This chapter used secondary data to discuss the case of innovative startup Thaely. The chapter focuses on understanding the problem identification, the innovative process of sneaker manufacturing from plastic bags & bottles, and the social and sustainable impact this startup is making on society.

INTRODUCTION

"Don't just do it, do it right," the Indian startup Thaely, with this slogan, has gained the attention of businessmen, social activists, social entrepreneurs, and the government. This startup manufactures sneakers from waste materials and plastics. This sneaker has received PETA's Best Vegan Sneaker Award 2021. The emergence of such social entrepreneurs is one of the prominent aspects of an Indian Startup ecosystem.

DOI: 10.4018/978-1-6684-6123-5.ch011

Business can be generalized as trade & commerce activity aiming toward higher revenue & profits. This phenomenon is widely accepted & implemented across the world. Every industrial revolution has changed the methods or procedures of manufacturing goods or providing services but the aim of making a profit was perpetual. It is perceived that it stands on the pillar of operations, finance, marketing, research & development sharing & contributing to the common goal of value addition with products resulting in high revenue & profits. Every manager works on maximizing the revenue & resulting in profiting through the efficient application of the management theories.

All these theories got strong in the mid of September 1970 by Friedman who argued that "The business profits can be increased by working for society, by becoming socially responsible. Many researchers delve into the area of corporate social responsibility & try to establish the relationship between making money from the business by spending money for society. In India, CSR practices are mandatory for companies if they comply with the benchmark established regarding revenue & profits as per the Company Act, 2013.

Apart from legal formalities, one's culture has a tremendous influence on the business activity & implementation of CSR policies. In India, the contributions of a few business houses are evident to society. Here we need to accept that legal provision has expanded the scope of CSR but still, it is limited only to the companies which are crossing the respective revenue/profit levels. The new age has highlighted that one cannot protect the environment & better society by building legal boundaries but requires a proactive approach to problem-solving.

Social entrepreneurship is the result of the churning of thoughts about solving societal problems and making the business sustainable & self-sufficient by applying the traditional management method for running the activity. The population composition of India supports the title of the world's youngest country, inspiring new ways and methods of solving the challenges. In this chapter, the author discusses the case of the social entrepreneur Mr. Ashay Bhave's venture "Thaely". "Thaely" means the bag in Hindi & Marathi, the native languages of India.

Research Approach

This chapter has adopted the qualitative research approach wherein case study research design is used based on secondary data sources (Reddy & Agrawal, 2012). The research method consists of how the researcher collects, analyzes, and interprets the data in the study (Creswell, 2009). Secondary analysis is a systematic method with procedural and evaluative steps, yet there is a lack of literature to define a specific process, therefore this chapter proposes a process that begins with the development of the research questions (Identification of the problem in the case), then the identification of the data sources, and thorough evaluation and organization the data to present the case of Thaely for addressing the issues of sustainability through innovation.

ABOUT THE STARTUP, THAELY

The Thaely is founded by the 23-year-old Mr. Ashay Bhave, resident of Maharashtra. Mr. Akshay noticed that every year 100 billion bags are produced & used after the consumption of 12 million barrels of oil leading to the death of 1,00,000 marine animals annually. It has triggered the idea of making sneakers made out of used plastic bags.

Sustainability Through Innovation

"Embarrassed I didn't know about this inspiring startup. These are the kinds of startups we need to cheer on—not just the obvious unicorns. I'm going to buy a pair today. (Can someone tell me the best way to get them?) And when he raises funds-count me in!" - this tweet from Anand Mahindra, Chairperson of Mahindra and Mahindra is evident to understand the success and need of the startups like Thaely.

Ashay worked on creating the plastic bag fabric for two years from 2017 to 2018 with the local shoe repair in Mumbai comparing the durability and workability with the available/working patterns of lasts & soles. Plastic bag fabric was developed by following practice continuous research & development to achieve the quality of the fabric used for traditional sneakers which strongly resembles the leather ThaelyTex (Patent Pending) This project won first place in Amity University Dubai's startup pitch award in 2019. The design team got the inspiration from the basketball sneakers of the early 2000s but maintain the uniqueness in their designs to attract customers.

PROBLEM IDENTIFIED

According to the (OECD Global Plastic Outlook, 2022), The world is producing twice as much plastic waste as two decades ago, with the bulk of it ending up in landfill, incinerated, or leaking into the environment, and only 9% successfully recycled. The report shows that rising populations and incomes are important drivers for this exponential increase in the number of plastics being used and flung into the environment leading to environmental issues.

The (OECD Global Plastic Outlook, 2022) also presents that COVID-19 has led to a 2.2% decrease in plastic use in 2020 as economic activities were slowed down however there was an increase in littering due to the food takeaway packaging and plastic medical equipment. Again, plastic consumption has also rebounded after economic activity resumed in 2021.

Some of the important key findings from the (OECD Global Plastic Outlook, 2022) are,

- Plastic intake has quadrupled during the last 30 years, pushed with the aid of using the boom in rising markets. Global plastics manufacturing doubled from 2000 to 2019 to attain 460 million tonnes.
- Plastic waste generation has doubled from 2000 to 2019. Nearly two-thirds of plastic waste comes from plastics with lifetimes of under five years, with 40% coming from packaging, 12% from consumer goods, and 11% from clothing and textiles.
- Only 9% of plastic waste is recycled (15% is collected for recycling but 40% of that is disposed of as residues).
- In 2019, 6.1 million tonnes (Mt) of plastic waste leaked into aquatic environments, and 1.7 Mt flowed into oceans. An estimated 30 million tons of plastic waste is currently in our oceans, with another 109 million tons accumulating in rivers. The accumulation of plastic in rivers means that the discharge into the ocean will continue for decades, even if we can significantly reduce mishandled plastic waste.
- Considering worldwide value chains and change in plastics, aligning layout procedures and the law of chemical substances can be key to enhancing the circularity of plastics. A worldwide technique to waste control ought to cause all to be had sources of financing, along with improvement

aid, to be mobilized to assist low and middle-earnings nations to meet anticipated charges of EUR 25 billion every 12 months to enhance waste control infrastructure.

The climate crisis is real. Yet, small changes in our lifestyle can make a big difference to the planet. something as casual as a pair of footwear—which is traditionally made from leather, plastic, rubber, or petroleum-based materials—does not degrade under natural conditions?

The Fashion Industry is the 2nd largest polluter in the world. Plastic bottles and bags are some of the main single-use waste components frequently leaking into the environment. From an ocean plastic perspective, they are among the most common beach litter.

100 billion plastic bags use 12 million barrels of oil and kill 100,000 marine animals annually. Plastics and petroleum-derived petrochemicals are one of the causes of unending wars, the makings, and the failings of economies leading to the current climate crisis. "Yet, there's no denying how the fashion industry is the biggest polluting industry, considering the millions of footwears thrown into landfills which take a few decades to decompose." (Mondal, 2022)

Sneakers alone account for 1.4% of global greenhouse gas emissions. Air travel is responsible for 2.5%. All this is swelling landfills adding to 300 million tons of plastic waste produced every year.

Plastic waste has numerous implications on the environment and health. The plastic waste dumped in landfills leaches into the ground and nearby water systems causing land and water pollution and ultimately reaching the food chain. The uncontrolled burning of waste, including plastic, causes air pollution. In addition, the clogged plastic waste in sewerage systems further pollutes rivers and groundwater. The plastic in food and water can cause severe health issues such as genetic disorders, and endocrine system damage. According to the United States Environmental Protection Agency, all the plastic waste ever generated is still present on Earth today, this makes sustainable management of plastic waste important.

Upcycling: In Search for Solutions to the Problem

To address the reduction of the carbon footprint from plastic waste, governments and environmentalists have encouraged the 3Rs of waste in the last few decades.

As per the Missouri Department of Natural Resources, "The three R's – reduce, reuse and recycle – all help to cut down on the amount of waste we throw away. They conserve natural resources, landfill space, and energy. Plus, the three R's save land and money that communities must use to dispose of waste in landfills. Siting a new landfill has become difficult and more expensive due to environmental regulations and public opposition."

Reducing your consumption is the first step and one of the most important. If we could all reduce the amount of waste that we create, then we can reduce the overall size of our landfills and ocean pollution. (Anchor Disposal, 2020)

The recycling of plastic was on the agenda of government institutes, NGOs, and social enterprises to protect the environment. Plastic recycling is the process of converting plastic waste into new products. If efficiently done plastic recycling is useful to conserve natural resources, reduce dependency on landfill, and in protecting the environment from greenhouse gas emissions. With more concerns about environmental protection recycling rates are increasing. The global recycling rate in 2015 was 9%, while 12% was incinerated and the remaining 79% disposed of in a landfill or the environment including the sea. From the beginning of plastic production in the 20th century, until 2015, the world has produced

Figure 1.
Source: (OECD Global Plastic Outlook, 2022)

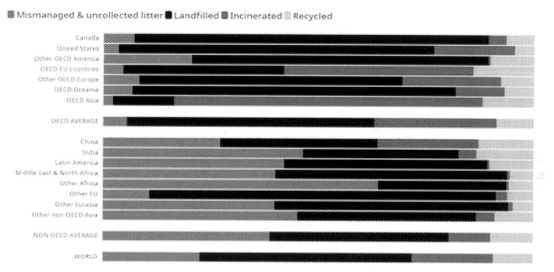

some 6.3 billion tons of plastic waste, only 9% of which has been recycled, and only 1% has been recycle more than once.

Almost all kinds of plastic are non-biodegradable and harmful to the environment hence the need for recycling is need of an hour. For example, approximately 8 million tons of waste plastic enter the Earth's oceans every year, causing damage to the aquatic ecosystem and forming large ocean garbage patches.

The Reuse from the 3Rs which was an unattended aspect of waste management has emerged as an alternative for a sustainable solution through the concept of upcycling plastic wastes. Reusing the products that we have obtained instead of throwing them directly into the garbage lightens the waste that we have. The concept of upcycling involves the process of converting wastes or objects to new utilities which can be used continuously. An example to understand upcycling is of conversion of scrapped tires into outdoor seating or toys made from old socks.

The process of upcycling differs from the process of recycling in many ways. The Cambridge English Dictionary defines upcycling as: "the activity of making new items out of old or used things". (https://www.cam.ac.uk/, 2019)

The difference between recycling and upcycling lies in the process of the conversion of inputs to outputs. The recycling process involves destroying the waste to produce new products whereas the process of upcycling involves using the waste to create something new from its current form of utility.

As upcycling converts waste into something new which increases the durability of the product hence need for recycling can be reduced and hence a good option for protecting the environment. Once a material can no longer serve any purpose, then it is more eco-friendlier to recycle it than it is to send it to a landfill.

According to (Sung, 2015), the process of upcycling provides better value to the used materials in the second life. It has been increasingly recognized as one promising means to reduce material and energy use and engender sustainable production and consumption.

Many large firms have started upcycling plastic waste to offer new products to the customer which are more sustainable especially in the fashion industry However, as shoppers become more educated about the environmental concerns associated with producing their clothes and shoes, brands become more incentivized to offer sustainable alternatives.

Upcycling for Sustainable Fashion in Modern Times

"Sustainable fashion in modern times is considered by looking at every stage in the supply chain: design, material procurement, processing and production, transportation, distribution, end of life, and understanding of companies' initiatives." (Bansal, 2022) How financial material factors are assessed based on the impact on nature - how polluting and intense the water usage and carbon emissions are during processing; labor - balancing the positive, such as poverty alleviation, while ensuring proper waste management at the pre/post-consumer stage, are all important.

Here are four factors that need to be addressed in the fashion industry to move towards sustainability: -

- using more sustainable materials,
- minimizing the carbon footprint of the industry and improving resource efficiency,
- reducing waste,
- and perhaps the least talked about, making the industry transparent and easy to understand for the consumer. (Agarwal, 2022).

VEGAN FOOTWEAR: THAELY'S SOLUTION TO THE PROBLEM OF PLASTIC WASTE

The Founder Ashay Bhave in his experimental design came up with the idea to use fashion as a force for good with a positive impact on the environment by upcycling used plastic bottles and bags. He with his teammates developed an innovative method of upcycling waste plastic bags that creates a strong and flexible fabric that resembles leather, named ThaelyTex fabric. This fabric is similar to leather in terms of its looks and feel.

Each pair uses a total of 15 plastic bags and 22 bottles. The manufacturing process of these sneakers doesn't use any additional chemicals and doesn't release any kind of toxic or harmful chemicals.

The outer layer of sneakers is manufactured from this upcycled plastic material i.e., ThaelyTex fabric made from plastic bags and the inside is manufactured by Eco polyester created from the recycling of plastic bottles.

The soles are made from ethically sourced natural rubber that causes no harm to rubber trees and forests and finally, the glue used would be Vegan, free of any animal products.

All components including the glue are 100% Vegan. Each pair is packaged in a reusable shopping tote made from 4 plastic bottles & a box made from recycled paper which is embedded with basil seeds & dyed with waste coffee grounds.

Figure 2. Thealy Sneaker Source: (www.thaely.com, 2021)

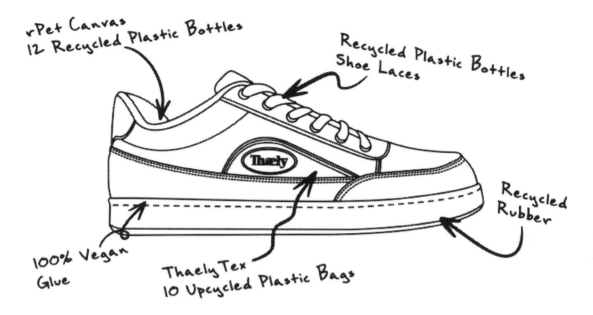

Understanding the Innovative Process of Manufacturing

Sustainable fashion in modern times is considered by looking at every stage in the supply chain: design, material procurement, processing and production, transportation, distribution, end of life, and understanding of companies' initiatives (Agarwal, 2022).

The first step in the process of manufacturing Thealy's footwear is the procurement of raw materials. The founder Ashay Bhave tied up with Triotap technologies, a waste management unit in Gurugam, Haryana to collect the wastes and segregates plastic bags from them.

They are then washed, sanitized, and hung to dry. The bags are cut into sheets and fused with heat and pressure to make ThaelyTex, fabric.

ThaelyTex sheets are shipped to the shoe manufacturer in Jalandhar and are dye cut according to the pattern.

Thaely also manufactures Recycled Polyethylene Terephthalate (rPET), a fabric from recycling waste PET Plastic Bottles. rPET is the same as virgin polyester in terms of quality, but its production requires 59 percent less energy. rPET is used in the toe box, lining, laces, and a tote bag in which sneakers are packed.

Furthermore, recycled rubber is used for the sole. The shoe soles for v Thaely Y2K Pro sneakers are made from the recycling of scrapped materials such as used shoe soles, tires, and other industrial waste. Recycling rubber means that millions of scrap shoe soles and tires are no longer dumped in landfills or open grounds.

All the pieces are stitched and glued together using a 100 percent vegan glue solution. A pair of sneakers is boxed in a plantable shoe box made by Agra-based firm Plantable, which is further packed in a tote bag made from recycled plastic bottles. This packaging material grows into a plant, if sowed in the ground. Making the entire packaging a zero product.

Figure 3. The process flow of Thaely Y2k Pro sneakers: From Waste to Vegan Footwear in the customer's hand

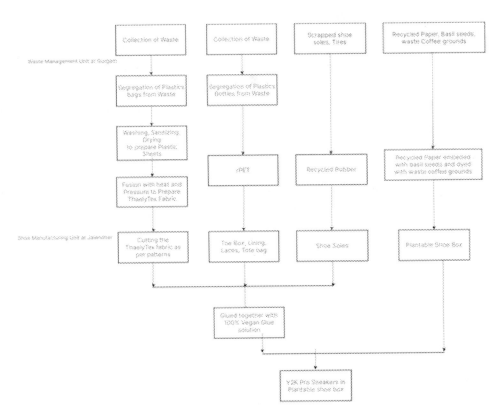

Each pair of sneakers utilize 10 plastic bags and 12 plastic bottles and is priced at $99 (Rs. 7,000).

Figure 3 depicts the process of Thaely Y2K pro sneakers manufacturing process and involved sustainable and recyclable materials.

Recycled Polyethylene Terephthalate, Recycled Rubber, and Recycled Paper (used for the packaging) create the most sustainable sneaker ever made.

ThaelyTex: Used for the Upper

The fabric used for the outer layer of these vegan sneakers is innovative and revolutionary in the fashion world. Thaely has filed an application for patenting it. The fabric can be used as an alternative to the leather as it looks and feels like it. This fabric is made entirely of plastic bags segregated from the waste collected. An important aspect of manufacturing of ThaelyTex fabric is that it doesn't produce any toxic or harmful chemicals or doesn't consume any chemicals in the manufacturing process.

The production process is designed to be safe, cost-effective, and have the lowest emissions possible. The Founder Ashay Bhave collaborated with the waste management unit run by TrioTap Technologies in Gurugram, India to collect the waste and segregate plastic bags from it. The waste plastic bags are collected from housing complexes, offices, and stores in and around Gurugram. "The collected plastic bags are then segregated, sanitized, and processed into ThaelyTex at TrioTap Technologies' Gurugram

The Thaely Y2K Pro box is made from recycled paper which is embedded with basil seeds and dyed with waste coffee grounds. Customers can simply cut it and plant it. Customers will have a basil plant in 10 days - ready for pesto. The box is sourced from Plantables in Agra, India.

waste management unit itself. (www.thaely.com, 2021)" This allows reducing carbon emissions. "All waste management professionals at TrioTap Technologies are paid fair wages, have access to safe working conditions, and are also provided with housing facilities (www.thaely.com, 2021)"

rPET: Used for the Toe Box, Lining, Laces, and Tote

rPET or Recycled Polyethylene Terephthalate is a fabric made from recycled waste PET Plastic Bottles. rPET is the same as virgin polyester in terms of quality, but its production requires 59 percent less energy compared to virgin polyester. Our rPET Canvas is sourced from Harper Group in Delhi, India. Our rPET Shoelaces are sourced from Dongguan ("Materials – Thaely Sustainable Sneakers")

Mingjia Weaving String Co. Ltd. in Guangdong, China. Unlike most other fashion brands, the rPET we use is 100% recycled and not blended with any other virgin materials.

Each sneaker uses rPET that's made using 12 Recycled waste plastic bottles. Each pair is also packaged in a reusable rPET shopping tote made from 4 plastic bottles.

Recycled Rubber: Used for the Sole

Recycled rubber is rubber that has been reclaimed from scrap materials such as used shoe soles, tires, and other industrial waste. Recycling rubber means that millions of scrap shoe soles and tires are no longer dumped in landfills, or left illegally in lakes, abandoned lots, along the side of the road, and in sensitive habitats. Recycling rubber saves impressive amounts of energy, which ultimately reduces greenhouse gas emissions. For example, recycling four tires reduces CO2 by about 323 pounds, which is equivalent to 18 gallons of gasoline. ("Materials – Thaely Sustainable Sneakers") Using recycled rubber in molded products like our sole, for example, creates a substantially smaller (by a factor of up to 20 times) carbon footprint as compared to using virgin plastic resins. These recycled rubber soles are sourced from Sambo Rubber Co. Ltd in Taichung City, Taiwan. The soles use about 40% Recycled rubber in their construction.

Printable Recycled Paper: Used for the Box

Social and Sustainable Impact of the Startup Thaely

The 2030 Agenda recognizes the role of the private sector and calls for its innovative and creative capacity to solve sustainable development challenges. However, it is unclear to what extent companies contribute to solving SDG issues. (Johannes, Wall, Thijssens, & Maas, 2021)

There is a stream of literature on sustainability-oriented innovation (SOI), which incorporates both the environmental and social aspects of sustainability covered by the SDGs (Adams, Jeanrenaud, Denyer, & Overy, 2016)

(Adams et al., 2016). of those two, the environmental aspect is the most directly actionable in business contexts, driven by market demand for cleaner products, increasing environmental regulations, and price savings, and thus dominant in a very broad stream of literature on industrial ecology and company eco-innovation. Responsible innovation adds to the present governance and responsibility aspect. there's no consensus about what sustainability exactly means, how it is achieved and what types of innovation will be called sustainable. Different terms and definitions are used when relating innovation to sustainability, with sustainable innovation and SOI being found most often. Both terms can discuss "innovations within which the renewal or improvement of products, services, technological or organizational processes delivers not only improved economic performance but also an enhanced environmental and social performance, both within the short and future and have the capacity to come up with positive social and environmental impacts.

Nowadays, talking about the United Nation's SDGs is in trend, but very few innovative entrepreneurs implement the same while running their businesses. If we visit the website of the organization under the tab of impact we can find that "Thaely" measure & report its impact based on the 5 SDGs

Goal 8 – Decent work & economic growth

Goal 8 narrates the close relationship between improving productivity & decent work for all. Overall economic growth can be achieved by improving the productivity of the organization as well as an individual by enhancing the work culture & environment.

The production process at "Thaely" mainly comprises manual work, therefore the sustainability of the worker on the assembly line reflects on productivity. The organization has adopted the fair wages system for maintaining a good lifestyle for workers including the appropriate working conditions.

Goal 9 – Industry, innovation & infrastructure

Goal 12 – Responsible consumption & production

Goal 13 – Climate Action - by inventorizing the greenhouse emission

Phenomenal study of 1972, "LTG - Limits to growth" gauge to understand the impact of the exponential economic & population growth on the finite resource available on the earth & concludes with the horrifying fact that either civilization or growth must end. This compels us to think about the cycle of the growth rate similarly we conduct for the product life cycle.

Goals 9,12,13 promotes the idea of expanding the growth cycle by becoming the responsible manufacturer and consumer towards environmental changes through innovative & forward-looking strategies. Here one needs to understand that path-breaking innovation can be implemented if we stop following the methods we were following in the past. As it said, " If you keep doing the same thing you used to do in the past, don't expect different results."

Thaley is following the principle of protecting the environment by following the path-breaking way of utilizing plastic waste as its core raw material for manufacturing. On the other hand, now it's appealing to the customer as well to get the opportunity for protecting the environment during the course of satisfying one's necessities.

Goal 15 – Life below water – By restricting the entry of the plastic bag into the seawater

Many studies in oceanography have stated the ill effects of dumping plastic waste on marine plants & animals. This has triggered many intellectuals to set a new trend in the business ecosystem which will not address the problem but help economic development as well. One can note the names like a start-up "river cleaning" where they claim that 80% of the plastic waste comes from the river.

On the same ground, "Thaely" has managed to recycle or reuse around 48,000 + plastic bottles & 40,000 + plastic bags for the production of sneakers as per the information available on their impact factor dashboard.

It's observed that "Thaely" may face strong competition from the big companies in this niche market. Adidas, the established brand in the shoe & apparel market, has announced a partnership with Parley for collecting plastic waste & converting it into high-performance apparel & footwear. Adidas aims to manufacture its shoes & apparel with 100% recycled polyester.

Acceptance of the Stakeholders for the Innovative Product

Different kinds of organizations are taking efforts to reduce the negative impact of plastic waste through reduction, reuse, and recycling. Many of them have created a significant social impact through their products and processes. One of the ways to understand the social impact of such efforts is the acceptance of stakeholders such as customers, government, community, employees, and non-governmental organizations working towards sustainable development. The Thaely has received overwhelming support and appreciation from various stakeholders. A few of them are mentioned below:

Amirlan Kurakov from the United Arab Emirates and a sneaker lover said, as a first model, it's a very impressive shoe and added, "Thaely is soft and light but a bit too flat." (Ahuja, 2021) It's comfortable for daily wear, but you can't play sports in them. I would recommend it to people who have a casual style.

One of the largest schools named The Swiss International Scientific School at Al Jaddaf became the first school to include the vegan sneakers manufactured by Thaely in their official school uniform. Ruth Burke, the school's headteacher and chief executive, said:

"It's fantastic to find a cool, local brand — created by a young student — which is sustainable and which our pupils like. We encourage our pupils to think out of the box and Thaely shows what's possible when you put your mind to something. We want our pupils to feel comfortable, and to look smart, but also, hope that every time they look at their feet, they are reminded of our collective, global responsibility to act sustainably, and the power of pursuing your dreams." (Maisey, 2022)

Thaely also received the award of Best Vegan Sneakers from PETA India in the month of Nov 2022. (PETA India, 2021)

The Upcoming Wave of Innovations in a Sustainable Fashion

The evolution of digital traceability and how it can increase transparency in the supply chain has become a reliable option. The firms can use contemporary technology to track down each movement in the process of value creation and value delivery which includes the movement of raw materials from suppliers and finished goods to the customers which enables firms to identify the potential risks and strategically target to mitigate the same (Bansal, 2022). Some firms are also working on unique digital passports that are tied to a specific garment. This might be a QR code or RFID that presents to both consumers and companies where something was manufactured, what the materials are, how many times it's been owned by someone and end-of-life information such as the most effective recycling practice. We're in the first innings; this digital infrastructure will take an incredible amount of investment and time. Also, the digital printing tech emergence has enabled designers to take their ideas from concept to garment in a much faster time frame and open new avenues of creativity.

CONCLUSION

In the global competitive business environment, more fashion brands and firms are taking initiatives for offering sustainable products through transformations in the manufacturing processes. The use of waste materials especially plastic bags, plastic bottles, used rubbers, tires, etc. is increased to cut done their carbon footprints. Firms are making a sufficient investment in R&D for new-age technology which will enable them to reduce the use of plastics and water consumption. The chapter helps the students, researchers, and practitioners to understand the reuse of plastic waste and encourage entrepreneurs to think of more sustainable ways to manufacture their product with the efficient use of natural resources and reduce the overall environmental impact.

REFERENCES

Adams, R., Jeanrenaud, S., Denyer, D., & Overy, P. (2016). Sustainability oriented innovation: A systematic review. *International Journal of Management Reviews, 18*(2), 180–205. doi:10.1111/ijmr.12068

Agarwal, G. (2022, October 26). Moving closer to sustainability. *Precious Kashmir.* https://preciouskashmir.com/2022/10/26/moving-closer-to-sustainability/

Ahuja, A. (2021, Nov 21). Waste Warriors of India. *Swach India.* www.swachindia.ndtv.com: https://swachhindia.ndtv.com/climate-warrior-23-year-old-recycles-10-plastic-bags-and-12-plastic-bottles-into-a-pair-of-sneakers-64173/

Anchor Disposal. (2020, April). Anchor Disposal. www.anchordisposal.com: https://anchordisposal.com/news/2020/4/1/whats-the-difference-between-upcycling-and-recycling

Bansal, R. (2022, September 5). *Indianretailer.com.* www.indianretailer.com: https://www.indianretailer.com/article/retail-people/the-global-eco-wakening-why-sustainability-is-now-the-key-driver-of-innovation.a8153/#:~:text=The%20Global%20Eco%2Dwakening%3A%20Why,the%20Key%20Driver%20of%20Innovation

Johannes, W. (2021, February 20). The innovative contribution of multinational enterprises to sustainable development goals. *Journal of Cleaner Production, 285,* 1–13. doi:10.1016/j.jclepro.2020.125319

Maisey, S. (2022, July 6). Lifestyle. *The National News.* www.thenationalnews.com: https://www.thenationalnews.com/lifestyle/fashion/2022/07/06/dubai-school-becomes-first-in-the-world-to-add-thaely-vegan-shoes-to-its-uniform/

Mondal, A. (2022). *Feet & fine: How vegan leather, biodegradable shoes are trying to save Earth*. MSN. www.msn.com: https://www.msn.com/en-in/news/other/feet-fine-how-vegan-leather-biodegradable-shoes-are-trying-to-save-earth/ar-AASOTG0

PETA India. (2021, November 8). PETA. www.petaindia.com: https://www.petaindia.com/blog/milind-soman-alia-bhatts-ed-a-mamma-and-sunny-leones-i-am-animal-among-winners-of-peta-indias-vegan-fashion-awards-2021/

ReddyK.AgrawalR. (2012). Designing Case Studies from Secondary Sources – A Conceptual Framework. doi:10.2139/ssrn.2167776

OECD. (2022) *Global Plastic Outlook*. OECD Library.

Sung, K. (2015). A Review on Upcycling: Current Body of Literature, knowledge gaps and a way forward. Venice Italy, 28-40.

APPENDIX

Responses to the revision:

Table 1. Evaluation 1

Queries	Author Response
It is not clear what methodological approach was adopted by the author. Aspects such as approach and data used can be useful for a better understanding of the readers	Research approach is added (P. 2)
One of the weaknesses of the present chapter is the presence of a few deviations in the writing	Reviewed and text is corrected (P. 2 & 3)

Table 2. Evaluation 2

Queries	Author Response
The document structure does not characterize scientific research	Research approach is added (P. 2)
The methodological procedures are fragile and superficial and tend to the case analyzed. The work does not bring together an adequate theoretical foundation oriented to the problem in question.	Research approach is added (P. 2)
The theoretical and practical gap was not clear to the reader. The authors made an effort to present the result, the case of success, but the minimum scientific structure required for academic work is superficial.	Research approach is added (P. 2) The case is considered to understand how Thaely is using innovations and sustainable business processes.

Chapter 12
Corporate Governance and Ethics for Sustainability:
The Case of the Company Mercur S.A. in Brazil

Leon Maximiliano Rodrigues
https://orcid.org/0000-0002-7132-3187
Universidade Estadual do Rio Grande do Sul, Brazil

Elis Shaida Raichande
Instituto Politécnico de Setúbal, Portugal

Mónica Filipa Nunes Carvalho Gomes
https://orcid.org/0000-0002-8385-5615
Instituto Superior de Educação e Ciências, Portugal

Mirian Benair Semedo
Universidade de Santiago, Cape Verde

ABSTRACT

This study analyzed, in the form of a case study, the experience of a large company (Mercur S.A) in the implementation of a new vision focused on sustainability. The target company is located in the south of Brazil, has almost a century of existence, and has been managed by the same family since its foundation. The study is based on the notions and concepts of corporate governance, ethics for sustainability, and social-ecological systems. The study shows that the changes implemented in the company are catalyzed by an initial change of the 'inner change' type, and are disseminated by the company and the community through social and cultural innovations. Important changes in the economic vision and social and environmental responsibility related to the structure/architecture and operation of the company, as well as trade-offs between financial and social and environmental aspects, were identified.

DOI: 10.4018/978-1-6684-6123-5.ch012

INTRODUCTION

The growing awareness about the impact of human activities on the environment and human development, especially since the 80's, has resulted in a growing social debate on the concept of sustainable development. One of the main triggers for this global trend is the Brundtland Report (WCED, 1987), which defines sustainability as "meeting the needs of the present without compromising the ability of future generations to meet their needs" (p. 8). The report also states that "sustainable development requires meeting the basic needs of all and extending to all the opportunity to satisfy their aspirations for a better life" (p. 44).

However, sustainable development is a non-consensual concept, upon which divergent views and interpretations may lead to different standpoints, goals, and strategies, especially regarding different political ideologies (Davidson, 2014; Hopwood et al., 2005). The concept implies the possibility of economic growth aiming at justice and social equity, as well as respect for the limits of nature, while, on the other hand, since the Industrial Revolution, economic growth has been considered associated with social and environmental problems (Aras & Crowther, 2008; Lele, 2013; Rist, 2007, 2008).

In fact, the Brundtland Report already warned that, in order to meet human needs, humanity has been fueling an unsustainable lifestyle which not only results in inequality within our generations but also compromises the future generations as we "borrow environmental capital from future generations with no intention or prospect of repaying" (WCED, 1987, p. 8). A study modeled human population dynamics by adding accumulated wealth and economic inequality to a predator-prey model of the human-nature relationship, identifying two possible paths to collapse: (1) due to scarcity of labor (following an inequality-induced famine; Type-L Collapse) or (2) due to scarcity of Nature (depletion of natural resources; Type-N Collapse) (Motesharrei et al., 2014).

The path to (un)sustainability finds support in the prevailing worldview in society. "The increasingly global capacity of civilization to manipulate natural and human capital has fueled faith in the economic concept that humankind can be freed from its dependence on nature" (Butler, 2000, p. 156). The fragmentation of knowledge — and the economy — in modern society reinforces this faith. Such fragmentation has led to the isolated development of disciplines and often to contradictions between them. And no issue more clearly separates the traditional economic view from the view of most natural scientists than the issue of sustainability (Ayres et al., 2001).

While economic growth still has a major place in the political agenda for human development, there is much evidence that alone, economic growth does not promote human development. In fact, although there is improvement on human development in the past few decades (WCED, 1987), the environmental impacts like climate change or biodiversity loss, as well as global poverty and inequality, especially between rich and poor countries, are growing (Lele, 2013; Maddison, 2008; Rist, 2008; WCED, 1987).

Therefore, integrating with nature means perceiving humans as part of the natural systems on which humanity depends, and perceiving the opportunities to prosper in a sustainable way, both economically, socially and environmentally - 'a better life'.

Considering this perspective, this study sought to investigate the changes implemented by a large company (to try) to achieve sustainability. The changes implemented over a period of more than a decade were evaluated from the perspectives of corporate governance (CG), ethics for sustainability and social-ecological systems.

The following topic presents the theoretical foundation. Next, the methodological approach is described. In the following topic, the case study is described and analyzed. Recommendations, considerations and future directions for studies and applications are suggested. Finally, the main conclusions and final considerations of the study are presented.

BACKGROUND

Corporate Governance and Sustainability

As introduced above, to meet human needs humanity has been fueling an unsustainable lifestyle which not only results in inequality within our generations but also compromises the right of future generations to access resources (tangible and intangible) to sustain their quality of life and aspirations. Often, sustainability has been taken from a 'myopic view', which tends to regard sustainability as efficiency on the resource usage, or that it simply implies business continuity, which can be ensured through technical improvements.

Whereas in the past, cost savings and reductions in waste and energy were the main motives to engage in sustainability activities, executives have more recently recognised the growth potential of sustainability with respect to reputational advantages and new value creation. However, only a minority of managers reported having clear sustainability goals and strategies in place. They still struggle with committing to the creation and application of new ideas for products, services, or processes that better serve customer needs and contribute to reduced environmental stress and improved social situations. (Globocnik et al., 2020, p. 2)

On the one hand, companies can't be profitable in a damaged environment with scarce resources. Therefore, environmental issues become a main worry as its problems impact business performance as well as prosperity (Aras & Crowther, 2008; Carroll, 2021). On the other hand, value creation is not only economic, and it is distributed among society's stakeholders, increasing social well-being, improving companies' legitimacy (Drempetic et al., 2020; Scherer & Voegtlin, 2020; Zaman et al., 2022), as well as employee's job meaningfulness, affective organizational commitment, and motivation, which will inevitably lead to increased productivity in the company (Aras & Crowther, 2008; Farooq et al., 2014; Glavas & Kelley, 2014).

For Turnbull (1997), CG is applied to all social institutions engaged in the production and sale of goods and services, and describes all the influences affecting the institutional processes. The author analyzed four different models of corporate control, which correspond to different dimensions of CG and bring some insights:

- *Firms are not alone in society,* and external stakeholders like community and government provide the necessary legal and market infrastructure for the firm's activities. Thus, firms must align their own interests with the interests of these critical stakeholders, engaging them in the firm's governance, and creating value to be distributed.
- The way it is done is *the allocation of corporate power, privileges and profits* between owners, managers and other stakeholders influences how governments favour their various constituencies.

- *Cultural values* such as trust and confidence and the type of relationships among internal stakeholders matters, and it can influence value consensus or resource dependency, affecting the performance of the company.
- Shareholders must provide an *appropriate separation of powers among management teams*, as knowledge and will to act is not sufficient to allow them to increase stakeholders' welfare.

Therefore, corporate actions and attitudes have impacts not only on its internal and external audiences, but on all its stakeholders, which makes it have a significant responsibility in the environment and, consequently, it should opt for assertive and careful approaches, as any hasty and inappropriate decision could jeopardize your entire reputation. From this perception emerges the notion of corporate social responsibility (CSR), which mediates profitability (Phillips et al., 2015).

CSR concept debate has grown since the second half of the 20th century, resulting in a significant proliferation of theories, approaches, and terminologies, sometimes with different meanings (Carroll, 2016, 2021; Garriga & Melé, 2004; Kolk, 2016), with the CSR concept changing over time. Initially the focus was on the economic aspect and shareholders' interests. Currently, in order for businesses to remain competitive and to survive within the spheres in which they operate, managers must implement management systems that reflect the realities of economic, social, and environmental issues, concerns, and conditions (Ávila et al., 2013).

One of the main theories on CSR is Carrolls' Pyramid of CSR, which considers that companies are not alone in their environment and, in order to survive they must address economic and legal obligations, while meeting ethical and philanthropic expectations from their stakeholders. In this scope, CSR dimensions relate to wider stakeholders' interests, while its nature is not static but dynamic, including intrinsic trade-offs between its different dimensions, as well as between short and long-term goals of the company (Carroll, 2016, 2021). Zaman et al. (2022) define CSR as a broad term that encompasses internal and external mechanisms resulting from policies, processes, and practices that companies implement to improve the social status and well-being of their stakeholders and society, whenever it's voluntarily conducted or required by rules, standards or customs. The authors enhance that CSR has internal mechanisms (company ethical codes, employee health and safety, work-life balance, training, human rights protection, equal opportunities, and diversity practices) and external mechanisms (partnerships with charitable organizations, philanthropy, environmental and community-focused practices).

In general, CSR seems to be a flexible concept, which is applied in different ways, according to geographical, cultural, governance systems, or other context factors (Carroll, 2016, 2021; Zaman et al., 2022). In most cases, CSR is a function of CG, which means that different configurations of CG systems, structures and processes impact CSR policies and practices (Scherer & Voegtlin, 2020; Zaman et al., 2022). Although, it should not yet be seen as an imposition, but rather as an obligation that companies assume and that, in the medium and long term, helps them to achieve beneficial goals for the company and the surrounding community (Martins, 2020). In that regard, according to Garriga & Melé (2004), CSR theories are focused on four main aspects: (1) achieving long-term profits, (2) using business power in a responsible way, (3) integrating social demands, and (4) simply doing what is ethically correct, while the great challenge is to overcome their overlapping limitations.

Even though there seems not to be a consensus on the nature of the relationship between sustainable CG and CSR (Zaman et al., 2022), they are concepts that reinforce each other (Aras & Crowther, 2008; Naciti et al., 2022), with a symbiotic relationship (Porter & Kramer, 2006; Zaman et al., 2022). At a global scale, there are many political frameworks regarding CG, as well as its relationship with sustainability concerns (e.g., EC & DG JUST, 2022; OECD, 2015; Ruggie, 2011). Business purposes and how companies should be run are principal issues of CG. More recently, trends on human development and environmental impact of human activities, in the past decades, made sustainability a main issue in business agenda, giving space for the concept of triple bottom line (TBL). TBL expresses the economic, social and environmental dimensions on which companies and other organizations create or destroy value (Elkington, 2006). Hence, the sustainability term, behind this tripod, has become ubiquitous both in the discourse of globalization and in the discourse of corporate performance (Aras & Crowther, 2008; Carroll, 2021) where costs and value created must be considered in the present and for the future of the business, as well as for society as a whole.

There are many studies addressing CG models, configurations, and mechanisms that contribute to sustainability goals attainment, and effective internal governance mechanisms help firms to meet sustainability goals (Herrera, 2015; Hussain et al., 2018; Scherer & Voegtlin, 2020). Some authors suggest gender diversity as relevant governance mechanisms (Hussain et al., 2018; Rosati & Faria, 2019; Zaman et al., 2022), as well as age (Rosati & Faria, 2019; Zaman et al., 2022) and race diversity of the council (Zaman et al., 2022), the existence of a Corporate Social Council committee, the board's independence (Hussain et al., 2018; Zaman et al., 2022), as well as the board's experience and networking capacity, greater expertise of the audit committee, and executive remuneration on CSR performance (Zaman et al., 2022).

Nonetheless, to achieve a sustainable path one must agree that it requires collective actions, and corporate stakeholders have much to learn and evolve in order to meet this ambition, considering their agency capability. In this sense, although it is recognised that a sustainable path will not be easy, and that fast and trade-offs must be made, academics and politicians address private enterprises, from small to multinational enterprises, as key-stakeholders to make changes and improvements through their governance frameworks, to foster innovation on their value chain and to create shared value — not only economic — for society (Ruggie, 2011; WCED, 1987).

Corporate Social Innovation

Social innovation can be defined as "a novel solution to a social problem that is more effective, efficient, or just than existing solutions and for which the value created accrues primarily to society as a whole rather than private individuals" (Phills et al., 2008, p. 36). The aim is to address unmet human and social needs, differing from usual business innovations which are driven by market rules (Lettice & Parekh, 2010; Phillips et al., 2015; Scherer & Voegtlin, 2020).

In the corporate arena, corporate social innovation (CSI) differs from CSR because the former implies a proactive approach to tackling social and environmental problems through the corporate value chain, shifting from a perspective of "giving" and "repairing" to a strategic alignment within the corporation goals (Popoli, 2016; Porter & Kramer, 2006). Porter & Kramer (2006) highlight that CSR can be a source of opportunity and innovation, if companies use the value chain to identify negative social impacts to clear away, as well as opportunities that may be addressed for social and strategic distinction, creating competitive advantage. Therefore, CSR paves the way for CSI.

Popoli (2016) narrows down some main differences between the two concepts (see Table 1).

Moreover, CSI dynamics lie in partnership and collaboration, which enable a better understanding of stakeholders' needs and expectations, improve knowledge transfer mechanisms and innovation processes (Mirvis et al., 2016; Popoli, 2016), and promote trust and loyalty from those stakeholders (Iglesias et

Table 1. Main differences between CSR and CSI

Corporate Social Responsibility (CSR)	Corporate Social Innovation (CSI)
Firms takes responsibility for their impact on society, even if beneficiaries become players in some social innovation processes	Firms and society pursue cross-fertilization between commercial and social efforts, in a two-sided co-creation of value from emergent opportunities
Society benefits from the actions carried out by firms	Firms and society take collaborative actions to benefit both sides
Beneficiaries are not tasked with carrying out anything or something that may be self-help	Beneficiaries are called upon to take a role and to be an active participant in achieving social innovation
Society benefits are mainly indirect and linked to improving corporate image and reputation	Benefits are direct, where firms exploit opportunities connected with social needs in order to get economic returns

Source: Adapted from Popoli (2016)

al., 2020). This is particularly relevant because CSI implies changes in companies' business models, which leads to higher complexity, especially concerning impact assessment and its effects on the whole business network (Evans et al., 2017).

A different concept that seems to stand in between these two concepts — perhaps integrating the two concepts — is *responsible innovation* (RI) which has been defined as "a transparent, interactive process by which social actors and innovators become mutually responsive to each other with a view on the (ethical) acceptability, sustainability and societal desirability of the innovation process and its marketable products" (Scherer & Voegtlin, 2020, p. 6; von Schomberg, 2012, p. 50).

RI is a broader concept for which CG has a main role, facilitating innovations that avoid harm and do good, thus contributing to sustainable development (Scherer & Voegtlin, 2020). Under this concept, firms take sustainability concerns into the innovation processes, even when they are for-profit organizations or do not pursue social goals. Regarding innovation for sustainability, whether it falls on RI or CSI, CG structures and configurations influence the way these innovations evolve and result in effective outcomes.

Despite its own challenges, Scherer, & Voegtlin (2020) highlights the importance of a reflexive governance in order to better address sustainable issues on corporate innovations processes, since it allows firms to reflect and reinvent themselves based on deliberation processes among stakeholders. The authors also point out some suggestions on CG configurations which may provide firms with the capacities for reflexivity and RI:

1. *Changing the ownership structure* involving investors with social interests, leading to increased reflexivity among shareholders and management about innovation contribution to sustainable development;
2. *Promoting sustainability reporting in the firm*, addressing the TBL performance, which creates transparency, inviting stakeholder dialogue on innovation goals;
3. *Including sustainability purposes on the legal statutes of firms* as means of legitimation for responsible innovations and social innovations, since it allows managers to allocate resources for it, without compromising their fiduciary duty to shareholders;
4. *Fostering stakeholder participation* in various ways (*e.g.*, stakeholder committees, citizen panels, focus groups), balancing the involvement of expertise in generating ideas and broad societal support to facilitate idea implementation;

5. *Creating modes of decision making* based on consensus and agreement among relevant stakeholders, since it leads to positive effects on defining goals and facilitating acceptance;
6. *Allocating resources for sustainable purposes*, even beyond firms' boundaries, which allow different forms of return on investment, including intangible capital.

Herrera (2015) proposes another perspective, arguing that social innovation is most powerful when firms institutionalize it within their systems and structures, and points out some critical factors in this process, which combines strategic and operational processes:

1. *Strategic alignment* – In order to warrant that social innovation assessment is strategic and will guide companies in the attainment of their goals, companies should consider social and sustainability elements at the strategy assessment, including evaluating the interests and influences of stakeholders, as well as characterize social concerns and stakeholder collaboration opportunities; evaluating companies' footprint, which involves evaluating the extended value chain, and evaluating companies' core values, philosophies, resources, and competencies;
2. *Institutional enablers of social innovation* – Companies must promote active stakeholder engagement, fostering collaboration and co-creation; create operational structures and processes (e.g. innovations programmes and resources to facilitate creativity and new sources of innovation opportunities); and, last but not least, companies must nourish an organizational culture, including corporate values, organizational norms, and employee attitudes that encourage social innovation through corporate values supporting experimentation, risk-taking, and collaboration;
3. *Clarity in intent regarding social goals* – which can fall within three main areas: governance and society (e.g. community involvement, education and culture, livelihood programs, ethical and governance practices), product responsibility and consumer (e.g., fair marketing, product labeling, product safety, and sustainable consumption) and environment and value chain management for optimizing the social and environmental footprint (e.g., host community engagement and environmental protection practices).

These two perspectives converge towards changes in the way of organizing corporate architecture and values, including those related to socio-environmental issues, and institutional goals and objectives.

Concluding this topic, sustainability is not just about companies' growth and development, in a strict sense. There's a trend towards increasing CG configurations and CSR expressions that corroborate the idea that CG is not only about maximizing shareholder value, but also about relationships between various stakeholders, such as investors, employees, and society, creating responsibility and accountability for the impact of corporate actions on the community and the environment (Scherer & Voegtlin, 2020; Zaman et al., 2022). And companies can adopt different levels of complexity in terms of CG for sustainability, from levels of complexity that are not concerned with social and environmental issues, to levels in which new values and issues that emerge from social and environmental issues are as much a part of corporate culture and architecture as economic issues, with variations between these two extremes.

Corporate Ethics for Sustainability

One of the major challenges in the market today is to protect and improve environmental quality through the use of standards and instruments that do not jeopardize the environment and ecological life support systems. This implies an ethical dimension of business relationships. The word 'ethics' comes from the Greek word *ethos*, which means a person's way of being, conduct or character, and etymologically corresponds to the Latin word *morale*, which has the same meaning (Souza, 2009). Over time, the term came to mean attitude towards the place of residence, housing and, later, to mean the attitude of Man in society and his spiritual values in relation to the world. From a contemporary perspective, ethics means rules of conduct for man in the preservation of his home, that is, the space of nature that we take care of to be our habitat (Macêdo et al., 2015).

However, as far as business is concerned, there is no special kind of ethics, that is, there is no such thing as a separate business ethic or a set of principles specific only for the conduct of business because ethics is concerned with the principles of good and evil, which are universal and eternal (Vallance, 1995). On the other hand, business ethics evolves from a critical attack on capitalism, which aims to make a profit, to a more productive and constructive analysis of the rules and practices that underlie commerce (Solomon, 1993).

Sustainability aims to balance social, environmental, and economic aspects. Its ethical aspect is related to combined and permanent life attitudes, which are built internally and externally, commanded by reason and managed according to ethical principles and virtues that integrate these three dimensions. Therefore, talking about sustainability can imply reflecting and giving a break to spirituality, as it implies looking inward and reflecting, leaving individualism aside so that there is a participatory democracy and, consequently, a mature and conscious discussion about global problems (Garcia, 2020).

Regardless of our management, technical and technological capacity, behind the machine and the decision-making table, there are people pushing the buttons and signing papers, that is, behind the market and governments operate people with their worldviews. Whatever the dialectical path taken, it will lead us to the social unit: the people. They constitute the functional unit of society. Therefore, we need to change the way we see reality in order to change the way we interact with nature (Ericson et al., 2014). And this concern makes sense, because as Ericson et al. (2014) highlight, the way we perceive reality influences our attitudes and behaviors, and a shift from materialist to post-materialist values, from anthropocentric to ecological worldviews is needed. Such a shift is important because our beliefs (ideologies) influence our stance on sustainability (Davidson, 2014). Thus, it is expected that people support a particular development model and the actors that work in favor of this model. Therefore, "inner changes" can be an important step towards sustainability. Davidson (2014) also demonstrates, when discussing the technique of "mindfulness", that a person can intentionally change/improve their way of seeing reality, with consequences in the way they relate to others and to nature.

Another important aspect to be addressed when talking about sustainability and the relationship with nature is the biological-ecological dimension. Biological diversity, from an ethical point of view, suggests that the biological resources of nations must be shared and exploited, taking into account non-"biopiracy", since the existing resources must be distributed in a way that guarantees equal rights to all citizens and without harming the culture of the people who have been exploiting these resources for several centuries, such as indigenous peoples, such as the Caiçara communities on the Brazilian coast (Bartholo et al., 2002).

Considering the human-nature relationship, and the inherent spiritual aspect, each individual in a community must be responsible for the consequences of their actions, instead of leaving everything to the mercy and good will of others. Thus, it is legitimate to say that most problems related to sustainability and ecosystems are related to ethical issues. In this sense, Boff (2006) proposes four principles that can serve to create policies aimed at protecting nature (Table 2). In other words, from the perspective

Table 2. The four principles of a new ethics for sustainability according to Boff (2006)

Principle	Description
Principle of affectivity	The crisis we are currently experiencing is one of sensitivity and affection. We ignore and are indifferent to a large part of the population that lives in poverty and misery, to the degradation of ecosystems, air and soil pollution, the slow extinction of species. We must work on the ability to feel indignant and sensitized towards others, because without this, the utilitarian ethic prevails over us, where the isolated individual seeks to survive or enjoy the natural and cultural benefits alone. The new sustainability that must be created to last, implies a greater sensitivity to values, solidarity, care, love and compassion, as these dimensions give more value and meaning to life.
Principle of care/ compassion	The first manifestation of sensitivity and pathos is care for life. All life must be cared for or it dies. Everything we take care of lasts longer. The ethics of compassion reminds us of the care we must have with the earth, so that it does not weaken or degrade it because of the wounds we inflict on it. Care for life, for the human being; for those who are most threatened; care for ecosystems, care for spirituality and care even for death, so that we can say goodbye with gratitude for this life.
Principle of cooperation	Cooperation, as a principle for sustainable ethics, consists of the objective logic of the evolutionary process and of life. All energies and beings need to cooperate with each other to have dynamic equilibrium, guarantee diversity, and for everyone to co-evolve. The cooperation that is required nowadays cannot exist only as a logic of the evolution of life, but must exist in us consciously and as a life project. Otherwise, we will not be able to save lives or guarantee a future commitment with humanity.
Principle of responsibility	To be responsible is to respond and assume the consequences of our actions. If we manage to destroy human lives, profoundly disrupt the life systems, and produce a disaster of unimaginable and irreversible proportions, we must be able to assume our responsibility for the Common Home and for the shared future. We must act responsibly so that the consequences of our actions are not so harmful to life and our future. Our actions must result in the promotion of life, care, cooperation, and love.

Source: Boff (2006)

of sustainability, ethics is a set of values and norms that regulate and direct the behavior and attitudes of people in the community or within groups, aiming at the preservation of the physical, emotional and spiritual integrity of people, environment, and ecological life support systems.

As seen above, ethical considerations are embedded in CG frameworks and, especially, in CSR activities, representing behaviors and ethical norms that society expects businesses to follow, beyond what is required by the law (Carroll, 1999, 2021). Ethical principles should be reflected in the organization's code of ethical conduct, formal and informal controls, strategy, policies, processes and procedures (Bonn & Fisher, 2005). Regarding CSR and its relation to ethical foundations and motivations, Schaltegger & Burritt (2018) identify four main kinds of business cases, that may coexist within a single large organization, simultaneously (e.g., different motivations between different managers) or over time (e.g., according to the company's maturity and structure) (Table 3).

While the collaborative business case appears to better meet sustainability goals, it is unclear whether it sufficiently reflects the ethical principles necessary to do so. Becker (2012) argues that a new type of ethics is needed for sustainability, as existing approaches do not fully respond to all dimensions in an integrated way, that is, between contemporary humans, between humans and future generations, and between humans and nature. The author criticizes the current metastructures (science, technology and economics) that drive development, and how they limit the development of the sustainable person and sustainable relationships. In the sphere of the economy, nature occurs as an object of economic rationality, being a means to meet human needs. Happiness is increasingly related to the level of income and consumption, and the entire economic system is based on self-interest and focused on short-term results, preventing the development of an attitude of responsibility and care for future generations.

Table 3. Four main kinds of business cases for sustainability according to Schaltegger & Burritt (2018)

Model	Description
reactionary business case	Occurs when CSR activities are perceived to increase costs and are implemented to protect the conventional profit-driven business if it is endangered.
reputational business case	Occurs when CSR activities only address reputation, brand value, and growth of sales. CSR activities can be accompanied by social and environmental positive outcomes, but when they're not, they might be scrutinized in public and have a negative impact on companies.
responsible business case	Occurs when CSR is seen as a means to increase the efficiency, quality, and performance of companies' processes and products. Thus, it is incremental, within the existing operational and business model, and does not shift into a real engagement with societal and political stakeholders to change the market framework and business environment.
collaborative business case	It is the results from dialogue-based management and stakeholders' engagement, including the vulnerable. Companies contribute substantially to solving problems of unsustainability, offer social innovations and create a structural change in the market and society. The rationale behind this perspective is that sustainability is an opportunity to enhance societal and environmental well-being, including a firm's financial viability.

Schaltegger & Burritt (2018)

Thus, Becker (2012) suggests guiding principles for the development of *ethics for sustainability*, which requires a broader understanding of the human being, while being relational and interdependent in relation to nature. In this context, the paradigm of economic growth must be replaced by the principle of sustainability, stability, and simplicity, based on a new social contract, assuming that more knowledge and more material goods do not necessarily mean a better life.

Increasingly, customers tend to look for companies that act ethically, with a good image in the market and that are ecologically responsible and sustainable. These characteristics are fundamental for companies to achieve sustainable development and, at the same time, increase their profitability in business (Kameyama, 2004). Business ethics is also concerned with good and bad, or right and wrong, behavior and practices that take place within a business context, and has increasingly been interpreted today to include the more difficult and subtle questions of fairness, justice, and equity (Carroll & Buchholtz, 2008).

There was usually a widespread belief that the impact on the environment was part of the company's social responsibility, implying purely legal or even ethical and moral obligations on the part of companies (Naciti et al., 2022). Over time, companies were pushed to review their investments in sustainability practices, increasing them and introducing social innovations to their business models to succeed (Porter & Kramer, 2006). In this scope, CSR might be a CSI predictor (Phillips et al., 2015; Popoli, 2016; Porter & Kramer, 2006). Indeed, companies can cooperate for the common good in many ways, by creating well-being, providing products and services in an honest and reliable way, and respecting the dignity and unshakable and fundamental rights of individuals (Garriga & Melé, 2004; Varela & António, 2012). Porter & Kramer (2006) provide a wide range of actions which can be taken by firms to implement CSR through their value chain (Table 4).

Companies as Social-Ecological Systems

Recognizing that humanity is approaching (or have already reached) the limits of the planet —— the Planetary Boundaries of Rockström et al. (2009) ——, society must face environmental dilemmas not only as questions of justice and ethics and morals. The realization that we need to deal with physical limits and scarce resources imposes the need for an ecological vision of sustainability. Approaches from an ecological perspective can contribute to understanding the factors or relationships that determine the

Table 4. Scope of CSR opportunities through firm value chain

Support Activities	Firm Infrastructure	● Financial reporting practices ● Government practices ● Transparency ● Use of lobbying
	Human Resource Management	● Education & job training ● Safe working conditions ● Diversity & discrimination ● Health care & other benefits ● Compensation policies ● Layoff policies
	Technology Development	● Relationships with universities ● Ethical research practices (e.g., animal testing) ● Product safety ● Conservation of raw material ● Recycling
	Procurement	● Procurement & supply chain practices (e.g. child labor, pricing for farmers) ● Uses of inputs (e.g., animal fur) ● Utilization of natural resources
Primary Activities	Inbound Logistics	● Transportation impacts (e.g., emissions, congestion, logging roads)
	Operations	● Emissions & waste ● Biodiversity & ecological impacts ● Energy & water usage ● Worker safety & labor relations ● Hazardous materials
	Outbound Logistics	● Packaging use and disposal ● Transportation impacts
	Marketing & Sales	● Marketing & advertising (e.g., truthful advertising, advertising to children) ● Pricing practices (e.g., price discrimination among customers, anticompetitive pricing practices, pricing policy for the poor) ● Consumer information ● Privacy
	After-Sales Service	● Disposal of obsolete products ● Handling of consumables ● (e.g., motor oil, printing ink) ● Customer privacy

Source: adapted from Porter & Kramer (2006)

human ecological condition on Earth when we include the social dimension in ecosystem dynamics (Folke, 2006).

When considering the relationship of social structures (businesses, governments, etc.) with nature, it is necessary to take into account the implications for the ecological structures involved, and the very ecological nature of social systems. An ecological view of social structures is aligned with the notion of *strong sustainability*, in which society is a part of and sustained by the ecological system —— human beings depend on natural capital and ecosystem services that can be irreplaceable (Ekins et al., 2003). Thus, governance for sustainability implies an ecological dimension. The concept of a *social-ecological system* provides a conceptual framework for interpreting human-nature relationships.

In a study which reviewed 1289 publications to assess the use of the socio-ecological system concept, it was found that the term is still a concept under construction, integrating many currents of thought, originated in different disciplines, but with common elements, such as the analysis of resilience, ecosystem services, sustainability, governance and adaptive management (Herrero-Jáuregui et al., 2018). Social-ecological systems have been defined as complex adaptive systems where the social and environmental factors and agents closely interact, and where such interactions among components drives change (Martínez-Fernández & Banos-González, 2021). This concept considers that the delineation between social and ecological systems is artificial and arbitrary (Folke, 2006).

An important concept that emerges from the study of social-ecological systems is that of resilience. The concept of resilience has evolved considerably since Holling's seminal 1973 paper (Walker et al., 2004). Resilience is concerned with the capacity of a system to persist and develop in its regime without shifting into another regime (Holling, 1973; Li et al., 2018; Walker et al., 2004). In the context of a firm, resilience is the ability to embrace change through innovation to persist over time (Carpenter et al., 2015; Folke, 2016; Li et al., 2018). Thus, simply put, resilience "is the capacity of an ecosystem to buffer disturbance and surprise and thereby conserve future options and opportunities" (Ekins et al., 2003, p. 170).

From this perspective, a company is closely connected with the community and the environment and, therefore, establishes a relationship of mutual influence with them, and needs to consider synergisms between social and ecological systems to promote resilience.

METHODOLOGY

The collection of Mercur's information and data required different methodologies. Open-ended and semi-structured interviews were conducted with company members using video call resources. Specific questions were asked to different members in the company via email. The data was obtained directly from the company's website.

The emissions, water use and renewable raw materials data were provided by the company 'Mercur S.A.'. The emissions data were obtained from the calculation of the emission of greenhouse gasses generated by the company's operation in all scopes. The methodology and source of emission factors are from the GHG Protocol Brazil (FGVces, 2011) (Table 5).

Table 5. Types of greenhouse gas emissions in the three company scopes

	Emission Type	Control Level	Description
Scope 1	Direct emissions	Sources controlled by the company	Fuels that are necessary for the operation of the company
Scope 2	Indirect emissions		Electricity purchased
Scope 3	Indirect Emissions	Sources not controlled by the company	Transport movements (inputs, outputs, imports and exports), trips and displacements of internal and external collaborators

Source: Mercur S.A

THE MERCUR S.A. CASE

In this topic, we present, based on the data and information obtained, a description of the changes implemented by the company Mercur S.A aiming at sustainability. Mercur is a large company managed since its founding, almost a century ago, by the same family. Since 2009, the company implemented changes that led to a restructuring in all sectors of the company since it began the process known as the "turn of the key".

The trigger for change, according to Jorge Hoezel Neto[1], came from him, starting with a question during a course on sustainability. This characterizes an inner change type of transformation (Pisters et al., 2020). However, what started as an individual transformation (inner change) has spread throughout the company and beyond into the community. In Annex 1, a timeline of the main measures and changes implemented by the company since the *turn of the key* in 2009 is presented.

According to Hoezel, the "main positive aspect was giving people a voice", and the "main negative aspect was the loss of revenue, loss of profit due to the economic model installed". In this case, it seems that an important trade-off was the exchange of financial capital gains by human and social capital gains (Carroll, 2016, 2021). This trade-off signals a value shift in the company's culture. This change will become clear in the course of this topic.

Socio-Environmental Responsibility

The "turn of the key" process started to introduce in the mentality of the Mercur staff a concern that transcends the company's internal production processes. As will be seen in this topic, the company seeks to promote not only internal changes, but has been intentionally seeking to promote changes both in the community and in the production chain, and in the local economy, by valuing suppliers in the region. The new approaches to company management characterize, as we will see, standards that satisfy the principles of good CG (e.g., Turnbull, 1997; Zaman et al., 2022) CSR (e.g., Carroll, 2021; Garriga & Melé, 2004; Martins, 2020), and CSI (e.g., Evans et al., 2017; Iglesias et al., 2020; Porter & Kramer, 2006).

Mercur's vision is to be a "company committed to building relationships that value life" (Mercur, [s.d.]). And to achieve this vision, some 'directions' (organizational objectives) were adopted, which aim to promote social issues, especially those aimed at valuing the local economy (Alexandre Antinarelli[2] and João Carlos Vogt[3]). In addition to valuing local suppliers and reinforcing the valorisation of the local production and economy, there is the "Import Reduction" guideline, which contributes to the generation of jobs, education and qualification of local workers, creation of new occupations and innovation in products and services, which add value to the local production chain, by making it possible to offer products with greater added value, which increases the wealth of the region and, if there are fairer productive relations, it boosts the redistribution of wealth and the reduction of inequalities social. For Mercur:

Buying locally [and reducing imports] has important impacts on environmental issues, notably in the reduction of greenhouse gas emissions, and also makes it possible to qualify this environment through the relationships that occur through meaningful conversations, through the identification of opportunities to 'create and do with', generating the protagonism of all local actors in the formulation of adequate solutions for each location (Alexandre Antinarelli[2] and João Carlos Vogt[3], personal communication).

To follow the direction "Replace imports, valuing local production", two "Strategic Indicators" are adopted (Table 6).

In addition to seeking to enhance the local economy, Mercur implemented important internal changes, which, among other aspects, promoted social justice and reduction of inequalities within the company. On March 1, 2010, the working hours were reduced from 44 hours a week to 40 hours a week. This

Table 6. Caption should be sentence case with no ending punctuation if only one sentence

Strategic Indicator	Goal	Target for 2022
Representation of the Billing of Imported Products on the Total Billing	To foment the reduction of imports; value and encourage local economy and production; promote generation of jobs, qualification of local workers and the creation of innovation in products and services	Representativeness of billings of products imported by MERCUR < or = 18% of the company's total sales
Total Purchases of Materials and Services from Suppliers in Rio Grande do Sul state vs. Brazil	To promote the selection of local suppliers and the relationship in its local value chain; to meet the needs for materials and services of the areas and processes involved, promoting actions that enable the restoration of socio-environmental aspects, such as the reduction of GHG emissions, occupancy and income and engagement on site	Commitment that the percentage of representativeness of Mercur purchases from suppliers in the state of Rio Grande do Sul – where the company's plants are located – is > or = 64% of the general total of purchases made considering all company's suppliers

Source: Mercur S.A

decrease did not affect Mercur's production targets in the market and there was no financial loss for employees (Jorge Hoezel[1]).

Night work was necessary for some production processes, and the equipment was kept on for 24 hours. When analyzing the process, it was noticed that night work was related to negative indicators, especially with regard to physical and mental health, with impacts on family life (Bianco Marques[4] and João Carlos Vogt[3]). So, according to Marques and Vogt, the following question was raised: "Do we need to work at night? Why do we work the third shift? Do we want to produce profits/dividends like people's malaise?" In 2014, the third shift (night work) was ended and the staff was allocated to other activities.

In November 2015, the collective bargaining agreement pointed to a salary adjustment in the order of 10%. On the other hand, the calculation carried out by the "Production Planning and Control" sector indicated that 34 people were idle in production. The common practice in the market would be the dismissal of idle people to regulate the necessary framework for the operation. The company understood that it could seek other solutions and not fire these people, entering into an agreement with the Union. The alternative found and approved was the reduction of the working day by 10%, that is, from 40 hours per week to 36 hours, instead of the readjustment in the order of 10%.

Based on the understanding that the company's (human) "architecture" influences the relationship and participation of employees and publics that interact with the brand, the company started to adopt a horizontal hierarchy and a participatory decision-making process. The company's employees "understand that this enables a more open and participatory relationship" (Fabiani Elisabet Spiegel[5] and Camila Lima[6]).

All people, as long as they have built and agreed on the decision process together with others involved in the situation, are empowered to make decisions at Mercur. … The company has experienced that, by sharing autonomy and responsibilities, with leaderships that rotate between each person's expertise, everyone is more focused on common goals than on the dispute for hierarchical growth (Fabiani Elisabet Spiegel[5] and Camila Lima[6]).

Aiming to support the changes implemented, the company created a "Learning" sector, which seeks to work with the employees on the necessary skills. Work in this sector is based on concepts such as *collaboration*, *protagonism* and *responsibility*. Through the co-creation process, "Fazer Com" (doing with), implemented through the Diversity on the Street (*Diversidade na Rua*) Project and the LAB (Laboratory of Social Innovation), products and services began to be developed in partnership with the community. This new way of developing products and services leads to an important change in the company, that is, from doing "things for" people to doing "things with" people. With these changes, the company seeks to stop encouraging consumerism through its products and services, to build these products and services together with people, from the understanding of your needs (Dayani Rabuske[7] and João Vogt[3]).

The "Fazer Com" process of co-creation came to be considered a way of being and doing with people, in which, through collaboration, intelligence and collective construction, it makes it possible to create solutions for the challenges/needs that emerge from the relationships between people in the company and in the community. It is present in relationships and is translated into Mercur's institutional commitment to "unite people and organizations to create sustainable solutions and referrals" (Mercur, [s.d.]).

For the "doing with" process to happen, the need arose for a space of its own. The LAB, already mentioned above, "is a space that opens its doors in order to serve as an instrument for the promotion of significant moments of teaching and learning and, also, for the creation of solutions that help to improve people's lives, from your legitimate needs and in life in common with them" (João Vogt[3]).

LAB puts people at the center of the process, giving them the possibility to build solutions that meet their needs and help them to develop their skills. It is believed that this space makes possible the experience of collaborative processes of co-creation, in order to build dynamics of open social innovation for new products, processes and services that meet the real needs of people, seeking the constitution of an autonomous and empowered subject, strengthening self-esteem and social participation, through creative activities and experiences (João Vogt[3]).

The development and implementation of this way of "doing with people", according to Rabuske and Vogt, is influenced by different methodologies, such as Circle Pedagogy (see Freire, 2015), Design Thinking (see Brown, 2008) and other collaborative methodologies, which contributed to the construction of this form of relationship and which, in turn, were taken into the company's day-to-day activities. Some principles and practices that are part of this way of "doing with people", according to Rabuske and Vogt, are: leadership rotation/alternation, encouraging people to contribute, shared responsibility, trust in a greater purpose, speak with intent, listen carefully and with the heart, and to make collective knowledge visible. This new cultural vision promotes opportunity for collective creativity, with more employee engagement (Turnbull, 1997; Mirvis et al, 2016; Popoli, 2016). And it is the social capital that comes to offset losses in the company's profitability. This trade-off is justified in a vision of social-ecological systems, in which social capital is important to sustain adaptive capacity, through flexibility and innovative creativity, from the system to changes (Folke et al. 2005, 2006).

In addition to the aforementioned, Mercur expands its actions beyond the limits of the community and the company (Turnbull, 1997; Popoli, 2016; Scherer & Voegtlin, 2020). The company maintains the "Borracha Nativa" (native rubber) project, which aims to contribute to the preservation of culture and the consolidation of extractive reserves and indigenous lands by encouraging the resumption of production and marketing of natural rubber. Native Amazonian rubber (latex), a natural raw material used in part of the company's products, is obtained from rubber tappers (*seringueiros*) in two protected areas: Rio Iriri Extractive Reserve and Xipaya Indigenous Land, located in the municipality of Altamira (Pará, Brazil)[8]. Through this project, the company intends to maintain "a commercialization model focused on fair remuneration, in addition to helping to improve local infrastructure and improve extractive processes" (Mercur, [s.d.]). They also intend "to value the work of the traditional populations of the forest, deserved for the service they provide to the planet, in addition to making them seen, recognized and reclaiming their rights as citizens in our society" (Mercur, [s.d.]).

The option to buy from extractivists and indigenous people goes beyond the purely lucrative vision. We are committed to a long-term strategy, where what counts is the standing forest and the way of life of the traditional populations that help to conserve it. We want to generate income for the local communities through the resumption of production and consequent commercialization of natural rubber by the rubber tappers, in a place whose culture was dying out, in addition to learning and experiences for the organization. For Mercur, the value of extracting latex is much greater than what results in the final product. It must contemplate and value the environmental services provided by extractivists, family farmers and indigenous peoples (Mercur, [s.d.]).

It is clear that the company's management adopts a holistic view, considering its sphere of action and decision beyond the physical limits of the company. As a social-ecological system, the company establishes a relationship of mutual beneficial influence with the social and ecological systems in which it operates, establishing bidirectional positive feedback loops (Motesharrei et al., 2016, Elsawah et al., 2020). However, the current condition of the company clearly emerges from the intentional and conscious action of the management (Scherer & Voegtlin, 2020). The direction of change, that is, whether it takes place from the inside out or from the outside in, is still an open question. In the case of Mercur, as already mentioned, the transformation can be considered an "inner change" process (Pisters et al., 2020). However, we shouldn't expect all companies to implement change from the inside out. This doesn't seem to be the trend. In this sense, the direction of change is clear from its starting point.

The future implementation of new actions, projects and/or programs aimed at promoting sustainability and improving environmental quality, according to João Vogt[3], is a permanent goal of the company, which goes beyond the environment (Herrera, 2015; Popoli, 2016; Scherer & Voegtlin, 2020). The company's position of "uniting people and organizations to build referrals and create sustainable solutions" carries the commitment to continuously seek the improvement of environmental quality.

Environmental Impacts

Regarding the environmental dimension, a set of changes were also implemented to improve the company's environmental efficiency. To exemplify the physical effects of the changes implemented by the company from 2009 to 2021, the variation over time of three indicators is presented in Figure 1, including: greenhouse gas emissions, water use, and amount of renewable raw materials in the company's total production. These results clearly show signs of positive environmental impacts (Porter & Kramer, 2006). As main measures taken by the company to reduce emissions, the following can be mentioned: stimulation of carpooling and use of public transport; use of ethanol in the company's vehicles; elimination of the boiler that operated with BPF oil; Replacement of incandescent and fluorescent lamps with LED lamps; decreased air travel; selection of nearest suppliers; reduction of imports; use of intermodality in transport of raw material inputs and outputs of finished products, replacing trucks with trains or ships; conversations with suppliers on the subject (an action that led some to make an emissions inventory and offset their emissions). Measures were also adopted to offset emissions. Trees of various native species were planted according to the target biome. Considering the areas planted by Mercur and its partners,

Figure 1. Temporal variation in (a) annual greenhouse gasses emissions (in tons per year), (b) annual water use (in cubic meters per year), and (c) relative share of renewable raw materials in the total production of the company (in percentage) (Source of data: Mercur S.A)

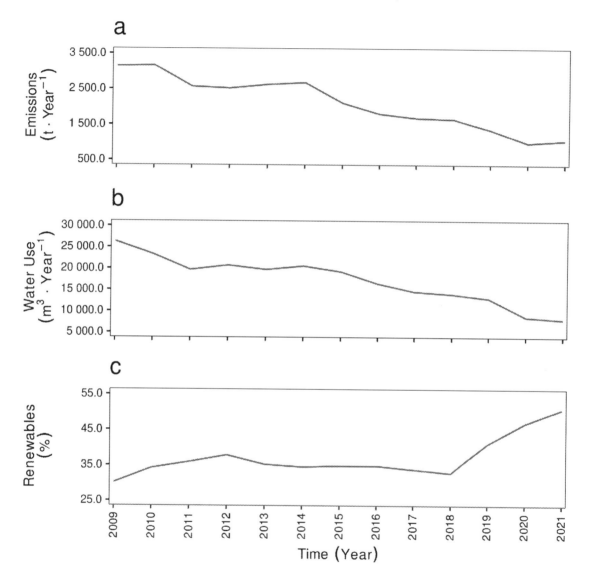

a total of approximately 93 thousand tree seedlings have been planted up until the date of this study (Mercur, [s.d.]).

In order to promote reductions in the use of water, some measures were adopted, including, mainly: closing the company's artesian wells, considered as water for use by the community; consuming only water from the supply company and rainwater harvesting; and intensification of water reuse in the company. In addition to measures to reduce water consumption, wastewater and effluent treatment is now being carried out so that the water returns to the water bodies in appropriate conditions. Mercur maintains two types of ETEs (Effluent Treatment Station) in operation: one industrial and one cloacal.

As for energy consumption, all the electricity purchased comes from the free energy market, including: solar, wind and Small Hydroelectric Power Plants (SHPs). The company's photovoltaic plant is also scheduled to start operating in 2022, which will generate at least 50% of the company's electricity needs. As for solid waste generated by the company, its nature is very diverse, leading to the adoption of different paths for each type of waste. For the recovery of certain types of waste, complete reverse logistics of some materials are carried out. An example is textile waste, for which part of the recovered yarn is used again in the form of recycled cotton.

Mercur is also associated with the Association of Reverse Logistics for Packaging (*ASLORE – Associação de Logística Reversa de Embalagens*), which aims to recover part of the packaging placed on the Brazilian market. The current target in Brazil is 22%, according to the "Sectoral Agreement for the Implementation of the Reverse Logistics System for Packaging in General", signed in 2015 between voluntary private companies and the government, supported by the "National Solid Waste Policy" (*PNRS – Política Nacional de Resíduos Sólidos*), established by Law No. 12,305/2010 and regulated by Decree No. 7,404/2010.

The packaging recovery model involves contracting collector cooperatives and specific projects from recycling companies. Then, the solid waste generated is separated and sent to companies specialized in each type of material.

- *Recyclable waste*, such as paper, cardboard and plastic, are donated to a local collector cooperative, as a partnership, which aims to generate employment and income, as well as enabling company employees to get to know the cooperative.
- *Hazardous waste* is sent for co-processing (waste blending activity for the production of fuel for power generation plants) together with Fupasc (Santa Cruz do Sul Environmental Protection Foundation), a local entity which Mercur has been associated with since 2004.
- *Metals, drums, contaminated plastic packaging, light bulbs, boiler ash, health waste*, etc. are sent to specialized companies according to their characteristics.

In addition to resource and waste management, a series of innovations in the company's products were implemented. Table 6 lists these innovations, the incremental changes they represent, and the numbers in terms of renewable materials and reduction of non-renewables. These innovations result from the company's work to replace non-renewable or renewable materials in its products.

Regarding the food offered to the company's employees, Mercur has been seeking to offer healthier foods, free of agrochemicals. The introduction of healthier eating was a gradual process that is still ongoing. Initially, organic rice and beans were introduced (Table J). In addition to organic rice and beans[9], other organic food items are also offered, including: brown rice, sweet potatoes, cassava, lettuce, brown sugar, soy and wheat. Some foods depend on the supply and are eventually organic, such as: onions, cabbage, chayote, zucchini, carrots, beets, cabbage, broccoli, cornmeal, watermelon, bananas, oranges and bergamot (a variety of tangerine). This diversity of foods comes from local producers, stimulating the local rural economy.

To enable the introduction of organic food in the company, partnerships were established and started the process of learning about healthy eating and conversations about the institutional positioning with partners: Refeições ao Ponto (company in the collective food sector), Center for Support and Promotion of Agroecology (*Centro de Apoio e Promoção da Agroecologia – CAPA*) and Regional Cooperative of Ecological Family Farmers (*Cooperativa Regional de Agricultores Familiar – Ecovale*).

Table 6. Examples of product innovations that have led to reductions in solid waste generation

Product	Process	Renewable Raw Materials in the Composition	
		Before	After
Mercur Natural Erasers	Replacement of dolomite with a renewable filler from a vegetable source (cassava starch)	12%	75%
Natural Thermal Bag	Produced with raw materials grown in an agroecological way, contributing locally to conserve biodiversity, taking advantage of a by-product that was discarded, locally stimulating the preservation of the Atlantic Forest and the juçara palm (*Euterpe edulis*)	First product created with 100% renewable raw materials	
Natural Thermal Bag with Recycled Cotton	Like the Natural Thermal Bag, it is filled with juçara palm kernels grown by agroecological processes, but with recycled cotton fabric	Product created with 100% renewable raw materials and partially made with recycled material	
Fixed Canadian Crutch	Developed in Brazil, with a reduction in the amount of raw material in the product and packaging, in addition to the replacement of non-renewable raw material for renewable in parts of the product	Body with 24% less aluminium (non-renewable raw material) than the previous version	
		Adhesive film packaging involving only the tube, providing more visibility to the product and a 75% reduction in plastic consumption in packaging	
		Articulated tip made of natural rubber, with silica of vegetable origin in the composition, extracted from the ash of the rice husk, making it about 70% renewable	
Total production	Change in relative amount of renewable raw material from 2009 to 2021	31.1%	50.8%

Source: Mercur S.A.

All the changes described so far demonstrate that Mercur has incorporated the environmental issue into the entire structure and functioning of the company, which represents a global shift towards sustainability (Porter & Kramer, 2006; Popoli, 2016). This is relevant, because while larger companies are under more pressure to disclose more information to gain legitimacy, higher scores on CSR reports do not mean that these companies couple their sustainability reporting with their business practices (Drempetic et al., 2020). Larger companies seem to better approach sustainability reporting (Drempetic et al., 2020; Rosati & Faria, 2019) as they have more resources and higher levels of intangible assets to achieve sustainability goals (Rosati & Faria, 2019), while at the same time they can have a great impact on the local economy.

From the reports of the company's employees, it is possible to perceive that, in addition to the changes in the company, these are accompanied by changes in the employees' consciousness. For example, in addition to starting to serve healthier foods in the company, employees developed an awareness of the role of healthier eating, but also of the role of food production for soil quality, environment and society.

The idea of increasing the consumption of organic food at Mercur, as well as other initiatives, is positively supported by all the actors involved, who claim to understand the importance of an "environmentally friendly agricultural production model that is important for the regeneration of the planet" (João Carlos Vogt[3] and João Carlos Kipper Trinks[10]). According to Vogt and Trinks, such a model "does not pollute the air, water and soil, and also contributes to improving integral health, in a healthy and natural way, in harmony with people and the environment" (personal communication). This vision is also in line with the "Vision 2050" adopted by Mercur: "Mercur is committed to building relationships that value life, based on local wellness initiatives" (Mercur, [s.d.]).

SOLUTIONS AND RECOMMENDATIONS

The study of human/social institutions from the perspective of sustainability can offer execution challenges, as companies considered to be in disagreement with the principles of sustainability and social responsibility may be resistant to providing information about the company's operation. There is, therefore, a social barrier that depends on the interest and goodwill of the institution studied. Finding ways to study institutions that adopt different management models should enrich the debate on corporate governance in general, and governance for sustainability, in particular.

Furthermore, advances in studies in the field of governance for sustainability are important to contribute to the improvement of practices in institutions. Perhaps the study of new experiences can be a stimulating factor for self-criticism in companies that are more resistant to change. It can also be a stimulus for other companies to seek to join studies as they perceive that these add value to the company. In a broader perspective, the experience of Mercur S.A can serve as an example to new experiences, pointing to what can work and the risks and benefits of adopting governance for sustainability.

FUTURE RESEARCH DIRECTIONS

The disruptive changes implemented by Mercur point to possibilities considered impractical by the business world until recently. However, this is a very recent experience that needs to be better understood in the long term. In this sense, future studies may be interesting to assess the evolution of Mercur S.A since the implemented changes.

Comparisons with other experiences will also be important to understand the issue of governance for sustainability in a broader and deeper way. Therefore, an obvious direction will be to study new experiences in corporate governance for sustainability in businesses of different natures from those of Mercur, and in different cultural, environmental and economic contexts to understand the particularities of governance for sustainability in arenas with different configurations and operating forces. Also, from a holistic standpoint of the topic, the study of other types of institutions (governments, public companies, NGOs, etc.) has the potential to provide new insights and broaden the understanding of the diversity of nuances of the topic.

In addition to the study on governance for sustainability itself, an issue that needs to be addressed is divulgation and communication. It is urgent not only to understand the governance of social structures, but mainly to seek ways to catalyze the dissemination of values and cultures that emerge from new successful experiences.

DISCUSSION AND CONCLUSION

The case of Mercur seems to be an exception in relation to what is expected and observed from the conventional behavior of companies. Family property and duality of the CEO are negatively related to CSR (Hussain et al., 2018; Zaman et al., 2022). Family owners may be concerned about accumulating family financial wealth and consider investments in CSR as unnecessary additional costs (Zaman et al., 2022), while CEO duality may compromise the board's monitoring process of CEO decisions on CSR (Hussain et al., 2018). And the evidence seems to reinforce this, if only the financial aspect is considered.

Corporate Governance and Ethics for Sustainability

A study that examines the effect of a stricter environmental protection regime on industrial performance in China found evidence which suggest that pollution-intensive firms increased profits in response to the environmental regulation, whereas clean yet energy-intensive firms decreased their profits in response to the environmental regulation (Jefferson et al., 2013).

On the other hand, the case of Mercur seems to be in line with a new vision of development that has gained strength in recent decades. van den Bergh (2009) argued that removing the information flaw that GDP represents will lead to decisions and developments more in line with improving human well-being. Moreover, ignoring GDP information is consistent with a perfectly neutral stance regarding economic (GDP) growth, and would eliminate an unnecessary constraint on our search for human progress (van den Bergh, 2009).

On the other hand, as already mentioned, an institution does not exist alone in society, but rather coexists with other social and ecological structures with which it interacts. The media effect on CSR, for example, differs between firms. Companies with higher public ownership are more reactive to media coverage and significantly increase their corporate social engagement, investing in CSR as an instrumental mechanism to accumulate reputation gains and improve the corporate image. This means that companies with concentrated ownership are driven by their own values in their search for CSR (Zaman et al., 2022).

Therefore, for an accelerated transformation towards sustainability it is necessary that the different sectors of society do their part, and consider the natural support systems for life and financial activities. This implies a systemic vision that includes both social and natural systems, evoking the concept of socio-ecological systems (Herrero-Jáuregui et al., 2018). Such a holistic view is embraced by the approach of social-ecological systems as complex adaptive systems (Herrero-Jáuregui et al., 2018). Governance, from the perspective of complex adaptive systems, is the structures and processes by which social actors make decisions and share power, resolve trade-offs and provide vision and direction for sustainability (Folke et al., 2005). To realize the governance vision, management (operationalization) and monitoring (feedback on the evolution of changes) are important dimensions (Boyle et al., 2001), which are evident in the company's new culture.

Finally, the Mercur CG experience studied reveals a process that can be considered disruptive in relation to the "status quo" in terms of business management. But, it is also disruptive with regard to the objectives of the changes, which is the search for a genuine value of the company in the community and for the environment, beyond the internal economic and financial aspects. Instead of a dominantly economic value, that is, concerned with financial capital, other capitals become relevant, notably social and human capital, and environmental and ecological capital. This is more evident by the fact that the company gave up important contracts, with loss of revenue, to seek social and environmental gains (Jorge Hoezel[3]). And the fact that the company has prospered practically for a century, passing through different social contexts over time, indicates a good adaptive and innovative capacity. And this ability seems to be confirmed given the recent changes implemented, which are in line with the new challenges faced by humanity in the Anthropocene.

ACKNOWLEDGMENT

We thank Jorge Hoelzel, João Carlos Vogt, Cassiano Ricardo Kappaun and other employees of Mercur for their efforts in collaborating with the collection of data and information for this study.

REFERENCES

Aras, G., & Crowther, D. (2008). Governance and sustainability: An investigation into the relationship between corporate governance and corporate sustainability. *Management Decision*, *46*(3), 433–448. doi:10.1108/00251740810863870

Ávila, L. V., Hoffmann, C., Corrêa, A. C., da Rosa Gama Madruga, L. R., Schuch Júnior, V. F., de Sousa Júnior, A. F., & Zanini, R. R. (2013). Social Responsibility Initiatives Using ISO 26000: An Analysis from Brazil. *Environmental Quality Management*, *23*(2), 15–30. doi:10.1002/tqem.21362

Ayres, R. U., van den Bergh, J. C. J. M., & Gowdy, J. M. (2001). Strong versus Weak Sustainability: Economics, Natural Sciences, and "Consilience". *Environmental Ethics*, *23*(2), 155–168. doi:10.5840/enviroethics200123225

Bartholo, R., Ribeiro, H., & Bittencourt, J. (2002). *Ética e Sustentabilidade* [*Ethics and Sustainability*.]. Editora E-papers.

Becker, C. U. (2012). *Sustainability Ethics and Sustainability Research*. Springer., doi:10.1007/978-94-007-2285-9

Boff, L. (2006). *Ética e Sustentabilidade* [*Ethics and Sustainability*.]. Ministério do Meio Ambiente.

Bonn, I., & Fisher, J. (2005). Corporate Governance and Business Ethics: Insights from the strategic planning experience. *Corporate Governance*, *13*(6), 730–738. doi:10.1111/j.1467-8683.2005.00466.x

Boyle, M., Kay, J., & Pond, B. (2001). Monitoring in support of policy: an adaptive ecosystem approach. Encyclopedia of global environmental change, 4(14), 116-137.

Brown, T. (2008). Design Thinking. *Harvard Business Review*, *86*(6), 84–92. PMID:18605031

Butler, C. (2000). Inequality, Global Change and the Sustainability of Civilisation. *Global Change & Human Health*, *1*(2), 17. doi:10.1023/A:1010029222095

Carpenter, S. R., Brock, W. A., Folke, C., van Nes, E. H., & Scheffer, M. (2015). Allowing variance may enlarge the safe operating space for exploited ecosystems. *Proceedings of the National Academy of Sciences of the United States of America*, *112*(46), 14384–14389. doi:10.1073/pnas.1511804112 PMID:26438857

Carroll, A. B. (1999). Corporate Social Responsibility: Evolution of a Definitional Construct. *Business & Society*, *38*(3), 268–295. doi:10.1177/000765039903800303

Carroll, A. B. (2016). Carroll's pyramid of CSR: Taking another look. *International Journal of Corporate Social Responsibility*, *1*(3), 1–8. doi:10.118640991-016-0004-6

Carroll, A. B. (2021). Corporate Social Responsibility: Perspectives on the CSR Construct's Development and Future. *Business & Society*, *60*(6), 1258–1278. doi:10.1177/00076503211001765

Carroll, A. B., & Buchholtz, A. K. (2008). *Business & society: Ethics and stakeholder management* (7th ed.). South-Western Cengage Learning.

Davidson, K. (2014). A Typology to Categorize the Ideologies of Actors in the Sustainable Development Debate: A Political Economy Typology of Sustainability. *Sustainable Development*, *22*(1), 1–14. doi:10.1002d.520

de Macêdo, I. I., Rodrigues, D. F., Chevitarese, L. P., & Feichas, S. A. Q. (2015). *Ética e sustentabilidade*. Editora FGV.

Drempetic, S., Klein, C., & Zwergel, B. (2020). The Influence of Firm Size on the ESG Score: Corporate Sustainability Ratings Under Review. *Journal of Business Ethics*, *167*(2), 333–360. doi:10.100710551-019-04164-1

EC. E. C., & DG JUST, D.-G. for J. and C. (2022). *Proposal for a directive of the European parliament and of the council on Corporate Sustainability Due Diligence and amending Directive (EU) 2019/1937* (Final COM/2022/71; p. 76). European Union. https://eur-lex.europa.eu/resource.html?uri=cellar:bc4dcea4-9584-11ec-b4e4-01aa75ed71a1.0018.02/DOC_1&format=PDF

Ekins, P., Simon, S., Deutsch, L., Folke, C., & De Groot, R. (2003). A framework for the practical application of the concepts of critical natural capital and strong sustainability. *Ecological Economics*, *44*(2–3), 165–185. doi:10.1016/S0921-8009(02)00272-0

Elkington, J. (2006). Governance for Sustainability. *Corporate Governance*, *14*(6), 522–529. doi:10.1111/j.1467-8683.2006.00527.x

Elsawah, S., Filatova, T., Jakeman, A. J., Kettner, A. J., Zellner, M. L., Athanasiadis, I. N., Hamilton, S. H., Axtell, R. L., Brown, D. G., Gilligan, J. M., Janssen, M. A., Robinson, D. T., Rozenberg, J., Ullah, I. I. T., & Lade, S. J. (2020). Eight grand challenges in socio-environmental systems modeling. *Socio-Environmental Systems Modelling*, *2*, 16226–16226. doi:10.18174esmo.2020a16226

Ericson, T., Kjønstad, B. G., & Barstad, A. (2014). Mindfulness and sustainability. *Ecological Economics*, *104*, 73–79. doi:10.1016/j.ecolecon.2014.04.007

Evans, S., Vladimirova, D., Holgado, M., Van Fossen, K., Yang, M., Silva, E. A., & Barlow, C. Y. (2017). Business Model Innovation for Sustainability: Towards a Unified Perspective for Creation of Sustainable Business Models. *Business Strategy and the Environment*, *26*(5), 597–608. doi:10.1002/bse.1939

Farooq, O., Payaud, M., Merunka, D., & Valette-Florence, P. (2014). The Impact of Corporate Social Responsibility on Organizational Commitment: Exploring Multiple Mediation Mechanisms. *Journal of Business Ethics*, *125*(4), 563–580. doi:10.100710551-013-1928-3

FGVces. C. de E. em S. (2011). *Especificações de verificação do Programa Brasileiro GHG Protocol* (Programa Brasileiro GHG Protocol, p. 42). [*Verification specifications of the Brazilian GHG Protocol Program* (Brazilian GHG Protocol Program, p. 42).] Fundação Getulio Vargas (FGV). http://bibliotecadigital.fgv.br:80/dspace/handle/10438/30258

Folke, C. (2006). Resilience: The emergence of a perspective for social–ecological systems analyses. *Global Environmental Change*, *16*(3), 253–267. doi:10.1016/j.gloenvcha.2006.04.002

Folke, C. (2016). Resilience (Republished). *Ecology and Society*, *21*(4), art44. Advance online publication. doi:10.5751/ES-09088-210444

Folke, C., Hahn, T., Olsson, P., & Norberg, J. (2005). Adaptive governance of social-ecological systems. *Annual Review of Environment and Resources, 30*(1), 441–473. doi:10.1146/annurev.energy.30.050504.144511

Freire, P. (2015). *Educação Como Prática Da Liberdade [Education As A Freedom Practice.]*. Paz e Terra.

Garcia, D. S. S. (2020). Sustentabilidade e ética: Um debate urgente e necessário. [Sustainability and ethics: An urgent and necessary debate.] *Revista Direitos Culturais, 15*(35), 51–75.

Garriga, E., & Melé, D. (2004). Corporate Social Responsibility Theories: Mapping the Territory. *Journal of Business Ethics, 53*(1), 51–71. doi:10.1023/B:BUSI.0000039399.90587.34

Glavas, A., & Kelley, K. (2014). The Effects of Perceived Corporate Social Responsibility on Employee Attitudes. *Business Ethics Quarterly, 24*(2), 165–202. doi:10.5840/beq20143206

Globocnik, D., Rauter, R., & Baumgartner, R. J. (2020). Synergy or conflict? The relationships among organisational culture, sustainability-related innovation performance, and economic innovation performance. *International Journal of Innovation Management, 24*(01), 2050004. doi:10.1142/S1363919620500048

Herrera, M. E. B. (2015). Creating competitive advantage by institutionalizing corporate social innovation. *Journal of Business Research, 68*(7), 1468–1474. doi:10.1016/j.jbusres.2015.01.036

Herrero-Jáuregui, C., Arnaiz-Schmitz, C., Reyes, M. F., Telesnicki, M., Agramonte, I., Easdale, M. H., Schmitz, M. F., Aguiar, M., Gómez-Sal, A., & Montes, C. (2018). What do we talk about when we talk about social-ecological systems? A literature review. *Sustainability, 10*(8), 2950. doi:10.3390u10082950

Holling, C. S. (1973). Resilience and Stability of Ecological Systems. *Annual Review of Ecology and Systematics, 4*(1), 1–23. doi:10.1146/annurev.es.04.110173.000245

Hopwood, B., Mellor, M., & O'Brien, G. (2005). Sustainable development: Mapping different approaches. *Sustainable Development, 13*(1), 38–52. doi:10.1002d.244

Hussain, N., Rigoni, U., & Orij, R. P. (2018). Corporate Governance and Sustainability Performance: Analysis of Triple Bottom Line Performance. *Journal of Business Ethics, 149*(2), 411–432. doi:10.100710551-016-3099-5

Iglesias, O., Markovic, S., Bagherzadeh, M., & Singh, J. J. (2020). Co-creation: A Key Link Between Corporate Social Responsibility, Customer Trust, and Customer Loyalty. *Journal of Business Ethics, 163*(1), 151–166. doi:10.100710551-018-4015-y

JeffersonG. H.TanakaS.YinW. (2013). Environmental Regulation and Industrial Performance: Evidence from Unexpected Externalities in China. doi:10.2139/ssrn.2216220

Kameyama, N. (2004). Ética Empresarial [Business ethics]. *Revista Praia Vermelha [Red Beach Magazine], 11*, 148–166.

Kolk, A. (2016). The social responsibility of international business: From ethics and the environment to CSR and sustainable development. *Journal of World Business, 51*(1), 23–34. doi:10.1016/j.jwb.2015.08.010

Lebel, L., Anderies, J. M., Campbell, B., Folke, C., Hatfield-Dodds, S., Hughes, T. P., & Wilson, J. (2006). Governance and the Capacity to Manage Resilience in Regional Social-Ecological Systems. *Ecology and Society, 11*(1), 19. doi:10.5751/ES-01606-110119

Lele, S. (2013). Rethinking Sustainable Development. *Current History (New York, N.Y.), 112*(757), 311–316. doi:10.1525/curh.2013.112.757.311

Lettice, F., & Parekh, M. (2010). The social innovation process: Themes, challenges and implications for practice. *International Journal of Technology Management, 51*(1), 139–158. doi:10.1504/IJTM.2010.033133

Li, C.-Z., Crépin, A.-S., & Folke, C. (2018). The Economics of Resilience. *International Review of Environmental and Resource Economics, 11*(4), 309–353. doi:10.1561/101.00000096

Maddison, A. (2008). The West and the Rest in the World Economy: 1000-2030. *World Economy, 9*(4).

Martínez-Fernández, J., & Banos-González, I. (2021). *An integral approach to address socio-ecological systems sustainability.*

Martins, J. D. D. (2020). Função social e responsabilidade social empresarial: O princípio da solidariedade como marco jurídico-constitucional para uma nova empresa cidadã. [Social function and corporate social responsibility: The principle of solidarity as a legal-constitutional framework for a new citizen enterprise.]. *Revista de Direito Ambiental e Socioambientalismo, 6*(2), 38–52. doi:10.26668/IndexLawJournals/2525-9628/2020.v6i2.7124

Mercur, B. ([s.d.]). *Mercur - Desde 1924—O mundo de um jeito bom pra todo o mundo.* [*Mercur - Since 1924—The world in a good way for everyone.*] Mercur. https://mercur.com.br/

Mirvis, P., Herrera, M. E. B., Googins, B., & Albareda, L. (2016). Corporate social innovation: How firms learn to innovate for the greater good. *Journal of Business Research, 69*(11), 5014–5021. doi:10.1016/j.jbusres.2016.04.073

Motesharrei, S., Rivas, J., & Kalnay, E. (2014). Human and nature dynamics (HANDY): Modeling inequality and use of resources in the collapse or sustainability of societies. *Ecological Economics, 101*, 90–102. doi:10.1016/j.ecolecon.2014.02.014

Motesharrei, S., Rivas, J., Kalnay, E., Asrar, G. R., Busalacchi, A. J., Cahalan, R. F., Cane, M. A., Colwell, R. R., Feng, K., Franklin, R. S., Hubacek, K., Miralles-Wilhelm, F., Miyoshi, T., Ruth, M., Sagdeev, R., Shirmohammadi, A., Shukla, J., Srebric, J., Yakovenko, V. M., & Zeng, N. (2016). Modeling sustainability: Population, inequality, consumption, and bidirectional coupling of the Earth and Human Systems. *National Science Review, 3*(4), 470–494. doi:10.1093/nsr/nww081 PMID:32747868

Naciti, V., Cesaroni, F., & Pulejo, L. (2022). Corporate governance and sustainability: A review of the existing literature. *The Journal of Management and Governance, 26*(1), 55–74. doi:10.100710997-020-09554-6

OECD. (2015). *G20/OECD Principles of Corporate Governance.* Organisation for Economic Co-operation and Development. doi:10.1787/9789264236882-

Phillips, W., Lee, H., Ghobadian, A., O'Regan, N., & James, P. (2015). Social Innovation and Social Entrepreneurship: A Systematic Review. *Group & Organization Management*, *40*(3), 428–461. doi:10.1177/1059601114560063

Phills, J. A. Jr, Deiglmeier, K., & Miller, D. T. (2008). Rediscovering Social Innovation. *Stanford Social Innovation Review*, *6*(4), 34–43. doi:10.48558/GBJY-GJ47

Pisters, S. R., Vihinen, H., & Figueiredo, E. (2020). Inner change and sustainability initiatives: Exploring the narratives from eco-villagers through a place-based transformative learning approach. *Sustainability Science*, *15*(2), 395–409. doi:10.100711625-019-00775-9

Popoli, P. (2016). Social Enterprise and Social Innovation: A Look Beyond Corporate Social Responsibility. Em R. Laratta, Social Enterprise—Context-Dependent Dynamics In A Global Perspective. IntechOpen. doi:10.5772/62980

Porter, M. E., & Kramer, M. R. (2006). Strategy and society: The link between competitive advantage and corporate social responsibility. *Harvard Business Review*, *84*(12), 78–92, 163. PMID:17183795

Rist, G. (2007). Development as a buzzword. *Development in Practice*, *17*(4–5), 485–491. doi:10.1080/09614520701469328

Rist, G. (2008). *The history of development: From Western origins to global faith* (3. ed., 2. impr). Zed Books.

Rockström, J., Steffen, W., Noone, K., Persson, Å., Chapin, F. S. I., Lambin, E., Lenton, T. M., Scheffer, M., Folke, C., Schellnhuber, H. J., Nykvist, B., de Wit, C. A., Hughes, T., van der Leeuw, S., Rodhe, H., Sörlin, S., Snyder, P. K., Costanza, R., Svedin, U., & Foley, J. (2009). Planetary Boundaries: Exploring the Safe Operating Space for Humanity. *Ecology and Society*, *14*(2), art32. doi:10.5751/ES-03180-140232

Rosati, F., & Faria, L. G. D. (2019). Business contribution to the Sustainable Development Agenda: Organizational factors related to early adoption of SDG reporting. *Corporate Social Responsibility and Environmental Management*, *26*(3), 588–597. doi:10.1002/csr.1705

Ruggie, J. (2011). Report of the Special Representative of the Secretary-General on the Issue of Human Rights and Transnational Corporations and other Business Enterprises: Guiding Principles on Business and Human Rights: Implementing the United Nations 'Protect, Respect and Remedy' Framework. *Netherlands Quarterly of Human Rights*, *29*(2), 224–253. doi:10.1177/016934411102900206

Schaltegger, S., & Burritt, R. (2018). Business Cases and Corporate Engagement with Sustainability: Differentiating Ethical Motivations. *Journal of Business Ethics*, *147*(2), 241–259. doi:10.100710551-015-2938-0

Scherer, A. G., & Voegtlin, C. (2020). Corporate Governance for Responsible Innovation: Approaches to Corporate Governance and Their Implications for Sustainable Development. *The Academy of Management Perspectives*, *34*(2), 182–208. doi:10.5465/amp.2017.0175

Solomon, R. C. (1993). Business Ethics. In *A Companion to Ethics* (pp. 354–365). Blackwell Publishers.

Souza, M. C. G. D. (2009). *Ética no ambiente de trabalho*. Elsevier Editora.

Turnbull, S. (1997). Stakeholder Governance: A Cybernetic and Property Rights Analysis. *Corporate Governance*, 5(1), 11–23. doi:10.1111/1467-8683.00035

Vallance, E. (1995). *Business ethics at work*. Cambridge University Press. doi:10.1017/CBO9781139166461

van den Bergh, J. C. J. M. (2009). The GDP paradox. *Journal of Economic Psychology*, 30(2), 117–135. doi:10.1016/j.joep.2008.12.001

Varela, J. A., & António, N. S. (2012). *O Bem Comum e a Teoria dos Stakeholders*. [*The Common Good and the Stakeholder Theory*.] ISCTE-IUL: Business Research Unit.

von Schomberg, R. (2012). Prospects for technology assessment in a framework of responsible research and innovation. Em M. Dusseldorp & R. Beecroft (Orgs.), Technikfolgen abschätzen lehren: Bildungspotenziale transdisziplinärer Methoden (p. 39–61). VS Verlag für Sozialwissenschaften. doi:10.1007/978-3-531-93468-6_2

Walker, B., Holling, C. S., Carpenter, S., & Kinzig, A. (2004). Resilience, Adaptability and Transformability in Social–ecological Systems. *Ecology and Society*, 9(2), 5. doi:10.5751/ES-00650-090205

WCED. (1987). Our Common Future. Oxford University Press.

Zaman, R., Jain, T., Samara, G., & Jamali, D. (2022). Corporate Governance Meets Corporate Social Responsibility: Mapping the Interface. *Business & Society*, 61(3), 690–752. doi:10.1177/0007650320973415

ENDNOTES

1 Management Facilitator at Mercur
2 Procurement Area Coordinator at Mercur
3 Coordinator of Mercur's Social Innovation Laboratory
4 Human Resources Analyst at Mercur
5 Relationship Analyst at Mercur
6 Communications Manager at Mercur
7 Project Specialist at Mercur
8 As Mercur's demand for latex is greater than the production capacity of extractivists, the company also purchases natural rubber from other regions of the country.
9 Beans and rice, accompanied by other dishes they would make, are the basis of the main meals in Brazil.
10 Coordenador da área de Impactos da Atividade da Mercur
11 Inter-union Department of Statistics and Socioeconomic Studies (*Departamento Intersindical de Estatística e Estudos Socioeconômicos* – DIEESE)

[1] Inter-union Department of Statistics and Socioeconomic Studies (*Departamento Intersindical de Estatística e Estudos Socioeconômicos* – DIEESE)

APPENDIX

Table 7 shows a timeline with the main changes implemented in the company since the "turn of the key".

Table 7. Timeline of the actions and initiatives of the Mercur company from the "turn of the key"

Year	Event
2009	Mercur employees participated in an Ecological Literacy course
2009	Start of the "turn of the key"
2009	Start of the co-creation process
2009	Beginning of studies on the native rubber chain, with the recovery of the history of extractivists and their role in forest conservation
2010	Start of the application of readjustment percentages, aiming to reduce salary differences, according to the values of the salary ranges, that is, whoever receives the salary up to the value stipulated as a "cut-off point" (based on DIEESE[1]) earns a higher percentage of readjustment, and whoever receives a nominal salary greater than this "cut-off point" receives a slightly smaller readjustment.
2010	First purchase of native rubber from rubber tappers in Terra do Meio, Amazon, totaling 2 tons.
2012	To facilitate the introduction of organic food in the company, it began the learning about healthy eating and conversations about the institutional positioning with partners (Refeições ao Ponto, CAPA – Centro de Apoio e Promoção da Agroecologia and Ecovale – Cooperativa Regional dos Agricultores Familiares Ecologistas Ltda).
2013	This year, organic rice is now being served twice a week and organic beans once a week. A SIPAT (Semana Interna de Prevenção de Acidentes de Trabalho - Workplace Accident Prevention Internal Week) was held under the theme healthy nutrition, with nutritionists advising on the importance of the topic and the harm caused by pesticides. The format of the menu meetings was changed, being called "Talking about Food", with the proposal to improve the quality of what is served, reducing fried foods, lower salt consumption, reducing food waste, and demonstrating respect for food and for people who do not have access to adequate food.
2013	Mercur started to grant employees with reimbursements related to education and health, with different percentages, that is, those who receive a salary up to the amount stipulated as a "cut-off point" earn a higher percentage of reimbursement and those who receive a salary greater than this "cut-off point" receives a lower refund amount.
2013	Implementation of new ways of developing products and services through the *Diversidade na Rua* [Diversity of the Street] Project.
2014	Implementation of a space for the development of the co-creation process, the LAB (*Laboratory of Social Innovation*), which has an area of 1050 m^2.
2014	Expansion of the offer of organic rice and beans for employees in the company.
2014	Extinction of the third work shift (night work) and relocation of employees to other activities.
2015	Organic beans and rice are now served daily at the company.
2015	Reduction of the working week from 40 hours to 36 hours.
2018	Mercur Profit Sharing Program (*Programa de Participação dos Resultados da Mercur* – PROMEPAR). It is now distributed in equal parts to all employees, regardless of the salary they receive in the organization, sharing the company's results.
2019	The *Fazer Com* (Do With) co-creation process was implemented in the area of R&D (Research and Development) and it was defined that this is the process that matches the *Jeito de Ser* (Way of Being) of the company.

Surce: Mercur S.A.

Chapter 13
Symbiosis of Humanistic Leadership, Sustainability, and Circular Economy

Damini Saini
https://orcid.org/0000-0001-5734-5067
Indian Institute of Management, Raipur, India

Juhi Agarwal
https://orcid.org/0000-0001-9768-440X
IIT Roorkee, India

ABSTRACT

Corporate sustainability is presumed to be a business strategy that creates long-term value by focusing on innovative measures aimed towards the natural environment and aligning it with its external environment. Continuous measures at global, institutional, and individual level have to be taken up to ensure sustainability paving the way to bring concepts like circular economy, sustainable practices into limelight among the scholars, academicians, and even corporate houses. This chapter tries to establish a link between a leader's role in an organization and creating spaces for circular economy and sustainability in their already existing cycles. On the basis of the steps taken for creating a sustainable business by TATA rganization, the chapter explores their leadership style and suggests that the values such as responsibility, humanity, and empathy in leadership became more important and goes well with the vision of circular economy and how well it has been dwelled up with the business models of TATA.

INTRODUCTION

The conditions prevailing on the planet at present with the emergence of global pandemic, economic crisis and degrading environment has turned the interest area of researchers, academicians, scientist, industrialist and even common people towards the phenomenon of sustainability, environment protection, circular economy and other green philosophies. The situation is alarming because we are on the verge of

DOI: 10.4018/978-1-6684-6123-5.ch013

extinction of many valuable resources. With the population at its ever-growing rate (almost 8.5 billion by 2030 as per the forecast by United Nation Department of Economic and Social Affairs in 2017) even the need for basic amenities like food, shelter, clothing is at stake and paving its way for exploitation being done at a rapid rate causing severe depletion of the environment will make the planet a difficult place to live in. Along with that around 3 billion consumers (middle class) are expected to join the global markets by 2030 (Organization for Economic Co-operation and Development, 2010). Considering all these factors, in recent years, various green concepts have made an appealing impact on the people and influenced them to move towards a safer world.

Big business houses are exploring a business model (Bocken et al., 2016), to eradicate the risk of depletion, overutilization of resources, inflation and materialistic revenue increments. Henceforth, Sustainability is the only panacea in case applied can help out natural environment conservation, reducing overexploitation of resources, moving to circular use of resources, and many more. Thus, a sustainable approach will help in keeping the ecosystem operational, intact and maintaining the quality of life for future generations. There are many sources identified leading to environmental degradation (Andersen, 2007) and out of them the industrial sector has been the major stakeholder apart from consumers and public, which are affecting the environment and its sustainability with their anti-environmental acts and behavior. The industrial impact is though economical in the areas of employment, services, profits, growth, but it negatively influences the ecosystem and therefore at present, the companies are mandated with Corporate Social Responsibility (Rajeev & Kalagnanam, 2017), to assure that they are giving back to society and ensuring environmental sustainability.

The industries are now moving from the linear cycle to a circular cycle with respect to the goods and services, manufactured and delivered to the ultimate consumer groups, thereby ensuring the reusability, recycling, reduction and recovery of the environment and resources and maintaining economic prosperity for present as well future generations (Beccarello & Foggia, 2018). However there is still a disconnect between the industrial innovations and academics while it comes to create sustainable organisation and vice versa. The available research cannot recommend one best way to establish sustainability in the system. Apart from this the concept of circular economy is still at a developing stage; it is not that easy to make a shift and therefore, can be applied to different forms of industries only with the right leadership supporting it. The leaders play a vital role in making the organizations and its people adhere to the principles of circular economy as it requires round the world collaborations, cross portfolio co-ordinations and understanding of different cultures and businesses across the globe. The focus of the chapter is to establish the connection between leadership and circular economy and then identifying and establishing the right leadership (Robinson, Kleffner, & Bertels, 2011), which can create a trustable, sustainable and growing environment for people to work.

Thereby ensuring the environment conservation and protection as leaders are the ones helping the organizations in their continuous drive for growth and success by creating awareness about the circular economy and using the interpersonal skills to effectively spread out and communicate the broader intentions of it to the stakeholders (Doh & Quigley, 2014).

The next part of the chapter describes the related concept of sustainability like corporate social responsibility and circular economy and role of leadership and top management in the various areas of organization which directly or indirectly impact on sustainability. Further the chapter consists of a case of TATA organisation (based on secondary data) which provides the example of a leading style aligned with the concept of circular economy. After that the chapter focuses upon the challenges and opportunities associated with the proposed leadership and finally it talks about how it complements and is suitable in the present education scenario. It also talks about future requirements from research and academic perspectives and finally practical implications.

LITERATURE REVIEW

Leadership and Humanistic Leadership

"Leaders are born not made" – The Great man theory of leadership claims this with the examples of Martin Luther King, Mahatma Gandhi, Jack Welch, and many more great personalities from history. This cannot be denied but cannot be accepted to the fullest as there are people who have grilled themselves and developed to their fullest potential as a leader through their perseverance, hard work and compassion. Henceforth, leadership refers to the process of influencing others to follow what is being said and move in an aforesaid direction to accomplish ultimate goal or objective (Ulrich, Zenger, & Smallwood, 1999). It refers to uplifting and empowering people to revolutionize change and making people accept it wholeheartedly (Spreitzer & Quinn, 2001). Leadership can be conceptualized as a collective phenomenon where different individuals contribute to the organization (Pettigrew & Whipp 1991). Leadership competencies, in turn, refer to the knowledge, skills, or abilities that facilitate one's ability to perform a task (McClelland and Boyatzis, 1982). Leadership is seen more as a fluid consequence, naturally arising from, contributing to and being shaped by social practices (Day, 2001). According to Rauch and Behling (1985) leadership is a process of influencing the activities of an organized group towards goal achievement.

Based on the literature, the right one to discuss about humanistic leadership starts with the definition by Fu et al. (2020) humanistic leader considers people as unified humans with their needs and motives and provides opportunity for them to utilize their fuller potential, thereby being immersed with wealth maximization of the organization and overall societal good. Humanistic Leadership can often be understood as regarding human dignity (Pirson, 2017), captivating ethicality in decisions and engrossing stakeholders (Sharma, 2019). Some suggest humanistic leadership as a link to support individual good and community welfare through prioritizing long-term commitment, focusing on the community good, holding unified vision and being realistic (Fremeaux & Michaelson, 2017). It considers people, social relationships, community, and ethical values to be critical (Mele, 2016). Accordingly, it also stresses on employees being individuals mushrooming in the organizations (Fritz & Sorgel, 2017). It can be thus considered to be the one entailing interaction and apprehension for employees along with societal orientation.

Corporate Social Responsibility

Corporate social responsibility (CSR) is not only a word to be described; it is a mixture of emotions towards the betterment of the society. Achievers or Non-Achievers try to indulge in philanthropic activities to make this world a better place to live in, for those who are not resourceful. CSR is a commitment to improve community well-being through discretionary business practices and contributions of corporate resources (Kotler & Lee, 2005). It is the critical linkage between business strategy and sustainable development (Steurer et al., 2005). Later, Simpson and Taylor (2013) described CSR as an additional responsibility of businesses to local and wider communities apart from its core responsibility of profit maximization. With the rise in globalization and ecological issues, the perspective of corporate has been changed from profit maximization to profit maximization along with social welfare and thus they have broadened their ideologies of operating business in the social context. The World Business Council for Sustainable Development (WBCSD) in its publication "Making Good Business Sense" by Holme and Watts (2020) defined CSR as – the continuing commitment by business to behave ethically and contribute to economic development while improving the quality of life of the workforce and their families as well as of the local community and society at large.

In India, the concept of CSR has come a long way (Satapathy & Paltasingh, 2019) In the Pre – Independence Era, it was a charitable activity that was undertaken considering the various religious and cultural beliefs that were prevalent in the society. After Independence, because of the existence of the Gandhian Principle of Trusteeship, it was the responsibility of the Public Sector Units to undertake the responsibility of development. Then came the era of 90's, where, with the introduction of Liberalization, Privatization and Globalization policy, the world was emerging as a single market and thus now the business organizations have to look after the needs of the people so as to match with the global standards and thereby enhancing their brand value and goodwill (Mitra & Schmidpeter 2017). Further, in today's world, there are many countries like France, Netherlands, Norway, Swedeni which have implemented the Corporate Social Responsibility reporting laws, India (with effect from 1.4.2014) started mandating it for the organizations so as to ensure overall development with the emerging concepts like Triple Bottom Line and Sustainable Development and many more (Karnani, 2013).

SUSTAINABILITY

Human needs are fulfilled either directly or indirectly from the natural environment we live in. With the increase in these human needs and desires, exploitation and overutilization of resources started to take place. Henceforth, sustainability brings out conditions where humans and nature can coexist on the planet in harmony, fulfilling the present as well as future requirements (Robertson, 2017). United Nations Brundtland Commission (1987) defined sustainability as meeting the needs of the present without compromising the ability of future generations to meet their own needs.

So, sustainability refers to meeting the needs and requirements of the present without damaging the reserves for contingency and future. Sustainability is generally seen as based on the three main pillars that is, people, planet and profit which can be understood as social, environmental and economic parameters. Subsequently, the concept of Triple bottom Line was coined by Elkington and Fennell (1998) who is often referred to as one of the developers of the term sustainability. It refers to the generic development without compromising about the living and non-living components of the planet. With the growing indecency towards the environment, the term sustainability came into existence, for the world. The term 'sustainable development' was coined in 1987's report 'Our Common Future' by the World Commission of Environment and Development (WCED) (Redclift, 1989; Palmer, 1992). The Sustainable Development Goals were identified by the UN in the year 2015 so as to make this planet a better world to live in, by working on various parameters like poverty, inequality, environment conservation and many other concerns, by the year 2030.

Sustainability is a system that can evolve indefinitely toward greater human utility, greater efficiency of resource use, and a balance with the environment which is favorable to humans and most other species (Miller, Bersoff, & Harwood, 1990). Sustainability is an ideal end-state like democracy, it is a lofty goal whose perfect realization eludes us. It includes the well-being of people, nature, our economy, and our social institutions, working together effectively over the long-term (Fricker, 1998). Sustainability is anything more than a slogan or an expression of emotion; it must amount to an injunction to preserve productive capacity for the indefinite future (Solow 1991). Finally, sustainability takes the resources as being finite in nature, and should be used in a conservative manner, keeping in mind the long-term usage and priority, and wisely enough, understanding the consequences of how those resources are going to be utilized.

CIRCULAR ECONOMY

The growth in population, consumption, and human needs lead to the extraction and utilization of the natural resources abruptly, paving way for the environment to suffer, which eventually leads to human suffering. The disastrous experiences have made people rethink their attitude towards the environment and the planet. According to Ellen MacArthur Foundation (2013), the currently prevailing economic design has its roots in the historically uneven distribution of wealth by geographic region. The problem is thus rooted in here; the inequality has made people intolerant towards its right usage. Business houses and industries lay off the responsibility of using and disposing the products to the consumers and refrain themselves from having any accountability towards the impact of that on the environment altogether (Zink & Geyer 2017). To ensure that the future generations are also able to meet out their needs and well-being, people are adhering to the concept of Circular Economy, which tries to ensure that the resources, goods and services can be recycled to enhance its usability and they are not unnecessarily exploited or wasted (Kirchherr, Reike, & Hekkert, 2017). Circular Economy is often seen as an antonym of linear economy, also as a process which is cyclical in nature involving recycling as its core. Linear Economy is perceived as a cradle to waste economy, where the resources extracted, then put into manufacturing, consumed and ultimately thrown away just like that. This model works on the traditional make, use and waste concept and is referred to as cowboy economy (Boulding, 1966).

The origin of the term Circular Economy is often debated on, where some claim it to be a part of Chinese Literature, some say it originated from Germany, some argue with Western Literature as its origin. Boulding (1966) stated that man must find his place in a cyclical ecological system which is capable of continuous reproduction of material form, even though it cannot escape having inputs of energy. Yang and Feng (2008), argued that circular economy is the abbreviated version of closed material cycle economy or resources circulated economy, in their studies. A circular economy is a system in which the throughput of energy and raw material is reduced, where the materials are reduced, reused and recycled in the production, distribution and consumption processes (Cooper, 1999). Many parts of the world have embraced themselves well with this concept including Europe, France, China, UK, who are among the leading proponents of it. One such proponent is the NGO, the Ellen MacArthur Foundation, from Isle of Wight, UK, who commissioned McKinsey and Company, which has produced three reports on this concept (Ellen MacArthur Foundation 2012, 2013, 2014). They started with pointing towards the limitations of the linear economy and moved towards identification of the benefits that can be drawn from it.

It can also be expressed as a guiding principle for development that embraces three aims which must be accomplished instantaneously: Environmental quality, economic prosperity and social equity (Taylor, 2016; WCED, 1987). Therefore, it generates value by extending the lifetime of products, across all the stages of the life cycle, allowing them to remain in the economy for the longer period of time and recouping their material basics. It is a reformative system which slows down, closes and narrow down wastages, leakages and unnecessary emissions to enhance sustainability (Geissdoerfer et al., 2017) The adoption of a circular economy approach, therefore, requires creative and innovative business models which can help out in creating, capturing and delivering value, along with maintaining the efficiency of the resources and extending the useful life of the product.

Importance of Sustainable Practices

Sustainable practices involving green economy, corporate social responsibility, circular economy and many of these green measures provide an essentially global perspective to business doing as well as emphasizing on the problems at a planetary scale that has eventually lead to shared roles and responsibilities (Fulton, De Silva, & Anton, 2012). There are certain key drivers which often pushes the organizations to take up this role. There is a hidden value in the concept of sustainability and circular economy, which is making all the different sectors interested in it. The internal complexity of the processes in the system has drawn out arguments from different origins in defining it. The description of sustainable practices or circular economy is embedded with inter sectoral and inter organizational interests and preferences. The knowledge and database on these concepts are ever evolving; there is openness to make additions to the literature as very little has been explored till yet. When the organizations adopt sustainable business practices, there are certain benefits which come as a motivator (Pohl & Tolhurst,2010).

These benefits often influence the organizations to inculcate these green practices into the long-term plans. Various surveys have been done in this regard, one such was done by Natural Marketing Institute on the U.S. consumers, where it was identified that people do choose the products which are an outcome of green practices. So, it can be said that such practices lead to creation of brand image and goodwill for the organization. The sustainable practices streamline the efforts and the resources that are put into operations, and lead to costs and resource optimization and reduction in a way (Schmidt, Zanini, and Korzenowski, 2018). There are various guidelines given at national and international level with respect to the environment conservation and protection, adopting sustainable practices can help the organizations in complying with the global regulations and maintaining international standards. Another main concern for the environment is the waste that is not recyclable and is degrading the quality of the environment. Sustainable practices help in reducing the waste as well as using such facilities which are recyclable.

Moreover Mckinsey (2014) reported that according to the study by various organizations like Deutsche Bank and Carbon Disclosure Project, it was identified that the organizations adopting sustainable practices were able to make their stakeholders happy and content as well as they are able to attract and retain the employees. In general, the need and importance of sustainable practices, with respect to the environment at large can be many. The practices ensure a livable future be it for the humans on the planet of the organizations for a longer period of time. It helps in ensuring the sustainability of the businesses at large. Nature is the biggest designer for itself, it is ever evolving, transforming, and finding newer ways to provide for it and thus the business needs to cope up with its dynamicity to sustain (Collins, Roper, & Lawrence, 2010). Therefore, allowing the organizations to build the resilience to cope up with it, provides immense opportunities to the organizations, government and other stakeholders to make plans for the contingency. In this modern era of technological advancements, manufacturing redesigning, streamlined production, dynamicity and ever-changing natural environment, it is inevitable for the business and organizations of all scales and potential to induce quality, affordability and sustainability in their plans so that they may get a competitive edge as well as consider the environmental impact of their practices when making business strategic decisions.

Humanistic Leadership in Advancing Sustainable Practices

We have already established the significance of circular economy and sustainability in business through literature review in the above paragraphs. Corporate social responsibility of business is a wide-ranging array of strategies and operating practices that a company cultivates in its efforts to deal with and create relations with its many stakeholders and the biological atmosphere (Waddock, 2004). These efforts are the outcome of the ideological thinking of the firm, its top management, who are the potential leaders and drivers for bringing about a change and implementing it across the organization. People often resist the radical changes due to their own insecurities related to self; this is where the role of leader comes to play. Most of the leadership theories focus on the individualized or small group impact of leaders rather than on the organizational processes, like Leader member exchange theory or other leadership models. Going with the kind of crisis the environment is facing, right leadership is required at the various levels of the organization, especially at the strategic level, so as to make decisions and induce change in the overall system (Metcalf & Benn, 2013). With the increased research, various concepts like responsible leadership, crisis leadership (Maak, 2007) have come into existence lately. Further, study of Pasricha, Singh, and Verma (2017), established that the linkage between ethical leadership and CSR exists which can be either direct or indirect.

According to Maak and Pless (2006) responsible leaders become a coordinator and cultivator of relationships towards the group of different stakeholders. The leader has to come out of the traditional role of leader-subordinate relationship and start focusing on the leader stakeholder relationship as an outcome of responsible leadership. The top management plays a very crucial role as a crisis leader or as a responsible leader and embraces it by coming out with such strategies and policies which helps in manifesting the objective of sustainability and dealing with the issues of sustainability and corporate social responsibility (Quinn & Dalton, 2009; Hu, Chen & Wang, 2018). Many recent studies have tried to establish this fact with the humanistic leadership around the globe within various country contexts for example. Latin America (Davila & Alvira, 2012), UAE (Anadol & Behery, 2020) and China (Yang, Fu & Beveridge, 2020). In Indian context also we found a few studies where correlation between humanistic leadership and being socially responsible ensuring sustainability was established for example Singh & Singh (2020) and Sharma (2019) contributed a spiritual humanistic model of sustainable leadership by analysing the case from Indian organizations. Tripathi and Kumar (2020) analysed the role of humanistic leadership in adhering to the ethical values, passion for community development and Code of Conduct in the Indian organization.

This chapter proposes the use of a humanistic style of leadership where the focal point is both individuals and collective human beings. Subsequently, the philosophical and ethical values of the parent organization, guides the leadership behavior. This eventually helps in embarking such practices that highly impacts sustainability and CSR. Humanistic, eclectic, pragmatic, and collaborative leadership is required for long-term sustainability. Its goal is to promote the dignity and worth of humans along with environmental sustainability. It may contribute to making these concerns viable topics to discuss, as well as the issues included in the dialogue that develops the social policy, legal and regulatory frameworks (Colbert, Nicholson, & Kurucz, 2018). Research supports that a relationship exists between humanistic leadership with sustainability (Lawrence & Pirson 2015). A leader with a humanistic approach, works across the functional areas of the organizations; this requires a collaborative approach where there is a proper cross functional communication flow irrespective of the hierarchical structures from the leader to the employees (Waddock, 2016). Humanistic approach of leaders consists of the attributes identified for an effective leader for sustainability including –integrity with a high level of ethical decision making skills, willingness with open mindedness to confront resistance from the stakeholders, having a long term ethical perspective on the bottom line measures, demonstrating care for the people and stakeholders, being ethical in day to day behavior, effective communication skills and lastly playing the role of disseminator, spokesperson of the organization (Boeker, 1997). The responsibility lies on the shoulders of the leader to help the organization and its people to refute from the materialistic individual maximize approach and move towards a more group oriented utility maximization, so as to ensure and motivate the organizations well as the employees to contribute to the sustainable practices (Pirson, & Lawrence, 2010; Saini & Sengupta,2016).

CSR at Tata Consultancy Services – India

There are many organizations which have emerged out to be great players when it comes to the implementation of successful sustainable practices. One such organization is the TATA group, in India which is found out to be among the top position holders in community and environmental services. The case is about the counterpart of TATA group, that is, Tata Consultancy Services, which is coming up with promising sustainable practices across all its pillars of people, planet and profit, because of the strong responsible and ethical leadership of the founders and the top management. The case presents various green practices taken up for the welfare of the society and environment at large. CSR programs at TCS, promotes the ideology of creating a sustainable environment and wellbeing of all. Their guiding principle is "Impact through Empowerment", where the outcome of its activities and the impacts are conveyed through long-term efforts based on the agenda of sustainability. Utilizing its proficiency in the IT sector, it tries to utilize the strength of its large employee base which is globally present, to transform its initiatives and the way of delivering them. Various initiatives are taken up at the regional as well as corporate level across the various offices of TCS in India as well as Abroad. Regional Initiatives more or less have a commitment towards the fulfillment of the needs and requirements of that specific region or nearby regions and areas. Corporate Initiatives have much wider applicability as they are run on a much higher scale and cater to the needs of a larger group of society. Various initiatives which are taken up at the corporate level are;

1. Adult Literacy Program, which is an initiative started by TCS in the year 2000. It functions on TCS' proprietary Computer-based Functional Literacy (CBFL) platform, henceforth helping in the promotion of literacy and education through multiple innovative information and communications technology solutions. Empower is a program initiated to help the support staff of TCS to make them proficient in speaking and understanding English. Bridge IT is an alliance between the National Confederation of Dalit and Adivasi Organizations (NACDAOR), which is an Indian Non-Governmental Organization (NGO), committed towards the burgeoning of people across the nation and the organization.
2. IT Employability is an initiative by TCS for the rural based engineering college students to make them capable enough to pursue a career in IT industry. Almost around 2500 and more people benefitted from this in the financial year 2019, and were acquired by the organization itself.

Symbiosis of Humanistic Leadership, Sustainability, and Circular Economy

3. Launchpad and Insight is an initiative for the senior secondary level students to inculcate in them the skills required to be a part of this tech-oriented world by providing them a gamification approach to learn computer languages. In the financial year 2019, almost thirteen thousand students availed its benefit.
4. Various technological upgradations and advancements are being continuously researched and launched in the Cancer Institute at Chennai and Tata Medical Centre at Kolkata, which is working towards the heart diseases and cancer research continuously to serve the nation.
5. TCS also adopts the Environment Management System globally to ensure safety of the environment with respect to carbon, water, material, resources, energy and waste management. The below table (1) depicts the various environmental concerns and how it has been dealt by the organization effectively.

Table 1. Aspects and approaches used by TCS

Aspect	Approach
1. Biodegradable Waste	Treated onsite for biogas recovery or manure generation through bio-digesters or composting.
2. Food Wastage	In FY 2019, 42% of the total food waste generated was treated using onsite composting methods or bio-digester treatment.
3. Water Wastage	The new campuses have been designed for 50% higher water efficiency, 100% treatment and recycling of sewage, and rainwater harvesting.
4. Ozone Depletion	The new campus facilities at the campus are well equipped with the HVAC systems based on zero-ODP refrigerants.
5. Carbon Footprint	Investment in communication and Video conferencing Infrastructure to reduce carbon footprint from air by 59%, acquiring green buildings, data center/server room consolidation, rack cooling solutions, airflow management, UPS load optimization through modular UPS solutions and centralized monitoring to reduce it by 50%

Source: TATA Consultancy Services report (2018-19)

It can be outlined how the processes are being introduced across all the pillars of sustainability, that is, people, planet and profit by the organization. The out of the box thinking, innovative measures that are embedded is the system is making the organization one of the top companies in India to follow the sustainable practices and contributing towards Corporate Social Responsibility. The various CSR activities taken up by the Tata Consultancy Services, is reflecting the strong humanistic leadership of its top management and founding members, to indulge in the sustainable activities, ensuring that they have a concern towards their stakeholders as well as the environment at large. The right leadership qualities and skills had helped them to achieve this momentum right from the beginning, keeping intact its roots, ethical values and strong human prepositions without any compromise.

Humanistic Leadership at TATA

TATA always went ahead with remarkable ways to show passion for the community welfare with their campaigns and programs across their brands and businesses reflecting their humanitarian leadership approach. Right from the beginning, Jamshetji Tata in 1902, ensured that before beginning any operations, there are wide streets planted with shady trees, plenty of space for lawns and parks, large areas for football, hockey and gardens, areas for temples, mosques and churches`` (Lala, 2004), which laid down a strong foundation for the company to enact and implement the ideology. Then TATA established premier institutes like TATA Institute of Social Sciences, TATA Institute of Fundamental Research, Indian Institute of Sciences, Cancer Institute and many other medical and educational institutes to ensure they lead with the notions of humanistic approach. The focus is to count each and every step that can be taken, thus there is leadership with trust, with all the potential stakeholders of the company. The focus of the company is always societal welfare. For example in areas of Kunoor, Hosur and other districts they worked for the upliftment of the community. TATA group has been an exemplar for corporate responsibility right from the start, even before the mandatory rules, as they focus on grooming their business leaders and inculcating in them the ideology and generosity of the top leaders and their humanistic style of leadership. As we have discussed, TATA group is one of its kind when it comes to adherence of the principles by the founding members, with this scope and scale of business. They have synergistic and practical ways to carry on with the humanistic leadership forward not only in India but outside world (Cappelli et al.,2010).

CHALLENGES AND OPPORTUNITIES IN ADOPTING SUSTAINABLE PRACTICES WITH RESPECT TO LEADERSHIP

Irrespective of how promising the notions of corporate social responsibility or circular economy looks like, there are numerous challenges or limitations that are in the way of adopting the sustainable practices by the organizations. There are limitations with corporate social responsibility where organizations are actually not able to gather enough information about the business benefits that they can derive out of CSR implementation and indulge in inefficient and confusing strategies which leads them nowhere with respect to the standards given (Jumde and du Plessis, 2020). The objective or the rationale behind why the companies are adopting the sustainability measures remains vague within the organization itself, which eventually impacts the right implementation of the policies because of either lack of trained personnel, lack of proper funding or many other reasons and puts the corporate image at stake (Aswani, Chidambaran, & Hasan, 2020). When it comes to circular economy, there are differences with respect to the semantics, that is, the lack of clarity in the nomenclature where circular and linear are somewhere associated with economy and somewhere associated with flow of income, which can be confusing at times. Also, the concept of sustainability is based on the three main pillars of social, economic and environment, where generally the focus is either on environmental issues or the economic concerns, the societal aspect of it remains underlined. Another concern surrounding the concept of sustainability and corporate responsibility is the unintended consequences that become a part of the system. Many times the processes which are considered to be sustainable, actually lead to negative consequences for the environment. For example: the companies act, 2013 in India which prescribes mandatory provisions for Companies to fulfil their CSR.

Lack of judicious guidelines and laws, is again weakening the whole scenario, as whatsoever guidelines are being suggested differs with respect to different sectors which creates lack of uniformity (Gatti et al., 2019). Because of lack of transparency and validation, most of the sustainability practices have emerged out to be just looking after the areas which the organization is surrounded by, which is a major concern as the objective of global change is somewhat at stake. Many other issues like the standardization regarding sustainable practices and their proper evaluation and implementation creates a barrier for the success of sustainable practices. These limitations clearly define that the organizations are only emphasizing on the goals which seems to be too simplistic in carrying out the real objective behind the concept of sustainable practices (Aswani, Chidambaran, & Hasan, 2020). Irrespective of these limitations, there are various opportunities which pave the way for organizations to take measures and adopt sustainable practices. They help the organizations in fulfilling Intra and intergenerational commitments which can help in creation of goodwill for the organization. The organizations get access to many agencies for the multiple and coinciding pathways of development in the longer run.

Global models and measures can be adopted, which help in creating global imprint. Often these practices help the organizations to integrate non-economic aspects into development. The system changes and innovation is at the core of these practices which can help in continuous growth of the organization. This domain often opens up the opportunity for interdisciplinary research involving themes like Potential cost, risk, diversification, value co-creation and many more. It helps the organizations to get cooperation of different stakeholders in implementing the practices thereby giving them a space to establish and maintain the relations for long run. The organizations are often provided with the regulation and incentives as core implementation tools by the government, as well as from the national and international flag bearers of environment protection and conservation. The sustainability factors often highlight the central role of private business, due to resources and capabilities that they possess. It brings out the business model innovation as a key for industry transformation (Rajput and Singh, 2019). Various other opportunities that sustainable business practices can bring forth can be seen as what outcome it can draw for the organization as well as the community at large.

When it comes to organization, they can have enhanced brand loyalty and image, increased market share, access to more capital, dedicated personnel, high productivity and reliability, standardization of processes, lowering the costs and overall improved financial performance. When considering about the community and general environment and people, there is a lot of potential to do charitable programs, employee volunteering, community education, safety and quality of products, material recyclability, integration of environmental issues and concerns in overall business plans and many more (Korhonen, Honkasalo, & Seppälä, 2018). This way it can be concluded that no matter what challenges or limitations it has, the opportunities that come along acts as a great motivation for the organizations to get themselves embraced in the process.

CONCLUSION

The implementation and integration of the sustainable practices bring about various socio-cultural changes which is evident from the work practices that are being adopted by the organization's right from its inception. Leadership has always been condemnatory when it comes to corporate social responsibility and sustainability, it comes in varied forms and models, often stems out Corporate Social Irresponsibility when there is lack of leadership. The main factor which often influences the extent to which any organization gets integrated with the notion of sustainability is the skills, quality and expertise of the leader who is present either on the board or the top management of the organization. The intention of the chapter was to identify the link between the various sustainable practices like that of CSR or green economy or circular economy and the kind of leadership style that can be adopted in varied situations. The chapter tried to support this linkage with the case of TATA and the various sustainability practices adopted by TATA-TCS in the last few years, under the flagship and the great humanistic leadership of its parent company (TATA) and the top management. Over time the view on CSR has changed from "no harm" to "transforming the world a better place to live in", and the TATA-TCS has proved that by implementing this initiative in the leadership at top level management. A human touch is required to infuse CSR and sustainability into every aspect of the business and organization to obtain higher outcomes.

The implementation of a circular economy requires a lot of time, effort and energy to make a transition towards it at individual, regional as well as the global level, henceforth, the role of leadership outshines. While looking at the organizational front, with a case of Sustainable Practices at TATA-TCS, it can be said that though they pioneered, opportunities in this regard are going to expand only in the coming future. TATAs are always coming up with creative and innovative measures, and leading a way towards sustainability. Other business leaders can also volunteer as well as make the people of the organization volunteer towards the change that is expected to come by adopting the sustainable practices, and henceforth associate it with the long-term objectives and success of the organizations. The growing interest towards the concepts of sustainable practices and circular economy requires humanistic leadership like the TATA group, which is altogether being researched upon by the academicians, industrialists and researchers time and again.

Implementation for Academia

Despite the word sustainability, circular economy, social responsibility is heavily highlighted by great industrialists, scholars, academicians, environmentalists; it is still in its booming stage, it remains as a buzzword, where everyone is trying to figure out the rationale behind it. There should be more studies which can correlate and define the role of leaders and corporate social responsibility in organizations with the help of empirical studies. Leadership is associated with the organizations and the role they play in inducing any kind of change in the organization is foremost but when it comes to environment and sustainability, the similar leadership style and model can work or not, is still questionable and the most fascinating area for research. More studies are required to understand the role and impact of leadership in moving towards sustainable practices like CSR or circular economy. Especially to find out ways to assess responsible or humane leadership and to develop it in contemporary leadership practices.

Implementation for Industry

The organizations are coming up with various sustainable measures to keep up the pace with the rapidly changing business environment and adopting green practices with the help of distinct styles and models of leadership. Recently in the covid-19 pandemic various studies emphasised on responsible (Haque, Fernando, and Caputi, 2021; Pounder, 2021) and humane leadership (Hutagalung et al, 2020; Panebianco, 2021) as the best suitable leadership style to cope. Similarly we need responsible and humane factors in leaders for the organisations if we want our leaders to be more sensitive and empathetic about the planet and people. Therefore humanistic leadership can be recommended for the organizations to deal with the sustainability related issues and whether or not a combination of certain strategies can be implemented. However this chapter opens various questions with respect to the choice of practices and sustainable measures, the right choice of leadership style, and its overall impact in achieving the goal of sustainability, which can be answered through further research.

REFERENCES

Anadol, Y., & Behery, M. (2020). Humanistic leadership in the UAE context. *Cross Cultural & Strategic Management*, *27*(4), 645–664. doi:10.1108/CCSM-01-2020-0023

Andersen, M. S. (2007). An introductory note on the environmental economics of the circular economy. *Sustainability Science*, *2*(1), 133–140. doi:10.100711625-006-0013-6

Aswani, J., Chidambaran, N. K., & Hasan, I. (2021). Who benefits from mandatory CSR? Evidence from the Indian Companies Act 2013. *Emerging Markets Review*, *46*, 100753. doi:10.1016/j.ememar.2020.100753

Beccarello, M., & Di Foggia, G. (2018). Moving towards a circular economy: Economic impacts of higher material recycling targets. *Materials Today: Proceedings*, *5*(1), 531–543. doi:10.1016/j.matpr.2017.11.115

Behling, O., & Rauch, C. F. Jr. (1985). A functional perspective on improving leadership effectiveness. *Organizational Dynamics*, *13*(4), 51–61. doi:10.1016/0090-2616(85)90005-1

Bocken, N. M., De Pauw, I., Bakker, C., & Van Der Grinten, B. (2016). Product design and business model strategies for a circular economy. *Journal of industrial and production engineering*, *33*(5), 308-320.

Boeker, W. (1997). Strategic change: The influence of managerial characteristics and organizational growth. *Academy of Management Journal*, *40*(1), 152–170. doi:10.2307/257024

Boulding, K. E. (1966). *The economics of the coming spaceship earth*.

Brundtland, G. H. (1987). Our common future—Call for action. *Environmental Conservation*, *14*(4), 291–294. doi:10.1017/S0376892900016805

Cappelli, P., Singh, H., Singh, J. V., & Useem, M. (2010). Leadership lessons from India. *Harvard Business Review*, *88*(3), 90–97. PMID:20402052

Colbert, B. A., Nicholson, J., & Kurucz, E. C. (2018). Humanistic leadership for sustainable transformation. In *Evolving Leadership for Collective Wellbeing*. Emerald Publishing Limited. doi:10.1108/S2058-880120180000007004

Collins, E., Roper, J., & Lawrence, S. (2010). Sustainability practices: Trends in New Zealand businesses. *Business Strategy and the Environment*, *19*(8), 479–494. doi:10.1002/bse.653

Cooper, T. (1999). Creating an economic infrastructure for sustainable product design. *Journal of sustainable product design*, 7-17.

Davila, A., & Elvira, M. M. (2012). Humanistic leadership: Lessons from Latin America. *Journal of World Business*, *47*(4), 548–554. doi:10.1016/j.jwb.2012.01.008

Day, D. V. (2001). Leadership development: A review of industry best practices.

Doh, J. P., & Quigley, N. R. (2014). Responsible leadership and stakeholder management: Influence pathways and organizational outcomes. *The Academy of Management Perspectives*, *28*(3), 255–274. doi:10.5465/amp.2014.0013

Elkington, J., & Fennell, S. (1998). Partners for Sustainability. *Greener Management International*, (24).

Ellen MacArthur Foundation. (2012). Ellen MacArthur Foundation. Towards the circular economy vol. 1: an economic and business rationale for an accelarated transition.

Frémeaux, S., & Michelson, G. (2017). The common good of the firm and humanistic management: Conscious capitalism and economy of communion. *Journal of Business Ethics*, *145*(4), 701–709. doi:10.100710551-016-3118-6

Fricker, A. (1998). Measuring up to sustainability. *Futures*, *30*(4), 367–375. doi:10.1016/S0016-3287(98)00041-X

Fritz, S., & Sorgel, P. (2017). Recentering leadership around the human person: introducing a framework for humanistic leadership.

Fu, P. P., von Kimakowitz, E., Lemanski, M., Liu, L. A., & Pattnaik, C. (2020). Humanistic leadership in different cultures: Defining the field by pushing boundaries. *Cross Cultural & Strategic Management*.

Fulton, S. C., De Silva, L., & Anton, D. (2012, March). Twenty years after the rio earth summit: what is the agenda for the 2012 United Nations Conference on Sustainable Development? In *American Society of International Law. Proceedings of the Annual Meeting* (p. 91). Cambridge University Press.

Gatti, L., Vishwanath, B., Seele, P., & Cottier, B. (2019). Are we moving beyond voluntary CSR? Exploring theoretical and managerial implications of mandatory CSR resulting from the new Indian companies act. *Journal of Business Ethics*, *160*(4), 961–972. doi:10.100710551-018-3783-8

Geissdoerfer, M., Savaget, P., Bocken, N. M., & Hultink, E. J. (2017). The Circular Economy–A new sustainability paradigm? *Journal of Cleaner Production*, *143*, 757–768. doi:10.1016/j.jclepro.2016.12.048

Haque, A., Fernando, M., & Caputi, P. (2021). Responsible leadership and employee outcomes: A systematic literature review, integration and propositions. *Asia-Pacific Journal of Business Administration*, *13*(3), 383–408. doi:10.1108/APJBA-11-2019-0243

Holme, R., & Watts, P. (2001). Making good business sense. *Journal of Corporate Citizenship*, *2001*(2), 17–20. doi:10.9774/GLEAF.4700.2001.su.00005

Hu, Y., Chen, S., & Wang, J. (2018). Managerial humanistic attention and CSR: Do firm characteristics matter? *Sustainability*, *10*(11), 4029. doi:10.3390u10114029

Hutagalung, L., Purwanto, A., & Prasetya, A. B. (2020). The Five Leadership Style in Time of Pandemic Covid-19 throughout Industrial Revolution 4.0 as compared to Humane Leadership. *International Journal of Social. Policy and Law*, *1*(1), 79–87.

Jumde, A., & du Plessis, J. (2022). Legislated corporate social responsibility (CSR) in India: The law and practicalities of its compliance. *Statute Law Review*, *43*(2), 170–197. doi:10.1093lr/hmaa004

Karnani, A. (2013). Mandatory CSR in India: A Bad Proposal. *Stanford Social Innovation Review*, *16*, 20.

Kirchherr, J., Reike, D., & Hekkert, M. (2017). Conceptualizing the circular economy: An analysis of 114 definitions. *Resources, Conservation and Recycling*, *127*, 221–232. doi:10.1016/j.resconrec.2017.09.005

Korhonen, J., Honkasalo, A., & Seppälä, J. (2018). Circular economy: The concept and its limitations. *Ecological Economics*, *143*, 37–46. doi:10.1016/j.ecolecon.2017.06.041

Kotler, P., & Lee, N. (2005). Best of breed: When it comes to gaining a market edge while supporting a social cause, "corporate social marketing" leads the pack. *Social Marketing Quarterly*, *11*(3-4), 91–103. doi:10.1080/15245000500414480

Lala, R. M. (2004). *The creation of wealth: The Tatas from the 19th to the 21st century*. Penguin Books India.

Lawrence, P. R., & Pirson, M. (2015). Economistic and humanistic narratives of leadership in the age of globality: Toward a renewed Darwinian theory of leadership. *Journal of Business Ethics*, *128*(2), 383–394. doi:10.100710551-014-2090-2

Maak, T. (2007). Responsible leadership, stakeholder engagement, and the emergence of social capital. *Journal of Business Ethics*, *74*(4), 329–343. doi:10.100710551-007-9510-5

Maak, T., & Pless, N. M. (2006). Responsible leadership in a stakeholder society–a relational perspective. *Journal of Business Ethics*, *66*(1), 99–115. doi:10.100710551-006-9047-z

MacArthur, E. (2013). *Towards the circular economy, economic and business rationale for an accelerated transition*. Ellen MacArthur Foundation.

McClelland, D. C., & Boyatzis, R. E. (1982). Leadership motive pattern and long-term success in management. *The Journal of Applied Psychology*, *67*(6), 737–743. doi:10.1037/0021-9010.67.6.737

Melé, D. (2016). Understanding humanistic management. *Humanistic Management Journal*, *1*(1), 33–55. doi:10.100741463-016-0011-5

Metcalf, L., & Benn, S. (2013). Leadership for sustainability: An evolution of leadership ability. *Journal of Business Ethics*, *112*(3), 369–384. doi:10.100710551-012-1278-6

Miller, J. G., Bersoff, D. M., & Harwood, R. L. (1990). Perceptions of social responsibilities in India and in the United States: Moral imperatives or personal decisions? *Journal of Personality and Social Psychology*, *58*(1), 33–47. doi:10.1037/0022-3514.58.1.33 PMID:2308074

Mitra, N., & Schmidpeter, R. (2017). The why, what and how of the CSR mandate: The India story. In *Corporate Social Responsibility in India* (pp. 1–8). Springer. doi:10.1007/978-3-319-41781-3_1

Organisation for Economic Co-operation and Development (OECD). (2010). *Education at a glance 2010: OECD indicators*. OECD.

Panebianco, S. (2021). Towards a Human and Humane Approach? The EU Discourse on Migration amidst the Covid-19 Crisis. *The International Spectator*, *56*(2), 19–37. doi:10.1080/03932729.2021.1902650

Palmer, G. (1992). Earth summit: What went wrong at Rio. *Wash. ULQ*, *70*, 1005.

Pasricha, P., Singh, B., & Verma, P. (2018). Ethical leadership, organic organizational cultures, and corporate social responsibility: An empirical study in social enterprises. *Journal of Business Ethics*, *151*(4), 941–958. doi:10.100710551-017-3568-5

Pettigrew, A., & Whipp, R. (1992). Managing change and corporate performance. In *European industrial restructuring in the 1990s* (pp. 227–265). Palgrave Macmillan. doi:10.1007/978-1-349-12582-1_9

Pirson, M. (2017). *Humanistic management: Protecting dignity and promoting well-being.* Cambridge University Press. doi:10.1017/9781316675946

Pirson, M. A., & Lawrence, P. R. (2010). Humanism in business–towards a paradigm shift? *Journal of Business Ethics, 93*(4), 553–565. doi:10.100710551-009-0239-1

Pohl, M., & Tolhurst, N. (2010). *Responsible business: How to manage a CSR strategy successfully.* John Wiley & Sons.

Pounder, P. (2021). Responsible leadership and COVID-19: small Island making big waves in cruise tourism. *International Journal of Public Leadership.*

Quinn, L., & Dalton, M. (2009). Leading for sustainability: implementing the tasks of leadership. *Corporate Governance: The international journal of business in society.*

Rajeev, P. N., & Kalagnanam, S. (2017). India's mandatory CSR policy: Implications and implementation challenges. *International Journal of Business Governance and Ethics, 12*(1), 90–106. doi:10.1504/IJBGE.2017.085240

Rajput, S., & Singh, S. P. (2019). Connecting circular economy and industry 4.0. *International Journal of Information Management, 49,* 98–113. doi:10.1016/j.ijinfomgt.2019.03.002

Redclift, M. (1989). The environmental consequences of Latin America's agricultural development: Some thoughts on the Brundtland Commission report. *World Development, 17*(3), 365–377. doi:10.1016/0305-750X(89)90210-6

Robertson, M. (2017). *Dictionary of sustainability.* Taylor & Francis. doi:10.4324/9781315536705

Robinson, M., Kleffner, A., & Bertels, S. (2011). Signaling sustainability leadership: Empirical evidence of the value of DJSI membership. *Journal of Business Ethics, 101*(3), 493–505. doi:10.100710551-011-0735-y

Saini, D., & Sengupta, S. S. (2016). Responsibility, ethics, and leadership: An Indian study. *Asian Journal of Business Ethics, 5*(1), 97–109. doi:10.100713520-016-0058-2

Satapathy, J., & Paltasingh, T. (2019). CSR in India: A journey from compassion to commitment. *Asian Journal of Business Ethics, 8*(2), 225–240. doi:10.100713520-019-00095-2

Schmidt, F. C., Zanini, R. R., Korzenowski, A. L., Schmidt, R. Junior, & Xavier do Nascimento, K. B. (2018). Evaluation of sustainability practices in small and medium-sized manufacturing enterprises in Southern Brazil. *Sustainability, 10*(7), 2460. doi:10.3390u10072460

Sharma, R. R. (2019). Evolving a model of sustainable leadership: An ex-post facto research. *Vision (Basel), 23*(2), 152–169. doi:10.1177/0972262919840216

Simpson, J., & Taylor, J. R. (2013). *Corporate governance ethics and CSR.* Kogan Page Publishers.

Singh, R., & Singh, A. (2020). Socially Responsible Leadership as a Driver for Sustainable Growth in the World of Electronic Commerce. *E-Business: Issues and Challenges of 21st Century, 137.*

Solow, R. M. (1991). Sustainability: an economist's perspective.

Spreitzer, G. M., & Quinn, R. E. (2001). *A company of leaders: Five disciplines for unleashing the power in your workforce* (Vol. 3). Jossey-Bass.

Steurer, R., Langer, M. E., Konrad, A., & Martinuzzi, A. (2005). Corporations, stakeholders and sustainable development I: A theoretical exploration of business–society relations. *Journal of Business Ethics*, 61(3), 263–281. doi:10.100710551-005-7054-0

TATA Consultancy Services. (2019). *Corporate Sustainability Report 2018-2019*. TATA Consultancy Services, Mumbai, India (https://www.tcs.com/content/dam/tcs/pdf/discover-tcs/investor-relations/corporate-sustainability/GRI-Sustainability-Report-2018-2019.pdf)

Tripathi, R., & Kumar, A. (2020). Humanistic leadership in the Tata group: The synergy in personal values, organisational strategy and national cultural ethos. *Cross Cultural & Strategic Management*, 27(4), 607–626. doi:10.1108/CCSM-01-2020-0025

Ulrich, D., Zenger, J., & Smallwood, N. (1999). *Results-based leadership*. Harvard Business Press.

Waddock, S. (2004). Creating corporate accountability: Foundational principles to make corporate citizenship real. *Journal of Business Ethics*, 50(4), 313–327. doi:10.1023/B:BUSI.0000025080.77652.a3

Waddock, S. (2016). Developing humanistic leadership education. *Humanistic Management Journal*, 1(1), 57–73. doi:10.100741463-016-0003-5

Yang, B., Fu, P., Beveridge, A. J., & Qu, Q. (2020). Humanistic leadership in a Chinese context. *Cross Cultural & Strategic Management*, 27(4), 547–566. doi:10.1108/CCSM-01-2020-0019

Yang, S., & Feng, N. (2008). A case study of industrial symbiosis: Nanning Sugar Co., Ltd. in China. *Resources, Conservation and Recycling*, 52(5), 813–820. doi:10.1016/j.resconrec.2007.11.008

Zink, T., & Geyer, R. (2017). Circular economy rebound. *Journal of Industrial Ecology*, 21(3), 593–602. doi:10.1111/jiec.12545

Section 4
Social Engagement and Inclusion

Chapter 14
Challenges and Perspectives of Pinhão Production Considering the Dimensions of Sustainability:
A Study in a City in Southern Brazil

Jean Marcos da Silva
https://orcid.org/0000-0003-0331-3849
Federal Institute Sul-rio-Grandense, Brazil

Jordana Marques Kneipp
https://orcid.org/0000-0001-6982-994X
Federal University of Santa Maria, Brazil

Thiago Paulo Both
Federal Institute Sul-rio-Grandense, Brazil

Greice Eccel Pontelli
https://orcid.org/0000-0003-4643-478X
Federal University of Santa Maria, Brazil

ABSTRACT

The activity of collecting non-timber forest products (NTFPs) is a secular activity in Brazil, more specifically the sertão [Brazilian backlands]. Many of these activities have ceased to exist over the years, although pinhão (Araucaria angustifolia seeds) production still persists as an income generator. Given this context, this study sought to answer the following question: "What are the challenges and perspectives of pinhão production considering the context of the community of Barro Preto in the city of Arvorezinha (Rio Grande do Sul State, southern Brazil) from the dimensions of sustainability?" Using thematic analysis, this interpretative qualitative study employed conversational interviews by Boje and Rosile (2021). In mapping the challenges and perspectives of the extractive activity of pinhão, these findings showed that inserting actors from the base of the pinhão productive chain to induce them to tell their stories is not enough to build a narrative that contemplates all the dimensions of sustainability.

DOI: 10.4018/978-1-6684-6123-5.ch014

INTRODUCTION

The production chains of non-timber forest products (NTFPs) have been operating in Brazil for centuries since the commercialization of the so-called *drogas do sertão* [drugs of the *sertão*], which were commodities consisting of plants and spices grown in the region. These production chains were pointed out by Fausto (2006) as one of the productive clusters of Colonial Brazil in the seventeenth and eighteenth centuries. While gold was produced in Minas Gerais State and sugar in the mills of northeastern Brazil, the Amazon region was dedicated to harvesting non-timber forest products.

It is estimated that the Jesuit priests obtained revenues exceeding 2 million pounds sterling selling NTFPs to all of Europe in the seventeenth century (Fausto, 2006). In fact, the occupation of Brazil was strongly motivated by the search for the so-called spices widely found in the Indies in the period preceding the colonization of Brazil. According to Schwarcz and Starling (2015), in one of the expeditions in search of these products, the Portuguese 'accidentally' came across Brazil; therefore, the exploitation of NTFPs is closely associated with the country's history.

The Amazon biome was protected from deforestation mainly because the Portuguese wanted to exploit the so-called *drogas do sertão*, such as the Brazil nut (*Bertholletia excelsa*) and cacao (*Theobroma cacao*), the so-called NTFPs, products whose production demands the conservation of tree species. In other Brazilian territories, such as the south, the native vegetation slowly began being substituted by fields for livestock production. It is estimated that today there are 2–4% of the original Atlantic Forest (Guerra et al., 2002), while there is still at least 80% of the Amazon biome (INPE, 2020). Hence, it is not an exaggeration to say that stimulating the production of NTFPs can contribute to preserving native biomes, as has occurred since the colonization of Brazil in the Amazon region.

The *pinhão* is a type of NTFP commonly found in southern Brazil. The annual productivity of this product has varied considerably in recent decades, ranging from 4,396 tons in 2003 to 5,715 tons in 2010. As of 2010, production has risen, reaching 9,638 tons in 2012 (IBGE, 2017). In addition to these numbers, the sociocultural importance of NTFPs in general and *pinhão* in particular is a recurring agenda as many people who collect these products accumulate knowledge about the local flora and collection activities. This is especially true in communities where NTFP production is carried out in groups and taught since childhood (Barbosa et al., 2020, Silva-Jean et al., 2022; 2017; 2020).

Therefore, it is possible to note social, economic, and environmental aspects in *pinhão* production, inducing an analysis of its production chain in light of sustainability as a broad terminology, as per the definition by Sachs (2007). The concept of sustainability has been discussed and rediscussed numerous times; one of the most debated definitions in the literature is undoubtedly that of Ignacy Sachs, who conceptualizes sustainability as a guiding principle based on eight dimensions: social, economic, ecological, environmental, spatial, cultural, territorial, and political (Sachs, 2007). Throughout this text, these dimensions will be deepened.

The production chains of NTFPs, as a sequence of activities that involve a set of actors, have been reported as an important ally in this process of search for sustainability defined by Ignacy Sachs, that is, in the following aspects: social, cultural, ecological, environmental, territorial, economic, and political (Pedrozo et al., 2011; Silva-Jean et al., 2022; Barbosa et al., 2020).

Productive chains are complex because of the set of actors with which they must be involved, especially the NTFP chains that tend to be poorly structured and present a notion of technological productivity different from that of the traditional economy. Nonetheless, the various non-timber forest products have endured for centuries, contributing to income generation, cultural strengthening, and environmental preservation of native biomes.

This historical and conceptual context has sparked the interest of countless researchers from various fields of knowledge seeking to understand the potentialities and challenges of these productive activities. Given the above, we sought to answer the following question: "what are the challenges and perspectives of the activity of collecting *pinhão* considering the context of the community of Barro Preto in Arvorezinha (Rio Grande do Sul State, southern Brazil) from the dimensions of sustainability?". The community of Barro Preto was chosen as it is inserted in the biome formed by the Atlantic Forest, a vegetation with a predominance of *pinhão*.

In addition to this introduction, this chapter is structured into four parts: the theoretical references, in which we demonstrated some conceptual antecedents about the definition of sustainability and productive chains, the materials and methods describing the procedures that guided this study, the results, and discussion, in which we covered our findings from the field study, and lastly, the final considerations.

THEORETICAL FRAMEWORK

This topic presents the theoretical concepts that formed the basis of this article: sustainability and the production chain of non-timber forest products. The widely used term sustainable development refers to development that meets the needs of the present without compromising the ability of future generations to meet theirs (WCED, 1987). Undoubtedly, over all these years, since the concept was coined by Our Common Future in 1987, many definitions have been used as if they were sustainability. The report was written by a commission led by the Prime Minister of Norway, Gro Brundtland, at the invitation of the United Nations.

The title of this section, "What sustainability is not!" is borrowed from the article by Cairns Jr. (1998). This text focuses on a reverse path in the definition of sustainability, dedicating itself to demonstrating what sustainability is not rather than answering: what is sustainability? This section presents some of the primary authors and ideas prevailing in the literature on the theme. This is because a mistaken understanding of what sustainability is hinders the process of convincing society in the search for a more sustainable world. Here is a list that has been compiled to clarify the issue.

Sustainability is Not Just About The Triple Bottom Line!

The idea of the Triple Bottom Line (TBL), a term coined by Elkington (1992), is still one of the main concepts used in the literature to define sustainability. According to Elkington (1992), it is about understanding that life on Earth depends on the coexistence of three important fields: the social, economic, and environmental dimensions.

The social dimension covers issues related to people's quality of life regarding equity, inclusion, work, health, education, and food, among others (Elkington, 2006). Moreover, while the economic dimension is related to the care of assets, finances, and budgets to obtain profit, the environmental dimension focuses on ensuring more appropriate exploitation practices of natural resources, such as reducing the emission of pollutants (Elkington, 2006).

Sustainability in its social scope, which is meant to be emphasized, can still be broken down into three major areas: developmental, bridging, and maintenance sustainability. Developmental sustainability predicts that significant environmental advances will only occur when citizens' basic needs are met; such needs include clean water, healthy food, medicine, adequate housing, education, employment, equity, and justice. It is believed that with a more dignified quality of life, humans will be able to reflect and care more about the biophysical environment in which they live.

Bridging sustainability, however, thinks about ways to promote "greener" behavior or a stronger environmental ethic (Hobson, 2003; Linden & Carlsson-Kanyama, 2003; Bhatti & Church, 2004; Frame, 2004; Barr & Gilg, 2006; Boolaane, 2006; Lindenberg & Steg, 2007; Rutherford, 2007; Vlek & Steg, 2007). It can be divided into two subsets, transformative and non-transformative change, and it seeks to harness human potential with behavioral changes to bring about environmental results.

Transformative change aims to radically alter people's relationships with the environment; changes in beliefs, lifestyle, and philosophy of life, adhering to an intimate closeness with nature. Cairns (2003) promoted the notion of biophilia (having love for all kinds of life), a relevant and appropriate contribution to the present context. The non-transformative ones, in contrast, serve as support for the transformative ones. These are not radical changes in people's lifestyles but the adoption of technological innovations in favor of the environment. Examples that have already been implemented include hybrid and electric cars, solar energy, water recycling, and recycling facilities.

Lastly, we have maintenance sustainability, which works with sustaining traditions, practices, preferences, and places that people would like to see maintained or improved. Many cultural practices being maintained are environmentally friendly, such as low-density suburban living.

Although well known in the literature, the TBL has been criticized by Kuhlman and Farrington (2010) for encouraging a separation between social and economic aspects, as if they were dissociated. For Kuhlman and Farrington (2010), it is necessary to interrupt the belief that sustainability has 'three sides of the same coin.'

For Kuhlman and Farrington (2010), it is necessary to return to the meaning that originated sustainability, that is, "Development that meets the needs of the present without compromising the ability of future generations to meet theirs" (WCED, 1987, p. 40), but without losing sight of the fact that current generations also contribute social capital, infrastructure, technology, culture, and institutions that improve future well-being, survival, and even production.

As Lynch and Khan (2020) reported, understanding sustainability as a triple dimension has only served to bring psychological comfort and favorable conditions to maintain neoliberal capitalism. This is because, according to Lynch and Khan (2020), TBL intends to find a way in which companies can make winners of themselves, their customers, and the environment (i.e., a win-win relationship).

In Kuhlman and Farrington's (2010) work, Elkington's TBL concept is only useful if one considers Elkington's intention when elaborating the concept, that is, if the purpose is to operationalize sustainability within companies. Alternatively, as Lynch and Khan (2020) claimed, if the intention is to measure the principle of sustainability. Otherwise, for a more sustainable world, it is necessary "To the conventional result (profit) add care for the environment (the planet) and good for people, for example, providing facilities for the disabled and hiring minorities (social dimension)" (Kuhlman & Farrington, 2010, p. 3438).

The publications by Sachs (2007) lead to the conclusion that besides the TBL, it is necessary to think about three other dimensions: cultural, territorial, and political. The cultural dimension can unite people around a collective ideal in search of sustainability (Sachs, 2007). Culture is an ally in this sense, and there needs to be a balance between respect for tradition and the necessary innovations (Sachs, 2007). The territorial dimension considers the rural-urban configurations in terms of mutual respect in the allocation of public investments (Sachs, 2007).

For Sachs (2007), the territorial dimension still requires overcoming inter-regional disparities and creating differentiated strategies for ecologically fragile territories. The political dimension covers national and international aspects. It deals with the construction of a State capable of implementing projects involving different social actors, social cohesion, a system capable of preventing wars and guaranteeing international cooperation, and demercantilizing science and technology by making them part of the common heritage of humanity.

Another aspect being considered is that these dimensions cannot be separated from each other. An integral search for these principles is needed (Sachs, 2007). One of the dimensions commonly placed as different aspects, as two sides of the same coin, are the social and economic dimensions.

(In) Sustainability, Social and Economic Cannot be Separated!

According to Kuhlman and Farrington (2010), in sustainability, the distinction between social and economic dimensions is impossible in practice. According to Vallance et al. (2011), the social aspect has been neglected in studies that set out to study sustainability, leading to the conceptual chaos of the term.

The separation between the economic and the social originates in the view that the defense of more sustainable initiatives results in the loss of jobs in sectors crucial for generating income in many regions, such as the wood industry (Cairns, 1998). According to Cairns (1998), the argument for job losses does not consider that sustainability would also generate other jobs in other sectors, thus contributing to innovation by requiring a technological revolution and a substantial paradigm shift in how technology is produced.

Given the necessary change to meet the sustainability principles, several technological innovations would be necessary to promote the sustainable use of the planet. In recent years many of these initiatives have emerged; one of them was pointed out in the publications of Silva-Jean et al. (2022) on the capacity of income generation and quality of life for people working with NTFP production. The companies involved in the production chains of these products are investing time and money in technologies suitable for sustainability.

In these innovative initiatives, the industrial sector can be an essential partner in supporting NTFP production chains, considering the culture of those who produce them. Silva (2015) identified that the cosmetics company Natura acquires these products from Amazonian communities to manufacture cosmetics. The products derived from innovation in the way of looking at the industry, taking into account that until a few decades ago, the defense of sustainability went through solid waste recycling practices.

The proposal of NTFP as one aligned with sustainability seems to treat the social and economic aspects together, moving away from a bilateral perspective, as two sides of the same coin. Cairns (1998) argues that "[...] rather than a 'Them' versus 'Us' polarization of industry and sustainability stakeholders, there should be synergistic cooperation" (p. 4).

Opposing the social to the economic aspect in a capitalist society can make evident which dimension stands out in terms of dominance: the economic one, which, because it has a technical meaning, has a language — that of money and a presence in sociopolitical power in a way that only social justice cannot do (Lynch & Khan, 2020). Therefore, one cannot treat these two dimensions as separable, for they are not. Perceived technological advances in the industry can be used as aligned or not in promoting the so-called social and economic dimensions. Changes in social behavior will dictate whether they will contribute to a more sustainable planet. Therefore, major social changes are fundamental to achieving sustainability (WCED, 1987).

For Vallance et al. (2011), people can only think about environmental issues when their basic needs are met (food, protection against cold, safety, health, housing, drinking water, etc.). Therefore, poverty is a barrier to adopting sustainable technologies such as solar panels, energy-efficient appliances, and solid waste disposal.

Hence, in sustainability, the most appropriate is to internalize social and economic concerns into a single fundamental concern rather than configure them separately (Boulding, 1966). Part of this vision is preserving the environment, even if it is impossible to admit that sustainability is only about caring for natural capital.

Sustainability Is Not (Only) About Preserving The Environment!

Another aspect to be considered is that environment and sustainability are not synonymous. Although treated interchangeably by many, they are different terms (Sachs, 2007; Cairns, 1998; Vallance et al., 2011). Apparently, there are two contradictory visions in the relationship between man and the environment: one that defends harmonious coexistence and another that judges nature as a kind of resource to be conquered. There is something in common between these perspectives: the search for a better life. In light of this, many events and studies have been carried out over the years around a fundamental question: how to reconcile human aspirations for well-being with limited natural resources? Sustainability emerges as a possible way to solve these challenges.

For Cairns (1998), it is necessary to review the rules of the economic game, which, because they are negotiable, can be rethought to coexist with the non-negotiable rules of the environment. Although many people may believe that human intelligence to develop technologies would free humans from these harsh natural laws, events such as the COVID-19 pandemic indicate that these natural laws do not take this into account, as it caused the death of countless people despite 200 years of significant advances in science and technology.

Phenomena such as COVID-19 indicate that relying on others and the available technology is not advisable, leading one to believe that searching for solutions that reconcile the environment with economic aspects is a role for everyone. As Cairns (1998) argued, it seems that sustainability is closer to a search for avoiding environmental catastrophes for both present and future generations. This would explain the need to protect natural systems and the organisms that inhabit them.

Kates et al. (2001) suggested a science of sustainability that seeks to understand the fundamental character of the interactions between nature and society at regional and global scales. To achieve this goal, a core set of questions for sustainability science was created to keep research attention on the fundamental character of nature-society interactions and society's ability to guide these interactions along sustainable trajectories (Kates et al., 2001).

When advocating that preserving the environment is part of sustainability, many argue that proponents of the concept seek to stop humanity from evolving. However, this makes no sense because making the world more sustainable requires challenging the current status quo to implement the necessary changes in production and social practices that respect future generations. Therefore, it is a quest for the evolution of humanity, not the other way around.

Sustainability is Not Against the Evolution of Humanity!

Since sustainability seeks to challenge the status quo by proposing paradigm changes that respect the quality of life of current and future generations, the defenders of sustainable practices are pro-change (Cairns, 1998). Given the concepts addressed in sustainability, especially in terms of technological innovation it is evident that sustainability cannot be against the evolution of mankind.

This understanding has not always been present as social capital. Elkington (2006) envisioned the existence of 7 different moments for the future of sustainability: the first one referes to the market: driven by competition, it will be critical for businesses to be able to identify market conditions to survive, demanding a new approach; the second is driven by the global shift in human and social values, this phenomenon happens from generation to generation, and business adaptation will be key; the third refers to a revolution fueled by increasing international transparency with great pressure for companies to show how their business is dealing with the TBL dimensions and what they plan to execute; the fourth moment is characterized by Life Cycle Technology, companies are challenged with implications on production, the time of use of the product and its end of life, different from "the cradle to the grave" view (Elkington, 2006).

Going forward, the fifth moment is defined by partnerships and will dramatically accelerate the rate at which new forms of partnership emerge between companies and other institutions; the sixth is tied to time and will promote a change in the way we understand and manage time, increasingly scarce business leaders will have the difficult task of building a long-term dimension; and Finally, the seventh is tied to corporate governance, where new questions are being asked, such as what is business for? How are companies managed? What is the appropriate balance between shareholders and other stakeholders? And what balance should be achieved at the TBL level? (Elkington, 2006).

The future looks challenging. In this same line of thought are the texts written by Eller et al. (2020) and Tsales (2020). Tsales' work, for instance, focuses on a sustainability agenda for the future. Lynch and Khan (2020) wonder why policy principles toward sustainability fail to manage environmental impacts; in Lynch and Khan's (2020) words, there has been no real mitigating effect yet.

To address the challenges implied in the idea of sustainability, the 2030 Agenda for Sustainable Development was adopted at the United Nations Sustainable Development Summit in 2015. "This Agenda is a plan of action for people, planet, prosperity, peace, and partnerships" (United Nations Organization, 2018), forming 5 Ps. Table 1 lists the 17 Sustainable Development Goals.

The goals in Table 1 resulted from many years of discussions and experiences of different actors. They are a kind of update of publications such as the Earth Charter, Agenda 21, and the Millennium Development Goals. The 2030 Agenda lists these 17 goals, which in turn are subdivided into 169 targets into five major areas: people, planet, prosperity, peace, and partnership. Based on this agenda, the United Nations (2018) suggests that initiatives be adopted with different stakeholders to ensure sustainability in the coming years.

Production Chain Of Non-Timber Forest Products

The terminology *Filière*, known as "production chain," is a French term created in the 1960s. The term employs the sequence of operations that, in the end, generate a product that can be purchased by the final consumer, having as its beginning a certain unprocessed material. The operations are always interconnected, from material extraction to commercialization in retail or external markets. For Morvan

Table 1. Sustainable development goals by area

Goal	Description
Goal 1	End poverty in all its forms, everywhere.
Goal 2	End hunger, achieve food security and improved nutrition, and promote sustainable agriculture.
Goal 3	Ensure healthy living and promote well-being for all, at all ages.
Goal 4	Ensure inclusive and equitable quality education and promote lifelong learning opportunities for all.
Goal 5	Achieve gender equality and empower all women and girls.
Goal 6	Ensure availability and sustainable management of water and sanitation for all.
Goal 7	Ensure reliable, sustainable, modern, and affordable access to energy for all.
Goal 8	Promote sustained, inclusive, and sustainable economic growth, full and productive employment, and decent work for all.
Goal 9	Build resilient infrastructure, promote inclusive and sustainable industrialization, and foster innovation.
Goal 10	Reduce inequality within and between countries.
Goal 11	Make cities and human settlements inclusive, safe, resilient, and sustainable.
Goal 12	Ensure sustainable production and consumption patterns.
Goal 13	Take urgent action to combat climate change and its impacts.
Goal 14	Conservation and sustainable use of the oceans, seas, and marine resources for sustainable development.
Goal 15	Protect, restore and promote the sustainable use of terrestrial ecosystems, sustainably manage forests, combat desertification, halt and reverse land degradation and halt biodiversity loss.
Goal 16	Promote peaceful and inclusive societies for sustainable development, provide access to justice for all, and build effective, accountable, and inclusive institutions at all levels.
Goal 17	Strengthen the means of implementation and revitalize the global partnership for sustainable development.

Source: Gregolin et al. (2019) from the United Nations (2015, p. 18).

(1985) and Pedrozo, Estivalete, and Begnis (2004), *filière* is a sequence of activities that leads to the production of goods and shapes the relations of the agents in the chain, being them dependent on each other or complementary. The hierarchical strength of the chain also determines these relations. In the end, a system is obtained, which can guarantee its transformation.

For Alves (2010), the lack of scientific knowledge about the NTFPs and the non-exploitation of these products by companies, since they are considered secondary products in relation to wood, contribute to a vast concept of NTFPs.

One can analyze the following concepts for NTFPs: 1) products of plant origin found in a native forest, except for timber; 2) products of native forest or cultivated species, except for timber 3) plant products originating from native forest or cultivation, also contemplating living beings and animals 4) not giving importance to the environment that the material was collected (Alves, 2010).

With given contributions in the literature, we can classify NTFPs with the concept elaborated by Chamberlain et al. (1998), which consists of all products that are collected from a native forest or from cultivated species, provided that they are not classified as wood, and may be from various parts of a plant and have various uses.

NTFPs are often characterized as secondary, minor, special, and non-traditional forest products, and these characterizations end up obscuring the potential of these products. In reality, NTFPs have always been important to mankind, being just as traditional as timber. In ancient times, many gatherers enjoyed the benefits generated by non-timber forest products, even before they had the necessary technology to extract and manipulate wood.

In 1992, the sale and processing of Pacific Northwest mushrooms in the United States contributed 40 million USD to the economies of the states involved. In 1998, James Chamberlain and A. L. Hammett believed that the trade and use of NTFPs would grow substantially. Today, by considering the participation of these products in the gross domestic production (GDP) of Brazil, we see a scenario where industries exploit these materials very little, yet the potential still exists and can be very well exploited.

In developing countries such as those in Latin America, where a large part of the population lives in rural areas and family agriculture generates subsistence for individuals, the NTFPs have greater value, as reported by Zamora (2001 apud Soares et al., 2008). According to the author, in Latin America, NTFPs are used mainly as medicinal, food, aromatic, coloring, energy, industrial, craft, and ornamental products.

According to Chamberlain and Hammett (1998), NTFPs are classified into four general product lines: edible, specialty wood products, floral greens, and medicinal and dietary supplements. It is vital to note that specialty wood products are not derived from cutting the tree's trunk but from branches and vines, used mainly for handicrafts.

At this point, it is noted that NTFPs are important for small communities, but their economic potential is not exploited. According to Pastore Junior and Borges (1998 apud Fiedler et al., 2008), even in regions such as the Amazon, where these products employ over 1 million people, the contribution to the Brazilian GDP is only 1.85%.

These products, unfortunately, do not arouse the interest of companies and the federal government. According to Paes-de-Souza (2011), the governmental organs of fomentation hardly observe and work together with the economic interests of the collectors, producers, and exporters. With this lack of attention from the entities, the productive chain of the NTFP ends up being more precarious. Brazil is rich in NTFPs, and we can mention some of the most widespread ones: the cashew nut, the Amazon nut, the pine nut, the açaí, and the babaçu, among others. Our focus will be on the pine nut, whose main producer is the well-known *Araucaria* (*Araucaria angustifolia*). The species is native to Brazil, has a long lifespan, is wind pollinated, and its pine cones take about two years to mature, from which the pine nut is extracted.

In the last century, the Atlantic Forests, or Araucaria Forests, lost much space mainly due to agricultural expansion. Of the 35% of the Araucaria Forest at the beginning of the twentieth century, there is only about 2 to 4% of the original area (Guerra et al., 2002). This data worries many researchers. For Sachs (1986), sustainability is a strong way to promote the preservation of these forests. Still, according to the author, preservation alone is not enough. The social, economic, cultural, political, environmental, and human dimensions must be carefully observed. The *pinhão*, along with the other NTFPs, generates income for families and represents a form of subsistence, besides having cultural and social purposes. In recent years, these products have been gaining strength, and the focus of this study is to leverage the relevance of such products from the reality of the production chain of the *pinhão*. Next, the method for developing the study is presented.

MATERIALS AND METHODS

Considering that qualitative research is more appropriate for understanding the process of life experiences and future perspectives, as is the case of this study, this research has a qualitative approach. To understand the challenges and perspectives of the extractive activity of the *pinhão* from the dimensions

Figure 1. Study Locus, Barro Preto Community, Arvorezinha, Rio Grande do Sul
Source: *Adapted from IBGE (2022).*

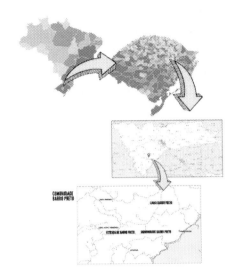

of sustainability, this study is situated in the interpretative paradigm through an analysis of the life history of actors in the community of Barro Preto, in Arvorezinha.

Therefore, the unit of analysis was the life stories of the participants interviewed through a process of identifying the meanings of this analyzed family unit composed of 3 individuals called in this study: Participant A, Participant B, and Participant C, obtained through a thematic analysis.

Characterization of the Study Site

The focus of this study was on actors that have great contact and knowledge about the *pinhão* NTFP. Specifically, the community of Barro Preto, in the municipality of Arvorezinha, in Rio Grande do Sul State. The municipality is surrounded by an Atlantic Forest biome, which favored the research (Figure 1).

The municipality of Arvorezinha is located in northern Rio Grande do Sul State, with an area of 271,643 square kilometers and 10,421 inhabitants, at a distance of 210 km from the capital Porto Alegre (IBGE, 2020). Part of the municipality is the Barro Preto community, 6 km away from the municipal seat, located in the Rural Zone. The QFC property (fictitious name) is located in this community. This region constitutes the study location of this study.

For data collection, contact was made with three extractors (Participants A, B, and C), all male and all older than 50 years, with experience in *pinhão* extraction activity. The interviewees collected *pinhão* for subsistence and income since childhood, as well as dealing with yerba mate production. The interviewees belong to the same family, working in the rural property inherited by their parents, where they currently live.

The choice to analyze the reality of the same family was due to the limited time and resources necessary to conduct interviews with a greater diversification of the surveyed extractors. Therefore, the conclusions verified in this study are not generalizable.

In addition, the justification for selecting the participants is based on the criterion of convenience described in Flick (2009) because the region where the selected family lives have a high incidence of *pinhão* in Rio Grande do Sul State. However, little has been studied in this aspect of productive chains of non-timber forest products besides the expectation of identifying other individuals to corroborate future research.

Data Collection

Primary data were collected from a field interview. A field trip was made in May 2022 to the Barro Preto Community in the municipality of Arvorezinha, where the extractors live. The interview lasted an hour and a half. The interview was guided by a script prepared strategically in advance. The questions ranged from personal data to specific issues related to the *pinhão*. In the end, important narratives about the central issue under study were acquired.

We used the concept of conversational storytelling interviews (CSI), according to Boje and Rosile (2021), which allows the interviewee to tell stories without interruption; the researcher and interviewee share stories in an interactive relationship. For Boje and Rosile (2021), the CSI is an alternative to data collection instruments based on structured or semi-structured interviews. In CSI, there is mutual sharing between the subjects involved, in which the researcher and interviewee share stories rather than a one-way interrogation; it is dialogical.

For Boje and Rosile (2020), semi-structured interviews are a form of colonization, manipulation, and cultural invasion since it starts from premises directed to a group (interviewees), taking into account concepts from another group. The definition of 'banking education' described by Freire (1975) illustrates this domain, as the expert/teacher deposits knowledge in the empty minds of students whose life experiences are neglected. They are untold stories, and they are people without a world.

For this reason, the interviews had only a few guiding questions, the main ones being: Tell me about yourself (name, age, profession). Tell me about your experience and what you do. How are you involved with collecting the *pinhão* NTFP? In your opinion, what are the challenges of collecting *pinhão*? Do you have any stories to tell about the process of collecting *pinhão*? Other questions were also asked to understand the interviewees' experiences, meanings, and reflections.

During the interview, numerous notes were taken to contribute to the analysis. Another strategy used was the audio recording of every interview. In addition, a cell phone camera was used to record all the statements and records in audio and photographs.

The secondary data are from various literary texts such as poems, books of short stories, and experience reports of the extractors interviewed. This secondary and primary data allowed us to gather a considerable amount of data.

Data Analysis

A thematic analysis was used for data analysis, focusing on understanding what was said, seeking to understand the interviewee's narrative. To this end, the interview was transcribed, and the data analysis was carried out, looking for common statements that made it possible to group them. The information was analyzed based on the dimensions of sustainability based on an a priori definition made considering the theoretical studies on the subject, as shown in Table 2.

Table 2. Dimensions of analysis of the interviews

Term	Variable	Constitutive definition	Applied definition	Indicator
Sustainability dimensions	Social	It deals with social inequality and the poor distribution of income. The labor scarcity and lack of access to resources and public services must be combated (SACHS, 2007). The State is responsible for ensuring social stability and full human rights (FREITAS, 2011).	It is linked to a lifestyle in which inequalities must be inhibited, people must have a dignified quality of life, and their constitutional rights must be truly assured. That which the human being is characterized.	Conversational storytelling interviews.
	Economic	The need for balanced intersectoral economic development promotes food security, capacity for continuous modernization of production instruments, reasonable condition of autonomy in scientific and technological research, and sovereign inclusion in the international economy. Such topics fuel a necessary change in production and consumption (SACHS, 2007).	It concerns income, production, and guaranteeing life's material conditions.	Conversational storytelling interviews.
	Environmental	It seeks to consciously enjoy nature's services and goods without causing harmful impacts. It involves ecological diversity, and human socio-cultural diversity must be preserved (SACHS, 2007).	Condition for harmonious coexistence between man and nature.	Conversational storytelling interviews.
	Cultural	The cultural traditions and, therefore, the culture of various societies have been minimized. Globalization and unbridled capitalist development end up passing over many cultures and traditions, which may be missed in the future, according to Sachs (2007). The preservation of cultures is essential, as it translates into respect and ethics in the face of differences.	It is what human activity produces, and it is the existing tradition.	Conversational storytelling interviews.
	Territorial	It is a dynamic and complex process based on the interaction and organization of several elements, material and immaterial: the material elements are the concrete processes and components, such as economy, production, and consumption; and the immaterial are cultural, ideological, symbolic, identity and natural aspects (SAQUET and GALVÃO, 2009). For Sachs (2007), the balance between urban and rural space fits this field's strategy.	They are the interactions between the physical and symbolic elements. The relationship between these elements characterizes a given community.	Conversational storytelling interviews.
	Political	It involves governmental strategies and actions, such as the promotion of democracy, as a means of appropriation of human rights, a reasonable degree of social cohesion, and, essentially, the State's ability to articulate itself in the implementation of a national project (SACHS, 2007).	Government entities corroborate to guarantee human rights and democracy and encourage changes to promote sustainability. The perspective of change.	Conversational storytelling interviews.

Source: Survey data.

From the stories told in the field study and using the Atlas ti software, 17 themes that emerged in the interviews were mapped, as follows: 1) social dimension: Generational succession, History, Change in the extractive activity, Personal profile for the agro-extractive activity; 2) economic dimension: Benchmarking, Commercialization channels, Quality control, Innovation in the property, Innovation in the Productive Process, Marketing, and Sale; 3) territorial dimension: Product characteristics, Relevance of the *pinhão*; 4) environmental dimension: Contribution to the environment; 5) cultural dimension: Extractive Activity, *pinhão* production process; 6) political dimension: Perspective for the extractive activity, Relationship with the community.

The mapped themes were subjected to a contextual and predictive analysis based on pre-narrative processes. As Boje and Rosile (2021) demonstrated, there are six pre-narrative processes at the core of the narrative paradigm, the so-called 6Bs: before, beneath, beyond, between, becoming, and bets, seeking the theory-method-praxis. The 6Bs are described in Figure 2.

Figure 2. The 6 Bs of narrative science
Source: *Elaborated from Boje and Rosile (2020)*

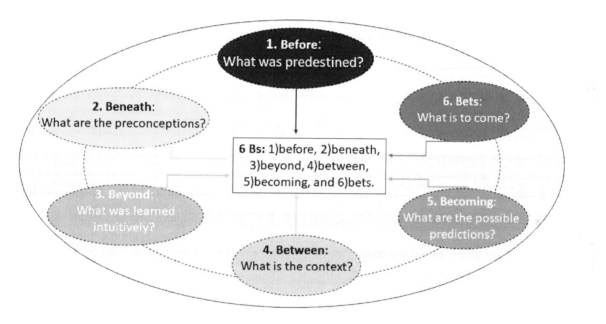

The 6Bs help to understand what the person said and what was not said or could not be said without some help. It is an attempt to read between the lines because what is being stated in narrative results from a set of multiplicities that come together to build meaning.

Taking into account the dimensions of sustainability proposed by Sachs, having the 6Bs of Boje and Rosile (2020) as a guideline, and collecting data through the CSI, this study sought to answer the following question: what are the challenges and prospects of the extractive activity of *pinhão* considering the context of the community of Barro Preto in Arvorezinha from the dimensions of sustainability? The results are presented below.

RESULTS AND DISCUSSIONS

The rural property QFC, where the participants live, is a productive unit based on pluriactivity, with production mainly of *pinhão* and yerba mate. They have been recently exploring tourism based on demonstrating their history as the main product to stimulate visitors. The individuals have lived on the property since the 1950s, exercising agro-extractivism. The results presented herein are based on *pinhão* collection, production, and commercialization activity.

The data analysis considered the context of QFC. As a result of this analysis, there is a distribution of themes among the sustainability dimensions pointed out in Sachs (2007). This analysis is listed in Table 3. The content of these themes presents a set of challenges and perspectives for the activity of collecting *pinhão* nuts in the community of Barro Preto in Arvorezinha.

A productive chain, such as the *pinhão* chain, is composed of actors with some level of interaction among themselves. In the case of this study, the data highlighted the perspective of the primary actors of the chain (i.e., the agroextractivists). The results presented in Table 3 constitute the meanings and stories told by these individuals.

We affirm that in the data collected, there is an appreciation by the interviewees of the patriarchs of the family that emerged since they came from Italy to Brazil in the 1940s. The patriarchs have attributed aspects such as courage, strength, and initiative. The research participants use previous generations' teachings as an anchor to continue devising strategies for the *pinhão* extraction activity as they seek means of production and expansion in the market.

The participants use the stories of the patriarch and their own life stories to encourage the younger generations to continue with the activity of collecting *pinhão* as an income opportunity. A clear prenarrative movement is defined as the 'before' of the narrative and as a 'bet' that something transformational will occur due to this narration (Boje, 2001).

Phrases such as 'Maybe even those young people who have abandoned land could invest in a plantation (of *pinhão*), for income. Then we can even see our pine nuts there' (Participant A). Right after, the narrative of his personal stories indicates the expectation placed on the new generations.

This analysis emerges from the theme 'Generational succession,' a social perspective that appears in the field data. In addition to the interest in a generational succession, the term 'tourism' can be found continuously throughout the interview, as in: 'Here maybe if you continue these tourism routes it will start to be visited a lot, because it is very beautiful, not only here at home but also in the neighbors' (Participant A); 'Even the name of our tourism thing here is QFC, which then honors the two surnames that are no longer with us' (Participant B) (once again, reference to the patriarchs).

Table 3 presents this analysis and the economic, environmental, territorial, cultural, and political aspects identified in the interviews.

Table 3 shows that the economic dimension stands out since it has the highest number of statements. It can be seen that the *pinhão* nut has a significant economic value, far from being a secondary product in relation to wood in this microenvironment. However, it is also noticeable from the data analysis that the potential in the macroenvironment is not well exploited. The phenomenon can be exemplified in the low sales prices of *pinhão*, the lack of industrialization, and the restriction of consumption to the region and the harvest season.

In the first instance, the territorial field revealed the interesting fact that in the last decades, the region had evolved substantially in the social aspect, taking several families out of poverty: "I think that maybe today the countryside is richer than it used to be. You have to say what is right, there used to be a lot of poor people, and there still are today, but it is much less."

As for the social whole, the territorial dimension appears to demonstrate the relevance of the *pinhão* for the interviewees and the region, since both the interviewees and their friends and neighbors enjoy a lot of the benefits related to the *pinhão*. On the other hand, the amount of *pinhão* trees in the region has decreased a lot, as it was pointed out: "Actually, there used to be more pine trees." In addition, the

Challenges and Perspectives of Pinhão Production Considering the Dimensions of Sustainability

Table 3. Thematic analysis

Dimensions	Statements illustrating the interview
Social	
Generational succession	Participant A: "Maybe even this youth who has dropped land could invest in a plantation for that income. Then we can even see our *pinhões*."
History	Participant A: "And then he even lost his arm in the mill. He had a loose sleeve that got caught in the belt and he couldn't pull it out." Participant B: "Our great-grandfather came from Italy when he was 7. Only Canton is from Spain, actually. Our last name is from Spain, and mom's is Italian from Italy."
Change in extractive activity	Participant B: "Climbing a *pinhão* tree anyone could climb. Now I don't know if you climb it." Participant B: "There was an interest, it seems, in the boys gathering the *pinhão*." Participant B: "Production has decreased a lot because they cut many male *pinhão* trees." Participant A: "We think that after people are using poison, good thing that our region here is not a soybean region, but we noticed that the (number of) *pinhão* in the pinecones has decreased a little."
Personal profile for agroextractive activity	Participant A: "Actually, I've already had several invitations to live in the countryside, there are plenty of opportunities, but many times I think that I wasn't born to be an employee. There are some that you can't climb up to the top because you are getting old. The brother is almost 70 years old and he is still climbing (the trees)."
Economic	
Benchmarking	Participant A: "We would have to have a way to peel the nuts like they do with the Brazil nuts, or do a similar process, or whatever. Maybe you could even study that." Participant B: "I don't know if I wouldn't preserve the pine nuts with those chambers to put cattle meat in; I think it's too cold."
Commercialization channels	Participant A: "Here, we have two markets we sell in the city. And often, the tourists come to buy; they buy yerba mate too. We produce weed here in the old system." Participant B: "The *pinhão* from Arvorezinha, actually, some people buy them here and take them to the big centers, to Porto Alegre, to Lajeado. Botucaraí is the region from Soledade upwards. Here is Alto Taquari and there is Botucaraí."
Quality control	Participant A: "If you leave it too long on the pinecone, it will spoil." Participant C: "Here, usually, if the *pinhão* tree is big, they use ladders and climb the ladder from there. If it's a little low, then no, even you can climb up" (labor quality, occupational safety).
Innovation on the property	Participant A: "Here, maybe if you continue these tourism routes, it will start to be visited a lot because it is very beautiful, not only here at home, but in the neighbors as well." Participant A: "We don't have a cabin yet to stay in, but the people from Araucária do. The Araucaria Park often sends people here, and they send the people here and have lunch, and there is a landing and a beautiful place."
Innovation in the productive process	Participant A: "We would have to have a way to peel the nuts like they do with the Brazil nuts, or do a similar process, or whatever. Maybe you could even study that."
Marketing and sales	Participant C: "Even the name of our tourism business here is QFC, which pays homage to the two surnames that are no longer with us." Participant A: "The 'perau do facão' is right there, about 3 km from here, and it is very beautiful."
Territorial	
Product features	Participant B: "There are male and female *pinhão*. I don't know if you know. The male makes a cigar, and he makes a cigar in place of the pinecone. In September, he releases pollen to pollinate the pinecone on the seed." Participant B: "There's the black one, there's the white *pinhão*, there's the red *pinhão*. It has several colors, sizes..." Participant B: "We know that in the field, it comes before and after. Since September, the pinecones are already being formed for the next year." Participant B: "The azure jay helps with the planting. It even has music."
Relevance of the *pinhão*	Participant A: "There is a video I talk about me having to gather *pinhão* to buy my first guitar." Participant A: "I used to buy shoes."
Environmental	
Contribution to the environment	Participant A: "Yes, that's the problem. Maybe it would even be a way for the people to leave the pine tree standing just to extract the fruit, keep the forest standing more."
Cultural	
Extractive activity	Participant A: "No matter where you are, the smart guy, you have the gift. You can be in the capital or deep in the grotto, as Baitaca says, but you will..." Participant A: "I based this on a story that Dad used to tell. If I like a phrase or a word, music often comes out of nowhere." Participant B: "We write and sing a lot about chimarrão: sometimes, the *pinhão* is forgotten. Do you know the *jaracatiá*? There's a *jaracatiá* tree outside, you can grate it and make candy out of it. I don't know if there is one down there, but you can take a picture to get an idea of what it is."
The *pinhão* production process	Participant A: "We work mostly with organic and natural things. We don't like poison very much." Participant A: "More in this season, we collect *pinhão*. It's from April until the end of May, beginning next month. Most of them are that little, and then others are left out, those from June and August."
Political	
Perspective for the extractive activity	Participant A: "We would have to have a way to peel it as they do with the Brazil nuts, do a similar process, or whatever." Participant C: "It would have to be industrialized into flour, like an almond."
Relationship with the community	Participant A: "I occasionally go to some lecture." Participant B: "here is a *pinhão* festival in Fontoura Xavier."

Source: survey data

participants cited characteristics of the seed in the locality. Another point that corroborates with the

Figure 3. The poem 'Araucaria'
Source: *Research data.*
*Translation: At the top of Serra do Botucaraí, the native pine made its home here. Pine tree and the king of our forest Fontoura Xavier made pinenuts the party. In the tents on br 386, don't go unnoticed, there are pine nuts in the pine cone, threshed and cooked. Pine nuts in the weather vane already leaves pine nuts in the plate and well hammered. Pine nuts in water with honey or salted pacca jerky in pounded mortar.

political dimension is that there is little dialogue and partnership among political actors in the region when the issue is *pinhão*."

One of the main cultural points that should be mentioned is the interest in organic products rather than those that use pesticides, which negatively impact the regional environment. The interviewees also associate the *pinhão* with cultural traditions, such as the burning of the *pinhão*, which consists in burning the dried leaves of the *pinhão* tree, heating the *pinhão* and preparing it for consumption.

The *pinhão* is an inspiration for local literature, appearing in poems by local residents, as in the poem 'Araucária' by Canton (Figure 3). In the poem's first verse, it is possible to observe a territorial aspect. The author points out characteristics of the local vegetation by mentioning the presence of the *pinhão* in the Fontoura Xavier mountains, attributing to the tree species the role of forest resident in a position of 'Royalty': '*Pinhão* is the King of our Forest [...]' (Canton, 2018). This description imbues the notion that *Araucaria* exists beyond the production of *pinhão*; its role has a territorial imprint.

Through the cultural dimension, we argue the presence of a defense of the territory with a political positioning by the author. By describing the Araucaria as 'King,' the author resorts to the relevance of the species as a constitutive element of the territory they inhabit, highlighting his position in defense of the continuity of the tree species, therefore, arguing against the deforestation of this vegetation.

The poem also reflects aspects of the cultural dimension by describing stages of the productive process of collecting the *pinhão*. Barbosa et al. (2020) found that the process of collecting the *pinhão* involves climbing the *Araucaria* tree, requiring tools such as bamboo to perform 'clashing' movements between the extractivist and the *pinhão* in order to cause the product to fall, as described in the fourth verse of the poem (Figure 3).

Returning to the thematic analysis, one can note that the productive chain of *pinhão* is structured in such a way as to promote harmony between Sachs' (2007) social and economic dimensions, especially in the primary link of the chain (i.e., extractivism). At this stage, extractors believe they experience '[...] a decent quality of life with income' (Participant A), besides demanding innovations that can generate jobs (Participant B).

Adopting the *pinhão* as an ally to sustainability has similarities with the cosmetics company Natura, which works with products from Amazonian communities to manufacture cosmetics, with the important characteristic of considering the culture of these communities. However, unlike the cosmetics company, it was not possible to map the presence of large companies commercially exploiting the *pinhão* in the Arvorezinha region.

The political field did not encapsulate representative themes in the thematic analysis, leading to the conclusion that it is not a factor of great participation in the *pinhão* chain from the perspective of the stories told by the participants. Political interaction even exists for actions related to tourism, but when it comes to aspects related to the *pinhão*, the participation of political entities proved to be weak. What, unfortunately, translates into practice the contributions of Paes-de-Souza (2011), when he states that government agencies do not act in conjunction with the interests of collectors, producers, and exporters.

In the environmental field, it is emphasized that the interviewees give great importance to the environment, emphasizing the appreciation and preservation of the *Araucaria* tree. However, in the stories told, the participants emphasize this dimension little. This concern and interest in leaving the forest standing are of utmost importance for contributing to reducing the deforestation rates of the Atlantic Forest, a high rate as emphasized by Guerra et al. (2002).

Suppose the *pinhão* were industrialized and gained space in the market. In that case, the extractivists believe that more people would invest in planting *pinhão* trees: "Maybe it would even be a way for people to leave the pine tree standing just to extract the fruit/seed, to keep the forest standing" (Participant B). The assumption of the change in thinking regarding the pine tree is strongly linked to and exemplifies the transformative changes in the social field of sustainability.

In the social field, the role of *pinhão* stands out for the quality of life, for guaranteeing subsistence and food for the animals. On the other hand, the effort can be too much due to the use of rudimentary equipment and the difficulty of the process, which can put the lives of the extractors at risk.

During the interview, a demand for more modern extraction equipment emerged, such as improved ladders and a kind of "bamboo" that could evolve into a tool that seeks to avoid climbing the tree to collect the *pinhão*. From the social point of view, once again, participants seek to support new generations by stimulating them to become interested in the *pinhão* as an actor in social movement.

The possible technological advances indicated the improvement of the *pinhão* chain, as the following statements portray: "We would have to have a way to peel the nuts like they do with the chestnut, do a similar process, or I don't know. Maybe you can even study that"; "We would have to invent something to collect the pine nuts from the ground; nobody has invented that yet." These notes refute the common sense and the view of Cairns (1998), in which he defended the idea that sustainable actions result in the loss of jobs, especially in sectors of great importance for income generation in several regions. In the context of the *pinhão*, such a view is refuted precisely by the demand for innovations, which enable job generation and labor qualification.

Moreover, the desired industrialization of *pinhão* production contributes to achieving goal 9 of the 2030 Agenda for Sustainable Development, which consists of building resilient infrastructure, promoting inclusive and sustainable industrialization, and fostering innovation.

Pinhão, along with the other NTFPs, also contribute to another aspect of the agenda created by the UN, which is goal 15: 'Protect, restore, and promote the sustainable use of terrestrial ecosystems, sustainably manage forests, combat desertification, halt and reverse land degradation and halt biodiversity loss.' The goal can be fulfilled through reforestation by keeping the flora standing to capture the natural resources in favor of humanity. This is exemplified in the report of one of the interviewees: "Maybe even this youth that has land could invest in a plantation for that [...]. Then we can even see our *pinhão*."

The data analysis carried out corroborates Zamora (2001) and Soares et al. (2008), who see Latin American countries as countries that value NTFPs considerably. This is because, despite the *pinhão* production chain having shortcomings, the interviewees said that the product brings satisfaction, considering the income generated, the subsistence, the guarantee of food for the property's animals, the preservation of the environment, and the conservation of traditions. Nevertheless, the valorization is done only by actors who have direct contact with these products, and there is no interest from government agencies or the industry in general.

The productive chain of the *pinhão* is clearly a valuable alternative to contribute to sustainability. Despite all the precariousness, the lack of political interest in strengthening the chain, and the small product participation in the national and international markets, *pinhão* production has existed for centuries. With an improvement in some links of this chain, as the participants of this study believe (e.g., the extraction and industrialization stages), *pinhão* cultivation can become an even more powerful ally to sustainability.

CONCLUSION

This study shed light on the current challenges and perspectives of the extractive activity of *pinhão*. This study was carried out in the context of the Barro Preto community in Arvorezinha (Rio Grande do Sul State, southern Brazil) and based on the dimensions of sustainability. Throughout the data analysis, it was possible to observe that the challenges and perspectives tend to be organized in themes with convergence for each sustainability dimension.

Among the challenges, it is possible to list them based on data from interviews and readings carried out in the literature: 1) the lack of interest of the younger generations in continuing with *pinhão* extraction; 2) the arcadian nature of the production process, remaining the same compared to centuries of activity; 3) the need for a personal profile for the work, almost always based on physical strength and sensibility for the practice of initiatives to control the quality of the production; and 4) the feeling of acting alone without political support to implement actions to add value to the product. Among the perspectives, the participants reported: 1) the hope that innovations will arise that can make the work of collecting *pinhão* less arduous, especially from the inspiration of other NTFP productive chains; 2) the reappearance of interest among young people in becoming acquainted with the agro-extractive activity of *pinhão*; and 3) the insertion of *pinhão* in tourism activities as a means of implementing income and attracting market interest.

Although *pinhão* production can be aligned with the preservation of the *Araucaria* forest, our findings allow us to conclude that the theme is little addressed by the participants of the study, even though it is a recurring theme and pointed out in the scientific literature as being eminently linked to the agroextractive activity. This aspect has been shyly taken advantage of by the agroextractivists participating in this study.

This study highlights how relevant the art of storytelling can be to a society if one considers the context, who, where, and when this narrative is being constructed. It is important to say that narratives are powerful means of persuasion, constituted from stories. Furthermore, the narrative told in this study shows that the themes most explored by the participants involve economic, cultural, territorial, and social dimensions, while environmental and political dimensions remain in the background.

This narrative is of concern because this study aimed to identify challenges and perspectives of *pinhão* extraction; themes such as environmental and political aspects were little argued by the participants, who seemed busier discussing continuity in terms of generational succession. We believe that participants have the motivation to discuss the succession theme vehemently since they: i) show concern with the continuity of the extractive activity, ii) propose innovation initiatives to stimulate new generations to dedicate themselves to the activity, and iii) resort to the stories of the patriarchs to motivate a behavior change.

We conclude that the narrative that keeps the political and environmental dimensions in a peripheral discussion is dangerous for the assurance of Sachs' (2007) integral sustainability by mischaracterizing the essence of sustainability and by making the achievement and advances for the *pinhão* productive chain unfeasible due to the lack (or little) political adherence of the participants on the subject.

The contribution of this study is based on the methodological aspect because it allowed actors little considered in the *pinhão* productive chain (the agroextractivists) to present their vision in terms of production challenges and perspectives for this activity. History is polyphonic (Boje & Rosile, 2020); therefore, many individuals who, in many cases, are silenced need to participate in searching for solutions to sustainability challenges.

The present study also suggests a theoretical contribution to the literature by unveiling the concept that extractive activity contributes to sustainability. Its implementation may be compromised if not seen through the multiple dimensions of integral sustainability, and it proves that the narrative can be constructed from relevant actors within this process. Yet, dimensions such as the 'environmental' and 'political' seem to be little explored in the discussion.

We suggest, as future research opportunities, diversifying the interview participants by investigating the stories of other actors in the production chain (e.g., government actors, universities, and research institutes) to perform a polyphonic context analysis and find meanings and contradictions between the stories of the individual respondents.

REFERENCES

Alves, R. V. (2010). *Estudo de caso da comercialização dos produtos florestais não madeireiros (PFNM) como subsídio para restauração florestal.* [Case study of the commercialization of non-timber forest products (NMF) as a subsidy for forest restoration.] [Master's thesis, Universidade Federal de Viçosa].

Barbosa, C. S., Silva-Jean, M., Luz, J. P., Leandro, G., & Bohn, D. P. (2020). Processo Produtivo do PFNM Pinhão das araucárias: O caso do extrativista JDZ no Rio Grande do Sul. [Pfnm Pinhão Production Process of araucarias: the case of the JDZ extractive in Rio Grande do Sul.]. *Revista de Administração e Negócios da Amazônia, 2*(1), 4–17. doi:10.18361/2176-8366/rara.v12n1p4-17

Barr, S., & Gilg, A. (2006). Estilos de vida sustentáveis: Enquadrando a ação ambiental dentro e ao redor da casa. [Sustainable lifestyles: framing environmental action in and around the house.]. *Geoforum, 37*(6), 906–920. doi:10.1016/j.geoforum.2006.05.002

Bhatti, M., & Igreja, A. (2004). O lar, a cultura da natureza e os significados dos jardins na modernidade tardia. [The home, the culture of nature and the meanings of the gardens in late modernity.]. *Estudos de Habitação, 19*(1), 37–51.

Boje, D., & Rosile, G. A. (2020). *How to use conversational storytelling interviews for your dissertation.* Camberley: Edward Elgar Publishing.

Boje, D. M. (2001). *Narrative methods for organizational & communication research.* Sage Publications. doi:10.4135/9781849209496

Boolaane, B. (2006). Restrições à promoção de abordagens centradas nas pessoas na reciclagem. [Restrictions on promoting people-centred approaches to recycling.]. *Habitat International, 30*(4), 731–740.

Boulding, K. (1966). The economics of the coming spaceship Earth. In H. Jarrett H. (Ed.), Environmental Quality in a Growing Economy, pp. 20-30. Baltimore: John Hopkins University Press.

Cairns, J. Jr. (1998). What sustainability is not! *International Journal of Sustainable Development and World Ecology, 5*(2), 77–81. doi:10.1080/13504509809469972

Cairns, J. (2003). Materialfilia, biofilia e uso sustentável do planeta. [Materialphilia, biophilia and sustainable use of the planet.]. *Revista Internacional de Desenvolvimento Sustentável e Ecologia Mundial, 10*(1), 43–48.

Canton, O. F. (2018). *Contos de Arrepio.* [Tales of Chill.] Lajeado: Univates.

Chamberlain, J., Bush, R., & Hammet, A. L. (1998). Non-Timber Forest Products: The other forest products. *Forest Products Journal, 48*(10), 10–19.

Elkington, J. (2006). Governance for Sustainability. *Journal Compilation, 14*(6), 522–529.

Eller, F. J., Gielnik, W., Thölke, C., Holzapfel, S., Tegtmeier, S., & Halberstadt, J. (2020). Identifying business opportunities for sustainable development: Longitudinal and experimental evidence contributing to the field of sustainable entrepreneurship. *Business Strategy and the Environment, 29*(3), 1387–1403. doi:10.1002/bse.2439

Fausto, B. (2006). *História do Brasil*. [*History of Brazil.*] São Paulo: EdUSP.

Fiedler, N. C., Soares, T. S., & Silva, G. F. (2008). Produtos Florestais Não Madeireiros: Importância e Manejo Sustentável da Floresta. [Non-Timber Forest Products: Importance and Sustainable Forest Management.]. *Revista Ciências Exatas e Naturais, 10*(2), 263–278.

Flick, U. (2009). *Qualidade na pesquisa qualitativa* [*Quality in qualitative research.*]. Artmed.

Frame, B. (2004). The big clean up: Social marketing for the Auckland region. *Local Environment, 9*(6), 507–526. doi:10.1080/1354983042000288030

Freire, P. (1975). *Pedagogia do oprimido* [*Pedagogy of the oppressed.*]. Afrontamento.

Gregolin, G. C., Gregolin, M. R. P., Triches, R. M., & Zonin, W. J. (2019). Desenvolvimento: do unicamente econômico ao sustentável multidimensional. [Development: from the only economic to the multidimensional sustainable.] *Rev. Eletrônica de Humanidades do Curso de Ciências Sociais da UNIFAP, 12*(3), 51–64.

Guerra, M. P., Silveira, V., Reis, M. S., & Schneider, L. (2002). Exploração, manejo e conservação da Araucária (Araucaria angustifolia). [Exploration, management and conservation of Araucaria (Araucaria angustifolia).] In L. L. Simões, & C. F. Lino (Orgs.). Sustentável Mata Atlântica: a exploração de seus recursos florestais [Sustainable Atlantic forest: the exploitation of its forest resources] (pp. 55-80). São Paulo: SENAC São Paulo.

Hobson, K. (2003). Thinking habits into action: The role of knowledge and process in questioning household consumption practices. *Local Environment, 8*(1), 95–112. doi:10.1080/135498303200041359

Instituto Brasileiro de Geografia e Estatística [IBGE]. (2020). *Estimativas populacionais para os municípios e para as Unidades da Federação brasileiros em 01.07.2020*. [*Population estimates for the municipalities and for the Brazilian Federation Units as of 07.01.2020.*]. IBGE.

Instituto Nacional de Pesquisas Espaciais [INPE]. (2020). *Projeto Prodes Digital: Mapeamento do desmatamento da Amazônia com Imagens de Satélite*. [*Prodes Digital Project: Mapping deforestation of the Amazon with Satellite Images.*] São José dos Campos: Instituto Nacional de Pesquisas Espaciais.

Kates, R. W., Clark, W. C., Corell, R., Hall, J. M., Jaeger, C. C., Lowe, I., McCarthy, J. J., Schellnhuber, H. J., Bolin, B., Dickson, N. M., Faucheux, S., Gallopin, G. C., Grübler, A., Huntley, B., Jäger, J., Jodha, N. S., Kasperson, R. E., Mabogunje, A., Matson, P., ... Svedin, U. (2001). Sustainability Science. *Science, 292*(5517), 641–642. doi:10.1126cience.1059386 PMID:11330321

Kuhlman, T., & Farrington, J. (2010, November 01). What is sustainability? *Sustainability*, *2*(11), 3436–3448. doi:10.3390u2113436

Linden, A., & Carlsson-Kanyama, A. (2003). Environmentally friendly disposal behaviour and local support systems: Lessons from a metropolitan area. *Local Environment*, *8*(3), 291–301. doi:10.1080/13549830306664

Lindenberg, S., & Steg, L. (2007). Normative, gain and hedonic goal frames guiding environmental behaviour. *The Journal of Social Issues*, *63*(1), 117–137. doi:10.1111/j.1540-4560.2007.00499.x

Lynch, T., & Khan, T. (2020). Understanding what sustainability is not – and what it is. *The Ecological Citizen*, *3*(B), 55-65.

Morvan, Y. (1985). *Filière de production: fondementes d'economie industrielle* [Production chain: foundations of industrial economics.]. Economica.

Neumeier, S. (2012). Why do Social Innovations in Rural Development Matter and Should They be Considered More Seriously in Rural Development Research? Proposal for a Stronger Focus on Social Innovations in Rural Development Research. *Journal of the European Society for Rural Sociology*, *52*(1), 48–69. doi:10.1111/j.1467-9523.2011.00553.x

Organização das Nações Unidas [ONU]. (2018). *Objetivos de Desenvolvimento Sustentável*. [Production chain: foundations of industrial economics.] ONU.

Paes-de-Souza, M., Silva, T. N., Pedrozo, E. A., & Souza Filho, T. A. (2011). O Produto Florestal Não Madeirável (PFNM) amazônico açaí nativo: Proposição de uma organização social baseada na lógica de cadeia e rede para potencializar a exploração local. [The Native Amazonian Non-Timber Forest Product (PFNM): proposition of a social organization based on chain and network logic to enhance local exploration.]. *Revista de Administração e Negócios da Amazônia*, *3*(2), 44–57.

Pedrozo, E. A., Silva, T. N., Sato, S. A. S., & Oliveira, N. D. A. (2011). Produtos Florestais Não Madeiráveis (PFNMS): As Filières do Açaí e da Castanha da Amazônia. [Non-Timber Forest Products (PFNMS): the Filières do Açaí and Castanha da Amazônia]. *Revista de Administração e Negócios da Amazônia*, *3*(2), 88–112.

Rutherford, S. (2007). Green governmentality: Insights and opportunities in the study of nature's rule. *Progress in Human Geography*, *31*(3), 291–307. doi:10.1177/0309132507077080

Sachs, I. (1986). *Ecodesenvolvimento: crescer sem destruir* [Eco-development: grow without destroying.]. Editora Vértice.

Sachs, I. (2007). *Rumo à ecossocieconomia: teoria e prática do desenvolvimento* [Towards eco-economics: theory and practice of development.]. Cortez.

Santos, A. J., Corso, N. M., Martins, G., & Bittencourt, E. (2002). Aspectos produtivos e comerciais do Pinhão no estado do Paraná. *Revista Floresta*, *32*(2), 163–169. doi:10.5380/rf.v32i2.2281

Schwarcz, L. M., & Starling, H. (2015). *Brasil: uma biografia*. Companhia das Letras.

Silva, J. M. (2015). *Políticas públicas para composição de custos e formação de preços da atividade extrativa da castanha-da-Amazônia*. [Master's thesis, Fundação Universidade Federal de Rondônia].

Silva-Jean, M. (2017). Custos e preços da castanha-da-amazônia nos Estados do Acre e Rondônia. [Costs and prices of the Amazon nut in the states of Acre and Rondônia.] *Custos e @gronegócio on-line, 13*(2), 421-447.

Silva-Jean, M., Paes-De-Souza, M., Souz-Filho, T. A., Riva, F. R., & Barbosa, C. S. (2022). Public Policies Guarantee for Minimum Prices on Products of Sociobiodiversity (PGPMBio): Composition of the extration cost of Amazonian chestnut in Rondônia and Acre. *Revista de Administração UFSM, 15*(1), 62–82. doi:10.5902/1983465965906

Silva-Jean, M., Souza, M. P., & Filho, T. A. S. (2020). Cadeia produtiva da Castanha-da-Amazônia nos Estados do Acre e Rondônia. [Production chain of the Amazon Nut in the states of Acre and Rondônia.]. *Brazilian Journal of Development, 6*(11), 91277–91297. doi:10.34117/bjdv6n11-512

Soares, T. S., Fiedler, N. C., Silva, J. A., & Gasparini Junior, A. J. (2008). Produtos Florestais Não Madeireiros. [Non-Timber Forest Products.] *Revista Científica de Engenharia Florestal, 11*.

Tsalis, T. A., Malamateniou, K. E., Koulouriotis, D., & Nikolaou, I. E. (2020). New challenges for corporate sustainability reporting: United Nations' 2030 Agenda for sustainable development and the sustainable development goals. *Corporate Social Responsibility and Environmental Management, 27*(4), 1617–1629. doi:10.1002/csr.1910

Vallance, S., Perkins, H. C., & Dixon, J. E. (2011). What is social sustainability? A clarification of concepts. *Geoforum, 42*(3), 342–348. doi:10.1016/j.geoforum.2011.01.002

Vlek, C., & Steg, L. (2007). Human behavior and environmental sustainability: Problems, driving forces, and research topics. *The Journal of Social Issues, 63*(1), 1–19. doi:10.1111/j.1540-4560.2007.00493.x

World Commission On Environment And Development [WCED]. (1987). *Our Common Future*. Oxford: Oxford University Press.

Chapter 15
Engaging People on E-Participation Through Social Media Interactions

Daielly M. N. Mantovani
Universidade de São Paulo, Brazil

Kleber Rodrigues Santos
https://orcid.org/0000-0003-2123-5426
Universidade de São Paulo, Brazil

Thaisa Barcellos Pinheiro Nascimento
https://orcid.org/0000-0002-8360-8649
Universidade de São Paulo, Brazil

Celso Machado Jr.
https://orcid.org/0000-0003-3835-2979
Universidade Municipal de São Caetano do Sul, Brazil

ABSTRACT

The chapter aims to analyze how social media engages citizens in issues related to municipal management in Brazilian capital cities (27 cities). For that, Twitter data was collected, and descriptive analysis, text mining, and social network analysis were carried out. Results show the most frequent interactions regarded sharing posts, replies, and reactions were less frequent. Text mining suggested behavior on Twitter is related on the hot news, so discussions tend to be superficial; network analysis showed mayor accounts have more connections with users than the cities' official accounts, which suggests a necessity for personification on the conversation. Interactions are both centralized (started by the city) and decentralized (start by the citizen), but consist merely of information transmission and opinion sharing, and more complex kinds of participation, such as co-creation and decision-making were not observed. These findings show the potential of social media communication for public management and give insights on how to develop a successful policy to participate in social media.

DOI: 10.4018/978-1-6684-6123-5.ch015

Engaging People on E-Participation Through Social Media Interactions

INTRODUCTION

Citizen participation is interpreted as a fundamental tool to strengthen democratic processes and public governance mechanisms and to foster sustainable development (Stratu-Strelet et al., 2021). Participation implies the citizen's voluntary involvement with the intention to influence public decision making and management (Tejedo-Romero et al., 2022), in other words, it means giving citizen voice in the governance structure (Callahan, 2007) or may be understood as citizen power (Arnstein, 1969). Citizen participation is widely recognized as positive and fundamental to democracy as it permits citizen taking part on decisions that affect their fades, however, it may be a source of conflict and uncertainty, since citizens usually do not have technical competency to influence some decisions, for example public security issues (Burke, 1968). Despite of that, it is an intelligence tool that enables data collection at the local level, making it possible to explore and set political priorities, as well as allocate resources more efficiently and transparently. Citizen participation involves a two-way communication, which establishes the interaction of civil society, represented by the citizen individually or organized in groups and communities, with the political class or the administrative sphere (Stratu-Strelet et al., 2021).

Citizen participation can occur with or without the mediation of technology, the first case being called e-participation. The United Nations defines e-participation as a key factor in governance and one of the pillars of sustainable development (UN, 2020), as it enables society to participate in decision-making on public issues, strengthening democratic processes (Stratu- Strelet et al., 2021).

The use of technology has direct effects on the elaboration of public policies, as it provides the opportunity for the common citizen to participate in decision-making spheres, generating empowerment of the individual and, consequently, of society itself. However, even with the collaborative use of technology, the process of creating an effective partnership between government and citizen is challenging so that, in fact, the democratic structure can be advanced (Bouzguenda, Alalouch, & Fava, 2019).

Citizen participation and engagement in public administration issues have been positioned over decades as challenging. However, the advancement of technologies, especially ICTs, allowed the creation of channels of multi-way interaction between the various actors involved, which changed the form of communication between institutions and citizens, thus enhancing the emergence of various initiatives and models encouraging citizen participation (Carvajal Bermúdez & König, 2021). Thus, ICTs have great potential to help overcome barriers to citizen participation, as they provide user-friendly and low-cost forms of citizen-government interaction. In this sense, Figure 1 shows the interaction between the ICTs and citizen participation.

To make urban development more democratic and participatory, it is necessary to create effective mechanisms for citizen engagement in the design of the urban structure, with emphasis, for example, on the design of streets and neighborhoods where they live (Carvajal Bermúdez & König, 2021). Current forms of citizen participation show low citizen engagement. However, digital media and web 2.0 technologies, especially social platforms, the focus of this chapter, create possible ways to overcome such barriers. Additionally, these new resources enable the establishment of new forms of governance, in which public entities not only transmit information, but also collect it from citizens. Despite the benefits that technology enables citizen participation, the impact of social media on citizen engagement has not been systematically measured (Carvajal Bermúdez & König, 2021). In an aggregate context, it is important to highlight that e-participation is one of the fundamental elements of e-government, presenting different levels of participation, namely: e-information, e-consultation, e-decision-making (Stratu-Strelet et al., 2021), e-petition and e-co-creation (Carvajal Bermúdez & König, 2021). Thus, it is possible to establish

Figure 1. Relation between ICTs and Participation
Source: the authors

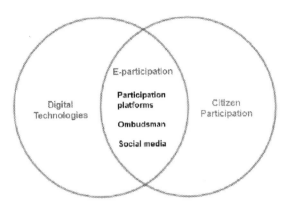

that citizen participation can vary from a restricted context to a broader and more inclusive one. Citizen co-production created a new interface for government-citizen interaction, however the effectiveness of this relationship in providing better public services is still little explored. The web 2.0 tools adopted by smart cities, such as mobile applications, social media, among others, have influenced how public administrations manage their services (Allen et al., 2020). The use of insights from the citizen in the provision of public services has the potential to raise the quality of the service, from the citizen's point of view, he observes that it is possible to participate in public management, through collaboration or co-production through the use of platforms similar to those used in everyday life for online shopping, banking, or social networks.

Governments in their turn, benefit from knowing their citizens in order to offer better services, which, in turn, contributes to achieving the objectives of providing better quality of life and sustainability to citizens. Data collected from participatory platforms is a relevant source of big data for smart cities and smart governments. The authors studied the relationship between co-production and service performance. The digitization of government brings great opportunities for citizen co-creation, the development of ICTs allows for a deeper engagement between citizen and government while bringing the opportunity for citizens to do good for themselves, by obtaining better public services. Citizen participation is also seen as a potential for developing more cost-effective services and increasing government accountability, in addition to playing a central role in the development of open government (Allen et al., 2020).

The ubiquity of ICTs in people's daily lives enhances the debate on their suitability for the challenge of identifying responses aimed at promoting sustainable development. Regarding digital participation, the focus of the literature is on technological alternatives, with little attention to issues of empowerment, inclusion, equity, access and belonging. These variables are fundamental for the social pillar of sustainability and, consequently, for achieving sustainable development (Bouzguenda, Alalouch & Fava, 2019).

The study by Stratu-Strelet et al. (2021) showed that, when citizens perceive the effectiveness of their participation, they feel more motivated to continue participating. The authors' research results show that: the use of ICTs increases citizen participation; the policies for the use of ICTs, by the public power, have a greater influence on e-participation than the available urban infrastructure or the individual demands of the citizen; the levels of democracy present a positive relationship with the levels of e-participation, with a direct relationship between democracy and participation; democracy and the digitization of government must be achieved together in order to achieve high levels of e-participation.

Since social media is a type of technology present in the daily life of the common citizen and with great prominence among the youngest, they could be adopted by public administrations as a platform for participation. The use of social media by the government in its relationship with the citizen is presented as potentially adequate, as this technology is widely used daily by practically the entire society for a diverse set of purposes. It is worth mentioning, as an example of the use of social media by society, the practices of buying products and services, expressing opinions, communicating with companies and personal contacts. In this sense, the chapter aims to analyze how social media engages citizens in issues related to municipal management, discussing how these interactions can feed municipal smartness. The empirical study used secondary data extracted by the API of the social network Twitter, considering interactions in the profiles of Brazilian municipalities.

Brazil has a fairly developed digital government, getting the 49th place out of 193 countries, on the EGDI 2022 (UN, 2022), which makes it an interesting country to analyze. Moreover, it is not of our knowledge a similar study considering Brazilian cities, being found only studies that focused on the 2018 presidential elections, which had a massive usage of social media during the campaigns.

This research intends to advance in knowledge of how to make use of current technology on participation and e-government, aiming to offer relevant insights to public management on how to improve their policies to obtain better outcomes and advance in governance and transparency. Bonsón et al., (2019) pointed out Twitter is the best social media for public issues participation, nevertheless studies on the use of this platform by local government are scarce, focusing mostly on international or national levels and exploring elections and political campaigns engagement. In this sense the present chapter tries to bring a municipal point of view on nonpolitical/election situations but on daily city issues. Another contribution may be specifically for Brazilian local governments, as the chapter presents local data which reflects the country's specificities.

GOVERNMENT, SOCIAL MEDIA, AND PARTICIPATION

Citizen political participation is beneficial for both the government and the society. For the first, participation provides information that allows better decisions and for citizens, in turn, it is a form of empowerment and citizenship. Traditionally, forms of participation were time-consuming and tedious, such as visits to public agencies, phone calls and letters; however, with the advances in technology and the emergence of new communication solutions, the so-called e-participation and therefore the e-citizen emerged. The interaction potential of social networks and their wide reach has drawn the attention of governments that seek to create policies and strategies for their presence in these networks (Arshad & Khurram, 2020).

The use of social media in the strategic communication of public management has become a current practice in governments around the world. In this case, the corporate use of social media platforms aims to increase population engagement, increase participation, collaboration and transparency in public management, as well as provide closer contact with citizens, which would be unlikely without using these platforms (Mergel, 2013). Thus, social media, in addition to changing the forms of interaction with citizens, have changed the way public institutions work (Criado et al., 2013).

Social media is part of the so-called Web 2.0, which is characterized by technologies that allow interaction between users and the creation and sharing of content in an open and democratic way, insofar as basic knowledge of navigation on the World Wide Web allows participation. These characteristics result in greater socialization among users dispersed anywhere in the world, creation of communities and individual empowerment, as posted messages can be disseminated instantly and at low cost (Magro, 2012). Among the main types of Web 2.0 used in public management, blogs, microblogs, social networks, sharing platforms, virtual reality and crowdsourcing stand out (Criado et al., 2013).

Social media websites, the focus of this chapter, are part of social media and are characterized by the set of technological tools that allow the creation of a user profile, the development of content by the user and connection with other users. Due to these characteristics, social networking websites, in addition to enabling public management to communicate with the user, that is, sending information in real time, allow citizens to discuss among themselves and offer ideas and their views on various issues of public management (Mossberger et al., 2013).

Similar to the term e-commerce, e-government (or electronic government) appears in the early 1990s under the internet boom, representing the union of the words government and internet (Grönlund & Horan, 2004). When we specifically deal with the insertion of e-government in the context of Web 2.0, and therefore, of social media, Magro (2012) points out that this phenomenon began in 2007 in the USA. One of the pioneering and most advanced countries in forms of e-government, Estonia, allows its citizens to enroll their children in school, file taxes, create digital identities and signatures, and even vote from their homes via the internet, for example, the country went from 0.3% of the population voting online in 2009 to 46% in 2019 (e-Estonia, 2022).

The UN (2020) defines e-government as: "(...) the use of communication and information technologies to effectively and efficiently deliver public services to citizens and businesses". Critically, Schönberger (2011) points out that most e-government initiatives focus on gaining efficiency and improving user experience while using public services, but it has deficiencies in promoting civil society participation and engagement. Antiroikko (2004, p.40) complements with the idea of democratic electronic governance, in which: "(...) citizens have the chance to participate and effectively influence relevant issues through different institutionally organized and legitimate modes of participation".

Web 2.0, in turn, is one of the great foundations for the insertion of electronic government from a more democratic and participatory perspective. Kes-Erkul and Erkul (2009, p. 5) point out that "(...) Web 2.0 tools, such as social media, have the ability to modify the relationship between the internet and its users, altering power structures and providing opportunities for users to have greater engagement with their respective communities."

Magro (2012) carried out an extensive literature review from 2007 to 2011 on the use of social media in e-government. The author points out that this practice began in the USA around 2007, without major strategic planning. As a result, there was, on the one hand, low citizen participation; on the other hand, the population was favorable of this initiative, increasing the government's credibility. From this, social media application initiatives evolved in several countries around the world, being used in electoral campaigns, creating a closer contact between candidates and voters.

In this sense, it can be considered that social media represent a new stage of e-government policies – known as government 2.0 (Guo et al., 2016) –, which, at first, sought to offer online services to citizens, that is, at this stage there is a static relationship of State service provision to citizens. With the use of social media, we seek to offer additional channels of communication, providing a democratic participation of the citizen in the formulation of public policies and decisions of the spheres of government (Mergel, 2013).

Social media, unlike traditional e-government platforms, are a two-way citizen-government interaction tool and consist of an effortless process on the part of the citizen to interact, communicate and engage with public management, which can encourage more effective citizen participation in decision-making. Despite the potential, citizen participation in e-government services delivered by social media is still low (Khan et al., 2021). The key to social media use is to make communication faster and transparent enabling citizens to engage in new projects or policies, which means informing them but also gathering information from them. Gathering information may assume different forms, such as collecting reports regarding infrastructure problems but also idea collection which is a way of co-creating along with the citizens (Carvajal Bermúdez & König, 2021).

Trust is a central element in this participation process. The study by Khan et al. (2021) used the TAM technology acceptance model to assess antecedents of intention to participate in e-government on social media, using data from Pakistan. They observed that privacy (in citizen-government interaction), security (of shared data), infrastructure (providing secure networks for exchanging information on social media), information quality and ease of use significantly positively influence trust in e-government via social networks. Similarly, trust influences the citizen's intention to participate. Pakistan is a developing country with low development of digital government solutions and low adoption of e-government solutions by the population. It occupies the 150th position in the EGDI 2022, while Brazil occupies the 49th position in the ranking, which illustrates the degree of Brazilian development in e-government. However, when it comes to social networks, have governments managed to engage the participation of the population? Have you been able to offer services? Or even, the focus question of this work, to stimulate the joint creation of the citizen for municipal management issues?

In developing countries, the use of social media is still at the simplest level, the informational, which aims to disseminate information and news, with little activity at two-way collaborative levels. Social media allows the government to speak directly to the citizen without the intermediation of traditional media which makes the process faster and simpler. While in developed countries social media are used at different levels of government with clear and formal policies and strategies for use, in developing countries the use is informational and poorly structured, without the presence of a strategic framework. It is common for governments and government agencies to have profiles on social networks, but only to have a digital presence, without an adequate communication strategy with the public (Arshad & Khurram, 2020).

Another important situation where social media comes to place are the extreme events. In extreme events (i.e., natural disasters or large-scale man-made harm), the rapid and accurate delivery of information from government is crucial to maintain political stability, public safety and trust in government. In their study, Chatfield and Reddick (2018) focused on New York City's network co-production during the response phase of disaster management of Sandy Hurricane. They analyzed Twitter posts #sandy, to assess how NYC public agencies used the social media to respond to the disaster. They point out Twitter is a powerful tool for risk management during crisis, for its speed and reach, however, the co-production does not happen naturally, it is necessary to develop a social media policy which enables the dissemination of high quality and fast information, as well as the monitoring and reacting to the inputs made by the users. In addition, their analysis also shown that identifying influencer users and mass media profiles, which have a relevant number of followers, is important to make the strategy succeed. In their perspective, co-production implies helping the government disseminate important information and bringing relevant information from population. Other important reflection brought by the authors is the importance of developing a strong social media policy in non-crisis times, then when extreme events happen, the communication structure will be already in operation.

Arshad and Khurram, (2020) release a survey with Pakistani citizens that showed the dissemination of quality information by the public agency (food agency) is an antecedent of the citizen's online political participation, since the perception of information quality raises the perception of the agency's transparency and the citizen's trust in it. Disclosure of information on networks reduces the likelihood of noise and misunderstandings, in addition to providing an understanding of how decisions are made and what work has been done by the agency, which also contributes to increased trust. Social networks make it possible for citizens to connect with the government, which also makes them assess their responsiveness, the better the perception of quality, the better the perception of responsiveness. Trust in the agency had no direct relationship with the citizen's political participation, as citizens tend to manifest themselves only when there is something complaint about. The authors also found the better the perception of responsiveness, the lower the online participation, because in this case, the citizen believes that the agency is already taking care of the relevant issues, making it unnecessary to interact with public management.

For Bonsón et al. (2019) Twitter is used by local governments as a communication tool for citizen engagement, understanding engagement as the population's participation in social matters of public interest.

In their work, they analyze Twitter posts from 29 Spanish municipalities and observed that 28 of them had an official profile on this network, but there is a heterogeneous behavior in use, with profiles with large numbers of posts and followers, others with very low activity. Engagement is measured by the number of replies, retweets and likes obtained by the post. The most frequent use of Twitter by city halls is for the cultural promotion of the city, with other functions, for example, issues related to mobility, health, education, among others, being less frequent. Regarding the communication format, the use of links to external websites is more frequent than the use of text, image and video, possibly due to the limitations of the platform itself regarding the allowed size of messages. The authors did not observe a significant relationship between the size of the municipality and engagement on Twitter, but they obtained significance in relation to the type of media, with image and video being the forms that receive the most retweets and likes; text messages getting more responses, ie, debate. In this way, it is suggested that the message format has an impact on the generation of engagement, being more visual forms suitable for the dissemination of information by retweet and text suitable for discussions (Bonsón et al., 2019).

Stratu-Strelet et al., (2021) investigated the institutionalization of e-participation policy as a relevant part of a strong e-government in European countries. They found that government leadership on driving ICT policies is more relevant than technology infrastructure to promote participation. Democracy, legitimacy and government efficiency were also found significant in order to promote e-participation.

Although the high interactivity on social media platforms can bring benefits to communication between public management and citizens, the mere creation of online channels to unilaterally disseminate information can bring problems to the reputation of governments. Mergel (2013) points out that, in order to have effective results from the use of social media, it is necessary to invest in human and social capital, which will make government processes more efficient. However, just disclosing information and messages without reacting to the comments and feelings expressed by the user can generate immense dissatisfaction.

On the other hand, the article by Wukich and Mergel (2016) suggests that several public management entities have increasingly adopted strategies to monitor posts on social media platforms. Monitoring allows the diagnosis of citizens' feelings, but also aims to identify relevant issues that can be reproduced for other recipients, that is, it has the function of identifying relevant information that can be disseminated. However, public agents are very cautious in this regard, with a preference for the use of official institutional information to the detriment of inputs provided by the citizen, which suggests a limitation of trust between the parties (Wukich & Mergel, 2016).

Different applications of social media can be identified in public management: dissemination of information, feedback on the quality of public services, citizen participation and internal collaboration of public servants. Studies by Oliveira and Welch (2013) propose that the evolution of the use of social media by public management depends on the level of organizational innovation, the influence of stakeholders, organizational culture and the skills and abilities of the teams involved (Oliveira & Welch, 2013).

The work by Guo et al. (2016) suggests three fundamental variables for the success of e-government interactions via social media: feeling of belonging (level of personal involvement in which the individual feels an integral part of that environment); motivation (the needs met by interactions via social networks motivate the use of these platforms); and flow state (feelings experienced when being fully involved in a certain activity). These variables, in the case of communication via social networking platforms, lead to the individual's loyalty to this communication channel.

Linders (2012), in turn, proposes three levels of citizen participation in the digital media age:

- Citizen sourcing: type of Citizen-Government interaction, in which the citizen shares opinions with the government and tries to improve the representation of his interest group. In this way, the citizen offers feedback on public services and other relevant information, such as the occurrence of crimes, signs of corruption, among others (Linders, 2012).
- Government as a platform: type of Government-Citizen interaction (Government to Citizen), in which the government provides relevant information for citizens to make their decisions, such as the development of crime maps (Linders, 2012), designed either by statistics developed with internal data, or together with data provided by citizens themselves.
- Do-it-yourself government: Citizen-to-Citizen interaction, in which communities self-organize and discuss issues related to public management, with little or no government intervention (Linders, 2012).

An important paradox concerns the relationship between digital inclusion and citizen participation through social media platforms: although social media have a high potential to allow participatory discussion of public issues, there will, in fact, only be those individuals who have access to the Internet and domain of use of social media. Thus, the following question is raised: Is there, in fact, the participation of citizens as a whole in discussions regarding Brazilian public management?

Among the spheres of power, municipal governments have the greatest proximity to the citizen. Thus, citizen participation tends to be more intense at the local level. In addition, large cities tend to be driving the adoption of innovations in the area of e-government (Mossberger et al., 2013). In this way, the municipal governments of Brazilian capitals were selected as units of analysis for the research.

Using text mining, Abu-Shanab and Yousra Harbanalisou (2019) analyzed more than 2000 articles related to e-government and identified that, within the technology theme, the keyword "social network" was the most mentioned. There are several episodes that illustrate the power of mobilization and engagement of social networks, see the Arab Spring that took place in 2011 and even the protests in Brazil in 2013. This process of civil engagement through information technologies is called e-participation, which is defined by the UN (2014, p. 61) as: "the process of involving citizens through ICT [Information and

Communication Technologies] in policy and decision-making, in order to make public administration participatory, inclusive, collaborative and deliberative (...)"

Metallo et. al (2020) identify two groups of determinants that can influence the level of public engagement in social networks:

1. Activity of the municipality (amount of information, type of posts and time of publication); and
2. Sociodemographic determinants (population size, average age of the population, level of education, citizens' income and Internet penetration).

Additionally, Choia and Song (2020) suggest that citizens with greater social capital – a commitment to the community and trust in government – are more likely to be present in electronic participation. As a response to digital transformations, government institutions are increasingly using different forms of electronic participation channels, from interactive online platforms, social media channels and mobile applications. (Fenney & Welch, 2014)

Fenney and Welch (2014) also point out that "Social media technology is expected to improve participation, learning and knowledge production in government settings, aligning traditional structural and authority boundaries while challenging them". Despite of the progress of electronic government in the universe of social networks, several authors point to limiting factors that inhibit the realization of what Antiroikko (2004, p.40) calls Democratic Electronic Governance. In order for the potential to promote citizen participation in public issues to be truly explored, it is essential that internet access has a wide reach and that participants have mastery over the use of social media.

Considering the literature reviewed, citizen participation through ICT tools is still a developing field. The authors reviewed suggest the usage of technology, including social media adds transparency, access, agility and engagement to participation, however, this process is not organic and must be well planned and led by government according to well established policies. The empirical studies found in the literature expose there is still room for improvement in the relation citizen-government through social media, then the present chapter propose a model to assess the interactions between citizen and government on social media web sites. The adopted analysis model is shown in Figure 2, and takes those levels into consideration. On the left side of the model there are the centralized and on-way communication, whereas on the right side there is the decentralized and participative interactions.

METHODOLOGICAL PROCEDURES

To meet the research's objective, analyze how social media engages citizens in issues related to municipal management, it was developed an empirical descriptive research based on social media data. Data collection occurred in January 2022 was done through the python language and the use of the tweepy library to access and consume the data provided by the Twitter API, gathering posts from 2019 to 2021. Initially, a pilot study was carried out with data from four municipalities and based on the insights generated, the study was replicated in other municipalities. Four API calls were made, specifying the search for the desired keyword (in this case, the name of the official accounts of the analyzed municipalities). In addition, a rule was set up so that only the records that were within a radius of 100 kilometers were gathered, having as reference the latitude and longitude of the city. After loading and storing, the data of interest was processed through the NetworkX library and the Node XL software to generate network

Figure 2. Analysis model
Source: Elaborated by the authors, based on Carvajal Bermúdez and König, 2021, OCDE 2001, and Stratu-Strelet et al., 2021.

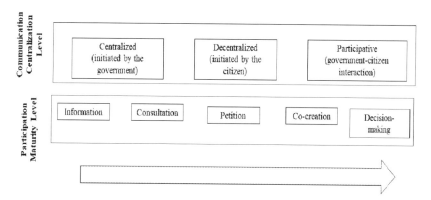

graphs. This process included the following steps: data extraction using Twitter API, storing data in csv file, creating networks' nodes and edges, storing the graphs and networks. It is noteworthy that the data collected consists of posts on Twitter that refer to the official profiles of the municipalities, which includes the posts made by the municipalities and posts by users mentioning the municipalities, which allowed evaluate the interactions between the parties.

Twitter was chosen as the social network platform to obtain the data, since it presents itself as one of the most used social media both in Brazil and in the world, and in addition, it makes its API (Application Programming Interface) available to developers and researchers so they can benefit from information relevant to their contexts (Twitter, 2020). The literature review itself pointed to a majority use of Twitter data in published articles, as it is the only social network that provides access to data in recent days.

There are different ways to extract data from the social media, performing searches by keywords, mentions of hashtags or profile (Recuero et al., 2020). The network analysis proposed in this study defined as nodes, those users who tweeted or retweeted publications mentioning the official pages of city halls and their respective mayors. The comments made in the tweets published by the city halls were also captured, as well as the responses given to these comments.

Additionally, to the network analysis text mining was developed in order to find out the topics discussed by municipalities and users. Text mining was carried out consisting of evaluating the frequency of keywords, so that it is possible to analyze their content. The process of structuring, mining and visualizing the obtained data consisted of the following steps:

- 1st Step: standardization of words to lowercase

It consisted of putting all the words in lowercase. In this step the same word written in different fonts or sizes are standardized.

- 2nd Stage: elimination of punctuation

It consisted of eliminating text punctuation. The objective of this step is to avoid noise caused by irrelevant terms.

- 3rd Step: elimination of additional irrelevant terms

At this stage, irrelevant words, called stopwords, were excluded, such as: "so", "thus", "because", "how", "beyond", among others. The objective of this step is to reduce the analysis to words that have meaning to the problem studied.

- 4th Step: removal of spaces left by previous steps

In this step, the spaces left by the elimination of punctuation, numbers and stopwords were removed.

- 5th Step: transformation of terms into a frequency matrix

In this step, the words resulting from the previous steps were first transformed into a matrix, that is, into a table with their frequencies.

- 6th Step: sorting and visualization of the data obtained

Sorting the matrix in descending form and selecting the top N words for analyzing the results. In addition, through this matrix, frequency graphs (clouds of words) were made, also considering the terms with greater frequency.

Both network analysis and text mining were carried out for each Brazilian municipality considered in the study. In addition, correlations among the interactions metrics (replies, retweets, quotes and likes) were calculated.

RESULTS

Initially, the profiles of Brazilian capitals, as well as their mayors in, were collected for the networks Twitter, Facebook and Instagram, the main social networks in use in Brazil. The number of followers of city halls in each of these networks was also surveyed (Appendix A). Except for the city of Brasília, which does not have a profile on Facebook, all capitals are present in the three networks, as well as their mayors (only the current mayors of Maceió and Teresina do not have a profile on Twitter). Table shows the number of followers of the municipalities in each of the networks. It is observed that for São Paulo, Rio de Janeiro and Curitiba, the largest concentration of followers occurs on the Facebook network, with a large discrepancy in relation to Twitter. For the other cities, the distribution of the number of followers is more balanced among the three platforms. In this way, it is important to emphasize that the use of data only from Twitter is a limitation of the research, because for some municipalities (in the case of SP, RJ and PR) this network is the less active. This also raises opportunities for a research agenda, with the comparative assessment of interactions in these networks.

Twitter data was extracted from 2019 to 2021, including the official profiles of city halls and mayors, totaling 7,647,561 interactions in the three years. It is noteworthy that interactions include posts, shares, responses and likes in posted messages. Descriptive analysis of the post shows that the highest frequency of mentions regards the mayors of Rio de Janeiro (current), Belo Horizonte (current), SP (2017-2020 administration). The most mentioned city in the period was Rio de Janeiro. This exploratory evaluation

Table 1. Municipalities followers on social media

State	City	Followers - Facebook	Followers - Twitter	Followers - Instagram
AC	Rio Branco	26514	943	27424
AL	Maceió	225955	26075	195726
AP	Macapá	114029	32006	89392
AM	Manaus	301196	303758	19272
BA	Salvador	296476	233837	649523
CE	Fortaleza	18026	241866	196633
DF	Brasília	0	326138	157756
ES	Vitória	103062	1915	119481
GO	Goiânia	112172	3484	237104
MA	São Luis	213473	357104	69645
MT	Cuiabá	90427	634	58489
MS	Campo Grande	42382	142	37532
MG	Belo Horizonte	269377	232235	317928
PA	Belém	167023	83648	187442
PB	João Pessoa	110956	3486	147702
PR	Curitiba	1021487	360952	508958
PE	Recife	242549	310277	243944
PI	Teresina	120353	10746	196677
RJ	Rio de Janeiro	682285	328242	539126
RN	Natal	174145	59055	172677
RS	Porto Alegre	161454	283244	98232
RO	Porto Velho	62752	1296	57986
RR	Boa Vista	58987	717	98353
SC	Florianópolis	242897	51855	148958
SE	Aracaju	42203	22047	147663
TO	Palmas	57135	43846	82938
SP	São Paulo	655578	9225	214426

suggests a peculiar characteristic in relation to the behavior on social networks in Brazil, the interaction with the figure of the mayor itself has greater engagement than the institutional interaction, with the city hall profile, suggesting a necessity for personification of the communication. This phenomenon opens opportunities for research, such as qualitative analysis, through interviews or surveys with participants in interactions on social networks, to understand their behavior pattern and the reasons why there are more interactions with mayors than with city halls.

Due to the computational constraints, it was decided to process the data for the second semester of 2021, to evaluate the city-citizen interactions. Table 2 shows the number of posts that mention the city halls (including those made by the city hall profile and those made by other users), the number of replies to the posted messages (replies), the number of message shares, the number of quotes (consist

Table 2. Posts on Twitter

City	Posts (a)	Reply (b)	Share (c)	Quote (d)	Reactions (e)	b/a	c/a	d/a	c/a
Rio Branco	4792	775	223511	527	12238	0,16	47	0,11	2,55
Maceió	4859	1205	393938	128	19299	0,25	81	0,03	3,97
Macapá	16797	4895	179274	1888	35941	0,29	11	0,11	2,14
Manaus	54949	9665	67233764	2262	100915	0,18	1224	0,04	1,84
Salvador	104646	28459	15849073	11864	327891	0,27	151	0,11	3,13
Fortaleza	29774	11149	433911	2036	117905	0,37	15	0,07	3,96
Vitória	1649	851	1105	68	2597	0,52	1	0,04	1,57
Goiânia	14050	4692	72992	1053	29129	0,33	5	0,07	2,07
São Luis	42878	11983	3806172	3508	135016	0,28	89	0,08	3,15
Cuiabá	2851	1081	6390	89	2843	0,38	2	0,03	1,00
Campo Grande	1982	753	1229	40	1722	0,38	1	0,02	0,87
Belo Horizonte	81063	26915	62469037	3602	228105	0,33	771	0,04	2,81
Belém	105596	37750	5260616	7865	387452	0,36	50	0,07	3,67
João Pessoa	12585	2797	74114	508	25129	0,22	6	0,04	2,00
Curitiba	40981	12299	715575	3612	117503	0,30	17	0,09	2,87
Recife	62784	19559	31728918	8650	219428	0,31	505	0,14	3,49
Teresina	4311	1517	30779	307	11477	0,35	7	0,07	2,66
Rio de Janeiro	916573	295478	35546018	94777	3828148	0,32	39	0,10	4,18
Natal	11724	2634	85541	633	28393	0,22	7	0,05	2,42
Porto Velho	8727	1629	820501	223	7688	0,19	94	0,03	0,88
Boa Vista	3276	1219	24771	255	7544	0,37	8	0,08	2,30
Florianópolis	94304	35491	161419131	17177	465039	0,38	1712	0,18	4,93
Aracaju	19137	7840	200732	2924	94581	0,41	10	0,15	4,94
Palmas	22220	8289	189648	939	63146	0,37	9	0,04	2,84
São Paulo	40697	17183	1664489	2769	97497	0,42	41	0,07	2,40

of messages shares with some comment added) and the number of reactions (also known as likes). In addition, the ratios between the number of replies, shares, quotes, and reactions by the total number of posts were calculated.

The municipalities with the highest number of posts are Rio de Janeiro, Salvador and Belém. The highest frequencies for sharing posts are for the municipalities of Florianópolis, Manaus and Belo Horizonte. The highest averages of replies per post are for Vitória, São Paulo, and Aracaju. These descriptive statistics demonstrate a heterogeneity in the behavior of the considered municipalities, regarding Twitter.

A correlation analysis (Table 3) was carried out between the interaction metrics. A positive, moderate to strong correlation is observed between the number of quotes and replies, that is, the more responses to a post, the greater the number of commented shares (quotes), indicating that answered posts tend to be shared, and lead to an active posture of users. There is also a significant and positive relationship between replies and reactions and between quotes and reactions, indicating that posts that receive more

Table 3. Correlation analysis

	Posts (a)	Reply (b)	Share (c)	Quote (d)	Reactions (e)	b/a	c/a	d/a	c/a
Posts (a)	1								
Reply (b)	0,999	1							
Share (c)	0,223	0,228	1						
Quote (d)	0,993	0,994	0,273	1					
Reactions (e)	0,998	0,999	0,228	0,996	1				
b/a	0,004	0,032	0,007	0,023	0,022	1			
c/a	0,030	0,028	0,954	0,067	0,026	-0,101	1		
d/a	0,203	0,214	0,449	0,280	0,220	0,088	0,343	1	
c/a	0,348	0,363	0,393	0,386	0,362	0,135	0,275	0,687	1

responses also receive more reactions from users, causing more active behavior. There is no significant correlation between replies and shares, between quotes and shares, and between shares and reactions. These findings are consistent with Bonsón et al., (2019) results, indicating retweeting and liking the posts are more numerous since they are simpler procedures, while replying requires more time, attention and writing abilities.

After the descriptive analysis, text mining of the posts, was carried out and the following patterns are observed:

The content posted on Twitter is dynamic and strongly considers the topics in evidence real time. As the data analyzed date from the second half of 2021, in most municipalities, the dominant terms are vaccination, vaccine, COVID-19, dose (of the vaccine), get (the vaccine), health, that is, terms related to the COVID-19 pandemic. Figures 3 and 4 exemplify this content in posts from the city hall of Rio Branco and Macapá. The same behavior is observed in the municipalities of Salvador, Fortaleza, Goiânia, São Luís, Belém, João Pessoa, Curitiba, Teresina, Rio de Janeiro, Natal, Porto Velho, Aracaju, Palmas, and Brasília. Some municipalities had a predominance of messages with content different from those related to health and the pandemic, for example, the municipality of Maceió had a predominance of posts related to the dismissal of the Municipal Secretary of Tourism of the city, who was involved in a polemic situation, as seen in Figure 5. The same occurs with the municipalities of Manaus, Recife and Florianópolis, in which there was a predominance of discussions about facts that had occurred locally. This pattern of intense discussion about current news can help to disseminate information, which consists of the initial level of citizen participation (sending communications), however, it can be an obstacle for the more sophisticated levels of participation that require long-term contribution and co-creation. Considering our theoretical model (Figure 2), the interactions observed may be started by both citizens and the municipalities, on the first case the messages are mostly of petition type, for the second, messages are mostly informational. An example of an informational post from Maceió on June 18th, 2021 was "Maceió communicates the dismissal of our A Municipal Secretary of Tourism (free translation)". This post had 616 retweets, 601 quotes and 9,362 likes, although there was intense activity regarding the post, there was no responses from the municipality profile to the reactions of the public, which means, the city used Twitter to disclosure the information, but not to discuss about issues related to the dismissal. Reactions from citizen vary from support to the dismissal to criticism and doubts about the Secretary

Figure 3. Wordcloud Rio Branco

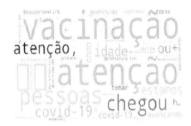

being hired to another position in the future. Figure 6 illustrates this kind of communication. Considering interactions started by citizens, Figure 7 illustrates the kind of requests made by them to the City Hall. In this post the citizen asks for a public vet hospital in São Luís, limping the City Hall profile. In short, the patterns identified on our data suggests two predominant styles of interaction:

1. Centralized: started by the public agent, aiming to disseminate relevant information, no debate pattern identified (responding to citizens posts). For those posts it was observed behaviors of support, criticism and irony from citizens towards city's posts, in cases the message concerned to polemic issues. In cases where the content of the message regarded general information, the most frequent behavior is sharing the post, with no responses.
2. Decentralized: started by the citizen, aiming to demand a service or attention to an issue by the municipality. For this kind of post, debate pattern was not either observed, which means the municipality does not reply publicly to the citizen. The most common activity for these posts is sharing or reacting (like or dislike), by other users.

Figure 4. Wordcloud Macapá

The most frequent kind of interaction on this platform are shares and replies and frequently using referring (using @) to other user profiles. It is quite common to find interactions in which no content is written, but a user referred (@). The Vitória City Hall cloud exemplifies this pattern; interactions mentioning this city hall only limp users and do not actually post comments, in general the users referred are mayors and governors of other cities and states. The same is observed in Campo Grande, Rio de Janeiro, Boa Vista and Brasília.

Figure 5. Wordcloud Maceió

Figure 6. Interaction for Communication Purpose, started by the city

Although social networks are platforms for everyday use by the population, regarding their use as a platform to facilitate e-participation, some obstacles emerge. The content posted on social networks can consist of text, image, video, audio and emojis and to collect and analyze all this content it is necessary to develop processing algorithms. In the case of textual content, text mining is relatively simple and efficient, however image, video and audio processing is more complex. Emojis for example, which must have its meaning interpreted according to the context of the interaction, which makes data processing more complex. Posts that include images are not captured by the Twitter API, which leaves this content out of data extraction and analysis.

Regarding the type of communication, the data show that interaction is initiated both by the city hall and by the citizen, but in the first case, with the objective of information, that is, to disseminate relevant official information. In the case of citizens, interactions aim to express opinions and report facts that require action from the municipality (petition level), which is consistent with the analysis framework proposed for the research. An example of a user's post referring to the city hall, which demonstrates a criticism and implicitly asks for an action: "everyone is satisfied in the wonderworld of the city of Rio Branco", responding to a post from the city hall and mocking cases of flooding due to the rain in the city. The processing of content produced on social networks presents, therefore, two challenges: 1) the volume is high, there can be thousands or even millions of interactions in a short period of time, which makes

Figure 7. Interaction for Petition Purpose

manual monitoring and processing unfeasible; 2) the processing and automatic extraction of interactions in social networks presents difficulties, such as image, video and audio processing, the identification of fake news and the recognition of language expressions, such as irony, which can compromise the understanding of the message content.

In order to comprehend the network structure of the interactions between city-citizen on Twitter, social network analysis was used. Network graphs were developed to verify the dynamics of interaction between cities/mayors and users. For this purpose, the Fruchterman-Reingold graph was used, usually used in studies involving data from social networks (Hansen, Shneiderman, and Smith, 2020). The software used, Node XL, plots the images related to the user's profile on Twitter as nodes in the network and the lines connecting the nodes are called edges, indicating that there was interaction between users. The wider the line, the greater the amount of interactions between nodes. It was observed that, in general, the number of nodes that interact with the mayors is much higher than the nodes that interact exclusively with the city hall or with both. The graph presented in Figure 6 presents the network of interactions for the city of Porto Alegre, considering users who had 30 or more interactions in the period, a minimum of interactions was used so that it was possible to have a legible visual representation. This network totalized 26,775 interactions and it can be seen that it focuses on a small number of users, which makes the representation of population groups questionable, as suggested in the literature.

Figure 8. Network for the city of Porto Alegre

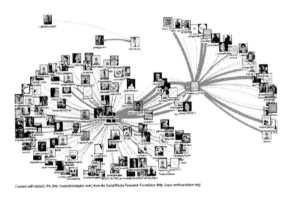

The city of São Paulo presents a distinct behavior of interactions, the only exception among the analyzed cities, in which the interactions with the city hall surpass the interactions with the mayor. An interesting fact about this network is the reduced number of interactions, totaling only 15595 interactions in the period, which can be explained by the more intense presence of this city hall in social networks other than Twitter, as mentioned earlier in the chapter.

It was observed that the city halls of Brazilian capitals, although being present in different social network platforms, still make incipient use of these technologies regarding citizen participation. In general, city halls restrict themselves to using social networks to send communications to citizens (for example, lack of energy in the neighborhood and storm alert). Citizens, in turn, use social media to express their opinions, various complaints and as a call to action towards the municipality on specific issues (for example, repairs of holes in the roads, pruning of trees, among others). The results indicate the existence of an interface between the public power and the citizen, however, this process has not yet evolved towards the effective engagement of the population, so the more sophisticated forms of participation were not observed, co-creation for example.

Social networks are used merely for institutional communication, it means only for the dissemination of information. Observing Brazilian legal system, participation can officially occur through legal mechanisms such as the participative budget. Contributions to the budget may happen only through the official established processes which do not consider what are demanded and discussed on social media. Another remark regarding social media content is the differentiation between what ideas, opinions and actual proposals, which makes it even more complicate to include these inputs into public decision making, finally, social media users may be anonymous or use a fake identity, which compromises transparency and governance. This ambiguity is reduced by following the legal process for participation which usually includes a user registration and a structured.

In short, legally, citizen participation in Brazil aims to obtain proposals from the population to compose the budget law, so that participation platforms are able to establish a flow of receiving and analyzing proposals, with transparency. Social media data is created spontaneously by the user, and it is not possible to classify this material automatically or simply as a proposal, opinion or idea. The population's proposals go through a process of prioritization, voting and analysis, which requires an identification of authorship, which does not occur in social networks, where an anonymous profile can be created, which has implications for the legitimacy of the process.

The results suggest that social media usage by Brazilian cities is not mature yet, although all the cities are present at these platforms, their strategies lies in the simpler levels of our research model, which is convergent to the arguments of Arshad and Khurram, (2020). It was observed both government and citizen initiate communication, however there is a difficult in advancing to a more sophisticate process of interaction, where government provokes, and citizen fully engage in a specific topic of discussion. On the side of the government news dissemination was the dominant type of message whereas on the side of the citizen petition was the dominant type. For the citizen, using Twitter to express their feelings and reporting their problems is a rapid, efficient and effortless way, which explains the dominance of this type of messages (Arshad & Khurram, 2020). The most concerning results regard the government behavior, restraining the communication to just publishing relevant news. As Chatfield and Reddick, (2018) propose, social media may be powerful if its implementation follows a transparent policy which will ultimately foster democracy development and social inclusion (Carvajal Bermúdez & König, 2021), but this will not happen spontaneously or organically.

It seems Brazilian cities do not have a strategic social media policy, aiming to engage people in participation for co-creation and decision making, nor have a policy to react to critics and population requests, leaving the users with no answers. It reminds us of the traditional one-way communication media used for many years: government disclosures information on newspapers, TV and print media and citizens report their problems, needs and suggestions by mail, e-mail and phone call; in all of these options, no interaction was possible and the same pattern is observed on social media nowadays.

Thus, it is necessary that government understand social media as an e-government tool that might help or harm its relationship with the population, that is why developing a social media policy with well-defined objectives is essential. Transparency and governance are essential as well, so citizens become aware of how their contributions are being used, creating a trust relationship (Khan et al., 2021).

CONCLUSION

Social networks are platforms have changed the way of relating and communicating. By generating material in large volume and in real time, they present great potential for identifying opportunities and problems related to private and public institutions. In the context of participation, the focus of this chapter, it was observed that the municipalities of Brazilian capitals are present on social networks, sometimes with more intense action on a specific platform (for example, the municipality of São Paulo has greater activity on the Facebook network), however, the interactions are focused on disseminating relevant information and guidance to the population. When analyzing Twitter data, it is observed that city halls disclose information starting the communication process, in the same way the population starts the communication process, but with a different objective, in general the user makes "petitions", that is, they report opinions and complaints that aim to sensitize public management. Thus, in the framework of analysis (Figure 7), centralized and decentralized communication are observed, but participatory communication does not occur. Likewise, there are no levels of participation in consultation, co-creation and decision-making on these platforms. The official tools of participation are built within the mechanisms of participation related to open government policies and construction of municipal budgets, and social media is not included in this process.

Figure 9. Research framework analysis

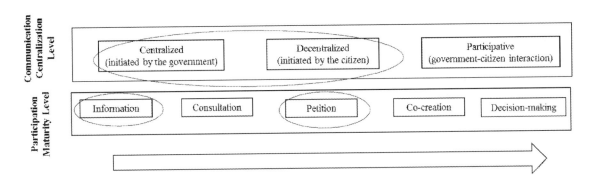

Although social networks are present in people's daily lives, the way these platforms are used may not be in line with what is necessary to make the process of citizen participation viable as a public policy. The anonymity, restriction of characters that can be published and the dynamism of these platforms make the processes of co-creating and giving voice in decisions difficult. Additionally, the data show that the interactions in the networks are concentrated in small numbers of users, which leads to the reflection if the networks would really raise the level of citizen participation. On the other hand, the networks' ability to disseminate information is powerful and efficient, so that its use to disseminate information has great potential. Additionally, the use of networks to monitor the approval of policies in progress can be interesting.

As limitations of the research, there can be highlighted the study of only Twitter data and capital cities. As a suggestion for future studies, deepening the analysis of communication between mayors and citizens in the networks may be interesting to identify whether these interactions have different characteristics from those carried out with city halls. In addition, machine learning and sentiment analysis models can be applied to Twitter data in order to identify the citizen's perception of city' policies. Finally, the study of fake news can be interesting to identify how this type of content impacts the relationship between citizens and cities on social networks.

ACKNOWLEDGMENT

This research was supported by the Conselho Nacional de Desenvolvimento Científico e Tecnológico (CNPQ) [grant number - 431484/2018-0] and the University of Sao Paulo [CFP USP-Humboldt 2019].

REFERENCES

Abu-Shanab, E., & Harb, Y. (2019). E-government research insights: Text mining analysis. *Electronic Commerce Research and Applications*, *38*(C), 2019. doi:10.1016/j.elerap.2019.100892 12p

Allen, B., Tamindael, L. E., Bickerton, S. H., & Cho, W. (2020). Does citizen coproduction lead to better urban services in smart cities projects? An empirical study on e-participation in a mobile big data platform. *Government Information Quarterly*, *37*(1), 101412. doi:10.1016/j.giq.2019.101412

Antiroikko, A. V., Malkia, M., & Savolainen, R. E. (2004). *Transformation in Governance: New Directions in Government and Politics*. Idea Group Publishing.

Arnstein, S. R. (1969). A Ladder Of Citizen Participation. *Journal of the American Institute of Planners*, *35*(4), 216–224. doi:10.1080/01944366908977225

Arshad, S., & Khurram, S. (2020). Can government's presence on social media stimulate citizens' online political participation? Investigating the influence of transparency, trust, and responsiveness. *Government Information Quarterly*, *37*(3), 101486. doi:10.1016/j.giq.2020.101486

Bonsón, E., Perea, D., & Bednárová, M. (2019). Twitter as a tool for citizen engagement: An empirical study of the Andalusian municipalities. *Government Information Quarterly*, *36*(3), 480–489. doi:10.1016/j.giq.2019.03.001

Bouzguenda, I., Alalouch, C., and Fava, N. (2019) 'Towards smart sustainable cities: A review of the role digital citizen participation could play in advancing social sustainability', *Sustainable Cities and Society*, *50*(November 2018), p. 101627. . doi:10.1016/j.scs.2019.101627

Burke, E. M. (1968). Citizen Participation Strategies. *Journal of the American Institute of Planners*, *34*(5), 287–294. doi:10.1080/01944366808977547

Callahan, K. (2007). Citizen Participation: Models and Methods. *International Journal of Public Administration*, *30*(11), 1179–1196. doi:10.1080/01900690701225366

Carvajal Bermúdez, J. C. and König, R. (2021) The role of technologies and citizen organizations in decentralized forms of participation. A case study about residential streets in Vienna. *Technology in Society*, *66*(October 2020). . doi:10.1016/j.techsoc.2021.101604

Chatfield, A. T., & Reddick, C. G. (2018). All hands on deck to tweet #sandy: Networked governance of citizen coproduction in turbulent times. *Government Information Quarterly*, *35*(2), 259–272. doi:10.1016/j.giq.2017.09.004

Choia, J., & Song, C. (2020). Factors explaining why some citizens engage in E-participation, while others do not. *Government Information Quarterly*, *37*(4), 101524. doi:10.1016/j.giq.2020.101524

Criado, J. I., Sandoval-Almazan, R., & Gil-Garcia, J. R. (2013). Government innovation through social media. *Government Information Quarterly*, *30*(4), 319–326. doi:10.1016/j.giq.2013.10.003

e-Estonia. (2022). e-Democracy and open data. Available in: https://e-estonia.com/solutions/e-governance/e-democracy/

Grönlund, Å., & Horan, T. A. (2005). Introducing e-Gov: History, Definitions, and Issues. *Communications of the Association for Information Systems*, *15*, 39.

Guo, J., Liu, Z., & Liu, Y. (2016). Key success factors for the launch of government social media platform: Identifying the formation mechanism of continuance intention. *Computers in Human Behavior*, *55*, 750–763.

Hansen, D., Shneiderman, B., & Smith, M. A. (2010). *Analyzing social media networks with NodeXL: Insights from a connected world*. Morgan Kaufmann.

Kes-Erkul, A., & Erkul, R. (2009). Web 2.0 in the Process of e-participation: The Case of Organizing for America and the Obama Administration. *National Center for Digital Government Working Paper Series*, *9*(1), 1-19.

Khan, S., Umer, R., Umer, S., & Naqvi, S. (2021). Antecedents of trust in using social media for E-government services: An empirical study in Pakistan. *Technology in Society*, *64*, 101400. https://doi.org/10.1016/j.techsoc.2020.101400

Linders, D. (2012). From e-government to we-government: Defining a typology for citizen coproduction in the age of social media. *Government Information Quarterly*, *29*(4), 446–454.

Magro, M. J. (2012). A Review of Social Media Use in E-Government. *Administrative Sciences*, *2*(2), 148–161.

Mergel, I. (2013). A framework for interpreting social media interactions in the public sector. *Government Information Quarterly*, *30*(4), 327–334.

Metallo, C., Gesuele, B., Guillamón, M., & Ríos, A. (2020). Determinants of public engagement on municipal Facebook pages. *The Information Society*, *36*(3), 147–159.

Mossberger, K., Wu, Y., & Crawford, J. (2013). Connecting citizens and local governments? Social media and interactivity in major U.S. cities. *Government Information Quarterly*, *30*(4), 351–358.

Oliveira, G. H. M., & Welch, E. W. (2013). Social media use in local government: Linkage of technology, task, and organizational context. *Government Information Quarterly*, *30*(4), 397–405.

Organização das Nações Unidas. (2020). *UN government survey 2020*. UN. https://publicadministration.un.org/egovkb/en-us/Reports/UN-E-Government-Survey-2020

Recuero, R., Bastos, M., & Zago, G. (2015). *Análise de redes para mídia social* [Network analysis for social media.]. Editora Sulina.

Stratu-Strelet, D., Gil-Gómez, H., Oltra-Badenes, R., & Oltra-Gutierrez, J. V. (2021). Critical factors in the institutionalization of e-participation in e-government in Europe: Technology or leadership? *Technological Forecasting and Social Change*, *164*, 120489.

Tejedo-Romero, F., Araujo, J. F. F. E., Tejada, Á., & Ramírez, Y. (2022). E-government mechanisms to enhance the participation of citizens and society: Exploratory analysis through the dimension of municipalities. *Technology in Society*, *70*, 101978. https://doi.org/10.1016/j.techsoc.2022.101978

United Nations. (2020). *E-Government*. UN. https://publicadministration.un.org/egovkb/en-us/about/unegovdd-framework

Wukich, C., & Mergel, I. (2016). Reusing social media information in government. *Government Information Quarterly*, 1–8.

ADDITIONAL READING

Amores, J. J., Blanco-Herrero, D., Sánchez-Holgado, P., & Frías-Vázquez, M. (2021). Detectando el odio ideológico en Twitter. Desarrollo y evaluación de un detector de discurso de odio por ideología política en tuits en español. [Detecting ideological hatred on Twitter. Development and evaluation of a hate speech detector for political ideology in tweets in Spanish.]. *Cuadernos.Info*, (49), 98–124. doi:10.7764/cdi.49.27817

Cantador, I., Cortés-Cediel, M. E., & Fernández, M. (2020). Exploiting Open Data to analyze discussion and controversy in online citizen participation. *Information Processing & Management*, *57*(5), 102301. doi:10.1016/j.ipm.2020.102301

Certomà, C., Corsini, F., and Frey, M. (2020) Hyperconnected, receptive and do-it-yourself city. An investigation into the European "imaginary" of crowdsourcing for urban governance. *Technology in Society*, 61(April 2019). . doi:10.1016/j.techsoc.2020.101229

Feeney, M., & Welch, E. (2012) Electronic participation technologies and perceived outcomes for local government managers. Public Management Review. Volume 14, 2012 - Issue 630 p. doi:10.1080/14719037.2011.642628

KEY TERMS AND DEFINITIONS

Citizen participation: Citizen involvement in public decision-making
e-Government: Delivering public services to society making use of ICTs.

e-Participation: Stimulating engagement in public decision-making employing ICTs

ICT: Information and communication technologies consist of sources (hardware and software) employed to enable access, transmission and processing of information.

Network analysis: A set of techniques employed to investigate social structures behind interactions among entities (people or organizations) through graphs and network metrics.

Social media: Websites and other tools that allows users to generate and share content and engage in social networks.

Web 2.0: Second generation of the Web, which focuses on interaction and creation and sharing of information by the users.

Chapter 16
The Social Challenge of Migrant Integration:
The Role of Mobile Apps

Leonilde Reis
https://orcid.org/0000-0002-4398-8384
Polytechnic Institute of Setubal, Portugal

Marcelo Pereira
Polytechnic Institute of Setubal, Portugal

Clara Silveira
https://orcid.org/0000-0003-2809-4208
Polytechnic Institute of Guarda, Portugal

ABSTRACT

Information and communication technologies can be a driving factor towards the resolution of social challenges and assertively contribute to the development of people and regions. The objective of this chapter is to present a mobile application designed to contribute to the inclusion of Migrants in the city of Viseu in Portugal. The methodology used to support the study was a design science research of creating the artifact to involve the various stakeholders in the process. The main results emphasize the relevance in organizational context of the use of information and communication technologies to bridge the gaps underlying the various social challenges and to develop innovative solutions so that the access to information is global and ubiquitous. In this sense, it is considered that the IntegraBrasil mobile application contributes to the well-being and integration of migrants, given the difficulties they face when they arrive in Portugal.

INTRODUCTION

Currently, the concerns underlying the integration of people are of particular interest given the current

DOI: 10.4018/978-1-6684-6123-5.ch016

situation. Portugal is particularly sensitive to the issue of integration and inclusion of people from diverse backgrounds. This development is of particular interest, especially in what concerns cities and territories of low population density. In this sense, Information Systems and Information and Communication Technologies (ICT) have specific characteristics implicit to enable the design of innovative strategies. Due to their innate characteristics, technologies can contribute to the real inclusion of sustainability measures, namely in the technical, human, social and economic dimensions. They are also instruments of added value with respect to the Sustainable Development Goals (SDGs).

According to the latest report from SEF - Foreigners and Borders Service - in 2021, it is estimated that about 698,887 thousand foreigners will be living in Portugal. Among them, 204,694 thousand are Brazilians living in the country, which, in terms of percentage, represent 29.3% of foreign residents. The migration of Brazilians to Portugal and the consequent choice of Viseu as a new address is based on a survey conducted by DECO (2021), where it mentions that the city of Viseu is considered a "garden", which received the title again. According to the study cited, Viseu is the Portuguese city with the highest quality of life. The district of Viseu is in the center of the country, three hundred kilometers from the Capital of the country, Lisbon, and about one hundred kilometers from the city of Porto. The quality of life is one of the first items that weighs on the choice of migrants and Viseu even geographically not being favorable due to the distances of the districts, has been chosen from the more than three thousand Brazilians who arrived in Portugal.

The main motivation to develop this work focuses on the opportunity to present the development of a mobile application that promotes the integration of Brazilian migrants in the city of Viseu. The purpose of the chapter is to present the development of an Android mobile application that will be available in the Google Play Store for download. It is intended to promote the information it makes available to create conditions to meet as many Brazilian migrants as possible.

The Associação Casa Brasil (ACB), is a non-profit association that promotes a set of services in order to contribute to the full integration of migrants who arrive in Viseu and need to provide steps ranging from offering employment to finding a school for their children. Connect Brazilian migrants to the services provided by Association Casa Brasil (ACB), through the app and thus enhance access to the information they need instantly. This access will be made in an agile manner, with autonomy and directed to support the processes of public agencies. Migrants constantly need information that is found in various services and many of them are confused or do not allow understanding due to the difference in the language Portuguese.

This access will be done in an agile manner, with autonomy and oriented to support the processes of public agencies. Migrants constantly need information that is found in various services and many of them are confused or do not allow for understanding due to the difference in legal frameworks.

This chapter is organized into six sections. In the first section, the introduction to the problem is presented, in which the need for a tool of this nature is identified as a contribution to the integration of Brazilian Migrants. Section two presents the methodology adopted - the Scientific Design Research. Section three presents the state of the art in the field of the theme under study. In section four, it presents the design of the application for the integration of Migrants. The discussion and future work is presented in section five. Finally, the conclusion is presented in section six.

The Social Challenge of Migrant Integration

METHODOLOGY

This section describes the methodology used during the development of the Mobile Application given its relevance in the Information Systems (IS) scientific field, and the fact that it enhances the successive iterations underlying the process.

The design of a mobile artifact that enhances social results in relation to employment, health, education, rights and duties, has as its basic characteristic directed to social innovation. The chapter presents the underlying problem of the development of an informational artifact, namely mobile application for ACB.

The Design Science Research (DSR) methodology, (Wieringa, 2009) was developed with the assumption of being a suitable methodology for the design of an artifact. The DSR is a type of research that aims to account for two types of problems: "practical problems", which enhance a change in the installed paradigm and that contribute to the objectives of decision-makers related to the problem, and contextual changes.

The authors, Baskerville, et al (2018), mention that DSR involves the creation of an artifact and/or theory and design to improve the current state of the practices used, but also the knowledge of existing development. Through the possibility of several iterations and given the phases of the methodology this enhances the creation of the artifact. According to Hevner, et al (2004) the criteria to be followed according to the proposed model is illustrated in Figure 1.

Figure 1. Design Science Research methodology
Source: adapted from Peffers, Tuunanen, Rothenberger, & Chatterjee (2007)

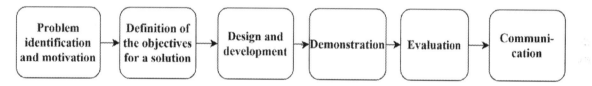

The DSR goes beyond a simple aid in the construction or development of the artifacts, it expresses itself through a rigorous process to design and solve problems, where its contribution is to create knowledge. For Simon (1996), it is a paradigm of pragmatic research that allows the creation of innovative artifacts to solve real-world problems.

To apply the DSR research methodology, particularly in the initial stages, the next section presents the state of the art of the topic under study that allows the development of the artifact.

BACKGROUND

The DSR, as a research methodology in the IS/ICT area allows the development of artifacts. In this sense, the importance of ICT and the Internet is highlighted. The popularization of the Internet has allowed society to become increasingly connected, technological and digital, allowing information to be disseminated through Information and Communication Technologies (ICT), thus reducing the distance between people and presenting benefits for a new society. The evolution of the New ICT, Pompeo (2016) has led to a new configuration of social relations, covered by the concept of networked society. using information technologies, a new communication model emerges, the digital informational age. It is considered that by using information technologies, these enhance the emergence of new communication models, framed in the digital information age.

Based on this principle, with the development and improvement of smartphones and the mobile network, access to information quickly and accurately allowed the dissemination of information to gain greater reach and be disseminated. In 2019 year, 96.8% of Portuguese have mobile phones and about 75% access the internet through it (ANACOM, 2019).

It is considered that a tool that accelerated this access to information by the smartphone is the application or application for software (app). For Song (2014), the growing use of mobile apps is an important trend in mobile information technology. This technological solution has brought a new routine to everyday life, namely online shopping, appointments, location to a specific address through maps, games, etc. In this sense, only by downloading contributes to the way people communicate, seek and exchange information among them including how to socialize and acquire knowledge (Ungerer, 2013).

According to Marketeer (2020) in 2014, only 16.4% of the Portuguese population downloaded free apps, a figure that rose to 20.3% the following year and to 29.6% in 2016. Since then, it has increased every year, reaching the current 48.4%.

Third Sector

The Social Organization (SO) is part of the so-called third sector, with the State as the first sector and the Private Initiative as the second sector. Despite the division, the Third Sector has in its course, acting where the State sometimes leaves social hiatus, practically acting in its place. The third sector is always another partner of the State in meeting social requests. In this sense, Community networks continue to fill the gaps that policies never fully fill (Schmidt, 2018). According to Coutinho (2006), among the activities that can be object of social organizations stands out: defense and incentive to culture; defense, preservation, and conservation of the environment; education; health; scientific research and technological development.

The SO, according to Da Silva Pereira et al. (2013), can be foundations, institutes, community associations, assistance and philanthropic entities established for public purposes, created by people with the ideal of offering improvement for society. Non-Governmental Organizations (NGOs) are also part of the SO, which have no ties to the State, where their proposal is to meet the requests of the population who are unassisted by the government. Its specificity is that they do not distribute their profit among their managers and need a more efficient management of their own resources, so that they can achieve the purpose of their assistance to the disadvantaged.

According to Cazzolato (2009), to be able to continue its activities, the SO needs to have a greater concern with the management process, as it relates the objective to which it proposes with the availability of resources it has. This need for management control is due to the number of resources they receive from donor entities, which request the NGOs support to direct the projects developed to their final goal. For Silva (2011), despite having a different purpose from profitable organizations, NGOs have an organizational structure similar to a company and, as such, need institutional management tools in order to achieve the desired objectives.

Social Responsibility

Social responsibility is part of their organizational commitment, however, there is no support or incentive for migrants to the level of rights and duties. To understand social responsibility, we must understand what the practice of work and employment was before it and how this inclusion brought improvements. In the late 1970s and early 1980s, several scandals arose that damaged the images of companies in a continuous strand of corporate problems, due to the inhuman issues that were generated by capitalism at the time.

For Antunes (2015), with the objective of framing the competitive standards of the market, at certain times there is the risk of becoming, consequently, an exclusionary and marginalizing society. To modify this scenario, several companies bet on a model of behavior and actions that promoted the satisfaction of their employees and customers, thus emerging social responsibility. With the change in the way man produced, leaving the artisanal method, to the industrial one, where respect and rights did not exist, with workdays of up to twelve hours, low wages and absence of laws, companies were pressured to change this method so that the best conditions were implemented. Therefore, in this first period, with the problems arising from industrialization, the understanding of the company's obligations in relation to social agents began to change (Tenório, 2015).

The objective of the company that adopts this philosophy based on strategies internally for its employees, as well as externally for its customers, partners, the environment and the community in general, is not to assume the role of the State and its obligations, but rather, to be an aspect of volunteerism. Tenório (2015) mentions that the approach of corporate action emerged in the early twentieth century, with philanthropism. Then, with the exhaustion of the industrial model and the development of post-industrial society, the concept evolved to incorporate the desires of social agents in the business plan of corporations.

After its natural evolution, social responsibility, which used to be only addressed as such, is now defined more broadly as follows: Corporate Social Responsibility, and Environmental Social Responsibility. To better understand the role of each one, the Corporate Social Responsibility is more directed to their processes, business and employees, the Corporate Social Responsibility is directed to the quality of life of its employees, but still act in the community and environment, and finally the Corporate Social Responsibility, and Environmental Social Responsibility that shows to be the most complete, because their concern is based on people, human values and the genuine value of the environment.

For the purposes of the return provided by the adoption of social responsibility, the company assumes an important role with society, and society in turn, perceives this relationship, with the successful initiatives and consequently the acceptance, and brand identification for something greater than its economic success. Ajmal (2018) advocates that, this responsibility with sustainability aspects contributes beneficially to organizations, while their values and commitment to society in general are highlighted.

Social Innovation

To understand social innovation (SI), we must understand its conception, as much as it seems to be a recent theme, SI is present in the performance with society throughout the ages, from the evolution in medical care to the less assisted, to the creation of community schools (Mulgan, 2007). Despite the apparent novelty of social innovation as a construct or set of discourses, history shows that humans have repeatedly experienced and achieved social changes that are disruptive and durable. Innovations permeate feelings of change to something that brings evolution or set of improvements to what we already know.

Nowadays, when the term social innovation is mentioned, technology is associated as the factor of this change. Technology and its means have been a unique factor in accelerating change. The driving force of innovation, previously carried out by the production line, changed with the advent of technology, as men began to use more processes carried out through thoughts, ideas, and research. For Kon (2018) the development of technology in the period, allowed new paradigms and new ideas to flourish, which was made possible by the emphasis on knowledge as a relevant input in socioeconomic processes and by the fact that they became available to both the private and government sectors.

Social innovation aims to respond to social problems through people's development and growth. It aims to enhance capabilities, devise strategies, value cooperation, and thus promote quality of life, well-being in their daily lives, and achieve the desired social transformation. Thus, a scenario of social, environmental, political, and economic ruptures emerges with a focus on structural changes that foster solutions to society's problems, along with the desire to promote new ways of doing things (Martinez, 2017).

In 2010, the European Union created a commission to examine actions to meet social needs, called social innovation. Its objective was to create new ideas/projects for economic and social growth, and to generate employment. In these studies, documents were generated that guided governments and private groups to finance the projects presented.

Since then, SI has been seen differently, due to the possibility of promoting improvements in the actors involved. As an example of this impact, after the 2008 crisis, Portugal suffered the intervention of the International Monetary Fund (IMF) to adjust its public accounts and saw the need to reformulate its capacity to generate jobs, education, and reduce poverty. Such need looked to the SI to generate new actions on the same themes. For unemployment, several training courses were provided to help people with less educational levels and thus allow them to re-enter new areas. It is considered that the educational process has been undergoing changes. However, it has been observed that students' disinterest and dropout has become a reality. In this sense, the existence of new tools to support teachers can be an added value to promote student motivation (Portugal, 2022).

Mobile Applications

Nowadays, mobile applications have gone beyond the barriers of being used only on smartphones. Smart devices such as smart TVs, and smartwatches make use of this artifact to make people's lives easier. For Couto (2016), with the intense presence of mobile technologies in our lives, apps have become popular and help thousands of people who are connected to organize their daily lives. Commonly called app, they can come installed on cell phones or be downloaded in specialized digital stores, they can be paid or free and are used for endless purposes, messaging, sending files, editing images, video calls or health control, from entertainment to their use in business, for example.

Mobile apps are software programs that have been developed for mobile devices and with the opening of Apple's online store for developers to submit their apps, their use became massive. Soon after, Google, realizing the potential of apps, also launched its online app store. Currently, to develop a mobile application for the ICT professional, it is necessary to have knowledge in programming languages, designer, back-end and front-end development. As an example of the programming languages used for native application development, for Android, which makes use of Java and for iOS with Objective-C. The languages used for the Windows phone system, C♯, for Symbian and Palm OS, C++, these less popular today.

Besides the native apps, which get access to the physical components and software of the cell phone, two other types of mobile apps. First, the hybrid apps, which has behavior of the native apps, but were developed with HTML5 and JavaScript programming, even though the user does not identify the difference in use. Second, the web app, accessed through a shortcut on the cell phone's main screen, which does not get access to the cell phone's hardware. Still for app development, the ICT market makes available methods that facilitate its creation, such as the sites/programs that are known as no-code, where it is not necessary to have a very deep knowledge in programming and the use of codes. The possibility of developing with this method becomes easier due to the simplicity of creation, with clicks and connections between the screens created, because it is quickly possible to have an application currently ready.

To ensure the quality of the application, tests are performed before it is launched in the stores. Such tests are responsible for presenting the developer with errors and needs for adjustments. Molinari (2018) advocates that a system may have no apparent errors and may be considered to have been checked correctly at first. However, this system may not serve the user, that is, it may be considered incorrect in the validation.

In this sense, it is considered that the quality of an application can be measured further according to the International Organization for Standardization, (ISO). ISO in this knowledge domain was initially as defined as ISO 9126 - ISO 9126 Software Quality Characteristics, which was updated in 2011 to ISO/IEC 25010:2011 - Systems and software engineering - Systems and software Quality Requirements and Evaluation (SQuaRE) - System and software quality models. The standard is aimed at quality analysis in the systems and software engineering domain. However, the standard was updated in 2017 ISO/IEC TS 25011:2017 Information technology - Systems and software Quality Requirements and Evaluation (SQuaRE) - Service quality models emphasizing on the quality of the models.

The Quality of an application can be measured internally and externally (ISO/IEC, 2017) and about:

- Functionality, where we essentially evaluate Adequacy; Interoperability; Access Security and Compliance.
- Reliability where the underlying issues of Maturity; Fault Tolerance; Recovery and Compliance are emphasized.
- Usability is also one of the components in which Intelligibility; Comprehensibility; Operability; Attractiveness and Conformity are evaluated.
- Efficiency enables us to analyze Performance Behavior over Time; Resource Utilization and Conformance.
- Portability where we analyze Adaptability; Ability to be installed; Coexistence; Ability to replace and Compliance.

After ISO 9126 was updated to ISO 25010, the characteristics, and sub-characteristics, listed in Table 1, have been increased from six to eight.

This update added security and compatibility features where it seeks to promote software protection and compatibility with other equipment. Each characteristic should be used to evaluate the software and its sub-characteristics and thus promote the quality of its artifact. For Natale (2011), each characteristic can be measured by defining an algorithm, a specific method and an expected value that will need to be achieved depending on the context in which it is used.

The identified features will be a quality factor in the mobile application design described in the next section.

Table 1. Product quality model

Maintainability	Performance and efficiency	Usability	Security
- Modularity - reusability - Parsability - Modifiability - Stability - Testability	- Behavior in relation to time - accuracy - Behavior in relation to resources	- Intelligibility - Apprehensibility - Operability - Error protection - Interface aesthetics of user	- Reliability - Integrity - Contestability - Accountability - Authenticity
Functionality	Reliability	Compatibility	Portability
- Functional completeness - Functional correctness - Functional adequacy	- Maturity - Availability - Fault tolerance - recoverability - accuracy - Behavior in relation to resources	- Coexistence - Interoperability	- Adaptability - Ability to be installed - Ability to to replace

Source: (ISO 25010:2017)

DESIGN OF THE MOBILE APPLICATION

This section aims to present the process of design of mobile applications. The practical component of the project is presented as well as the services and courses that the ACB makes available, how the information is accessed, the presentation of the problem, the analysis of requirements, the architecture of the system and, finally, the prototype of the application.

Association Casa do Brasil

The Association Casa do Brasil in Viseu (ACB) is an independent, non-partisan and non-profit association, with the beginning of its activities in the year 2019, which aims to welcome and integrate Brazilian migrants through services that can be facilitators in their adaptation in Viseu. Namely, the services are legal orientation, psychological orientation, job counseling at the Employment and Professional Training Institute, modeling courses (cutting and sewing), painting, music (guitar), and judo. It also provides guidance to public bodies, namely the Foreigners and Borders Service (FBS), Social Security, Parish Council and Health, about the documents needed to fulfill rights and duties with these bodies. It also aims to define and provide guidance on education for dependents in the first cycle, second cycle and third cycle regarding criteria, documentations, and internship (exclusive for the third cycle).

ACB participants register for free and have no decision-making rights at meetings. To have the right to make decisions, a monetary contribution is made, paid through a single quota, valid for 12 months. This annual contribution gives participants' access to the benefits of the ACB Card, such as discounts of various percentages on pharmacy products, discounts on English courses and on residential locksmith services.

Currently the ACB's physical address is an establishment in the city of Viseu and it provides digital access to its information through a platform at https://associacaocasadobrasil.pt/, where it presents, among other things, its mission, vision and values.

Currently, migrants who want information/guidance must go to the address of the ACB and register or search the organizational website for what they want. You can also come in person to use a computer

The Social Challenge of Migrant Integration

or to access the information you need. This limitation to accessing ACB information was analyzed, thus generating a need to mitigate it, and a proposal for a new model that allows a greater reach of its information. As a contribution to solving the problem, the various phases are presented to be able to build the artifact. Through the needs survey it is possible to identify the problem under study as well as the organizational requirements.

The immigration of Brazilians to Portugal has been evolving, past the economic crisis, and with the possibility of employment and security, the entry of men, women and students has grown. Two periods of immigration are perceived, the first between the years 1980 and 1990 called the first wave and from the beginning of the year 2000 to 2010, called the second wave, where work and economic security were the factors of entry and exit. For Egreja et al (2011), in both waves, the labor character of this flow is highlighted, which is inserted in specific niches, both by virtue of the expressive presence in the labor market and the concentration in areas where the labor market is more dynamic, in the cities of Lisbon and Porto. The impact of this immigration process can be perceived in the agencies directly linked to the government, such as in schools, health centers, social security, and parish councils.

Aiming to promote access to information, it was outlined to develop a mobile application for cell phones with the android operating system, which would serve the migrants arriving in Portugal, specifically in the district of Viseu, through the ACB and thus promote the integration of migrants in the Viseu community. It is considered that they can be an added value for regional development given the shortage of people in the interior of Portugal. These people are crucial for progress and for population settlement, especially in these low-density territories.

The interior territories in general have specificities that can enhance a better quality of life. On the other hand, it is a fact that technologies have been making these territories more intelligent. As a basis for the research, we consulted journals, books, works available on google academic, the websites of the Portuguese government bodies, and vast documents and articles on ICT.

In this sense, the responsibility to develop a software with quality that meets the needs, has no flaws, and is delivered in the proposed period should be a determination for the developer. Machado (2016) emphasizes this concern; in recent years the evolution of techniques for the software development process is a constant in the search for the construction of more reliable systems, within reasonable deadlines, and with quality that meets the needs of the end customer.

According to Arruda, et al (2014), the requirements have a primary function in the software process, by presenting the function of identifying the requirements of the parties involved, defining the features, restrictions and among others, thus being considered a decisive factor for the success or failure of a project. On the other hand, for Machado (2016), requirements are goals or restrictions established by customers and users of the system that define the various properties of the system. Moreover, Sommerville (2016) considers that requirements analysis is the step in which requirements are organized into categories, to explore the relationships between requirements and rank their importance according to the needs of stakeholders.

In this regard, Davis (1993) presents another definition for requirement, suggesting that it is a user need or a necessary feature, function, or attribute of the system. However, (Sommerville, 2016), defines requirements according to functionality into two types, the functional or non-functional:

- Functional requirements describe what the system should accomplish and its benefits to the customer. They can be general, which cover what the system can do, and specific, which reflect the systems and the work routine of an organization.

Figure 2. Use Case Diagram

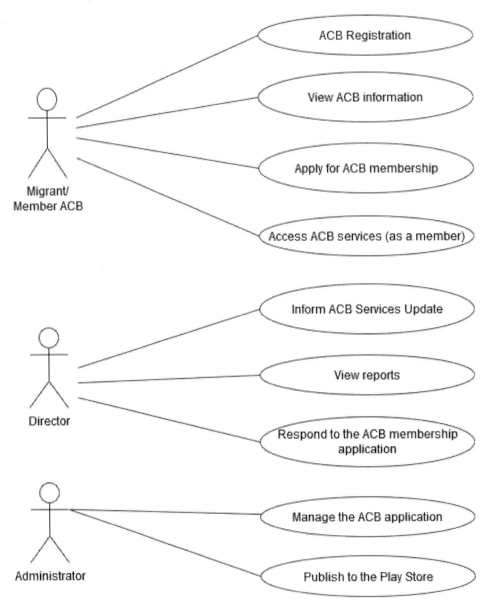

- Non-Functional requirements are not directly related to the specific services offered by the system to its users. They may be related to emergent system properties such as reliability, response time, and footprint.

In this sense, Figure 2 presents the Use Case in which the actors involved in the system are represented. Figure 2 emphasizes the various valences of actors. Thus, the actors in the process are:

- Director: is the person in charge of the ACB to signal to the application administrator the needs for inclusion, exclusion and updating of ACB services and information.

Figure 3. System Architecture

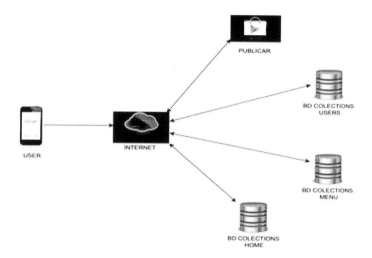

- Migrant: is an end user of the system. The migrant profile allows any person, with validation through registration with login and password to have access to the application and to ACB information. If so requested, the Migrant can also be a member of the ACB.
- Administrator: is responsible for managing the application. Its functions are to make the changes requested by the Director, manage the versions of the application, and keep the publication active in the Play Store for downloads by users.

Figure 3 shows the architecture of the ACB app, consisting of a no code platform, that is, without programming code; it is accessed by the browser and allows application development, for the web or cell phone, where through the components provided by it and settings made on these components and has as a result the desired application.

Figure 3 shows that the access is requested by the user through a cell phone with internet access. the ACB application is available for download in Google's virtual store, playstore, if this is the first contact with the application. The databases, called collections by the platform itself, are stored in a cloud service. The ACB app's databases are the home, menu, and user.

Application Prototypes

To facilitate communication with stakeholders the language used should be easy to understand and independent of the technology. In this sense, the VerbPhraseName pattern (Adolph & Bramble, 2003) was used to name the use cases, that is, to start with a verb in the infinitive representing the goal of the main actor.

Figure 4 shows the menu structure of the application.

It should be noted that this study aims to enhance the use of new technologies at the service of people, particularly in the integration of migrants. Therefore, the development of the mobile application takes into consideration the specific needs of the Community.

Figure 4. Menu Structure

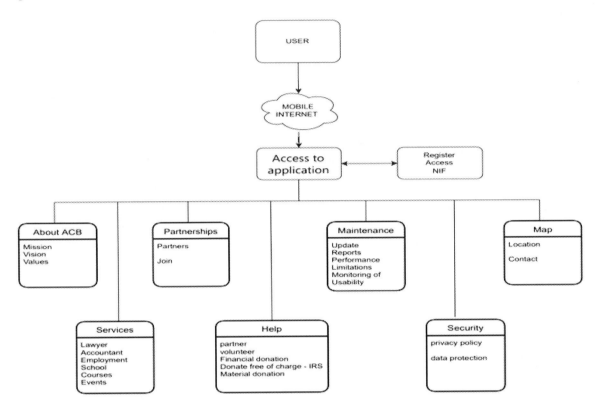

It is important to note that the artifact presented in this study, the App "ACB Migrante", has the potential to promote the SDGs, namely in the human and social dimensions. At the social sustainability level, the development of the "ACB Migrante" App includes features to promote the creation of a sustainable community, namely in helping social inclusion and integration; at the human sustainability level, the app is designed to promote the improvement of individual well-being, altruism and helping others.

Technology used in the social sphere, where through local knowledge and in conjunction with scientific knowledge, it can contribute to solving old problems with new approaches. Costa (2013) refers that in this perspective, the proposal of social technology advocates the development and use of ICT for social inclusion, based on the understanding that men and women should be involved in a constant process of action and reflection, so that the interaction between the individual and technology allows the expression of actions that enhance a more just, inclusive, and sustainable society.

Figure 5 shows the menu tree to make the entire structure of the designed mobile application visible.

Figure 5 consists of the various components representing the valences of the App "ACB Migrante" as it relates to the integration of Migrants. The main menu of the application presents menu options that include the presentation of the ACB, as in the valence that refers to the aid, as well as the partnerships. It also presents the services they provide to their associates, such as security policies and the physics of the ACB. Figure 6 shows the main screen menu of the Application.

Figure 7 shows the Volunteer activity, as well as the advantages of being a Partner. It also shows the Partner area.

The Social Challenge of Migrant Integration

Figure 5. Mobile application structure

ACB's stakeholders include partners, volunteers, and members. Figure 8 presents the psychological support component that the App "ACB Migrante" makes available to Migrants.

Psychological support is a relevant service for ACB members. Figure 9 presents the value-added component in this context as Partnerships are key to the full integration of Migrants.

Figure 6. Main menu

Figure 7. Stakeholders involved in the ACB

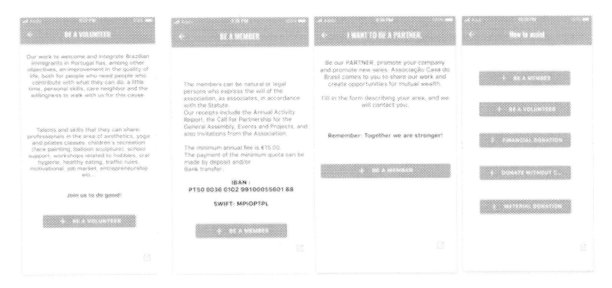

Figure 8. Psychological support

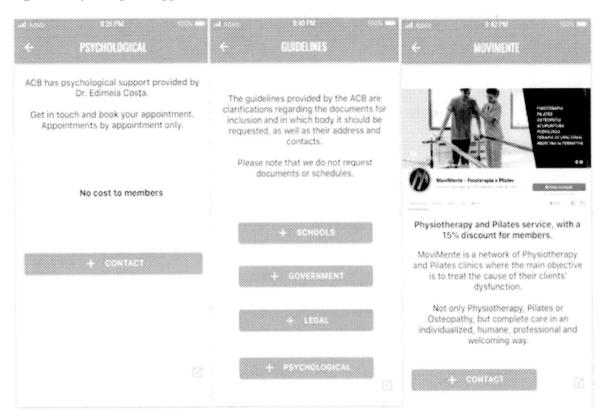

The Social Challenge of Migrant Integration

Figure 9. Integration activities

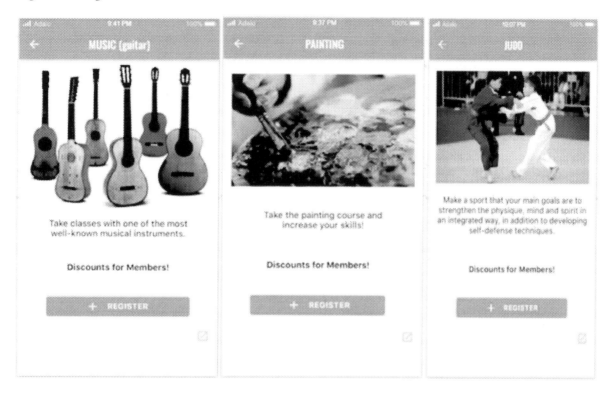

Figure 10 shows the Services within Courses that Members have access to. It also presents the Modeling Course and the English Course to contribute to the enrichment of the Migrants.

Figure 11 shows the types of donations in order to contribute to social well-being.

Figure 12 shows the connection with the Government Entities, schools, and legal support to speed up the interconnection with these entities.

Figure 13 shows the privacy policy, emphasizes the integration of Brazilians as well as the Registration form.

The application enhances the dissemination of information crucial for the integration of Brazilian Migrants, allowing the integration of sustainability factors, particularly with regard to human and social dimensions. It should be noted that the design of the mobile application was built according to the steps of the DSR. However, as the iterative process that it is, it allows us to enter a cycle of continuous improvement that future work perspectives will identify.

FUTURE RESEARCH DIRECTIONS

It is considered that the challenges underlying Society 5.0, namely as regards the dissemination of ICT use in the sense of increasingly putting them at people's service, implies the need to use more and more mobile applications. In this sense, it is advocated that the use of the mobile application in the ACB should be seen as a way to disseminate information and become a means of communication between Associations and citizens, facilitating integration into the Community.

Figure 10. Integration activities

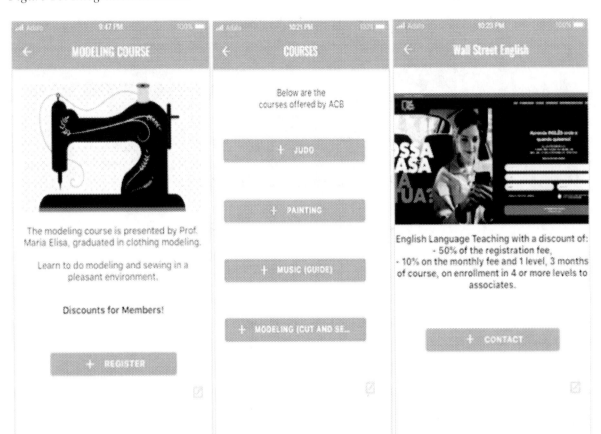

Also important was the awareness of the SDGs, both for those who develop the software and for those who use it thus contributing to several SDGs towards the full integration of Migrants. The mobile application developed incorporates sustainability measures, namely in the technical, social, and human dimensions. An improvement to be implemented in the next iteration will be to explicitly incorporate the environmental and economic dimensions.

Future work also includes improvements to the "ACB Migrant" application after obtaining feedback from users who download the application from the Google Play Store.

CONCLUSION

This study has fulfilled its objective by strengthening the use of new technologies for the integration and inclusion of migrants in the ACB through the construction of the artifact - the App "ACB Migrante". Therefore, it is contributing to the implementation of the principles of the European Pillar of Social Rights, thus helping the human well-being advocated by Society 5.0.

Figure 11. Types of donations

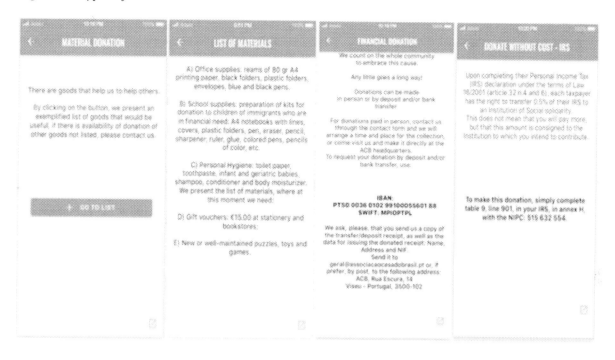

Figure 12. Institutional Entities

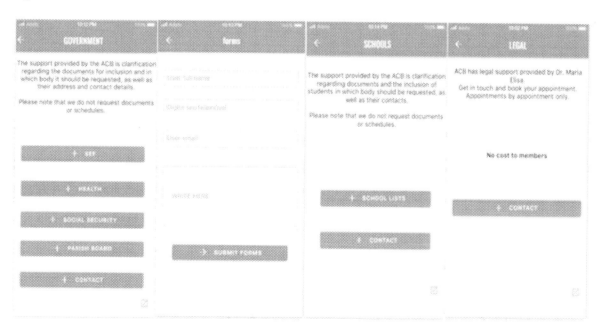

Figure 13. ACB Sign UP

The design of the mobile application includes sustainability concerns in order to constitute a solution that incorporates systemic thinking, also constituting a strategic opportunity within the scope of Society 5.0 and the strategy of increasingly developing solutions for the well-being of People. The incorporation of sustainability factors during the process of the DSR methodology was an important contribution.

It was outlined as a strategy to use the DSR which proved to be fundamental in the research process. The development of the artifact, the mobile application, enhances the search for results regarding the various social dimensions, namely in relation to employment, health, education, rights, and duties. It is an initiative with basic characteristics directed towards social innovation.

It is also considered that, regarding the dimensions of sustainability (Human/Individual, Economic, Environmental, Technical and Social), ICT can enhance the incorporation of these dimensions to include concerns of software reuse, open source, cloud computing, virtualization (Reis, et al., 2021).

REFERENCES

Adolph, S., & Bramble, P. (2003). *Patterns for Effective Use Cases*. Addison-WesleyPearson Education.

Ajmal, M. E., Khan, M., Hussain, M., & Helo, P. (2018). Conceptualizing and incorporating social sustainability in the business. *International Journal of Sustainable Development and World Ecology*, 25(4), 327–339. doi:10.1080/13504509.2017.1408714

ANACOM. (2019). *Comunicações eletrónicas - 2.º e 3.º trimestres de 2021*. [Electronic Communications - 2nd and 3rd quarters of 2021.] https://www.anacom.pt: https://www.anacom.pt/render.jsp?categoryId=370555

Antunes, R. (2015). *A sociedade da terceirização total* [The total outsourcing society.]. Revista da ABET.

Arruda, D., Soares, R., Vieira, D., Ferreira, R., Cabral, T., & Lencastre, M. (2014). Engenharia de Requisitos: Um Survey Realizado no Porto Digital, Recife/Brasil. [Requirements Engineering: A Survey Conducted in Porto Digital, Recife/Brazil.] Recife.

Baskerville, Baiyere, A., Gergor, S., Hevner, A., & Rossi, M. (2018). Design Science Research Contributions: Finding a Balance between Artifact and Theory. *Journal of the Association for Information Systems, 19*(5), 358–376. doi:10.17705/1jais.00495

Cazzolato, N. (2009). *As Dificuldades de Gestão das Organizações Não-Governamentais* [*The Management Difficulties of Non-Governmental Organizations.*]. Faculdade de Administração e Economia. doi:10.15603/2176-9583/refae.v1n1p66-81

Costa, A. B. (2013). *Tecnologia social & políticas públicas* [*Social technology and public policy.*]. Instituto Polis.

Coutinho, N. (2006). *As organizações sociais e o contrato de gestão* [*Social organizations and the management contract.*]. REVISTA DE DIREITO PÚBLICO. doi:10.5433/1980-511X.2006v1n2p25

Couto, E. E. (2016). *App Learning - Experiências de pesquisas e formação [App learning—research and training experiences]*. EDUFBA.

Da Silva Pereira, R., de Moraes, F. C. C., Junior, A. B. M., & Palmisano, A. (2013). Especificidades da gestão no terceiro setor. [Specificities of management in the third sector.]. *Revista Organizações em Contexto, 9*(18), 167–195. doi:10.15603/1982-8756/roc.v9n18p167-195

Davis, A. (1993). *Software requirements: objects, functions and states.*

DECO. (2021). *Viseu é a cidade do país com mais qualidade de vida. [Viseu is the country's city with the highest quality of life.]* Visit. Viseu. https://visitviseu.pt/sugestao?item=16

Egreja, C. P. (2011). Caminhos limitados ou mobilidade bloqueada? a mobilidade socioprofissional dos imigrantes brasilerios em Portugal. [Limited paths or blocked mobility? the socio-professional mobility of Brazilian immigrants in Portugal.] Coimbra.

Gupta, S. (2013). For mobile devices, think Apps, not ads. *Harvard Business Review.*

Hevner, A., & Chatterjee, S. (2004). *Design Science Research in Information Systems*. Springer. doi:10.2307/25148625

ISO/IEC. (2017). *TS 25011:2017 Information technology — Systems and software Quality Requirements and Evaluation (SQuaRE) — Service quality models*. ISO. https://www.iso.org/standard/35735.html

Kon, A. (2018). A inovação nos serviços como instrumento para a Inovação Social: Uma visão integrativa. [Service innovation as a tool for social innovation: an integrative view.]. *Revista de Economia Política, 38*(3), 584–605. doi:10.1590/0101-35172018-2814

Machado, F. (2016). *Análise e Gestão de Requisitos de Software – Onde nascem os sistemas [Analysis and Management of Software Requirements – Where systems are born.].* Érica - Saraiva.

Marketeer. (2020). *Marketeer.* https://www.marketeer.sapo.pt: https://www.marketeer.sapo.pt/numero-de-portugueses-que-descarregam-apps-gratis-triplica

Martinez, F., O'Sullivan, P., Smith, M., & Esposito, M. (2017). Perspectives on the role of business in social innovation. *Journal of Management Development, 36*(5), 36. doi:10.1108/JMD-10-2016-0212

Mulgan, G. (2007). *Gestão de organizações sem fins lucrativos: o desafio da inovação social* [Management of nonprofit organizations: the challenge of social innovation.]. Edições Vida Económica.

Natale, D. (2011). Complexity and data quality. *Commission UNINFO JTC1/SC7. Software Engineering*, 1–4.

Peffers, T., Tuunanen, T., Rothenberger, M. A., & Chatterjee, S. (2007). A Design Science Research Methodology for Information Systems Research. *Journal of Management Information Systems, 24*(3), 45–78. doi:10.2753/MIS0742-1222240302

Pompeo, W. (2016). *(R)Evolução Digital: Análises e perspectivas das novas tecnologias da informação e comunicação no direito, educação e gestão de negócios* [(R)Digital Evolution: Analysis and perspectives of new information and communication technologies in law, education and business management.]. Fadisma.

Portugal, 2. (17 de 04 de 2022). *Portugal Inovação social.* Portugal Inovação social. https://inovacaosocial.portugal2020.pt

Reis, L., Cagica Carvalho, L., Silveira, C., Marques, A., & Russo, N. (2021). *Inovação e Sustentabilidade em TIC* [Innovation and Sustainability in ICT.]. Silabo.

Schmidt, J. (2018). *Universidades comunitárias e terceiro setor – Fundamentos comunitaristas da cooperação em políticas públicas.* [Community universities and the third sector – Communitarian foundations of cooperation in public policies.] Santa cruz do sul: Edunisc.

Silva, M. C. (2011). *Sustentabilidade no Terceiro Setor: O desafio de armonizer* [Sustainability in the Third Sector: The challenge of harmonizing.]. Reuna.

Simon, H. (1996). *The sciences of artificial.* MIT PRESS.

Sommerville, I. (2016). *Software Engineering* (10th ed.). Pearson.

Song, J. J., Kim, J., Jones, D. R., Baker, J., & Chin, W. W. (2014). Application discoverability and user satisfaction in mobile application stores: An environmental psychology perspective. Decision Support Systems. [Decision Support Systems.]. *Decision Support Systems, 59*, 37–51. doi:10.1016/j.dss.2013.10.004

Tenório, O. (2015). *Responsabilidade Social Empresarial: Teoria e Prática* [Corporate Social Responsibility: Theory and Practice.]. FGV.

Ungerer, R. (2013). *Sociedade globalizada e mundo digital* [Globalized society and digital world.]. Artmed.

Wieringa, R. (2009). *Design Science Methodology For Information Systems And Software Engineering.* Springer.

KEY TERMS AND DEFINITIONS

Information Systems: This is the organized set of components such as people, processes of collection and transmission of data and material resources, automated or manual. The interaction of components enhances the processing and dissemination of information.

Information and Communication Technologies: A technological resource set used to process information and ensure communication. When used in an integrated way it enhances information transmission and communication processes.

International Standards: A set of technical standards establishing a quality management model for organizations in general, whatever their type or dimension, covering various areas within organizations.

Requirements Analysis: An iterative process to identify features and restrictions with a view to developing or changing a software product. Usually use cases are used.

Software Systems Development: A set of activities involved in the production of software. These activities are related to each other in an iterative and incremental process.

Sustainability: The ability to sustain life on the planet, considering the five dimensions: individual, social, economic, technical, and environmental.

Chapter 17
Youth Civic Engagement:
New Strategies for Social Innovation

Carlos Rodríguez-Hoyos
https://orcid.org/0000-0002-6949-6804
Universidad de Cantabria, Spain

Adelina Calvo-Salvador
https://orcid.org/0000-0002-9262-7905
Universidad de Cantabria, Spain

Aquilina Fueyo Gutiérrez
https://orcid.org/0000-0001-8668-923X
Universidad de Oviedo, Spain

Gloria Braga Blanco
https://orcid.org/0000-0003-4389-4908
Universidad de Oviedo, Spain

ABSTRACT

This chapter analyses the scientific bibliography on the civic engagement of young people published in English between 2014 and 2019. Sixty-nine articles published in international scientific journals were analyzed. The main objective of this meta-research is to understand what problems have been investigated and the data collection techniques used. The authors found that the most researched problem was the analysis and understanding of the impact that participation has on plans and programs designed by diverse socio-educational actors with regard to the civic engagement of young people. Based on the meta-research carried out, they propose three socially innovative intervention strategies aimed at improving youth civic engagement. Each strategy is designed to enhance young people's agency in order to increase their levels of social participation.

DOI: 10.4018/978-1-6684-6123-5.ch017

INTRODUCTION

Over the last few decades we have witnessed profound transformations that have, to a greater or lesser extent, revolutionized areas as diverse as the economy, education, leisure, commerce and the environment, among others. Some of these transformations have improved life on the planet. However, others have created significant problems that require new approaches in order to address them. Responding to the enormous challenges facing humanity from an innovative and sustainable perspective requires rethinking not only the technological developments designed over recent decades, but also the forms of participation and engagement of the different players involved in increasingly global societies. Social innovation aimed at sustainability cannot be understood exclusively as a way to develop cleaner technologies, but rather requires rethinking how people, citizens of a global world, engage and participate in the contexts they inhabit (Andrew & Klein, 2010; Ulitin et al., 2022). In short, it requires rethinking how the level of democracy in societies can be improved (Cahill, Wyn, & Borovica, 2019).

When we talk about democracy, we are talking about the government of the people. This concept has now acquired a new, more global dimension. In other words, it implies the government of all people who form part of a global society facing common problems. Therefore, it is necessary to seek civic strategies that encourage commitment to improving the living conditions of people on a planetary scale. The search for ways to improve commitment to the improvement of a society - of all people, including young people - is based on the idea that societies can govern themselves. This involves defining the type of sociopolitical agenda they want, managing their resources, organizing forms of communication and collaborating and defining their eco-social priorities, etc.

This chapter aims to provide policy strategies for social innovation aimed at increasing the agency of young people. These strategies are designed to improve the civic engagement of this group so that they can actively intervene in the response to current local and global challenges, especially problems related to improving sustainability. As suggested by some authors, social innovation is not only a technical response to diverse problems, but rather it must also have some characteristics that are related to the objectives of this book: it must be efficient, effective, fair and sustainable (Preskill & Beer, 2012). Social innovation strategies are being employed as a basic resource to respond to global ecological problems by various kinds of environmental organizations (Grimm et al., 2013). Such strategies aim to find local bottom-up solutions in the knowledge that it will be impossible for technological solutions alone to respond to the climate crisis (Seyfang & Smith 2007). Therefore, it is essential to rethink participation and civic action strategies by promoting the participation of all social agents, including young people.

BACKGROUND

The concept of civic engagement has been reviewed by various authors and institutions in the specialized literature from different disciplinary fields (Andolina et al., 2002). It is important to clarify that there is no standard definition of this term (Clement, Deering, Mikhael & Villa-García, 2014). The concept has been defined according to variables as diverse as the country in which a given bibliographic reference is generated, the discipline it comes from or the organization that promotes a document in which it is described (Meinzen-Dick, Di Gregorio & McCarthy, 2004). Nevertheless, some recurring aspects can be identified in the literature when defining the term. Broadly speaking, civic engagement refers to the actions developed by citizens seeking to participate in their community and trying to improve the living conditions of the people who belong to it (Adler & Goggin, 2005). Civic engagement can adopt more classical forms of involvement in traditional politics (participating in voting, belonging to political parties, organizing electoral campaigns, etc.) as well as other more community-based forms of civic participation (organizing marches, signing petitions, developing volunteer actions, etc.). More specifically, Checkoway and Aldana (2013) have identified four different ways of putting civic engagement into practice: grassroots organizations, intergroup dialogue, citizen participation and sociopolitical development.

Research carried out to date shows that civic engagement develops on two levels (Arvanitidis, 2017). On the one hand, each person participates individually (Gil de Zúñiga, 2015) in various causes as a consequence of several variables linked to their life experiences (values, existing resources, upbringing models, etc.). On the other hand, people participate civically according to the social structures (Warren, Sulaiman, & Jaafar, 2014) in which they live (for example, educational institutions, communities, families, etc.).

Generically, it can be considered that youth civic engagement is related to the development of diverse citizenship practices by young people in formal and informal contexts (Balsano, 2005; Harris, Wyn, & Younes, 2010; Chen, 2017; Gaby, 2017; Marshall, 2018) through activities developed outside peer groups or the family environment (Amná, 2012). Youth civic engagement in different social causes and institutions has been a growing concern in democratic countries over recent decades, reflected in the interest shown by their organizations in developing a feeling of affiliation among young people which is seen as a fundamental issue for the maintenance and improvement of democracy itself. This dynamic has encouraged the development of research based on different disciplines of social and human sciences such as education, politics, sociology or anthropology and so on. Likewise, other terms are used in the specialized literature that refer to variables relating to the diverse strategies through which young people intervene in issues that affect the interests of citizens in general: participation, politics, active citizenship, activism, youth volunteering, etc. (Shaw et al., 2014; Tsekoura, 2016a). In fact, youth civic engagement has some differential characteristics. It is relevant and specific to this age group. However, it cannot be understood as a predictor of future behavior. It also requires a level of engagement that may involve actions or psychological or behavioral states. Finally, the development of citizenship in young people involves participation in a specific community (Karakos, 2015).

While there is a certain degree of consensus when considering whether democratic societies should encourage and promote youth civic engagement, some authors have highlighted the dangers posed by the participation of young people in political life from authoritarian or extremist positions (Banaji, 2008, Pirro, & Róna, 2019). This terminological richness and complexity relates to the difficulties involved in separating what could be considered political life: it could be understood from a purely formal perspective related to involvement in formal institutions (for example, those belonging to the youth group of a political party), or from a civic perspective linked closely to other community problems that affect the expectations and experiences of young people more directly, such as issues related to sexual identity and gender (Youniss et al., 2002).

This study presents a meta-research aimed at finding out what has been investigated in this field and how it has been conducted, that is, the main lines of research and the data collection tools used. The results highlight topics that are underrepresented or directly absent in the specialized literature and identify the methodological assumptions on which research is developed. Consequently, through this analysis several socially innovative civic strategies for improving youth civic engagement have been suggested.

RESEARCH QUESTIONS

The main objectives of this work are to understand what has been investigated in the field of youth civic engagement and how this has been carried out, with the aim of gaining an insight into which topics require further investigation in order to improve the participation of young people in society.

To achieve these objectives, we have developed a meta-research of the specialized international literature in English produced between 2014 and 2019 (both included). This work has enabled us to understand the factors that lead young people to develop their civic engagement and identify the continuities and discontinuities produced in this process. In this meta-research we will address the following questions:

1. What are the main lines of investigation that are being developed? What absences are perceived?
2. What data-collection techniques are used?

Methodology

To carry out this meta-research we adopted a series of methodological decisions which we will explain. In order to guide the search we proceeded to elaborate a theoretical framework and defined research questions which would allow us to delimit and structure the search in the data bases used (Thomas & Pring, 2004). Following this we outlined the essential criteria used for delimiting the articles to include or exclude in our work (Evans & Benefield, 2001). Firstly, we limited the search to research articles published only in peer-reviewed scientific journals. Thus, other academic contributions such as those made in scientific meetings or conferences and chapters of books were not included. Secondly, we restricted the selection of articles to those published between 2014 and 2019 (both included). In order to identify the research which would be useful to our investigation we used the following international data bases: Scopus, Web of Science, RSCI and SciELO, using the following search terms: youth civic engagement, youth civic participation and youth online engagement. Thirdly, we carried out an analysis of the complete contents of each article, in contrast to previous works that have only analyzed the abstract of articles (Tzankova & Cicognani, 2019). Finally, although in the initial search we could identify articles written in other languages, we analyzed only those published in English.

Once all these criteria were applied, the first selection of articles returned a high number of works (187). To continue with the selection of articles, the research team carried out an initial thematic analysis with the aim of reducing their number. This thematic analysis allowed us to discard some works that were not directly related to our research objectives (participation processes in school activities, types of engagement developed by users in contact with digital applications, consumer products, etc.). Following this selection process we identified 69 articles which clearly addressed the objective of this meta-research. Finally, to facilitate the analysis of the selected works a coding system was developed linked to the research questions (topics investigated, research paradigm and data collection techniques). The system was developed ad hoc and agreed between the members of the team. The articles were then analyzed independently and arranged in each of the categories established in the coding system designed by the research team.

Table 1. Lines of research identified

Lines of research	Number of papers
Plans and programs to encourage civic engagement	28 (41%)
Political participation	19 (27%)
Online civic engagement	17 (25%)
Frameworks to understand civic engagement	5 (7%)

RESULTS OF META-RESEARCH

The main results of the meta-research are presented below. In total 69 articles that fulfilled the established criteria were analyzed. Due to the length of this chapter, we are going to cite and reference only some of the works included in each category with the aim of facilitating the reading and understanding of this document.

Lines of Research Found in the Literature

We have identified four main lines of research with different levels of importance: 1) Plans and programs to encourage civic engagement; 2) Political participation, 3) Online civic engagement and 4) Frameworks to understand civic engagement. Firstly, we found that 41% of the works analyzed (n=28) aimed to find out to what extent the participation of young people in certain plans or initiatives led to an improvement in the civic engagement in this group. Secondly, 27% of the works identified (n=19) are part of a line of research aimed at finding out what youth political participation is like, that is, what kind of engagement they develop in what could be considered civic participation. We also found that 25% of the studies analyzed (n=17) focused on understanding what type of mechanisms young people use to develop their civic engagement on the Internet. Finally, 7% of the works (n=5) aimed to identify frameworks related to the development of youth civic engagement.

Plans and Programs to Encourage Civic Engagement

In recent years, one of the main concerns of the scientific community has been to understand the impact of the participation of young people in plans and programs or experiences designed by diverse social, political, community and/or educational entities. To some extent, all these works suggest that the development of more or less institutionalized formal experiences can empower students and increase their agency. The literature analyzed allows us to verify that most of the works included in this line of research describe different processes which aim to look in-depth at youth civic engagement following a top-down logic. Research suggests that youth civic engagement is a significant concern to organizations of different sizes and influence, such as the UN (Cumiskey et al., 2015; Thew, 2018; Kwon, 2019), the governments of some countries (Narksompong & Limjirakan 2015; Boadu & Ile, 2018), regions or town halls (Cushing, 2014; Nwachukwu & Kang'ethe, 2014; Morciano et al., 2016; Davison & Russell, 2017; Sebti et al., 2019), as well as institutions that aim to improve the involvement of young people in their functioning (Dickson-Hoyle et al., 2018; Arunkumar et al., 2019; Cahill, Wyn, & Borovica, 2019).

The main results of these works show that including young people in decision-making makes it possible to identify a significant improvement in the civil engagement of the participants (Fernández & Shaw 2015; Cumiskey et al., 2015; Gal, 2017; Purdue, Peterson, & Deng 2018; Kwon, 2019). It also shows some (recurrent) variables related to the design and development of the types of initiatives that would limit or hinder their implementation.

Some works highlight the need to articulate plans and programs in which young people are involved in the design phase of the programs breaking, on the one hand, with unidirectionality when making decisions (Tsekoura, 2016b; Dickson-Hoyle et al., 2018) and, on the other hand, with the design of excessively rigid or formal operating dynamics (Morciano et al., 2016; Thew, 2018) whose sole objective is to promote tokenist (Narksompong & Limjirakan, 2015) or illusionary (Boadu & Ile, 2018) modes of intervention. Precisely, when this group perceives that their intervention is purely symbolic and they have no real capacity to make decisions, resistance behavior is produced (considered as disruptive on some occasions) which demonstrates an active opposition to both a decision-making process in which they can hardly intervene (Boldt, 2018) and merely ritual modes of intervention (Batsleer, Thomas, & Pohl, 2019).

We also found that the research analyzed suggests the need to rethink what role the diverse actors involved in these plans need to play if they want to improve youth civic engagement. Some of the works show the importance of designing processes in which adults can collaborate with young people, as long as this is always aimed at improving their agency (Tsekoura, 2016b; Krauss, 2018), identifying that it involves a characteristic that is socially mediated (Sebti et al., 2019). In this regard, the works analyzed reveal the importance that diverse agents (politicians, NGO workers, teachers, activists, etc.) acquire in the construction of supportive social contexts (Augsberger et al., 2019) which facilitate the improvement of civic engagement.

Political Participation

The second line of research aims to understand the political commitment of young people. In an effort to summarize the main themes of the works we can verify that they aim to analyze:

A) Factors that influence participation. What type of factors influence the political participation of young people (Henn & Foard, 2014; Soler-i-Martí & Ferrer-Fons, 2015; Määttä & Aaltonen, 2016; Vázquez et al., 2015; Mollica, 2017; Sahimi et al., 2018; Pirro, & Róna, 2019) and, on the other hand, (B) what levels of intervention they have in political life (Gozzo & Sampugnaro 2016; Goessling, 2017; Fowler, 2017; Gaby, 2017; Melles & Ricker 2017; Poteat, Calzo, & Yoshikawa, 2018; Sika, 2018). Firstly, some works have tried to identify what type of factors influence the commitment of young people in diverse political activities. These works show that one of the most influential variables is the perception they have about their own context. In contexts in which the economic and political situation is worse, young people's perception of their own reality is more critical and increases their possibilities of mobilization and involvement in public life (Soler-i-Martí & Ferrer-Fons, 2015; Sika, 2018). Other works suggest that the political participation of young people is motivated by issues that go beyond the transformation or improvement of the society in which they live. Among the most determining factors that lead young people to get involved in politics are motivations that could be considered endogenous such as sociability and professional progress (Vázquez et al., 2015), the impact of the personal relationships they develop (Body &

Hogg, 2019) or even the link between the type of activity carried out and their basic values which, in the long term, can end up becoming a way of weaving professional networks facilitating access to employment (Sahimi et al., 2018).

B) Level of intervention. Other investigations within this line of research analyse the levels of political commitment developed by young people in the contexts in which they live. Research within this framework addresses diverse topics such as young people's levels of intervention in organizations such as the gay-straight alliances and the influence of their activism in these organizations on their general civic engagement (Poteat, Calzo, & Yoshikawa, 2018), the impact of pre-registration in the increase of youth participation in electoral processes (Fowler, 2017) or the ways in which youth organizations resist and reinforce neoliberalism today (Goessling, 2017). These works show that young people increasingly choose modes of intervention in public life that could be considered unconventional (Gökçe-Kızılkaya & Onursal-Beşgül, 2017; Sika, 2018), thus making it necessary to broaden the notion of what is considered political in order to give meaning to diverse forms of youth participation (Tsekoura, 2016a) and question the hypothetical decrease in their levels of political commitment. For example, Gozzo and Sampugnaro (2016) analyse the participation of young Europeans and suggest their levels of intervention in public life are less in southern countries than northern ones. Although they question whether young people intervene in political issues, it was found that when they do intervene this is generally linked to specific problems which are often related to their daily or personal lives. A similar study in the North American context showed that if the data for youth participation between 1990 and 2009 is taken into account, it can be identified that youth civic engagement had increased and, therefore, the supposed decline in civic engagement had to be questioned (Gaby, 2017).

Online Civic Engagement

This section covers works that have specifically analyzed how young people develop their engagement through the Internet. These studies have attempted to show the complex continuities and discontinuities that lead young people to engage in political activities on and off the network, using various kinds of digital applications (Metzger et al., 2015; Banaji & Buckingham, 2016; Cicognani et al., 2016; Roque, Dasgupta, & Costanza-Chock, 2016; Kim, Russo, & Amnå, 2017; Van Haperen, Nicholls, & Uitermark, 2018; Banaji, Mejias, & de la Pava Velez, 2018; Onyechi, 2018). They show the important transformations that youth civic engagement has undergone as a consequence of the appearance of the Internet. The appearance of applications such as social networks has helped young people create hybrid spaces of engagement. Subsequently, they have developed a wide range of diverse practices and profiles in such a way that online political participation can become a gateway to offline political participation (Kim, Russo, & Amnå, 2017). While some groups mix their personal relationships with politics, others develop clear political uses of networks, including some forms of activism and citizen journalism (Mascheroni, 2017).

Other works have specifically analyzed the role online civic engagement can play when improving the participation of young people belonging to socially excluded or minority groups (El Marzouki, 2016; Simoes & Campos 2017; Shiratuddin et al., 2017; García, Fernández, & Jackson, 2019). Generally, these works show the great potential that digital applications have for allowing young people from these groups to participate more actively in decision-making processes of a social, political or economic nature (Shiratuddin et al., 2017). Likewise, it enables them to participate in the creation of content that encourages the empowerment of young people and break with their own community stereotypes (El Marzouki, 2016;

García, Fernández, & Jackson, 2019). Moreover, some works have attempted to understand how young people use digital technology in contexts produced in situations of political crises of varying degrees of severity (Zhu, Skoric, & Shen, 2017; Zhang & Lin, 2018). The results of these works suggest that in situations of high social tension in which special circumstances such as street protests occur, young participants do not use social networks such as Facebook to get information or learn about other points of view, but rather they try to protect themselves by hiding publications from friends who have different political opinions to their own (Zhu, Skoric, & Shen, 2017).

Frameworks for Understanding Civic Engagement

Finally, we have identified a smaller number of works (n=5) aimed at establishing frameworks focused on fostering a better understanding of youth civic engagement (Uršič, Dekker, & Filipovič Hrast, 2014; Themistokleous & Avraamidou, 2016; Cahill & Dadvand, 2018; Tzankova & Cicognani, 2019). Based on different perspectives or disciplines (such as videogame design, architecture or psychology), one of the main characteristics of the works in this line of research is that they try to identify some principles that enable or encourage the development of youth practices which are characterised by a high degree of civic engagement. On the one hand, this can be seen in the academic interest of youth participation in recent years (Tzankova & Cicognani, 2019). On the other hand, it is an enormously complex and controversial topic because it does not always produce social benefits and can replicate patterns of inequality (Cahill & Dadvand, 2018).

The work of Uršič, Dekker & Filipovič Hrast (2014) focused on understanding how spaces influence young university students' participation suggests that, currently, the trend towards the privatization and consumption in some of these places is evident, thus limiting their intervention. Another example is the work developed by Themistokleous & Avraamidou (2016) whose purpose is the definition of a series of theoretical principles aimed at designing games that allow improving the civic engagement of the young people who use these types of cultural artefacts. Their contributions show that in order to improve youth civic engagement using digital games, it is necessary to design gaming experiences in which the content is contextualised since, among other factors, this increases the likelihood that young people will use the skills acquired in games in their interventions in public life.

Data Collection Techniques Used in the Literature in this Scope

The authors analyzed the data collection techniques used most frequently by the scientific community to investigate youth civic engagement processes. Firstly, it should be noted that most of the research analyzed combines diverse data collection strategies, independent of the paradigm in which the research was developed. The most used techniques were questionnaires (n=28), interviews (n=26), and the analysis of various kinds of documents (n=25). Secondly, we found observation (n=11), focus groups (n=10), field notes (n=7), the analysis of web pages (n=3) and the analysis of virtual productions such as posts, tweets, etc. (n=3). However, in five of the works analyzed no information was provided on the data collection techniques used.

The questionnaire was the most widely used data collection technique in the research analyzed. It was used to carry out a descriptive analysis of the opinions of young people on aspects as varied as motivations for participating in voluntary associations (Vázquez et al., 2015), the impact of social networks on

Table 2. Data collection techniques

Data collection techniques	Number of papers
Questionnaire	28
Interview	26
Analysis of documents	25
Observation	11
Focus Group	10
Field notes	7
Analysis of web pages	3
Analysis of virtual productions (posts, tweets, etc.)	3
Not defined	5

civic engagement (Chen, 2017) or the impact of social media on political actions developed individually and collectively (Zhang & Lin, 2018).

Secondly, interviews were used recurrently to understand the opinions of the different agents involved in processes related to youth civic engagement. For example, they were used to ask young people about their experiences or reflections following participation in actions aimed at encouraging their agency in participatory spaces (Tsekoura, 2016b) or about the influence their participation in voluntary programs had on their civic engagement later on (Body & Hogg, 2019).

Thirdly, the technique of analyzing different kinds of documents was also used (Narksompong & Limjirakan, 2015; Augsberger et al., 2019). Other techniques included the observation of activities developed by young people in conferences, exchanges or participatory processes (Thew, 2018; Boldt, 2018), as well as field notes used in research clearly inspired by ethnography (Batsleer, Thomas, & Pohl, 2019; Schusler, Krings & Hernández, 2019).

A disaggregated analysis of the techniques used according to the paradigm allows us to identify that the most used technique in qualitative research is the interview (n=18), the analysis of diverse documents such as reports, posters, publications, etc. (n=18), observation (n=11) and field notes (n=7). Other techniques such as focus groups (n=5), the analysis of virtual productions (n=2) or web pages (n=1) are used less often. Some more theoretical works do not define data collection techniques (n=5).

On the contrary, the quantitative paradigm presents less variety of techniques. The questionnaire (n=17) is the most frequently used, while the analysis of documents (n=3) and virtual productions (n=1) are used to a lesser extent. Finally, in the mixed works the most used technique is the questionnaire (n=11), followed by the interview (n=8), the analysis of documents (n=4), the focus group (n=5) and the analysis of web pages (n=2).

Lessons Learned from the Meta-research for Innovative Intervention with Young People. Searching Form New Strategies

Research should be a source of inspiration for change and social innovation. Therefore, the lessons learned from the meta-research we carried out have enabled the authors to define intervention strategies inspired by the literature. Following the analysis of the specialized literature, the authors have proposed various socially innovative intervention strategies aimed at improving youth civic engagement. These strategies are designed to promote the voice of young people in society and create increasingly democratic and sustainable civic spaces (Checkoway, & Aldana, 2013). They are intended to give this group a voice and increase their level of participation in decision-making on global issues. These strategies should be understood as open-ended attempts that can be developed in different ways in each context of civic engagement. The main objective, meaning, and some procedural principles for implementing these strategies are explained below.

Strategy 1. Young People as Participation Detectives. Meta-participation

Objective

The aim of this strategy is to develop frameworks of thinking through which young people can understand how participation in social and organizational structures works (in formal, non-formal and informal contexts).

Meaning

In democratic societies, participation is a value and a process that is an integral part of life in common since, as we know, democracy means "the government of the people". Thus, in a democratic society, its organizations must be governed by these principles, and the value of participation and collective decision-making must be effective. Participation is based on dialogue and the search for collective agreements in which all social groups, especially young people with less voice, must participate. A participatory society needs spaces for dialogue and cooperative and civic action.

Goodman (2002) pointed out that maintaining a democratic system in today's societies, which he called critical democracy (based on the legacy of John Dewey), requires a difficult balance between the values of individuality and those of the community. At present, this balance is very fragile given that there is a clear predominance of individualism the values of which are securely maintained in the collective imagination. Research carried out with young people shows that in order to develop the attitudes, knowledge or values essential for the exercise of democratic citizenship, it is necessary to generate spaces in which participation is practised organically (McCowan, 2009). If young people perceive their intervention as symbolic, it generates ritual modes of intervention (Batsleer et al., 2019) or even resistance or disruptive behavior. This is why participation cannot be limited to a specific moment of their school life or the educational processes they experience, but rather it must go beyond and be present in any organizational dynamic.

As identified in the literature reviewed, in order to improve young people's levels of participation, firstly, it is necessary to understand the reasons that lead them to participate and, secondly, understand the ways in which this participation is carried out. This implies that creating spaces in which young people are encouraged to participate is insufficient. Young people need to be able to identify not only where they can develop civic action but also how they can improve it. This requires designing civic strategies aimed at understanding the complexity of the multiple dimensions inherent in any decision-making process or intervention (meta-participation), enabling them to initiate and propose changes in these types of processes themselves (Trilla & Novella, 2011).

Procedural Principles

- *Review existing participation channels in civic action contexts.* Meta-participation should be based on an exhaustive and systematic review of all formal channels (rules, laws, internal operating regulations, etc.) that regulate the contexts in which young people participate (e.g. in schools, neighborhood associations, political parties, student unions, etc.). The main objective is to help them understand the framework that currently shapes their opportunities to intervene in the contexts in which they carry out their activities (electoral processes, representative bodies, management committees, etc.).
- *Identify unregulated forms of participation.* In addition to the formal structures that organise activity in the contexts in which young people intervene, there are numerous mechanisms that articulate unregulated forms of participation. These alternatives are embedded in the culture of organizations and institutions and become informal or non-formal strategies of participation with enormous importance in the management of common things. Strategies need to be developed that allow young people to understand which agents can and cannot participate and the level of participation allowed in the institutions.
- *Identify variables that limit participation.* Based on the knowledge of formal and informal structures of participation, those elements that could be hindering or limiting young people's civic action need to be identified. To this end, young people will not only have to learn to identify the areas where they can participate but also: a) know what kinds of strategies are used to facilitate this intervention; b) know how to design mechanisms so that people who traditionally participate less, can participate.
- *Identify factors which promote participation processes in contexts of action.* Lastly, young people need to be able to understand and explain the mechanisms that encourage participation. This implies, on the one hand, recognising those civic strategies that are already in place and, on the other hand, envisioning others (formal or informal) that could facilitate or increase the power to make decisions in the management of common things, thus generating social contexts of mutual support.

Strategy 2. Hybridising Civic Action. Online and offline

Objective

This strategy aims to broaden spaces for youth civic engagement by establishing a continuum between the face-to-face and virtual contexts in which young people develop their civic action. Hybridizing civic action contexts involves recognizing forms of civic engagement that, until now, have either gone unnoticed or were considered secondary, especially those led by young people.

Meaning

The development of multiple communication systems and technical devices that allow for real-time and delayed modes of communication has generated profound transformations in all areas of human experience, from the personal to the political and from the intimate to the private. In recent years the growth of digital technologies, driven by large multinational technology companies, has led to increasing levels of penetration among citizens, even though there are still significant sectors of the population without access to them or with a cultural capital that conditions a very limited use of them (Álvarez-Icaza, Bustamante-Bello, Ramírez-Montoya, & Molina, 2022). Frequently these developments have been accompanied by discourses, driven by the multinational technology companies, suggesting that they have opened up new channels for public and community participation (Buckingham, 2017).

Studies of young people indicate that, at present, they demonstrate more open and flexible mechanisms of civic action, especially when it is linked to issues that concern them personally (Mascheroni, 2017). They prefer to get involved in actions that are less hierarchical than traditional ones which focus mostly on responding to specific problems (Amnå & Ekman, 2014). In order to do so, they use communication strategies and formulas that diverge from more conventional media (social networks, streaming communication systems, etc.). As some research suggests, the use of technologies by young people is transforming their forms of self-representation (Bell, 2019), the strategies they use to socialise, as well as their civic engagement practices (Gil de Zúñiga, 2015). It is therefore important to develop strategies that allow for the hybridisation of youth civic engagement by focusing, on the one hand, on the type of actions they carry out in virtual contexts and, on the other hand, by identifying more conventional patterns of engagement in face-to-face political participation.

Procedural Principles

This strategy is focused on linking the actions developed by young people in formal and informal contexts of political participation. In addition, it should enable young people to relate civic actions developed through technological devices with those carried out in face to face contexts. The main actions required by this strategy are outlined below:

- *Develop actions aimed at making young people aware of the civic uses they make of various applications and social networks.* Nowadays, technologies are used in all dimensions of human existence. For example, it is rare for young people to create different digital profiles aimed at structuring their digital identity according to the dimension to be addressed (personal, educational, political, etc.). Thus it is crucially important that the strategy allows young people to identify the political areas or social movements in which they are involved, to a greater or lesser extent (feminism, the fight against climate change, community work, etc.).
- *Redesign traditional forms of communication in organizations and institutions by promoting co-productive interactivity.* Some studies suggest the importance of designing online spaces that facilitate co-productive interactivity (Mainsah, Brandtzæg, & Følstad, 2016) where young people can manage content and intervene in an open way, enabling a high level of control over their own interventions. This type of interaction allows political participation spaces to be used to develop real-time encounters.
- *Provide spaces in which young people can experience and engage in traditional forms of civic participation.* While the expansion of information technologies has had a major impact on the way in which young people develop their civic engagement, it is still necessary to provide more traditional opportunities for community involvement. As some studies suggest, there is a clear relationship between the possibilities of participating in offline civic learning experiences (in various contexts such as family or school, among others) and the level of engagement developed (Bowyer & Kahne, 2020). Therefore, hybridising civic action among young people requires not only pro-

viding opportunities for civic engagement online, but also strengthening strategies in more formal or traditional structures and dynamics.

Strategy 3. Double Contextualisation of Civic Action. Between the Global and Local

Objective

Double contextualization should be focused and organized in a double movement. Firstly, through contextualization processes, it is necessary to direct young people's attention to their closest contexts in scenarios common to them, so that they are able to identify the challenges and problems they face (demographic, environmental, economic, etc.). Secondly, civic action must be recontextualized through strategies that facilitate the development of a more global or planetary outlook.

Meaning

The research analyzed suggests the improvement of youth civic engagement -and therefore the improvement of the societies in which young people live- are directly related to their knowledge and perceptions of the contexts they inhabit. In this regard, it has been suggested that there are some endogenous variables, such as peer relationships, values or potential professional benefits (Body & Hogg, 2019) that this group takes into account when it comes to civic engagement. Furthermore, the context is decisive. In other words, young people develop a more intense commitment when they analyze and care about their local contexts, especially those in a more precarious socio-economic situation (Sika, 2018). Therefore, it is advisable to develop processes based on the recognition of the scenarios in which young people live, with the objective of highlighting the impact that decision-making in the public sphere has on improving the living conditions of their community. As Percy-Smith, McMahon & Thomas (2019) suggest, involving young people in their daily contexts opens up possibilities for engaging them in civic actions crucially important to their interests or those of other groups close to them. Such contextualization is ultimately a way to offer young people a transition from analysis to effective social action (Watts & Hipólito-Delgado, 2015).

This first movement –contextualization– must be accompanied by a second movement that allows young people to recontextualize civic action and focus their perspective on more global issues and challenges. This task requires providing opportunities for young people to look beyond their daily contexts and ask themselves how complex globalization processes are affecting their lives. In short, it is a process that encourages young people to question the sociopolitical and economic consequences that globalization processes have on their lives, and identify the inequalities (gender-related, ethnic, economic, religious, etc.) caused by these profound social transformations.

Procedural Principles

The double contextualization of civic action requires the design of experiences that are characterized, on the one hand, by involving young people at the start of decision-making processes and, on the other hand, by their scalability. The actions that need to be implemented are outlined below:

- *Design civic action experiences so that they are articulated from a bottom-up logic*. As we have identified in this meta-research, the literature suggests that the development of initiatives conceived from a top-down logic and the lack of participation in their design are some of the variables that account for the lack of youth civic engagement in some experiences developed from the logic of adults. Some studies indicate that including young people in decision-making processes when designing experiences aimed at this group has positive effects on improving their civic engagement (Kwon, 2019). Thus, it is necessary to foster the leadership of young people and move away from less rigid organizational styles in favour of more flexible and emergent ones.

- *Ensure young people's interests are prioritised during the process of designing the experiences.* The literature reviewed strongly suggests that youth civic engagement is enhanced when specific life issues identified by young people are addressed (Dickson-Hoyle et al., 2018). In short, it is essential to design experiences in which young people can produce knowledge that offers a clear response to their life experiences. This requires viewing this group as human beings capable of researching and learning in order to improve the civic problems they experience in their lives (Rodríguez & Brown, 2009). Thus young people must be involved in the process of designing experiences so that they can respond to the specific needs they identify as their own.

- *Design scalable experiences to recontextualize their local civic issues with others of a more global nature.* The recontextualization of civic action should highlight the interdependence of human existence on a planetary scale and the undeniable challenge of providing collective responses to interconnected ecological and social issues, among others. This implies collaborating in the design of experiences that gradually deal with more complex issues (Ho, Clarke & Dougherty, 2015) and, as if it were a spiral curriculum, the issues need to be addressed repeatedly in order to understand them in an increasingly more complex and global way. As some studies suggest, improving youth civic engagement requires placing less emphasis on understanding formal political processes in favor of creating spaces that facilitate the development of more collaborative, exploratory or global knowledge (Davies et al., 2019). This recontextualization needs to provide analytical frameworks that help young people to respond in solidarity in a context that is uncertain, contingent, complex and defined by its difference (Andreotti, 2006).

CONCLUSION

This meta-research allows us to conclude that the analysis of the diverse ways through which youth civic engagement is expressed has been on the agenda of the scientific community in recent years. The presence of this theme in the journals of various academic disciplines should be interpreted as representing the concern of supranational and national organizations for improving the agency of young people, as well as their ability to intervene in the public life of democratic societies (Dickson-Hoyle et al., 2018; Arunkumar et al., 2019; Cahill, Wyn, & Borovica, 2019). The most important line of research analyses the impact that participation from different plans or programs has on youth civic engagement (Cumiskey et al., 2015; Morciano et al., 2016; Davison & Russell, 2017; Thew, 2018; Kwon, 2019).

A significant part of the research reviewed only analyses how civic engagement is modified in the short term. Although it is necessary to measure the impact of certain policies aimed at encouraging youth intervention in society, it should also be noted that this requires the design of longitudinal studies that analyse whether this civic engagement is maintained over the long term, the variables that influence the maintenance of engagement or its impact on political life in adulthood, among others. The analysis of the works included in this research also suggests that, although they use a wide variety of data collection techniques, young people have very few opportunities to actively participate in research design. Thus, it would be a good idea to review research design using more participatory research perspectives, so that the research objective is more coherent with epistemological assumptions and the techniques used (Åkerström, Aytar, & Brunnberg, 2015; Schelbe et al., 2015; Seymour et al., 2017).

It is important to note that the civic engagement of young people is in a continuous process of transformation and that their intervention in civic or political life presents new forms which were unknown until a few years ago (Tsekoura, 2016a; Gökçe-Kızılkaya & Onursal-Beşgül, 2017; Sika, 2018). These unconventional forms of intervention in public life have benefited from the development of phenomena such as social media, generating hybrid spaces of engagement. Regarding research focused on the analysis of virtual civic engagement, future research should incorporate other data collection techniques that allow understanding not only what young people say they do (through questionnaires, interviews or group discussion), but also what evidence they leave in the digital spaces they visit. This research has identified that there are very few works focused on what type of evidence young people leave in virtual spaces and how this online engagement interacts with that developed outside the Internet. Likewise, it is necessary to continue understanding what the continuities and discontinuities are between the civic engagement young people develop outside and inside the Internet.

This work has also allowed us to identify that there are very few studies which aim to understand how to improve the engagement of those young people who belong to minority groups or who experience situations of social exclusion, to a greater or lesser extent (Simoes & Campos, 2017; Shiratuddin et al., 2017; García, Fernández, & Jackson, 2019). Moreover, it has been shown that there are hardly any works aimed at analyzing the youth civic engagement from a gender perspective (Orloff & Palier, 2009; Cicognani et al., 2016; García, Fernández & Jackson, 2019). The scientific community must incorporate the gender perspective in their studies in order to ask themselves how gender could be influencing the development of mechanisms that enable a certain degree of civic engagement.

It is also evident that a comparative analysis between different age ranges is required, given that the term "young" in the research analyzed encompasses adolescents aged 11 – 12 years old to young people up to the age of 40. Thus, on the one hand, it is necessary to understand what particularities civic engagement presents in the different stages of youth and, on the other hand, how to make this sustainable in the long term.

In the final part of the chapter we have proposed three socially innovative civic strategies, inspired by this meta-research, aimed at improving young people's agency. Based on the specialized literature, the authors consider that improving young people's voice and participation in civic life requires imagining new strategies for youth civic action focused on three directions: a) improving their understanding of existing types of participation channels, as well as the limitations and possibilities for improvement; b) incorporating the new participation strategies they develop in digital spaces, combining them with those developed in person; c) recontextualizing local needs in a framework that is essentially planetary in its nature.

These three social innovation strategies will enable the objectives proposed in this chapter to be achieved. The analysis of the specialized literature provides evidence of the direct relationship between social innovation strategies and the improvement of sustainability. In fact, these strategies are being used to fight against climate crisis by organizations that protect the environment (Seyfang & Smith 2007). Improving youth civic engagement is an essential tool in sharing responsibility for the need to respond to the challenges of a global crisis. Increasing the participation of young people in civic life forces adults to regard them as subjects with full rights and responsibilities, that is subjects who are a model of active and committed citizenship. By doing this, we would be able to break through a particularly adult-centric culture that still does not sufficiently take into consideration the voice and perspectives of young people to understand and transform reality. The global response to current challenges (environmental, political, economic, etc.) requires the development of processes that deepen and improve existing levels of democracy in our societies. Thus, it is necessary to analyse which groups have had less voice, presence and participation in our societies and also demonstrate that these inequalities are not reproduced in exactly the same way in all contexts and at all times.

ACKNOWLEDGMENT

R+D+i project entitled Researching new socio-educational scenarios for the construction of global citizenship in the 21st century (Reference PID2020-114478RB-C21 financed by MCIN/AEI /10.13039/501100011033).

REFERENCES

Adler, R. P., & Goggin, J. (2005). What do we mean by "civic engagement"? *Journal of Transformative Education*, *3*(3), 236–253. doi:10.1177/1541344605276792

Åkerström, J., Aytar, O., & Brunnberg, E. (2015). Intra-and inter-generational perspectives on youth participation in Sweden: A study with young people as research partners. *Children & Society*, *29*(2), 134–145. doi:10.1111/chso.12027

Álvarez-Icaza, L., Bustamante-Bello, R., Ramírez-Montoya, M.S., & Molina, A. (2002). Systematic Mapping of Digital Gap and Gender, Age, Ethnicity, or Disability. *Sustainability*, *14*(1297). doi:10.3390/su14031297

Amná, E. (2012). How is civic engagement developed over time? Emerging answers from a multidisciplinary field. *Journal of Adolescence*, *35*(3), 611–627. doi:10.1016/j.adolescence.2012.04.011 PMID:22613043

Amnå, E., & Ekman, J. (2014). Standby citizens: Diverse faces of political passivity. *European Political Science Review*, *6*(2), 261–281. doi:10.1017/S175577391300009X

Andolina, M. W., Jenkins, K., Keeter, S., & Zukin, C. (2002). Searching for the meaning of youth civic engagement: Notes from the field. *Applied Developmental Science*, *6*(4), 189–195. doi:10.1207/S1532480XADS0604_5

Andreotti, V. (2006). *Soft versus critical global citizenship education in development education: Policy and practice*. Centre for the Study of Social and Global Justice, Nottingham University.

Andrew, C., & Klein, J. L. (2010). *Social Innovation: What is it and why is it important to understand it better* (Vol. ET10003). Ministry of Research and Innovation.

Arunkumar, K., Bowman, D. D., Coen, S. E., El-Bagdady, M. A., Ergler, C. R., Gilliland, J. A., & Paul, S. (2019). Conceptualizing youth participation in children's health research: Insights from a youth-driven process for developing a youth advisory council. *Children (Basel, Switzerland)*, *6*(1), 3. doi:10.3390/children6010003 PMID:30597913

Arvanitidis, P. A. (2017). The concept and determinants of civic engagement. *Human Affairs*, *27*(3), 252–272. doi:10.1515/humaff-2017-0022

Augsberger, A., Springwater, J. S., Hilliard-Koshinsky, G., Barber, K., & Martínez, L. S. (2019). Youth participation in policy advocacy: Examination of a multi-state former and current foster care youth coalition. *Children and Youth Services Review*, *107*, 104491. doi:10.1016/j.childyouth.2019.104491

Balsano, A. (2005). Youth civic engagement in the United States: Understanding and addressing the impact of social impediments on positive youth and community development. *Applied Developmental Science*, *9*(4), 188–201. doi:10.12071532480xads0904_2

Banaji, S. (2008). The trouble with civic: A snapshot of young people's civic and political engagements in Twenty-first Century Democracies. *Journal of Youth Studies*, *11*(5), 543–560. doi:10.1080/13676260802283008

Banaji, S., & Buckingham, D. (2016). Young people, the Internet, and civic participation: An overview of key findings from the Civic Web project. *International Journal of Learning and Media*, *2*(1), 15–24. doi:10.1162/ijlm_a_00038

Banaji, S., Mejias, S., & de la Pava Velez, B. (2018). The significance of ethnography in youth participation research: Active citizenship in the UK after the Brexit vote. *Sociální studia*, *15*(2), 97–115. doi:10.5817/SOC2018-2-97

Batsleer, J., Thomas, N. P., & Pohl, A. (2019). Who knows? Youth work and the mise-en-scene: Reframing pedagogies of youth participation. *Pedagogy, Culture & Society*, *28*(2), 205–221. doi:10.1080/14681366.2019.1627484

Bell, B. T. (2019). You take fifty photos, delete forty-nine and use one: A qualitative study of adolescent image-sharing practices on social media. *International Journal of Child-Computer Interaction*, *20*, 64–71. doi:10.1016/j.ijcci.2019.03.002

Bessant, J. (2016). Democracy denied, youth participation and criminalizing digital dissent. *Journal of Youth Studies*, *19*(7), 921–937. doi:10.1080/13676261.2015.1123235

Boadu, E. S., & Ile, I. (2018). The politics of youth participation in social intervention programs in Ghana: Implications for participatory monitoring and evaluation (PM&E). *Journal of Reviews on Global Economics*, *7*, 913–925. doi:10.6000/1929-7092.2018.07.89

Body, A., & Hogg, E. (2019). What mattered ten years on? Young people's reflections on their involvement with a charitable youth participation project. *Journal of Youth Studies*, *22*(2), 171–186. doi:10.1080/13676261.2018.1492101

Boldt, G. (2018). Condescension or co-decisions: A case of institutional youth participation. *Young*, *26*(2), 108–125. doi:10.1177/1103308817713620

Boulianne, S. (2016). Online news, civic awareness and engagement in civic and political life. *New Media & Society, 18*(9), 1840–1856. doi:10.1177/1461444815616222

Bowyer, B., & Kahne, J. (2020). The digital dimensions of civic education: Assessing the effects of learning opportunities. *Journal of Applied Developmental Psychology, 69*, 101162. doi:10.1016/j.appdev.2020.101162

Buckingham, D. (2017). Media theory 101: AGENCY. *Journal of Media Literacy, 64*(1-2), 12–15.

Cahill, H., & Dadvand, B. (2018). Re-conceptualising youth participation: A framework to inform action. *Children and Youth Services Review, 95*, 243–253. doi:10.1016/j.childyouth.2018.11.001

Cahill, H., Wyn, J., & Borovica, T. (2019). Youth participation informing care in hospital settings. *Child and Youth Services, 40*(2), 140–157. doi:10.1080/0145935X.2019.1567323

Checkoway, B., & Aldana, A. (2013). Four forms of youth civic engagement for diverse democracy. *Children and Youth Services Review, 35*(11), 1894–1899. doi:10.1016/j.childyouth.2013.09.005

Chen, J. (2017). Can online social networks foster young adults' civic engagement? *Telematics and Informatics, 34*(5), 487–497. doi:10.1016/j.tele.2016.09.013

Cicognani, E., Albanesi, C., Mazzoni, D., Prati, G., & Zani, B. (2016). Explaining offline and online civic engagement intentions between Italian and migrant youth. *Revista de Psicología Social, 1*(2), 282–316. doi:10.1080/02134748.2016.1143177

Clement, R., Deering, M., Mikhael, R., & Villa-García, C. (2014). *Youth civic engagement and leadership*. Retrieved from George Washington University website: https://elliott. gwu. edu/sites/elliott. gwu. edu/files/downloads/Youth% 20CE% 26L_FINAL. pdf.

Cumiskey, L., Hoang, T., Suzuki, S., Pettigrew, C., & Herrgård, M. M. (2015). Youth participation at the third UN world conference on disaster risk reduction. *International Journal of Disaster Risk Science, 6*(2), 150–163. doi:10.100713753-015-0054-5

Cushing, D. F. (2014). Promoting youth participation in communities through youth master planning. *Community Development (Columbus, Ohio), 46*(1), 43–55. doi:10.1080/15575330.2014.975139

Davies, I., Evans, M., Fülöp, M., Kiwan, D., Peterson, A., & Sim, J. B. Y. (2019). *Taking action for change: Educating for youth civic engagement and activism*. University of York.

Davison, K., & Russell, J. (2017). Disused religious space: Youth participation in built heritage regeneration. *Religions, 8*(6), 107. doi:10.3390/rel8060107

Dickson-Hoyle, S., Kovacevic, M., Cherbonnier, M., & Nicholas, K. A. (2018). Towards meaningful youth participation in science-policy processes: A case study of the Youth in Landscapes Initiative. *Elem Sci Anth, 6*, 67. doi:10.1525/elementa.327

do Valle Santos, W. C., Singh, D., Delgado Leandro da Cruz, L., de Carvalho Piassi, L. P., & Reis, G. (2019). Vertical gardens: Sustainability, youth participation, and the promotion of change in a socio-economically vulnerable community in Brazil. *Education Sciences, 9*(3), 161. doi:10.3390/educsci9030161

El Marzouki, M. (2016). Citizens of the margin: Citizenship and youth participation on the Moroccan social web. *Information Communication and Society*, *21*(1), 147–161. doi:10.1080/1369118X.2016.1266373

Evans, J., & Benefield, P. (2001). Systematic reviews of educational research: Does the medical model fit? *British Educational Research Journal*, *27*(5), 527–541. doi:10.1080/01411920120095717

Fernández, G., & Shaw, R. (2015). Youth participation in disaster risk reduction through science clubs in the Philippines. *Disasters*, *39*(2), 279–294. doi:10.1111/disa.12100 PMID:25440993

Fowler, A. (2017). Does voter preregistration increase youth participation? *Election Law Journal: Rules. Politics & Policy*, *16*(4), 485–494. doi:10.1089/elj.2017.0436

Gaby, S. (2017). The civic engagement gap(s): Youth participation and inequality from 1976 to 2009. *Youth & Society*, *49*(7), 923–946. doi:10.1177/0044118X16678155

Gal, T. (2017). An ecological model of child and youth participation. *Children and Youth Services Review*, *79*, 57–64. doi:10.1016/j.childyouth.2017.05.029

García, P., Fernández, C. H., & Jackson, A. (2019). Counternarratives of youth participation among black girls. *Youth & Society*, *00*(0), 1–22. doi:10.1177/0044118X18824738

Gil de Zúñiga, H. (2015). European public sphere| toward a European public sphere? The promise and perils of modern democracy in the age of digital and social media. *International Journal of Communication*, *9*, 3152–3160.

Goessling, K. P. (2017). Resisting and reinforcing neoliberalism: Youth activist organizations and youth participation in the contemporary Canadian context. *Mind, Culture, and Activity*, *24*(3), 199–216. doi:10.1080/10749039.2017.1313278

Gökçe-Kızılkaya, S., & Onursal-Beşgül, Ö. (2017). Youth participation in local politics: City councils and youth assemblies in Turkey. *Southeast European and Black Sea Studies*, *17*(1), 97–112. doi:10.1080/14683857.2016.1244239

Goodman, J. (2002). *La educación democrática en la escuela*. MCEP.

Gozzo, S., & Sampugnaro, R. (2016). What matters? Changes in European youth participation. *Partecipazione e Conflitto*, *9*(3), 748–776. doi:10.1285/i20356609v9i3p748

Grimm, R., Fox, C., Baines, S., & Albertson, K. (2013). Social innovation, an answer to contemporary societal challenges? Locating the concept in theory and practice. *Innovation (Abingdon)*, *26*(4), 436–455. doi:10.1080/13511610.2013.848163

Harris, A., Wyn, J., & Younes, S. (2010). Beyond apathetic or activist youth: 'Ordinary' young people and contemporary forms of participation. *Young*, *18*(1), 9–32. doi:10.1177/110330880901800103

Heinrich, A. J., & Million, A. (2016). Young people as city builders: Youth participation in German Municipalities. *disP-The Planning Review*, *52*(1), 56-71. DOI: doi:10.1080/02513625.2016.1171049

Henn, M., & Foard, N. (2014). Social differentiation in young people's political participation: The impact of social and educational factors on youth political engagement in Britain. *Journal of Youth Studies*, *17*(3), 360–380. doi:10.1080/13676261.2013.830704

Ho, E., Clarke, A., & Dougherty, I. (2015). Youth-led social change: Topics, engagement types, organizational types, strategies, and impacts. *Futures, 67*, 52–62. doi:10.1016/j.futures.2015.01.006

Karakos, H. L. (2015). *Understanding civic engagement among youth in diverse contexts*. Vanderbilt University.

Kim, Y., Russo, S., & Amnå, E. (2017). The longitudinal relation between online and offline political participation among youth at two different developmental stages. *New Media & Society, 19*(6), 899–917. doi:10.1177/1461444815624181

Krauss, S. E. (2018). Towards enhancing youth participation in Muslim-Majority countries: The Case of youth-adult partnership in Malaysia. *Pertanika Journal of Social Science & Humanities, 26*, 165–188.

Kwon, S. A. (2019). The politics of global youth participation. *Journal of Youth Studies, 22*(7), 926–940. doi:10.1080/13676261.2018.1559282

Määttä, M., & Aaltonen, S. (2016). Between rights and obligations–rethinking youth participation at the margins. *The International Journal of Sociology and Social Policy, 36*(3/4), 157–172. doi:10.1108/IJSSP-09-2014-0066

Mainsah, H., Brandtzæg, P. B., & Følstad, A. (2016). Bridging the generational culture gap in youth civic engagement through social media: Lessons learnt from young designers in three civic organizations. *The Journal of Media Innovations, 3*(1), 23–40. doi:10.5617/jmi.v3i1.2724

Marshall, K. (2018). Global education challenges: Exploring religious dimensions. *International Journal of Educational Development, 62*, 184–191. doi:10.1016/j.ijedudev.2018.04.005

Mascheroni, G. (2017). A practice-based approach to online participation: Young people's participatory habitus as a source of diverse online engagement. *International Journal of Communication, 11*, 4630–4651.

McCowan, T. (2009). A 'seamless enactment' of citizenship education. *Journal of Philosophy of Education, 43*(1), 85–99. doi:10.1111/j.1467-9752.2009.00669.x

Meinzen-Dick, R., Di Gregorio, M., & McCarthy, N. (2004). *Methods of Studying Collective Action in Rural Development*. International Food Policy Research Institute. doi:10.1016/j.agsy.2004.07.006

Melles, M. O., & Ricker, C. L. (2017). Youth participation in HIV and sexual and reproductive health decision-making, policies, programs: Perspectives from the field. *International Journal of Adolescence and Youth, 23*(2), 159–167. doi:10.1080/02673843.2017.1317642

Metzger, M., Erete, S., Barton, D. L., Desler, M. K., & Lewis, D. A. (2014). The new political voice of young Americans: Online engagement and civic development among first-year college students. *Education, Citizenship and Social Justice, 10*(1), 55–66. doi:10.1177/1746197914558398

Mollica, C. (2017). The diversity of identity: Youth participation at the Solomon Islands Truth and Reconciliation Commission. *Australian Journal of International Affairs, 71*(4), 371–388. doi:10.1080/10357718.2017.1290045

Morciano, D., Scardigno, F., Manuti, A., & Pastore, A. (2016). A theory-based evaluation to improve youth participation in progress: A case study of a youth policy in Italy. *Child and Youth Services*, *37*(4), 304–324. doi:10.1080/0145935X.2015.1125289

Narksompong, J., & Limjirakan, S. (2015). Youth participation in climate change for sustainable engagement. *Review of European, Comparative & International Environmental Law*, *24*(2), 171–181. doi:10.1111/reel.12121

Nwachukwu, P. T., & Kang'ethe, S. M. (2014). An insightful assessment on impact of local economic development programs in South Africa and Nigeria: Linking socio-cooperative connections and youths' participation. *Mediterranean Journal of Social Sciences*, *5*(23), 1621–1621. doi:10.5901/mjss.2014.v5n23p1621

Onyechi, N. J. (2018). Taking their destiny in their hands: Social media, youth participation and the 2015 political campaigns in Nigeria. *African Journalism Studies*, *39*(1), 69–89. doi:10.1080/23743670.2018.1434998

Orloff, A. S., & Palier, B. (2009). The power of gender perspectives: Feminist influence on policy paradigms, social science, and social politics. *Social Politics*, *16*(4), 405–412. doi:10.1093p/jxp021

Percy-Smith, B., McMahon, G., & Thomas, N. (2019). Recognition, inclusion and democracy: Learning from action research with young people. *Educational Action Research*, *27*(3), 347–361. doi:10.1080/09650792.2019.1577149

Pirro, A. L., & Róna, D. (2019). Far-right activism in Hungary: Youth participation in Jobbik and its network. *European Societies*, *21*(4), 603–626. doi:10.1080/14616696.2018.1494292

Poteat, V. P., Calzo, J. P., & Yoshikawa, H. (2018). Gay-Straight Alliance involvement and youths' participation in civic engagement, advocacy, and awareness-raising. *Journal of Applied Developmental Psychology*, *56*, 13–20. doi:10.1016/j.appdev.2018.01.001 PMID:29805190

Preskill, H., & Beer, T. (2012). *Evaluating social innovation*. Centre for Evaluation Innovation. doi:10.22163/fteval.2012.119

Purdue, S., Peterson, H., & Deng, C. (2018). The case for greater youth participation in monitoring and evaluation in international development. *Evaluation Journal of Australasia*, *18*(4), 206–221. doi:10.1177/1035719X18804401

Rodrigues, M., Menezes, I., & Ferreira, P. D. (2018). Effects of socialization on scout youth participation behaviors. *Educação e Pesquisa*, *44*, 1–16. doi:10.15901678-4634201844175560

Rodríguez, L. F., & Brown, T. M. (2009). From voice to agency: Guiding principles for participatory action research with youth. *New Directions for Youth Development*, *123*(123), 19–34. doi:10.1002/yd.312 PMID:19830799

Roque, R., Dasgupta, S., & Costanza-Chock, S. (2016). Children's civic engagement in the scratch online community. *Social Sciences*, *5*(4), 1–17. doi:10.3390ocsci5040055

Sahimi, A., Hamizah, N.A., Suandi, T., Ismail, I.A., & Hamzah, S.R. (2018). Profiling youth participation in volunteer activities in Malaysia: understanding the motivational factors influencing participation in volunteer work among malaysian youth. *Pertanika Journal of Social Sciences & Humanities, 26*(T), 49-62.

Schelbe, L., Chanmugam, A., Moses, T., Saltzburg, S., Williams, L. R., & Letendre, J. (2015). Youth participation in qualitative research: Challenges and possibilities. *Qualitative Social Work: Research and Practice, 14*(4), 504–521. doi:10.1177/1473325014556792

Schusler, T., Krings, A., & Hernández, M. (2019). Integrating youth participation and ecosocial work: New possibilities to advance environmental and social justice. *Journal of Community Practice, 27*(3-4), 460–475. doi:10.1080/10705422.2019.1657537

Sebti, A., Buck, M., Sanzone, L., Liduke, B. B., Sanga, G. M., & Carnevale, F. A. (2019). Child and youth participation in sexual health-related discussions, decisions, and actions in Njombe, Tanzania: A focused ethnography. *Journal of Child Health Care, 23*(3), 370–381. doi:10.1177/1367493518823920 PMID:30669864

Seyfang, G., & Smith, A. (2007). Grassroots Innovations for Sustainable Development: Towards a New Research and Policy Agenda. *Environmental Politics, 16*(4), 584–603. doi:10.1080/09644010701419121

Seymour, K., Bull, M., Homel, R., & Wright, P. (2017). Making the most of youth development: Evidence-based programs and the role of young people in research. *Queensland Review, 24*(1), 147–162. doi:10.1017/qre.2017.17

Shaw, A., Brady, B., McGrath, B., Brennan, M. A., & Dolan, P. (2014). Understanding youth civic engagement: Debates, discourses, and lessons from practice. *Community Development (Columbus, Ohio), 45*(4), 300–316. doi:10.1080/15575330.2014.931447

Shiratuddin, N., Hassan, S., Mohd Sani, M. A., Ahmad, M. K., Khalid, K. A., Abdull Rahman, N.L., Abd Rahman, Z.S., & Ahmad, N.S. (2017). Media and youth participation in social and political activities: development of a survey instrument and its critical findings. *Pertanika Journal of Social Sciences & Humanities, 25*(S), 1-20.

Sika, N. (2018). Civil society and the rise of unconventional modes of youth participation in the MENA. *Middle East Law and Governance, 10*(3), 237–263. doi:10.1163/18763375-01003002

Simoes, J. A., & Campos, R. (2017). Digital media, subcultural activity and youth participation: The cases of protest rap and graffiti in Portugal. *Journal of Youth Studies, 20*(1), 16–31. doi:10.1080/13676261.2016.1166190

Soler-i-Martí, R., & Ferrer-Fons, M. (2015). Youth participation in context: The impact of youth transition regimes on political action strategies in Europe. *The Sociological Review, 63*(2_suppl), 92–117. doi:10.1111/1467-954X.12264

Themistokleous, S., & Avraamidou, L. (2016). The role of online games in promoting young adults' civic engagement. *Educational Media International, 53*(1), 53–67. doi:10.1080/09523987.2016.1192352

Thew, H. (2018). Youth participation and agency in the United Nations framework convention on climate change. *International Environmental Agreement: Politics, Law and Economics*, *18*(3), 369–389. doi:10.100710784-018-9392-2

Thomas, G., & Pring, R. (2004). *Evidence-based practice in education*. Open University Press.

Trilla, J., & Novella, A. M. (2011). Participación, democracia y formación para la ciudadanía: Los consejos de infancia. *Review of Education*, *356*, 26–46.

Tsekoura, M. (2016a). Debates on youth participation: From citizens in preparation to active social agents. *Revista Katálysis*, *19*(1), 118–125. doi:10.1590/1414-49802016.00100012

Tsekoura, M. (2016b). Spaces for youth participation and youth empowerment: Case studies from the UK and Greece. *Young*, *24*(4), 326–341. doi:10.1177/1103308815618505

Tzankova, I., & Cicognani, E. (2019). Youth participation in psychological literature: A semantic analysis of scholarly publications in the PsycInfo Database. *Europe's Journal of Psychology*, *15*(2), 276–291. doi:10.5964/ejop.v15i2.1647 PMID:33574955

Ulitin, A., Mier-Alpaño, J. D., Labarda, M., Juban, N., Mier, A. R., Tucker, J. D., & Chan, P. L. (2022). Youth social innovation during the COVID-19 pandemic in the Philippines: A quantitative and qualitative descriptive analyses from a crowdsourcing open call and online hackathon. *BMJ Innovations*, *0*(3), 1–8. doi:10.1136/bmjinnov-2021-000887

Uršič, M., Dekker, K., & Filipovič Hrast, M., (2014). Spatial organization and youth participation: case of the University of Ljubljana and Tokyo Metropolitan University. *Annales, Series historia et sociología*, *24*(3), 433-450.

Van Haperen, S., Nicholls, W., & Uitermark, J. (2018). Building protest online: Engagement with the digitally networked #not1more protest campaign on Twitter. *Social Movement Studies*, *17*(4), 408–423. doi:10.1080/14742837.2018.1434499

Vázquez, J. L., Lanero, A., Gutiérrez, P., & García, M. P. (2015). Expressive and instrumental motivations explaining youth participation in non-profit voluntary associations: An application in Spain. *International Review on Public and Nonprofit Marketing*, *12*(3), 237–251. doi:10.100712208-015-0128-5

Warren, A. M., Sulaiman, A., & Jaafar, N. I. (2014). Social media effects on fostering online civic engagement and building citizen trust and trust in institutions. *Government Information Quarterly*, *31*(2), 291–301. doi:10.1016/j.giq.2013.11.007

Watts, R. J., & Hipólito-Delgado, C. P. (2015). Thinking ourselves to liberation? Advancing sociopolitical action in critical consciousness. *The Urban Review*, *47*(5), 847–867. doi:10.100711256-015-0341-x

Youniss, J., Bales, S., Christmas-Best, V., Diversi, M., Mclaughlin, M., & Silbereisen, R. (2002). Youth civic engagement in the twenty-first century. *Journal of Research on Adolescence*, *12*(1), 121–148. doi:10.1111/1532-7795.00027

Zhang, X., & Lin, W. Y. (2018). Hanging together or not? Impacts of social media use and organizational membership on individual and collective political actions. *International Political Science Review*, *39*(2), 273–289. doi:10.1177/0192512116641842

KEY TERMS AND DEFINITIONS

Agency: The ability of people to make decisions and participate in issues related to their own lives and the community in which they live, in contexts where they can be heard.

Co-Productive Interactivity: Characteristics of digital tools that allow for the easy management of content and active intervention in virtual spaces.

Informal Structures of Participation: Symbolic spaces where people can participate that are not regulated by formal or clearly structured rules.

Meta-Participation: Competence that allows the understanding of how participatory processes are being developed (content, format, actors, etc.).

Meta-Research: Systematic review work aimed at understanding how research is developing in a particular field of knowledge.

Online Civic Engagement: Strategies that people are using to participate through the various applications provided by the internet.

Youth Civic Engagement: The development of knowledge that allows young people to perceive themselves as being part of participation processes in common areas of their civil life.

Compilation of References

Aboelmaged, M., & Mouakket, S. (2020). Influencing models and determinants in big data analytics research: A bibliometric analysis. *Information Processing & Management*, 57(4), 102234. doi:10.1016/j.ipm.2020.102234

Aboelmaged, M., & Subbaugh, S. (2012). Factors influencing perceived productivity of Egyptian teleworkers: An empirical study. *Measuring Business Excellence*, 16(2), 3–22. doi:10.1108/13683041211230285

Abu-Elezz, I., Hassan, A., Nazeemudeen, A., Househ, M., & Abd-Alrazaq, A. (2020). The benefits and threats of blockchain technology in healthcare: A scoping review. *International Journal of Medical Informatics*, 142, 104246. doi:10.1016/j.ijmedinf.2020.104246 PMID:32828033

Abu-Shanab, E., & Harb, Y. (2019). E-government research insights: Text mining analysis. *Electronic Commerce Research and Applications*, 38(C), 2019. doi:10.1016/j.elerap.2019.100892 12p

Acs, Z. J., Audretsch, D. B., Braunerhjelm, P., & Carlsson, B. (2004). *The missing link: The knowledge filter and entrepreneurship in endogenous growth*. (Working Paper 4783). London Center for Economic Policy Research.

Acs, Z., & Armington, C. (2010). The Determinants of Regional Variation in New Firm Formation. *Regional Studies*, 36(1), 33–45. doi:10.1080/00343400120099843

Adams, R., Jeanrenaud, S., Denyer, D., & Overy, P. (2016). Sustainability oriented innovation: A systematic review. *International Journal of Management Reviews*, 18(2), 180–205. doi:10.1111/ijmr.12068

Aderaldo, I., Aderaldo, C., & Lima, A. (2017). Aspectos críticos do teletrabalho mu ma companhia multinacional [Critical aspects of teleworking in a multinational company]. *Cadernos EBAPE.BR*, 15(8), 511–533. doi:10.1590/1679-395160287

Adler, R. P., & Goggin, J. (2005). What do we mean by "civic engagement"? *Journal of Transformative Education*, 3(3), 236–253. doi:10.1177/1541344605276792

Adolph, S., & Bramble, P. (2003). *Patterns for Effective Use Cases*. Addison-WesleyPearson Education.

Agarwal, G. (2022, October 26). Moving closer to sustainability. *Precious Kashmir*. https://preciouskashmir.com/2022/10/26/moving-closer-to-sustainability/

Ahmed, A., Abubakari, Z., & Gasparatos, A. (2019). Labelling large-scale land acquisitions as land grabs: Procedural anddistributional considerations from two cases in Ghana. *Geoforum*, 105, 191–205. doi:10.1016/j.geoforum.2019.05.022

Ahuja, A. (2021, Nov 21). Waste Warriors of India. *Swach India*. www.swachindia.ndtv.com: https://swachhindia.ndtv.com/climate-warrior-23-year-old-recycles-10-plastic-bags-and-12-plastic-bottles-into-a-pair-of-sneakers-64173/

Compilation of References

Ajmal, M. E., Khan, M., Hussain, M., & Helo, P. (2018). Conceptualizing and incorporating social sustainability in the business. *International Journal of Sustainable Development and World Ecology*, 25(4), 327–339. doi:10.1080/13504509.2017.1408714

Åkerström, J., Aytar, O., & Brunnberg, E. (2015). Intra-and inter-generational perspectives on youth participation in Sweden: A study with young people as research partners. *Children & Society*, 29(2), 134–145. doi:10.1111/chso.12027

Aktas, F., Ceken, C., & Erdemli, Y. E. (2018). IoT-based healthcare framework for biomedical applications. *Journal of Medical and Biological Engineering*, 38(6), 966–979. doi:10.100740846-017-0349-7

Albino, V., Ardito, L., Dangelico, R. M., & Messeni Petruzzelli, A. (2014). Understanding the development trends of low-carbon energy technologies: A patent analysis. *Applied Energy*, 135, 836–854. doi:10.1016/j.apenergy.2014.08.012

Al-Breiki, M., & Bicer, Y. (2021, January 10). Comparative life cycle assessment of sustainable energy carriers including production, storage, overseas transport and utilization. *Journal of Cleaner Production*, 279, 1–16. doi:10.1016/j.jclepro.2020.123481

Aldrich, H., & Auster, E. R. (1986). Even dwarfs started small: Liabilities of age and size and their strategic implications. In B. M. Staw & L. L. Cummings (Eds.), *Research in Organizational Behavior* (Vol. 8, pp. 165–198). JAI Press.

Allen, B., Tamindael, L. E., Bickerton, S. H., & Cho, W. (2020). Does citizen coproduction lead to better urban services in smart cities projects? An empirical study on e-participation in a mobile big data platform. *Government Information Quarterly*, 37(1), 101412. doi:10.1016/j.giq.2019.101412

Almazroa, H., Alotaibi, W., & Alrwaythi, E. (2022). Sustainable Development Goals and Future-Oriented Teacher Education Programs. *IEEE Transactions on Engineering Management*, 1–14. doi:10.1109/TEM.2022.3165686

Alrabaiah, H. A., & Medina-Medina, N. (2021). Agile Beeswax: Mobile app development process and empirical study in real environment. *Sustainability*, 13(4), 1909. doi:10.3390u13041909

Álvarez-Icaza, L., Bustamante-Bello, R., Ramírez-Montoya, M.S., & Molina, A. (2002). Systematic Mapping of Digital Gap and Gender, Age, Ethnicity, or Disability. *Sustainability*, 14(1297). doi:10.3390/su14031297

Alves, R. V. (2010). *Estudo de caso da comercialização dos produtos florestais não madeireiros (PFNM) como subsídio para restauração florestal.* [*Case study of the commercialization of non-timber forest products (NMF) as a subsidy for forest restoration.*] [Master's thesis, Universidade Federal de Viçosa].

Alzamora, F. M., Carot, M. H., Carles, J., & Campos, A. (2019). Development and Use of a Digital Twin for the Water Supply and Distribution Network of Valencia (Spain). *Water Quality Models for Water Distribution Systems*. Pilar Conejos Aguas de Valencia.

Amabile, T., Hadley, C., & Kramer, S. (2002). Creativity under the gun. *Harvard Business Review*, 80(8), 52–61. PMID:12195920

Amar, I. A., Petit, C. G., & Tao, S. (2011). *Solid-state electrochemical synthesis of ammonia: a review*. Solid State Electrochem. doi:10.100710008-011-1376-x

Amburgey, T., & Rao, H. (1996). Organizational ecology: Past, present, and future directions. *Academy of Management Journal*, 39(5), 1265–1286. doi:10.2307/256999

Amná, E. (2012). How is civic engagement developed over time? Emerging answers from a multidisciplinary field. *Journal of Adolescence*, 35(3), 611–627. doi:10.1016/j.adolescence.2012.04.011 PMID:22613043

Amnå, E., & Ekman, J. (2014). Standby citizens: Diverse faces of political passivity. *European Political Science Review*, *6*(2), 261–281. doi:10.1017/S175577391300009X

Amore, M. D., & Bennedsen, M. (2016). Corporate governance and green innovation. *Journal of Environmental Economics and Management*, *75*, 54–72. doi:10.1016/j.jeem.2015.11.003

ANACOM. (2019). *Comunicações eletrónicas - 2.º e 3.º trimestres de 2021*. [Electronic Communications - 2nd and 3rd quarters of 2021.] https://www.anacom.pt: https://www.anacom.pt/render.jsp?categoryId=370555

Anadol, Y., & Behery, M. (2020). Humanistic leadership in the UAE context. *Cross Cultural & Strategic Management*, *27*(4), 645–664. doi:10.1108/CCSM-01-2020-0023

Anchor Disposal. (2020, April). Anchor Disposal. www.anchordisposal.com: https://anchordisposal.com/news/2020/4/1/whats-the-difference-between-upcycling-and-recycling

Andersen, M. S. (2007). An introductory note on the environmental economics of the circular economy. *Sustainability Science*, *2*(1), 133–140. doi:10.100711625-006-0013-6

Andolina, M. W., Jenkins, K., Keeter, S., & Zukin, C. (2002). Searching for the meaning of youth civic engagement: Notes from the field. *Applied Developmental Science*, *6*(4), 189–195. doi:10.1207/S1532480XADS0604_5

Andreotti, V. (2006). *Soft versus critical global citizenship education in development education: Policy and practice*. Centre for the Study of Social and Global Justice, Nottingham University.

Andrew, C., & Klein, J. L. (2010). *Social Innovation: What is it and why is it important to understand it better* (Vol. ET10003). Ministry of Research and Innovation.

Android. (2022). *Android*. https://www.android.com/intl/pt-BR_br/

Angraal, S., Krumholz, H. M., & Schulz, W. L. (2017). Blockchain technology: Applications in health care. *Circulation: Cardiovascular Quality and Outcomes*, *10*(9), e003800. doi:10.1161/CIRCOUTCOMES.117.003800 PMID:28912202

Anissimov, M. (2022). What are Different Types of Synthetic Fuels? *About Mechanics*. https://www.aboutmechanics.com/what-are-different-types-of-synthetic-fuels.htm

Anjum, H. F., Rasid, S. Z. A., Khalid, H., Alam, M. M., Daud, S. M., Abas, H., Sam, S. M., & Yusof, M. F. (2020). Mapping research trends of blockchain technology in healthcare. ieee. *Access*, *8*, 174244–174254. doi:10.1109/ACCESS.2020.3025011

Antiroikko, A. V., Malkia, M., & Savolainen, R. E. (2004). *Transformation in Governance: New Directions in Government and Politics*. Idea Group Publishing.

Antonioli, D., Mancinelli, S., & Mazzanti, M. (2013). Is environmental innovation embedded within high-performance organizational changes? The role of human resource management and complementarity in green business strategies. *Research Policy*, *42*(4), 975–988. doi:10.1016/j.respol.2012.12.005

Antunes, R. (2015). *A sociedade da terceirização total* [The total outsourcing society.]. Revista da ABET.

APA. (2022). *Waste*. Relatório do Estado do Ambiente. https://rea.apambiente.pt/environment_area/waste?language=en

Aras, G., & Crowther, D. (2008). Governance and sustainability: An investigation into the relationship between corporate governance and corporate sustainability. *Management Decision*, *46*(3), 433–448. doi:10.1108/00251740810863870

Compilation of References

Araújo, C. A. (2006). Bibliometria: Evolução histórica e questões atuais [Bibliometrics: Historical evolution and current issues]. *Questao*, *12*(1), 11–32.

Araújo, E. R., & Bento, S. C. (2002). *Teletrabalho e Aprendizagem: Contributos para uma problematização [Telework and Learning: Contributions to a problematization]*. Fundação Calouste Gulbenkian.

Arnstein, S. R. (1969). A Ladder Of Citizen Participation. *Journal of the American Institute of Planners*, *35*(4), 216–224. doi:10.1080/01944366908977225

Arruda, D., Soares, R., Vieira, D., Ferreira, R., Cabral, T., & Lencastre, M. (2014). Engenharia de Requisitos: Um Survey Realizado no Porto Digital, Recife/Brasil. [Requirements Engineering: A Survey Conducted in Porto Digital, Recife/Brazil.] Recife.

Arshad, S., & Khurram, S. (2020). Can government's presence on social media stimulate citizens' online political participation? Investigating the influence of transparency, trust, and responsiveness. *Government Information Quarterly*, *37*(3), 101486. doi:10.1016/j.giq.2020.101486

Arunkumar, K., Bowman, D. D., Coen, S. E., El-Bagdady, M. A., Ergler, C. R., Gilliland, J. A., & Paul, S. (2019). Conceptualizing youth participation in children's health research: Insights from a youth-driven process for developing a youth advisory council. *Children (Basel, Switzerland)*, *6*(1), 3. doi:10.3390/children6010003 PMID:30597913

Arvanitidis, P. A. (2017). The concept and determinants of civic engagement. *Human Affairs*, *27*(3), 252–272. doi:10.1515/humaff-2017-0022

Aswani, J., Chidambaran, N. K., & Hasan, I. (2021). Who benefits from mandatory CSR? Evidence from the Indian Companies Act 2013. *Emerging Markets Review*, *46*, 100753. doi:10.1016/j.ememar.2020.100753

ATAG. (2020, 9). Aviation Benefits beyond borders. *Global Fact Sheet*.

Audretsch, D. B. (2007). Entrepreneurship capital and economic growth. *Oxford Review of Economic Policy*, *23*(1), 63–78. doi:10.1093/oxrep/grm001

Audretsch, D. B., Belitski, M., Caiazza, R., Günther, C., & Menter, M. (2021). From latent to emergent entrepreneurship: The importance of context. *Technological Forecasting and Social Change*, 121356.

Audretsch, D. B., & Keilbach, M. (2005). Entrepreneurship capital and regional growth. *The Annals of Regional Science*, *39*(3), 457–469. doi:10.100700168-005-0246-9

Audretsch, D., & Fritsch, M. (2002). Growth Regimes over Time and Space. *Regional Studies*, *36*(2), 113–124. doi:10.1080/00343400220121909

Augsberger, A., Springwater, J. S., Hilliard-Koshinsky, G., Barber, K., & Martínez, L. S. (2019). Youth participation in policy advocacy: Examination of a multi-state former and current foster care youth coalition. *Children and Youth Services Review*, *107*, 104491. doi:10.1016/j.childyouth.2019.104491

Aujla, G. S., & Jindal, A. (2020). A decoupled blockchain approach for edge-envisioned IoT-based healthcare monitoring. *IEEE Journal on Selected Areas in Communications*, *39*(2), 491–499. doi:10.1109/JSAC.2020.3020655

Austria Glasrecycling. (2022). *Glas entsorgen. [Dispose of glass]*. AGR.. https://www.agr.at/glasrecycling/glas-entsorgen

Austrian Government. (2022). *Abfallentsorgung/Müllabfuhr.* [*Waste disposal/garbage collection*] Österreichs digitales Amt. oesterreich.gv.at https://www.oesterreich.gv.at/themen/bauen_wohnen_und_umwelt/umzug/5/Seite.180301.html

Austro Papier. (2022). *Positionen—Altpapier.* Austropapier. https://austropapier.at/positionen-altpapier/

Avdikos, V. (2014). Οι Πολιτιστικές και Δημιουργικές Βιομηχανίες στην Ελλάδα [The Cultural and Creative Industries in Greece]. Epikentro Ed. Athens.

Ávila, L. V., Hoffmann, C., Corrêa, A. C., da Rosa Gama Madruga, L. R., Schuch Júnior, V. F., de Sousa Júnior, A. F., & Zanini, R. R. (2013). Social Responsibility Initiatives Using ISO 26000: An Analysis from Brazil. *Environmental Quality Management, 23*(2), 15–30. doi:10.1002/tqem.21362

Ayres, R. U., van den Bergh, J. C. J. M., & Gowdy, J. M. (2001). Strong versus Weak Sustainability: Economics, Natural Sciences, and "Consilience". *Environmental Ethics, 23*(2), 155–168. doi:10.5840/enviroethics200123225

Ayuso, M. D., & Martínez, V. (2005). Gobierno electrónico. Contenidos y organización de las sedes webs de los parlamentos autonómicos. [E-government. Contents and organization of the headquarters websites of the autonomous parliaments.]. *Revista Espanola la de Documentacion Cientifica, 28*(4), 462–478.

Aziz, M., Wijayanta, A. T., & Nandiyanto, A. D. (2020). Ammonia as Effective Hydrogen Storage: A Review on Production, Storage and Utilization. *Energies*, (2022).

Aziz, M., Oda, T., Morihara, A., & Kashiwagi, T. (2017). Combine4d nitrogen production, ammonia synthesis, and power generation for efficient hydrogen dtorage. *Energy Procedia, 143*, 674–679. doi:10.1016/j.egypro.2017.12.745

Azogu, I., Norta, A., Papper, I., Longo, J., & Draheim, D. (2019, April). A framework for the adoption of blockchain technology in healthcare information management systems: A case study of Nigeria. In *Proceedings of the 12th International Conference on Theory and Practice of Electronic Governance* (pp. 310-316). ACM. 10.1145/3326365.3326405

Azzawi, M. A., Hassan, R., & Bakar, K. A. A. (2016). A review on Internet of Things (IoT) in healthcare. *International Journal of Applied Engineering Research, 11*(20), 10216–10221.

Babiniotis, G. (2002). Λεξικό της Νέας Ελληνικής Γλώσσας [Dictionary of the Modern Greek Language]. Lexicology Center. 2nd ed. Athens

Bahl, V. (2012), *Murder Capital to Modern Miracle? The Progression of Governance in Medellin.* Colombia. Development Planning Unit. The Bartlett, University College London. https://opendocs.ids.ac.uk/opendocs/bitstream/handle/20.500.12413/11792/Murder_capital.pdf?sequence=1&isAllowed=y (5/4/2022)

Bailey, D., & Kurland, N. (2002). A Review of Telework Research: Findings, New Directions, and Lessons for the Study of Modern Work. *Journal of Organizational Behavior, 23*(4), 383–400. doi:10.1002/job.144

Baines, T., Brown, S., Benedettini, O., & Ball, P. (2012). Examining green production and its role within the competitive strategy of manufacturers. *Journal of Industrial Engineering and Management, 5*(1), 53–87. doi:10.3926/jiem.405

Bai, Y., Song, S., Jiao, J., & Yang, R. (2019). The impacts of government R&D subsidies on green innovation: Evidence from Chinese energy-intensive firms. *Journal of Cleaner., 233*(9), 819–829. doi:10.1016/j.jclepro.2019.06.107

Balsano, A. (2005). Youth civic engagement in the United States: Understanding and addressing the impact of social impediments on positive youth and community development. *Applied Developmental Science, 9*(4), 188–201. doi:10.12071532480xads0904_2

Banaji, S. (2008). The trouble with civic: A snapshot of young people's civic and political engagements in Twenty-first Century Democracies. *Journal of Youth Studies, 11*(5), 543–560. doi:10.1080/13676260802283008

Banaji, S., & Buckingham, D. (2016). Young people, the Internet, and civic participation: An overview of key findings from the Civic Web project. *International Journal of Learning and Media, 2*(1), 15–24. doi:10.1162/ijlm_a_00038

Banaji, S., Mejias, S., & de la Pava Velez, B. (2018). The significance of ethnography in youth participation research: Active citizenship in the UK after the Brexit vote. *Sociální studia*, *15*(2), 97–115. doi:10.5817/SOC2018-2-97

Banerjee, C., Bhaduri, A., & Saraswat, C. (2022). Digitalization in Urban Water Governance: Case Study of Bengaluru and Singapore. *Frontiers in Environmental Science*, *10*(March), 1–12. doi:10.3389/fenvs.2022.816824

Bannister, F., & Connolly, R. (2011). The trouble with transparency: A critical review of openness in e-government. *Policy and Internet*, *3*(1), 1–30. doi:10.2202/1944-2866.1076

Bansal, R. (2022, September 5). *Indianretailer.com*. www.indianretailer.com: https://www.indianretailer.com/article/retail-people/the-global-eco-wakening-why-sustainability-is-now-the-key-driver-of-innovation.a8153/#:~:text=The%20Global%20Eco%2Dwakening%3A%20Why,the%20Key%20Driver%20of%20Innovation

Barbosa, C. S., Silva-Jean, M., Luz, J. P., Leandro, G., & Bohn, D. P. (2020). Processo Produtivo do PFNM Pinhão das araucárias: O caso do extrativista JDZ no Rio Grande do Sul. [Pfnm Pinhão Production Process of araucarias: the case of the JDZ extractive in Rio Grande do Sul.]. *Revista de Administração e Negócios da Amazônia*, *2*(1), 4–17. doi:10.18361/2176-8366/rara.v12n1p4-17

Bard, J., Gerhardt, N., Selzam, P., Beil, M., Wiemer, M., & Buddensiek, M. (2022). *The limitations of hydrogen blending in the european gas grid*. IEE- Fraunhofer Institute for Energy Economics and Energy System Technology. https://www.iee.fraunhofer.de/content/dam/iee/energiesystemtechnik/en/documents/Studies-Reports/FINAL_FraunhoferIEE_ShortStudy_H2_Blending_EU_ECF_Jan22.pdf

Bardin, L. (1996). *El análisis de contenido* [Content analysis.]. Ediciones Akal Universitaria.

Bardin, L. (2011). *Content analysis* (5th ed.). Edições.

Barhate, B., & Dirani, K. M. (2022). Career aspirations of generation Z: A systematic literature review. *European Journal of Training and Development*, *46*(1/2), 139–157. doi:10.1108/EJTD-07-2020-0124

Barros, A. & Silva, J. (2010). Percepções dos indivíduos sobre as consequências do teletrabalho na configuração home-office: estudo de caso na Shell Brasil [Perceptions of individuals about the consequences of telework in the home-office configuration: A case study at Shell, Brazil]. *Cadernos EBAPE.BR*, *8*(1) – artigo 5, 71-91.

Barros, M. V., Salvador, R., de Francisco, A. C., & Piekarski, C. M. (2020). Mapping of research lines on circular economy practices in agriculture: From waste to energy. *Renewable & Sustainable Energy Reviews*, *131*, 131. doi:10.1016/j.rser.2020.109958

Barr, S., & Gilg, A. (2006). Estilos de vida sustentáveis: Enquadrando a ação ambiental dentro e ao redor da casa. [Sustainable lifestyles: framing environmental action in and around the house.]. *Geoforum*, *37*(6), 906–920. doi:10.1016/j.geoforum.2006.05.002

Bartholo, R., Ribeiro, H., & Bittencourt, J. (2002). *Ética e Sustentabilidade* [Ethics and Sustainability.]. Editora E-papers.

Baskerville, Baiyere, A., Gergor, S., Hevner, A., & Rossi, M. (2018). Design Science Research Contributions: Finding a Balance between Artifact and Theory. *Journal of the Association for Information Systems*, *19*(5), 358–376. doi:10.17705/1jais.00495

Batsleer, J., Thomas, N. P., & Pohl, A. (2019). Who knows? Youth work and the mise-en-scene: Reframing pedagogies of youth participation. *Pedagogy, Culture & Society*, *28*(2), 205–221. doi:10.1080/14681366.2019.1627484

Baumol, W. J. (1993). *Entrepreneurship, management, and the structure of payoffs*. New York University.

Baumol, W. J. (1996). Entrepreneurship: Productive, Unproductive and Destructive. *Journal of Business Venturing*, *11*(1), 3–22. doi:10.1016/0883-9026(94)00014-X

Beaverstock, J. V., Smith, R. G., & Taylor, P. J. (1999). A roster of World Cities. In *Cities, 16* (6), 445-458. https://www.sciencedirect.com/science/article/pii/S0264275199000426?via%3Dihub (2/4/2022)

Beccarello, M., & Di Foggia, G. (2018). Moving towards a circular economy: Economic impacts of higher material recycling targets. *Materials Today: Proceedings*, *5*(1), 531–543. doi:10.1016/j.matpr.2017.11.115

Beck, K., Beedle, M., Bennekum, A. v., Cockburn, A., Cunningham, W., & Fowler, M. (2001). Princípios do Manifesto Ágil. *Agile Manifesto*. https://agilemanifesto.org/iso/ptpt/principles.html

Becker, M. C. (Ed.). (2008). Handbook of organizational routines. Cheltenham, UK: Edward Elgar Publishing Berkes, F., Folke, C., Colding, J. (eds.) (2003). Navigating Social-Ecological Systems: Building Resilience for Complexity and Change, Cambridge University Press. doi:10.4337/9781848442702

Becker, C. U. (2012). *Sustainability Ethics and Sustainability Research*. Springer., doi:10.1007/978-94-007-2285-9

Becker, C., Chitchyan, C., Duboc, R., & Easterbrook, L. (2015). Sustainability Design and Software: The Karlskrona Manifesto. *37th International Conference on Software Engineering (ICSE 15)*. IEEE. 10.1109/ICSE.2015.179

Behling, O., & Rauch, C. F. Jr. (1985). A functional perspective on improving leadership effectiveness. *Organizational Dynamics*, *13*(4), 51–61. doi:10.1016/0090-2616(85)90005-1

Bell, B. T. (2019). You take fifty photos, delete forty-nine and use one: A qualitative study of adolescent image-sharing practices on social media. *International Journal of Child-Computer Interaction*, *20*, 64–71. doi:10.1016/j.ijcci.2019.03.002

Belzunegui, A., & Erro-Garcés, A. (2020). Teleworking in the Context of the Covid-19 Crisis. *Sustainability*, *12*(9), 3662. doi:10.3390u12093662

Ben Fekih, R., & Lahami, M. (2020, June). Application of blockchain technology in healthcare: a comprehensive study. In *International Conference on Smart Homes and Health Telematics* (pp. 268-276). Springer, Cham. 10.1007/978-3-030-51517-1_23

Bencsik, A., Horváth-Csikós, G., & Juhász, T. (2016). Y and Z Generations at Workplaces. *Journal of Competitiveness*, *6*(3), 90–106. doi:10.7441/joc.2016.03.06

Bersi, D. (2019). *Πορτοκαλί Οικονομία και Ανοιχτά Συστήματα Καινοτομίας: Η Περίπτωση των Hubs στους Κλάδους Πολιτισμού και [Orange Economy and Open Innovation Systems: The Case of Hubs in the Culture and Creativity Sectors]* [Unpublished master's thesis, Harokopio University. Athens]. https://tinyurl.com/yc3c7m75

Bessant, J. (2016). Democracy denied, youth participation and criminalizing digital dissent. *Journal of Youth Studies*, *19*(7), 921–937. doi:10.1080/13676261.2015.1123235

Beuren, I. M., & Angonese, R. (2015). Instruments for determining the disclosure index of accounting information. *Revista Eletrônica de Estratégia e Negócios – REEN*, *8*(1), pp. 120-144.

Bevill, K. (2008). Building the 'Minnesota Model. *Ethanol Producer Magazine*, 114-120.

Bezerra, M. M., Lima, E. C., Brito, F. W., & Santos, A. C. (2019). Geração Z: Relações de uma Geração Hipertecnológica e o Mundo do Trabalho [Generation Z: Relationships of a hypertechnological generation and the world of work]. *Revista Gestão em Análise [Management in Analysis Magazine]*, *8* (1), 136-149.

Compilation of References

Bharadwaj, H. K., Agarwal, A., Chamola, V., Lakkaniga, N. R., Hassija, V., Guizani, M., & Sikdar, B. (2021). A review on the role of machine learning in enabling IoT based healthcare applications. *IEEE Access : Practical Innovations, Open Solutions*, *9*, 38859–38890. doi:10.1109/ACCESS.2021.3059858

Bhatti, M., & Igreja, A. (2004). O lar, a cultura da natureza e os significados dos jardins na modernidade tardia. [The home, the culture of nature and the meanings of the gardens in late modernity.]. *Estudos de Habitação*, *19*(1), 37–51.

Bian, Y., Wu, L., Bai, J., & Yang, Y. (2021). Does factor market distortions inhibit green economic growth? *World Econ Papers*, *2*, 105–119.

Birnholtz, J. P., Cohen, M. D., & Hoch, S. V. (2007). Organizational character: On the regeneration of Camp Poplar Grove. *Organization Science*, *18*(2), 315–332. doi:10.1287/orsc.1070.0248

BMK. (2021). *Die Bestandsaufnahme der Abfallwirtschaft in Österreich—Statusbericht 2021*. [*Taking Stock of Waste Management in Austria—Status Report 2021*] Bundesministerium für Klimaschutz, Umwelt, Energie, Mobilität, Innovation und Technologie. https://www.bmk.gv.at/dam/jcr:04ca87f4-fd7f-4f16-81ec-57fca79354a0/BAWP_Statusbericht2021.pdf

BMK. (2022a). *Allgemeines zur Abfallwirtschaft*. [*General information on waste management.*]. Österreichs digitales Amt. oesterreich.gv.at - https://www.oesterreich.gv.at/themen/bauen_wohnen_und_umwelt/abfall/1/Seite.3790060.html

BMK. (2022b). *Europäischer Green Deal*. BMK. https://www.bmk.gv.at/themen/klima_umwelt/eu_international/euop_greendeal.html

BMK. (2022c). *Kunststoffabfälle in Österreich*. [*Plastic waste in Austria*.]. BMK. https://www.bmk.gv.at/themen/klima_umwelt/kunststoffe/kunststoffabfaelle.html

Boadu, E. S., & Ile, I. (2018). The politics of youth participation in social intervention programs in Ghana: Implications for participatory monitoring and evaluation (PM&E). *Journal of Reviews on Global Economics*, *7*, 913–925. doi:10.6000/1929-7092.2018.07.89

Bocken, N. M., De Pauw, I., Bakker, C., & Van Der Grinten, B. (2016). Product design and business model strategies for a circular economy. *Journal of industrial and production engineering*, *33*(5), 308-320.

Bocken, N. M. P., Allwood, J. M., Willey, A. R., & King, J. M. H. (2012). Development of a tool for rapidly assessing the implementation difficulty and emissions benefits of innovations. *Technovation*, *32*(1), 19–31. doi:10.1016/j.technovation.2011.09.005

Bocken, N. M. P., Short, S. W., Rana, P., & Evans, S. (2014). A literature and practice review to develop sustainable business model archetypes. *Journal of Cleaner Production*, *65*, 42–56. doi:10.1016/j.jclepro.2013.11.039

Body, A., & Hogg, E. (2019). What mattered ten years on? Young people's reflections on their involvement with a charitable youth participation project. *Journal of Youth Studies*, *22*(2), 171–186. doi:10.1080/13676261.2018.1492101

Boeing, P. (2016). The allocation and effectiveness of China's R&D subsidies—Evidence from listed firms. *Research Policy*, *45*(9), 1774–1789. doi:10.1016/j.respol.2016.05.007

Boeker, W. (1997). Strategic change: The influence of managerial characteristics and organizational growth. *Academy of Management Journal*, *40*(1), 152–170. doi:10.2307/257024

Boje, D., & Rosile, G. A. (2020). *How to use conversational storytelling interviews for your dissertation*. Camberley: Edward Elgar Publishing.

Boje, D. M. (2001). *Narrative methods for organizational & communication research*. Sage Publications. doi:10.4135/9781849209496

Boldt, G. (2018). Condescension or co-decisions: A case of institutional youth participation. *Young*, *26*(2), 108–125. doi:10.1177/1103308817713620

Bonilla, C. A., Zanfei, A., Brentan, B., Montalvo, I., & Izquierdo, J. (2022). A Digital Twin of a Water Distribution System by Using Graph Convolutional Networks for Pump Speed-Based State Estimation. *Water (Basel)*, *14*(4), 514. doi:10.3390/w14040514

Bonn, I., & Fisher, J. (2005). Corporate Governance and Business Ethics: Insights from the strategic planning experience. *Corporate Governance*, *13*(6), 730–738. doi:10.1111/j.1467-8683.2005.00466.x

Bonsón, E., Perea, D., & Bednárová, M. (2019). Twitter as a tool for citizen engagement: An empirical study of the Andalusian municipalities. *Government Information Quarterly*, *36*(3), 480–489. doi:10.1016/j.giq.2019.03.001

Boolaane, B. (2006). Restrições à promoção de abordagens centradas nas pessoas na reciclagem. [Restrictions on promoting people-centred approaches to recycling.]. *Habitat International*, *30*(4), 731–740.

Bornmann, L., & Leydesdorff, L. (2013). Macro-Indicators of Citation Impacts of Six Prolific Countries: InCites Data and the Statistical Significance of Trends. *PLoS One*, *8*(2), 1–5. doi:10.1371/journal.pone.0056768 PMID:23418600

Bosch-Sijtsema, P., Ruohomaki, V., & Vartiainen, M. (2009). Knowledge Work Productivity In Distributed Teams. *Journal of Knowledge Management*, *13*(6), 533–546. doi:10.1108/13673270910997178

Bossle, M. B., Dutra De Barcellos, M., Vieira, L. M., & Sauvée, L. (2016). The drivers for adoption of eco-innovation. *Journal of Cleaner Production*, *113*, 861–872. doi:10.1016/j.jclepro.2015.11.033

Boubakri, N., Cosset, J. C., & Saffar, W. (2013). The role of state and foreign owners in corporate risk-taking: Evidence from privatization. *Journal of Financial Economics*, *108*(3), 641–658. doi:10.1016/j.jfineco.2012.12.007

Boulding, K. (1966). The economics of the coming spaceship Earth. In H. Jarrett H. (Ed.), Environmental Quality in a Growing Economy, pp. 20-30. Baltimore: John Hopkins University Press.

Boulding, K. E. (1966). *The economics of the coming spaceship earth*.

Boulianne, S. (2016). Online news, civic awareness and engagement in civic and political life. *New Media & Society*, *18*(9), 1840–1856. doi:10.1177/1461444815616222

Bouzguenda, I., Alalouch, C., and Fava, N. (2019) 'Towards smart sustainable cities: A review of the role digital citizen participation could play in advancing social sustainability', *Sustainable Cities and Society*, *50*(November 2018), p. 101627. . doi:10.1016/j.scs.2019.101627

Bowyer, B., & Kahne, J. (2020). The digital dimensions of civic education: Assessing the effects of learning opportunities. *Journal of Applied Developmental Psychology*, *69*, 101162. doi:10.1016/j.appdev.2020.101162

Boyle, M., Kay, J., & Pond, B. (2001). Monitoring in support of policy: an adaptive ecosystem approach. Encyclopedia of global environmental change, 4(14), 116-137.

Bradford, M., & Florin, J. (2003). Examining the role of innovation diffusion factors on the implementation success of enterprise resource planning systems. *International Journal of Accounting Information Systems*, *4*(3), 205–225. doi:10.1016/S1467-0895(03)00026-5

Braun, A., & Toth, R. (2020, December 15). Circular economy:national and global policy -overview. *Clean Technologies and Environmental Policy*, *23*(2), 301–304. doi:10.100710098-020-01988-8

Compilation of References

Bródka, P., Skibicki, K., Kazienko, P., & Musiał, K. (2011). A degree centrality in multi-layered social network. *Proceedings of the 2011 International Conference on Computational Aspects of Social Networks*, (pp. 237–242). IEEE. 10.1109/CASON.2011.6085951

Brouwer, M. T., Thoden van Velzen, E. U., Augustinus, A., Soethoudt, H., De Meester, S., & Ragaert, K. (2018). Predictive model for the Dutch post-consumer plastic packaging recycling system and implications for the circular economy. *Waste Management (New York, N.Y.), 71*, 62–85. doi:10.1016/j.wasman.2017.10.034 PMID:29107509

Brown, T. (2008). Design Thinking. *Harvard Business Review, 86*(6), 84–92. PMID:18605031

Brundtland, G. H. (1987). Our common future—Call for action. *Environmental Conservation, 14*(4), 291–294. doi:10.1017/S0376892900016805

Buckingham, D. (2017). Media theory 101: AGENCY. *Journal of Media Literacy, 64*(1-2), 12–15.

Buitrago, R., & Duque, M. (2013). *The Orange Economy: An Infinite Opportunity.* (Online). Available at: https://publications.iadb.org/el/orange-economy-infinite-opportunity (5/4/2022)

Burke, E. M. (1968). Citizen Participation Strategies. *Journal of the American Institute of Planners, 34*(5), 287–294. doi:10.1080/01944366808977547

Butler, C. (2000). Inequality, Global Change and the Sustainability of Civilisation. *Global Change & Human Health, 1*(2), 17. doi:10.1023/A:1010029222095

Cahill, H., & Dadvand, B. (2018). Re-conceptualising youth participation: A framework to inform action. *Children and Youth Services Review, 95*, 243–253. doi:10.1016/j.childyouth.2018.11.001

Cahill, H., Wyn, J., & Borovica, T. (2019). Youth participation informing care in hospital settings. *Child and Youth Services, 40*(2), 140–157. doi:10.1080/0145935X.2019.1567323

Caillier, J. (2017). Do Work-Life Benefits Enhance the Work Attitudes of Employees? Findings from a Panel Study. *Public Organization Review, 7*(3), 393–408. doi:10.100711115-016-0344-4

Cairns, J. (2003). Materialfilia, biofilia e uso sustentável do planeta. [Materialphilia, biophilia and sustainable use of the planet.]. *Revista Internacional de Desenvolvimento Sustentável e Ecologia Mundial, 10*(1), 43–48.

Cairns, J. Jr. (1998). What sustainability is not! *International Journal of Sustainable Development and World Ecology, 5*(2), 77–81. doi:10.1080/13504509809469972

Callahan, K. (2007). Citizen Participation: Models and Methods. *International Journal of Public Administration, 30*(11), 1179–1196. doi:10.1080/01900690701225366

Camarero, C. G. (2003). Las nuevas formas de comunicación de la administración con el ciudadano. [The new forms of communication of the administration with the citizen.] *Anales de documentación, 6*, 109-119.

Canton, O. F. (2018). *Contos de Arrepio.* [Tales of Chill.] Lajeado: Univates.

Cao, X., Deng, M., Song, F., Zhong, S., & Zhu, J. (2019). Direct and moderating effects of environmental regulation intensity on enterprise technological innovation: The case of China. *PLoS One, 14*(10), e0223175. doi:10.1371/journal.pone.0223175 PMID:31589643

Cappelli, P., Singh, H., Singh, J. V., & Useem, M. (2010). Leadership lessons from India. *Harvard Business Review, 88*(3), 90–97. PMID:20402052

Carpenter, S. R., Brock, W. A., Folke, C., van Nes, E. H., & Scheffer, M. (2015). Allowing variance may enlarge the safe operating space for exploited ecosystems. *Proceedings of the National Academy of Sciences of the United States of America*, *112*(46), 14384–14389. doi:10.1073/pnas.1511804112 PMID:26438857

Carree, M. A., & Thurik, A. R. (2010). The Impact of Entrepreneurship on Economic Growth. In Z. J. Acs & D. B. Audretsch (eds.) Handbook of Entrepreneurship Research, 557-594. Springer. doi:10.1007/978-1-4419-1191-9_20

Carriço, N., & Ferreira, B. (2021). Data and Information Systems Management for the Urban Water Infrastructure Condition Assessment. *Frontiers in Water*, *0*, 670550. Advance online publication. doi:10.3389/frwa.2021.670550

Carriço, N., Ferreira, B., Barreira, R., Antunes, A., Grueau, C., Mendes, A., Covas, D., Monteiro, L., Santos, J., & Brito, I. S. (2020). Data integration for infrastructure asset management in small to medium-sized water utilities. *Water Science and Technology*, *82*(12), 2737–2744. Advance online publication. doi:10.2166/wst.2020.377 PMID:33341766

Carrillo, L. L., Bergamini, T. P., & Navarro, C. L. C. (2014). El emprendimiento como motor del crecimiento económico. [Entrepreneurship as an engine of economic growth.] *Boletín económico de ICE. Información Comercial Española*, *3048*, 55–63.

Carroll, A. B. (1999). Corporate Social Responsibility: Evolution of a Definitional Construct. *Business & Society*, *38*(3), 268–295. doi:10.1177/000765039903800303

Carroll, A. B. (2016). Carroll's pyramid of CSR: Taking another look. *International Journal of Corporate Social Responsibility*, *1*(3), 1–8. doi:10.118640991-016-0004-6

Carroll, A. B. (2021). Corporate Social Responsibility: Perspectives on the CSR Construct's Development and Future. *Business & Society*, *60*(6), 1258–1278. doi:10.1177/00076503211001765

Carroll, A. B., & Buchholtz, A. K. (2008). *Business & society: Ethics and stakeholder management* (7th ed.). South-Western Cengage Learning.

Carvajal Bermúdez, J. C. and König, R. (2021) The role of technologies and citizen organizations in decentralized forms of participation. A case study about residential streets in Vienna. *Technology in Society*, *66*(October 2020). . doi:10.1016/j.techsoc.2021.101604

Carvalho, L., Gallardo, D., & Nevado, M. T. (2018). Local municipalities' involvement in promoting entrepreneurship: An analysis of web page orientation to the entrepreneurs in Portuguese municipalities. In L. Carvalho (ed.) Handbook of Research on Entrepreneurial Ecosystems and Social Dynamics in a Globalized World, 1-19. Évora, Portugal: IGI Global. doi:10.4018/978-1-5225-3525-6.ch001

Cazzolato, N. (2009). *As Dificuldades de Gestão das Organizações Não-Governamentais* [*The Management Difficulties of Non-Governmental Organizations.*]. Faculdade de Administração e Economia. doi:10.15603/2176-9583/refae.v1n1p66-81

Chaín, C., Muñoz, A., & Más, A. (2008). La gestión de información en las sedes webs de los ayuntamientos españoles. [The management of information in the headquarters websites of the Spanish municipalities.]. *Revista Espanola la de Documentacion Cientifica*, *31*(4), 612–638. doi:10.3989/redc.2008.4.662

Chakraborty, S., Aich, S., & Kim, H. C. (2019, February). A secure healthcare system design framework using blockchain technology. In *2019 21st International Conference on Advanced Communication Technology (ICACT)* (pp. 260-264). IEEE. 10.23919/ICACT.2019.8701983

Chamberlain, J., Bush, R., & Hammet, A. L. (1998). Non-Timber Forest Products: The other forest products. *Forest Products Journal*, *48*(10), 10–19.

Compilation of References

Chanchaichujit, J., Tan, A., Meng, F., & Eaimkhong, S. (2019). Blockchain technology in healthcare. In Healthcare 4.0 (pp. 37-62). Palgrave Pivot, Singapore. doi:10.1007/978-981-13-8114-0_3

Chandrasekhar, K., Kumar, A., Raj, T., Kumar, G., & Kim, S. (2021). *Bioelectrochemical system-mediated waste valorization. SMAB.* Systems Microbiology and Biomanifactturing. doi:10.100743393-021-00039-7

Chatfield, A. T., & Reddick, C. G. (2018). All hands on deck to tweet #sandy: Networked governance of citizen coproduction in turbulent times. *Government Information Quarterly, 35*(2), 259–272. doi:10.1016/j.giq.2017.09.004

Checkoway, B., & Aldana, A. (2013). Four forms of youth civic engagement for diverse democracy. *Children and Youth Services Review, 35*(11), 1894–1899. doi:10.1016/j.childyouth.2013.09.005

Chen, C. (2014). *The CiteSpace Manual,* 94. College of Computing and Informatics Drexel -- Drexel University. http://cluster.cis.drexel.edu/~cchen/citespace/

Chen, X., Chen, J., Wu, D., Xie, Y., & Li, J. (2016). Mapping the Research Trends by Co-word Analysis Based on Keywords from Funded Project. *Procedia Computer Science, 91*(Itqm), 547–555. doi:10.1016/j.procs.2016.07.140

Chen, C. (2006). CiteSpace II: Detecting and Visualizing Emerging Trends and Transient Patterns in Scientific Literature. *Journal of the American Society for Information Science and Technology, 57*(3), 359–377. doi:10.1002/asi.20317

Chen, C. C., Shih, H. S., Shyur, H. J., & Wu, K. S. (2012). A business strategy selection of green supply chain management via an analytic network process. *Computers & Mathematics with Applications (Oxford, England), 64*(8), 2544–2557. doi:10.1016/j.camwa.2012.06.013

Chen, C., & Song, M. (2019). Visualizing a field of research: A methodology of systematic scientometric reviews. *PLoS One, 14*(10), e0223994. doi:10.1371/journal.pone.0223994 PMID:31671124

Chen, H. L., & Hsu, W. T. (2009). Family ownership, board independence, and R&D investment. *Family Business Review, 22*(4), 347–362. doi:10.1177/0894486509341062

Chen, J. (2017). Can online social networks foster young adults' civic engagement? *Telematics and Informatics, 34*(5), 487–497. doi:10.1016/j.tele.2016.09.013

Chen, J., Wang, X., Shen, W., Tan, Y., Matac, L. M., & Samad, S. (2022). Environmental uncertainty, environmental regulation and enterprises' green technological innovation. *International Journal of Environmental Research and Public Health, 19*(16), 9781. doi:10.3390/ijerph19169781 PMID:36011417

Chen, Z., Zhang, X., & Chen, F. (2021). Do carbon emission trading schemes stimulate green innovation in enterprises? Evidence from China. *Technological Forecasting and Social Change, 168*, 120744. doi:10.1016/j.techfore.2021.120744

Chiru, C. (2017). Teleworking: Evolution and trends in USA, UE and Romania. Economics. *Management and Financial Markets, 12*(2), 222–229.

Chiu, Y. B., & Lee, C. C. (2020). Effects of financial development on energy consumption: The role of country risks. *Energy Econ, 90*, 104833. doi:10.1016/j.eneco.2020.104833

Choia, J., & Song, C. (2020). Factors explaining why some citizens engage in E-participation, while others do not. *Government Information Quarterly, 37*(4), 101524. doi:10.1016/j.giq.2020.101524

Cho, J. H., & Sohn, S. Y. (2018). A novel decomposition analysis of green patent applications for the evaluation of R&D efforts to reduce CO2 emissions from fossil fuel energy consumption. *Journal of Cleaner Production, 193*, 290–299. doi:10.1016/j.jclepro.2018.05.060

Chowdhary, C. L., & Acharjya, D. P. (2018). Singular Value Decomposition–Principal Component Analysis-Based Object Recognition Approach. In Bio-Inspired Computing for Image and Video Processing (pp. 323-341). Chapman and Hall/CRC. doi:10.1201/9781315153797-12

Chowdhary, C. L. (2020). Growth of financial transaction toward bitcoin and blockchain technology. In *Bitcoin and blockchain* (pp. 79–97). CRC Press. doi:10.1201/9781003032588-6

Cicognani, E., Albanesi, C., Mazzoni, D., Prati, G., & Zani, B. (2016). Explaining offline and online civic engagement intentions between Italian and migrant youth. *Revista de Psicología Social*, *1*(2), 282–316. doi:10.1080/02134748.2016.1143177

Çimşir, B. T., & Uzunboylu, H. (2019). Awareness Training for Sustainable Development: Development, Implementation and Evaluation of a Mobile Application. *Sustainability*, *11*(3), 1–17. doi:10.3390u11030611

Clauson, K. A., Breeden, E. A., Davidson, C., & Mackey, T. K. (2018). Leveraging Blockchain Technology to Enhance Supply Chain Management in Healthcare:: An exploration of challenges and opportunities in the health supply chain. *Blockchain in healthcare today*.

Clement, R., Deering, M., Mikhael, R., & Villa-García, C. (2014). *Youth civic engagement and leadership*. Retrieved from George Washington University website: https://elliott. gwu. edu/sites/elliott. gwu. edu/files/downloads/Youth%20CE% 26L_FINAL. pdf.

Clemons, A. (2016). From World's Murder Capital to the Medellin Miracle. *The culture trip*. https://theculturetrip.com/south-america/colombia/articles/from-world-s-murder-capital-to-the-medellin-miracle/ (2/4/2022)

Colbert, B. A., Nicholson, J., & Kurucz, E. C. (2018). Humanistic leadership for sustainable transformation. In *Evolving Leadership for Collective Wellbeing*. Emerald Publishing Limited. doi:10.1108/S2058-880120180000007004

Collins, E., Roper, J., & Lawrence, S. (2010). Sustainability practices: Trends in New Zealand businesses. *Business Strategy and the Environment*, *19*(8), 479–494. doi:10.1002/bse.653

Colterm. (2022). *Deseuri—Colterm S.A. [Waste—Colterm S.A.]*. Colterm. https://www.colterm.ro/anunturi/protectia-mediului/1078-deseur

Comisión Europea. (2003). *El libro verde. El espíritu empresarial en Europa. [The Green Paper. Entrepreneurship in Europe.]* Comisión Europea, Bruselas. 0027-final. https://eur-lex.europa.eu/legal-content/ES/TXT/?uri=celex:52003DC0027

Commission of the European Communities. (2007). Ανακοίνωση σχετικά με μια ευρωπαϊκή ατζέντα για τον πολιτισμό σ' έναν κόσμο παγκοσμιοποίησης *[Communication on a European Agenda for Culture in a Globalizing World]*. Brussels. https://eur-lex.europa.eu/legal-content/EL/TXT/PDF/?uri=CELEX:52007DC0242&from=DE

Conejos Fuertes, P., Martínez Alzamora, F., Hervás Carot, M., & Alonso Campos, J. C. (2020). Building and exploiting a Digital Twin for the management of drinking water distribution networks. *Urban Water Journal*, *17*(8), 704–713. doi:10.1080/1573062X.2020.1771382

Cooper, T. (1999). Creating an economic infrastructure for sustainable product design. *Journal of sustainable product design*, 7-17.

Compilation of References

Cooper, J. P., Jackson, S., Kamojjala, S., Owens, G., Szana, K., & Tomić, S. (2022). Demystifying Digital Twins: Definitions, Applications, and Benefits. *Journal - American Water Works Association*, *114*(5), 58–65. doi:10.1002/awwa.1922

Costa, C. C., Cunha, M. P., & Guilhoto, J. M. (2011). *The role of ethanol in the brazilian economy: three decades of progress*. MPRA- Munich Personal RePEc Archive.

Costa, A. B. (2013). *Tecnologia social & políticas públicas [Social technology and public policy.]*. Instituto Polis.

Courtial, J. P. (1994). A coword analysis of scientometrics. *Scientometrics*, *31*(3), 251–260. doi:10.1007/BF02016875

Coutinho, N. (2006). *As organizações sociais e o contrato de gestão [Social organizations and the management contract.]*. REVISTA DE DIREITO PÚBLICO. doi:10.5433/1980-511X.2006v1n2p25

Couto, E. E. (2016). *App Learning - Experiências de pesquisas e formação [App learning—research and training experiences]*. EDUFBA.

Criado, J. I., Sandoval-Almazan, R., & Gil-Garcia, J. R. (2013). Government innovation through social media. *Government Information Quarterly*, *30*(4), 319–326. doi:10.1016/j.giq.2013.10.003

Cumiskey, L., Hoang, T., Suzuki, S., Pettigrew, C., & Herrgård, M. M. (2015). Youth participation at the third UN world conference on disaster risk reduction. *International Journal of Disaster Risk Science*, *6*(2), 150–163. doi:10.100713753-015-0054-5

Cushing, D. F. (2014). Promoting youth participation in communities through youth master planning. *Community Development (Columbus, Ohio)*, *46*(1), 43–55. doi:10.1080/15575330.2014.975139

Da Silva Pereira, R., de Moraes, F. C. C., Junior, A. B. M., & Palmisano, A. (2013). Especificidades da gestão no terceiro setor. [Specificities of management in the third sector.]. *Revista Organizações em Contexto*, *9*(18), 167–195. doi:10.15603/1982-8756/roc.v9n18p167-195

Daniel, J., Sargolzaei, A., Abdelghani, M., Sargolzaei, S., & Amaba, B. (2017). Blockchain technology, cognitive computing, and healthcare innovations. *J. Adv. Inf. Technol*, *8*(3), 194–198. doi:10.12720/jait.8.3.194-198

Darshan, K. R., & Anandakumar, K. R. (2015, December). A comprehensive review on usage of Internet of Things (IoT) in healthcare system. In *2015 International Conference on Emerging Research in Electronics, Computer Science and Technology (ICERECT)* (pp. 132-136). IEEE. 10.1109/ERECT.2015.7499001

Dash, S., Gantayat, P. K., & Das, R. K. (2021). Blockchain technology in healthcare: opportunities and challenges. *Blockchain Technology: Applications and Challenges*, 97-111.

Dauda, L., Long, X., Mensah, C. N., Salman, M., Boamah, K. B., Ampon-Wireko, S., & Dogbe, C. S. K. (2021). Innovation, trade openness and CO2 emissions in selected countries in Africa. *Journal of Cleaner Production*, *281*, 125143. doi:10.1016/j.jclepro.2020.125143

Davidson, K. (2014). A Typology to Categorize the Ideologies of Actors in the Sustainable Development Debate: A Political Economy Typology of Sustainability. *Sustainable Development*, *22*(1), 1–14. doi:10.1002d.520

Davies, I., Evans, M., Fülöp, M., Kiwan, D., Peterson, A., & Sim, J. B. Y. (2019). *Taking action for change: Educating for youth civic engagement and activism*. University of York.

Davila, A., & Elvira, M. M. (2012). Humanistic leadership: Lessons from Latin America. *Journal of World Business*, *47*(4), 548–554. doi:10.1016/j.jwb.2012.01.008

Davis, A. (1993). *Software requirements: objects, functions and states*.

Davison, K., & Russell, J. (2017). Disused religious space: Youth participation in built heritage regeneration. *Religions*, *8*(6), 107. doi:10.3390/rel8060107

Day, D. V. (2001). Leadership development: A review of industry best practices.

de Macêdo, I. I., Rodrigues, D. F., Chevitarese, L. P., & Feichas, S. A. Q. (2015). *Ética e sustentabilidade*. Editora FGV.

de Morais Barroca Filho, I., Aquino, G., Malaquias, R. S., Girão, G., & Melo, S. R. M. (2021). An IoT-based healthcare platform for patients in ICU beds during the COVID-19 outbreak. *IEEE Access : Practical Innovations, Open Solutions*, *9*, 27262–27277. doi:10.1109/ACCESS.2021.3058448 PMID:34786307

DECO. (2021). *Viseu é a cidade do país com mais qualidade de vida. [Viseu is the country's city with the highest quality of life.]* Visit. Viseu. https://visitviseu.pt/sugestao?item=16

Delmas, M. A., & Montes-Sancho, M. J. (2010). Voluntary agreements to improve environmental quality: Symbolic and substantive cooperation. *Strategic Management Journal*, *31*(6), 575–601.

Deng, W., Liang, Q., Li, J., & Wang, W. (2020). Science mapping: a bibliometric analysis of female entrepreneurship studies. Gender in Management. doi:10.1108/GM-12-2019-0240

Deng, Q., & Ji, S. (2015). Organizational green IT adoption: *Concept and evidence. Sustainability*, *7*(12), 16737–16755. doi:10.3390u71215843

Desarrollo Sostenible, F. U. D. I. S. (2020). *Ods Research & Action*. https://odsresearch.com/

Dhir, S., Kumar, D., & Singh, V. (2019). Success and failure factors that impact on project implementation using agile software development methodology. *Software Engineering*, *731*, 647–654. doi:10.1007/978-981-10-8848-3_62

Dias, N. B., & Vieceli, N. (2018). Mechanical Biological Treatment. In Waste-to-Energy (WtE). NOVA.

Dickson-Hoyle, S., Kovacevic, M., Cherbonnier, M., & Nicholas, K. A. (2018). Towards meaningful youth participation in science-policy processes: A case study of the Youth in Landscapes Initiative. *Elem Sci Anth*, *6*, 67. doi:10.1525/elementa.327

DiMaggio, P., & Powell, W. W. (1983). The iron cage revisited: Collective rationality and institutional isomorphism in organizational fields. *American Sociological Review*, *48*(2), 147–160. doi:10.2307/2095101

Diniz, E. (2013). Editorial. *RAE - Revista de Administração de Empresas, 53*(3), 223.

Directives, E. (2014, outubro 22). Diretiva 2014/94/UE do Parlamento Europeu e do Conselho [Directive 2014/94/EU of the European Parliament and of the Council]. *Jornal Oficial da União Europeia*. https://eur-lex.europa.eu/legal-content/PT/TXT/?uri=CELEX:32014L0094

Djenna, A., & Saïdouni, D. E. (2018, October). Cyber-attacks classification in IoT-based-healthcare infrastructure. In *2018 2nd Cyber Security in Networking Conference (CSNet)* (pp. 1-4). IEEE. 10.1109/CSNET.2018.8602974

do Valle Santos, W. C., Singh, D., Delgado Leandro da Cruz, L., de Carvalho Piassi, L. P., & Reis, G. (2019). Vertical gardens: Sustainability, youth participation, and the promotion of change in a socio-economically vulnerable community in Brazil. *Education Sciences*, *9*(3), 161. doi:10.3390/educsci9030161

Doh, J. P., & Quigley, N. R. (2014). Responsible leadership and stakeholder management: Influence pathways and organizational outcomes. *The Academy of Management Perspectives*, *28*(3), 255–274. doi:10.5465/amp.2014.0013

Compilation of References

Doliente, S., Narayan, A., Tapia, J., & Samsatli, N. (2020, july 10). Bio-aviation Fuel: A Comprehensive Review and Analysis of the Supply Chain Components. *Fronties in Energy Research*.

Dolot, A. (2018). The characteristics of Generation Z. *The Characteristics of Generation Z. E-mentor*, *74*(2), 44–50. doi:10.15219/em74.1351

Dos Santos, R., & Kobashi, N. (2009). Bibliometria, cientometria, infometria: Conceitos e aplicações [Bibliometrics, scientometrics, infometrics: concepts and applications]. *Pesquisa Brasileira Em Ciência Da Informação [Brazilian research in information science]*, *2*(1), 155–172.

DRE - Diário da República Eletrónico. (2021). Decreto-Lei 8/2021, de 20 de Janeiro [Decree-Law 8-2021 of January 20]. *Diário da República n.º 13/2021, Série I de 2021-01-20. [Diario da republica no. 13/2021, Series I of 2021-01-20.]* https://dre.tretas.org/dre/4390632/decreto-lei-8-2021-de-20-de-janeiro

Drempetic, S., Klein, C., & Zwergel, B. (2020). The Influence of Firm Size on the ESG Score: Corporate Sustainability Ratings Under Review. *Journal of Business Ethics*, *167*(2), 333–360. doi:10.100710551-019-04164-1

Duan, G., Bai, Y., Ye, D., Lin, T., Peng, P., Liu, M., & Bai, S. (2020). Bibliometric evaluation of the status of Picea research and research hotspots: Comparison of China to other countries. *Journal of Forestry Research*, *31*(4), 1103–1114. doi:10.100711676-018-0861-9

Durand, C. (2012). Trois visions pour Nantes Métropole en 2030. [Three visions for Nantes Metropolis in 2030.] *Le courrier du pays de Retz journal*.

Dutta, P., Choi, T. M., Somani, S., & Butala, R. (2020). Blockchain technology in supply chain operations: Applications, challenges and research opportunities. *Transportation research part e: Logistics and transportation review*, *142*, 102067.

Eames, I., Austin, M., & Wojcik, A. (2022). Injection of gaseous hydrogen into a natural gas pipeline. *International Journal of Hydrogen*.

EC. E. C., & DG JUST, D.-G. for J. and C. (2022). *Proposal for a directive of the European parliament and of the council on Corporate Sustainability Due Diligence and amending Directive (EU) 2019/1937* (Final COM/2022/71; p. 76). European Union. https://eur-lex.europa.eu/resource.html?uri=cellar:bc4dcea4-9584-11ec-b4e4-01aa75ed71a1.0018.02/DOC_1&format=PDF

Edwards, F. (1991). The Banff Centre for Management's Recent or Impending Initiatives in Environmental Innovation. *Environmental Conservation*, *18*(4), 369–370. doi:10.1017/S0376892900022736

e-Estonia. (2022). e-Democracy and open data. Available in: https://e-estonia.com/solutions/e-governance/e-democracy/

Egreja, C. P. (2011). Caminhos limitados ou mobilidade bloqueada? a mobilidade socioprofissional dos imigrantes brasilerios em Portugal. [Limited paths or blocked mobility? the socio-professional mobility of Brazilian immigrants in Portugal.] Coimbra.

EIA- U.S. Energy Information Administration. (2022). Biofuels explained. *Biodiesel, renewable diesel, and other biofuels*. EIA.https://www.eia.gov/energyexplained/biofuels/biodiesel-rd-other-use-supply.php

EIA. (2022). *EIA- US- Energy Information Administration*. EIA. https://www.eia.gov/tools/faqs/faq.php?id=27&t=10#:~:text=The%20ethanol%20content%20of%20most,ethanol%20production%20capacity%20is%20located

Eiadat, Y., Kelly, A., Roche, F., & Eyadat, H. (2008). Green and competitive? An empirical test of the mediating role of environmental innovation strategy. *Journal of World Business*, *43*(2), 131–145. doi:. jwb.2007.11.012 doi:10.1016/j

Ekins, P., Simon, S., Deutsch, L., Folke, C., & De Groot, R. (2003). A framework for the practical application of the concepts of critical natural capital and strong sustainability. *Ecological Economics*, *44*(2–3), 165–185. doi:10.1016/S0921-8009(02)00272-0

El Marzouki, M. (2016). Citizens of the margin: Citizenship and youth participation on the Moroccan social web. *Information Communication and Society*, *21*(1), 147–161. doi:10.1080/1369118X.2016.1266373

Elhoseny, M., Ramírez-González, G., Abu-Elnasr, O. M., Shawkat, S. A., Arunkumar, N., & Farouk, A. (2018). Secure medical data transmission model for IoT-based healthcare systems. *IEEE Access : Practical Innovations, Open Solutions*, *6*, 20596–20608. doi:10.1109/ACCESS.2018.2817615

Elkington, J., & Fennell, S. (1998). Partners for Sustainability. *Greener Management International*, (24).

Elkington, J. (2006). Governance for Sustainability. *Corporate Governance*, *14*(6), 522–529. doi:10.1111/j.1467-8683.2006.00527.x

Elkington, J. (2006). Governance for Sustainability. *Journal Compilation*, *14*(6), 522–529.

Ellen MacArthur Foundation. (2012). Ellen MacArthur Foundation. Towards the circular economy vol. 1: an economic and business rationale for an accelarated transition.

Eller, F. J., Gielnik, W., Thölke, C., Holzapfel, S., Tegtmeier, S., & Halberstadt, J. (2020). Identifying business opportunities for sustainable development: Longitudinal and experimental evidence contributing to the field of sustainable entrepreneurship. *Business Strategy and the Environment*, *29*(3), 1387–1403. doi:10.1002/bse.2439

Elsawah, S., Filatova, T., Jakeman, A. J., Kettner, A. J., Zellner, M. L., Athanasiadis, I. N., Hamilton, S. H., Axtell, R. L., Brown, D. G., Gilligan, J. M., Janssen, M. A., Robinson, D. T., Rozenberg, J., Ullah, I. I. T., & Lade, S. J. (2020). Eight grand challenges in socio-environmental systems modeling. *Socio-Environmental Systems Modelling*, *2*, 16226–16226. doi:10.18174esmo.2020a16226

EMA. (2018). *Blockchain and the water industry—Smart contracts*. https://www.ema-inc.com/news-insights/2018/10/blockchain-and-the-water-industry-smart-contracts/

Emonts, B., Reuß, M., Stenzel, P., Welder, L., Knicker, F., Grube, T., Görner, K., Robinius, M., & Stolten, D. (2019). Flexible sector coupling with hydrogen: A climate-friendly fuel supply for road transport. *International Journal of Hydrogen Energy*, *44*(26), 12918–12930. doi:10.1016/j.ijhydene.2019.03.183

Engelhardt, M. A. (2017). Hitching healthcare to the chain: An introduction to blockchain technology in the healthcare sector. *Technology Innovation Management Review*, *7*(10).

EPA- U-S- Environemntal Protection Agency. (n.d.). *Economics of Biofuels*. EPA. https://www.epa.gov/environmental-economics/economics-biofuels

EPA. (n.d.). *Renewable Energy Certificates (RECs) | US EPA*. EPA. https://www.epa.gov/green-power-markets/renewable-energy-certificates-recs

Compilation of References

ePortugal. (2022). *Gestão de resíduos—EPortugal.gov.pt.* [*Waste management—EPortugal.gov.*]. ePortugal. *t*https://eportugal.gov.pt/cidadaos-europeus-viajar-viver-e-fazer-negocios-em-portugal/bens-e-mercadorias-em-portugal/gest ao-de-residuos

Ericson, T., Kjønstad, B. G., & Barstad, A. (2014). Mindfulness and sustainability. *Ecological Economics*, *104*, 73–79. doi:10.1016/j.ecolecon.2014.04.007

Erwin, K., & Shatto, B. (2016). Moving on From Millenials: Preparing for Generation Z. *Journal of Continuing Education in Nursing*, *47*(6), 253–254. doi:10.3928/00220124-20160518-05 PMID:27232222

Escamilla, S., Plaza, P., & Flores, S. (2016). Análisis de la divulgación de la información sobre la responsabilidad social corporativa en las empresas de transporte público urbano en España. [Analysis of the disclosure of information on corporate social responsibility in urban public transport companies in Spain.]. *Revista de Contabilidad*, *19*(2), 195–203. doi:10.1016/j.rcsar.2015.05.002

Etter, M., Ravasi, D., & Colleoni, E. (2017). Social Media and the Formation of Organizational Reputation. *Academy of Management Review*, *44*(1), 28–52. doi:10.5465/amr.2014.0280

Etzion, D. (2007). Research on organizations and the natural environment, 1992-present: A review. *Journal of Management*, *33*(4), 637–664. doi:10.1177/0149206307302553

EU Directives. (2018). *Directive (eu) 2018/2001 of the european parliament and of the council.* EU. https://eur-lex.europa.eu/legal-content/EN/TXT/PDF/?uri=CELEX:32018L2001&from=EN

European Comission. (2022). *Waste Framework Directive*. EC. https://environment.ec.europa.eu/topics/waste-and-recycling/waste-framework-directive_en

European Commisision. (2022). *Sustainable development*. EC. https://policy.trade.ec.europa.eu/development-and-sustainability/sustainable-development_en

European Commission. (2011). *Country factsheet Romania* [File]. EC. https://www.eea.europa.eu/themes/waste/waste-prevention/countries/romania-waste-prevention-country-profile-2021/view

European Commission. (2013). Entrepreneurship 2020 Action Plan. Reigniting the entrepreneurial spirit in Europe. EC. https://eur-lex.europa.eu/legal-content/EN/TXT/PDF/?uri=CELEX:52012DC0795&from=EN

European Commission. (2018). *Creative Europe program*. Brussels. https://eur-lex.europa.eu/legal-content/EN/TXT/?uri=COM%3A2018%3A366%3AFIN

European Commission. (2019). *The EU Environmental Implementation Review 2019 Country Report—Hungary*. EC. https://ec.europa.eu/environment/eir/country-reports/index_en.htm

European Commission. (2020). *A new Circular Economy Action Plan*. EUR-Lex. https://eur-lex.europa.eu/legal-content/EN/TXT/HTML/?uri=CELEX:52020DC0098&from=EN

European Commission. (2020). *Horizon 2020, Details of the EU funding program*. EC. https://ec.europa.eu/info/research-and-innovation/funding/funding-opportunities/funding-programmes-and-open-calls/horizon-2020_en#latest (3/4/2022)

European Commission. (2022). *Cultural and Creative Cities Monitor*. EC. https://composite-indicators.jrc.ec.europa.eu/cultural-creative-cities-monitor

European Commission. (2022). *Energy*. EC. https://energy.ec.europa.eu/topics/renewable-energy/bioenergy/biofuels_en#:~:text=By%202030%2C%20the%20EU%20aims,the%20achievement%20of%20this%20target

European Environment Agency. (2013). *Waste—National Responses (Hungary)* [SOER 2010 Common environmental theme (Deprecated)]. EEA. https://www.eea.europa.eu/soer/2010/countries/hu/waste-national-responses-hungary

European Environment Agency. (2021a). *Waste recycling in Europe*. EEA. https://www.eea.europa.eu/ims/waste-recycling-in-europe

European Environment Agency. (2021b). *Municipal waste management across European countries*. European Environment Agency. https://www.eea.europa.eu/publications/municipal-waste-management-across-european-countries

European Environment Agency. (2021c). *Waste: A problem or a resource?* European Environment Agency. https://www.eea.europa.eu/publications/signals-2014/articles/waste-a-problem-or-a-resource

European Environment Agency. (2022a). *Reaching 2030's residual municipal waste target—Why recycling is not enough—European Environment Agency* [Briefing]. EEA. https://www.eea.europa.eu/publications/reaching-2030s-residual-municipal-waste/reaching-2030s-residual-municipal-waste/

European Environment Agency. (2022b). *Europe is not on track to halve non-recycled municipal waste by 2030—European Environment Agency* [News]. EEA Europa. https://www.eea.europa.eu/highlights/europe-is-not-on-track

European Parliament & Council of the European Union. (2008). *Directive 2008/98/EC of the European Parliament and of the Council of 19 November 2008 on waste and repealing certain Directives (Text with EEA relevance)*. EUR-Lex. https://eur-lex.europa.eu/eli/dir/2008/98/oj/eng

European Parliament, & Council of the European Union. (2018). *Directive (EU) 2018/851 of the European Parliament and of the Council of 30 May 2018 amending Directive 2008/98/EC on waste*. https://eur-lex.europa.eu/legal-content/EN/TXT/?uri=celex%3A32018L0851

European Parliament. (2015). *Circular economy: definition, importance, and benefits*. European Parliament. https://www.europarl.europa.eu/news/en/headlines/economy/20151201STO05603/circular-economy-definition-importance-and-benefits

Europex. (2018). Renewable Energy Directive (RED II) -Directive (EU) 2018/2001 (recast) on the promotion of the use of energy from renewable sources.

Eurostat. (2022). *Statistics explained, Glossary*. EC. https://ec.europa.eu/eurostat/statistics-explained/index.php?title=Glossary:City (1/4/2022)

Compilation of References

Eurostat. (2022). *Statistics*. Eurostat. https://ec.europa.eu/eurostat/databrowser/view/sdg_11_60/default/table?lang=en

Evans, J., & Benefield, P. (2001). Systematic reviews of educational research: Does the medical model fit? *British Educational Research Journal*, *27*(5), 527–541. doi:10.1080/01411920120095717

Evans, S., Vladimirova, D., Holgado, M., Van Fossen, K., Yang, M., Silva, E. A., & Barlow, C. Y. (2017). Business Model Innovation for Sustainability: Towards a Unified Perspective for Creation of Sustainable Business Models. *Business Strategy and the Environment*, *26*(5), 597–608. doi:10.1002/bse.1939

Eveland, J. (2014). *Medellin transformed: from murder capital to modern city*. Lee Kuan Yew World City Prize. https://www.leekuanyewworldcityprize.gov.sg/resources/features/medellin-transformed/ (29/3/2022)

Faisal, M., Sadia, H., Ahmed, T., & Javed, N. (2022). Blockchain Technology for Healthcare Record Management. In *Pervasive Healthcare* (pp. 255–286). Springer. doi:10.1007/978-3-030-77746-3_17

FARM ENERGY. (2019). Biodiesel Cloud Point and Cold Weather Issues. Retrieved 2022, from https://farm-energy.extension.org/biodiesel-cloud-point-and-cold-weather-issues/#:~:text=In%20cold%20climates%2C%20it%20can,from%20which%20it%20is%20made

Farooq, O., Payaud, M., Merunka, D., & Valette-Florence, P. (2014). The Impact of Corporate Social Responsibility on Organizational Commitment: Exploring Multiple Mediation Mechanisms. *Journal of Business Ethics*, *125*(4), 563–580. doi:10.100710551-013-1928-3

Fausto, B. (2006). *História do Brasil*. [History of Brazil.] São Paulo: EdUSP.

FCH. (n.d.). Fuel Cells and Hydrogen - Green Hydrogen Guarantees of Origin (GO) now available on the market. Retrieved 2022, from https://www.fch.europa.eu/news/green-hydrogen-guarantees-origin-go-now-available-market

Feiertag, J., & Berge, Z. L. (2008). Training Generation N: How Educators Should Approach the Net Generation. *Education + Training*, *50*(6), 457–464. doi:10.1108/00400910810901782

Fernández, G., & Shaw, R. (2015). Youth participation in disaster risk reduction through science clubs in the Philippines. *Disasters*, *39*(2), 279–294. doi:10.1111/disa.12100 PMID:25440993

Ferreira, B., Carriço, N., Barreira, R., Dias, T., & Covas, D. (2022). Flowrate Time Series Processing in Engineering Tools for Water Distribution Networks. *Water Resources Research*, *58*(6), 1–20. doi:10.1029/2022WR032393 PMID:35813986

FGVces. C. de E. em S. (2011). *Especificações de verificação do Programa Brasileiro GHG Protocol* (Programa Brasileiro GHG Protocol, p. 42). [*Verification specifications of the Brazilian GHG Protocol Program* (Brazilian GHG Protocol Program, p. 42).] Fundação Getulio Vargas (FGV). http://bibliotecadigital.fgv.br:80/dspace/handle/10438/30258

Fiedler, N. C., Soares, T. S., & Silva, G. F. (2008). Produtos Florestais Não Madeireiros: Importância e Manejo Sustentável da Floresta. [Non-Timber Forest Products: Importance and Sustainable Forest Management.]. *Revista Ciências Exatas e Naturais*, *10*(2), 263–278.

Filoso, S., Carmo, J. B., Mardegan, S. F., Lins, S. M., Gomes, T. F., & Martinelli, L. A. (2015). Reassessing the environmental impacts of sugarcane ethanol production in Brazil to help meet sustainability goals. *Renewable & Sustainable Energy Reviews*, *52*, 1847–1856. doi:10.1016/j.rser.2015.08.012

Finn, P. (2021) Organising for entrepreneurship: How individuals negotiate power relations to make themselves entrepreneurial. *Technological Forecasting and Social Change, 166,* 120610. doi:10.1016/j.techfore.2021.120610

Fiolhais, R. (2007). Teletrabalho e Gestão dos Recursos Humanos [Telework and Human Resources management]. In A. Caetano, & J. Vala (Org.), Gestão de Recursos Humanos: Contextos, processos e técnicas [Human resource management: contexts, processes, and techniques] (3ª ed., pp. 235-262). Editora RH.

Flew, T. (2002). Beyond ad hocery: Defining Creative Industries. In *Cultural Sites, Cultural Theory, Cultural Policy. The Second International Conference on Cultural Policy Research*. Wellington. https://eprints.qut.edu.au/256/1/Flew_beyond.pdf (29/3/2022)

Flick, U. (2009). *Qualidade na pesquisa qualitativa* [*Quality in qualitative research.*]. Artmed.

Florida, R. (2002). The rise of the creative class. Basic Books Ed., New York

Folke, C. (2006). Resilience: The emergence of a perspective for social–ecological systems analyses. *Global Environmental Change, 16*(3), 253–267. doi:10.1016/j.gloenvcha.2006.04.002

Folke, C. (2016). Resilience (Republished). *Ecology and Society, 21*(4), art44. Advance online publication. doi:10.5751/ES-09088-210444

Folke, C., Carpenter, S., Elmqvist, T., Gunderson, L., Holling, C. S., & Walker, B. (2002). Resilience and Sustainable Development: Building Adaptive Capacity in a World of Transformations. *Ambio, 31*(5), 437–440. doi:10.1579/0044-7447-31.5.437 PMID:12374053

Folke, C., Hahn, T., Olsson, P., & Norberg, J. (2005). Adaptive governance of social-ecological systems. *Annual Review of Environment and Resources, 30*(1), 441–473. doi:10.1146/annurev.energy.30.050504.144511

Fonseca, J. D., Carmargo, M., Commenge, J.-M., Falk, L., & Gil, I. D. (2019). Trends in design of distributed energy systems using hydrogen as energy vector: A systematic literature review. Internaqtional Journal of hydrogen energy, 44, 9486-9504.

Fost Plus. (2022a). *Plastic verpakkingen. Fost Plus.* https://www.fostplus.be/nl/recycleren/plastic-verpakkingen

Fost Plus. (2022c). *Metalen verpakkingen.* [*Metal packaging.*] Fost Plus. https://www.fostplus.be/nl/recycleren/metalen-verpakkingen

Fost Plus. (2022d). *Papier—Karton.* Fost Plus. https://www.fostplus.be/nl/recycleren/papier-karton

Fowler, A. (2017). Does voter preregistration increase youth participation? *Election Law Journal: Rules. Politics & Policy, 16*(4), 485–494. doi:10.1089/elj.2017.0436

Frame, B. (2004). The big clean up: Social marketing for the Auckland region. *Local Environment, 9*(6), 507–526. doi:10.1080/1354983042000288030

Freire, P. (1975). *Pedagogia do oprimido* [*Pedagogy of the oppressed.*]. Afrontamento.

Freire, P. (2015). *Educação Como Prática Da Liberdade* [*Education As A Freedom Practice.*]. Paz e Terra.

Frémeaux, S., & Michelson, G. (2017). The common good of the firm and humanistic management: Conscious capitalism and economy of communion. *Journal of Business Ethics, 145*(4), 701–709. doi:10.100710551-016-3118-6

Fricker, A. (1998). Measuring up to sustainability. *Futures, 30*(4), 367–375. doi:10.1016/S0016-3287(98)00041-X

Friedrich Ebert Stiftung. (2022). *Monitor Social.* Monitor Social. https://monitorsocial.ro/

Compilation of References

Frischmuth, F., & Härtel, P. (2022). Energy - Hydrogen sourcing strategies and cross-sectoral flexibility trade-offs in net-neutral energy scenarios for Europe. *Energy, 238*, 121598. doi:10.1016/j.energy.2021.121598

Fritz, S., & Sorgel, P. (2017). Recentering leadership around the human person: introducing a framework for humanistic leadership.

Fuentelsaz, L., González, C., Maícas, J. P., & Montero, J. (2015). How different formal institutions affect opportunity and necessity entrepreneurship. *BRQ Business Research Quarterly, 18*(4), 246–258. doi:10.1016/j.brq.2015.02.001

Fu, H. Z., Wang, M. H., & Ho, Y. S. (2012). The most frequently cited adsorption research articles in the Science Citation Index (Expanded). *Journal of Colloid and Interface Science, 379*(1), 148–156. doi:10.1016/j.jcis.2012.04.051 PMID:22608849

Fujii, H. (2016). Decomposition analysis of green chemical technology inventions from 1971 to 2010 in Japan. *Journal of Cleaner Production, 112*, 4835–4843. doi:10.1016/j.jclepro.2015.07.123

Fujii, H., & Managi, S. (2016). Research and development strategy for environmental technology in Japan: A comparative study of the private and public sectors. *Technological Forecasting and Social Change, 112*, 293–302. doi:10.1016/j.techfore.2016.02.012

Fulton, S. C., De Silva, L., & Anton, D. (2012, March). Twenty years after the rio earth summit: what is the agenda for the 2012 United Nations Conference on Sustainable Development? In *American Society of International Law. Proceedings of the Annual Meeting* (p. 91). Cambridge University Press.

Fu, P. P., von Kimakowitz, E., Lemanski, M., Liu, L. A., & Pattnaik, C. (2020). Humanistic leadership in different cultures: Defining the field by pushing boundaries. *Cross Cultural & Strategic Management*.

Fusi, F., & Zhang, F. (2018). Social Media Communication in the Workplace: Evidence From Public Employees Network. *Review of Public Personnel Administration, 0*(0), 1–27. doi:10.1177/0734371X18804016

Gabrielova, K., & Buchko, A. (2021). Here comes Generation Z: Millennials as managers. *Business Horizons, 64*(4), 489–499. doi:10.1016/j.bushor.2021.02.013

Gaby, S. (2017). The civic engagement gap(s): Youth participation and inequality from 1976 to 2009. *Youth & Society, 49*(7), 923–946. doi:10.1177/0044118X16678155

Gadenne, D. L., Kennedy, J., & McKeiver, C. (2009). An empirical study of environmental awareness and practices in SMEs. *Journal of Business Ethics, 84*(1), 45–63. doi:10.100710551-008-9672-9

Gajendran, R., & Harrison, D. (2007). The Good, the Bad, and the Unknown About Telecommuting: Meta-Analysis of Psychological Mediators and Individual Consequences. *The Journal of Applied Psychology, 92*(6), 1524–1541. doi:10.1037/0021-9010.92.6.1524 PMID:18020794

Galindo, M. A., & Méndez, M. T. (2014). Entrepreneurship, economic growth, and innovation: Are feedback effects at work? *Journal of Business Research, 67*(5), 825–829. doi:10.1016/j.jbusres.2013.11.052

Galindo, M. Á., Ribeiro, D., & Méndez, M. T. (2012). Innovación y crecimiento económico: Factores que estimulan la innovación. [Innovation and economic growth: Factors that stimulate innovation.]. *Cuadernos de Gestión, 12*(Esp), 51–58. doi:10.5295/cdg.110309mg

Gal, T. (2017). An ecological model of child and youth participation. *Children and Youth Services Review, 79*, 57–64. doi:10.1016/j.childyouth.2017.05.029

Gandía, J. L., & Archidona, M. (2008). Determinants of web site information by Spanish city councils. *Online Information Review*, *32*(1), 35–57. doi:10.1108/14684520810865976

Gao, X., Lyu, Y., Shi, F., Zeng, J., & Liu, C. (2019). The impact of financial fac-tor market distortion on green innovation efficiency of high-tech industry. *Ekoloji*, *28*(107), 3449–3461.

Garcia, D. S. S. (2020). Sustentabilidade e ética: Um debate urgente e necessário. [Sustainability and ethics: An urgent and necessary debate.] *Revista Direitos Culturais*, *15*(35), 51–75.

García, P., Fernández, C. H., & Jackson, A. (2019). Counternarratives of youth participation among black girls. *Youth & Society*, *00*(0), 1–22. doi:10.1177/0044118X18824738

Garde, R., Rodríguez, M. P., & López, A. M. (2015). Are Australian universities making good use of ICT for CSR reporting? *Sustainability*, *7*(11), 14895–14916. doi:10.3390u71114895

Garrain, D., Herrera, I., Lechón, Y., & Lago, C. (2014). Well-to-Tank environmental analysis of a renewable diesel fuel from vegetable oil through co-processing in a hydrotreatment unit. Biomassa and Bionergy. doi:10.1016/j.biombioe.2014.01.035

Garriga, E., & Melé, D. (2004). Corporate Social Responsibility Theories: Mapping the Territory. *Journal of Business Ethics*, *53*(1), 51–71. doi:10.1023/B:BUSI.0000039399.90587.34

Gasparatos, A., Mudombi, S., Balde, B. S., Maltitz, G. V., Johnson, F. X., Romeu-Dalmau, C., ... Willis, K. J. (2022). Local food security impacts of biofuel crop production in southern Africa. *Renewable & Sustainable Energy Reviews*, *154*, 154. doi:10.1016/j.rser.2021.111875

Gatti, L., Vishwanath, B., Seele, P., & Cottier, B. (2019). Are we moving beyond voluntary CSR? Exploring theoretical and managerial implications of mandatory CSR resulting from the new Indian companies act. *Journal of Business Ethics*, *160*(4), 961–972. doi:10.100710551-018-3783-8

Geissdoerfer, M., Savaget, P., Bocken, N. M., & Hultink, E. J. (2017). The Circular Economy–A new sustainability paradigm? *Journal of Cleaner Production*, *143*, 757–768. doi:10.1016/j.jclepro.2016.12.048

Gezer, I., & Cardoso, S. P. (2015). Entrepreneurship and its impact on innovation and development. A multivariate analysis with socioeconomic indicators. Globalization, *Competitiveness & Governability*, *9*(2), 43–60. doi:10.3232/GCG.2015.V9.N2.02

Giddey, S., Badwal, S. S., & Kulkarni, A. (2013). Review of electrochemical ammonia production technologies and materials. *International Journal of Hydrogen Energy*, *38*(34), 14576–14594. doi:10.1016/j.ijhydene.2013.09.054

Giddey, S., Badwal, S. S., Minnings, C., & Dolan, M. (2017). Ammonia as Renewable Energy Transportation Media. *ACS Sustainable Chemistry & Engineering*, *5*(11), 10231–10239. doi:10.1021/acssuschemeng.7b02219

Gil de Zúñiga, H. (2015). European public sphere| toward a European public sphere? The promise and perils of modern democracy in the age of digital and social media. *International Journal of Communication*, *9*, 3152–3160.

Giovanis, E. (2018). The Relationship Between Flexible Employment Arrangements and Workplace Performance in Great Britain. *International Journal of Manpower*, *39*(1), 51–70. doi:10.1108/IJM-04-2016-0083

GitHub. (2022). *GitHub*. Obtido de https://github.com

Gkatzis, Th. (2020). *Η συμβολή της δημιουργικής οικονομίας στην αστική αναγέννηση και βιώσιμη ανάπτυξη των πόλεων: Θεωρητική προσέγγιση* [The contribution of the creative economy to urban regeneration and sustainable urban development: A theoretical approach], Το Βήμα των Κοινωνικών Επιστημών [Lexical characteristics of Greek language] [The Step of the Social Sciences]. V. 72. University of Thessaly. https://tinyurl.com/nxfzxxr2 (2/5/2022)

Compilation of References

Glaeser, E. L., Kerr, S. P., & Kerr, W. R. (2015). Entrepreneurship and urban growth: An empirical assessment with historical mines. *The Review of Economics and Statistics*, *97*(2), 498–520. doi:10.1162/REST_a_00456

Glänzel, W., & Moed, H. F. (2002). Journal impact measures in bibliometric research. *Scientometrics*, *53*(2), 171–193. doi:10.1023/A:1014848323806

Glas. (2022b). Fost Plus. https://www.fostplus.be/nl/recycleren/glas

Glass, A. (2007). Understanding Generational Differences for Competitive Success. *Industrial and Commercial Training*, *39*(2), 98–103. doi:10.1108/00197850710732424

Glavas, A., & Kelley, K. (2014). The Effects of Perceived Corporate Social Responsibility on Employee Attitudes. *Business Ethics Quarterly*, *24*(2), 165–202. doi:10.5840/beq20143206

Globocnik, D., Rauter, R., & Baumgartner, R. J. (2020). Synergy or conflict? The relationships among organisational culture, sustainability-related innovation performance, and economic innovation performance. *International Journal of Innovation Management*, *24*(01), 2050004. doi:10.1142/S1363919620500048

Goessling, K. P. (2017). Resisting and reinforcing neoliberalism: Youth activist organizations and youth participation in the contemporary Canadian context. *Mind, Culture, and Activity*, *24*(3), 199–216. doi:10.1080/10749039.2017.1313278

Gökçe-Kızılkaya, S., & Onursal-Beşgül, Ö. (2017). Youth participation in local politics: City councils and youth assemblies in Turkey. *Southeast European and Black Sea Studies*, *17*(1), 97–112. doi:10.1080/14683857.2016.1244239

Golden, T. (2006). Avoiding depletion in virtual work: Telework and the intervening impact of work exhaustion on commitment and turnover intentions. *Journal of Vocational Behavior*, *69*(1), 176–187. doi:10.1016/j.jvb.2006.02.003

González, B., & Ballesta, J. A. (2018). Caracterización del emprendimiento femenino en España: Una visión de conjunto. [Characterization of female entrepreneurship in Spain: An overview.] *REVESCO: Revista de estudios cooperativos*, (129), pp. 39-65.

Goodman, J. (2002). *La educación democrática en la escuela*. MCEP.

Górniak, A., Midor, K., Kaźmierczak, J., & Kaniak, W. (2018). *Advantages and Disadvantages of Using Methane from CNG in Motor Vehicles in Polish Conditions*. MAPE- Multidisciplinary Aspects of Production Engineering. doi:10.2478/mape-2018-0031

Goulart, J. (2009). *Teletrabalho: Alternativa de trabalho flexível [Telework: flexible work alternative]*. Editora Senac.

Govindan, K., Nasr, A. K., Saeed Heidary, M., Nosrati-Abarghooee, S., & Mina, H. (2022). Prioritizing adoption barriers of platforms based on blockchain technology from balanced scorecard perspectives in healthcare industry: A structural approach. *International Journal of Production Research*, 1–15. doi:10.1080/00207543.2021.2013560

Gozzo, S., & Sampugnaro, R. (2016). What matters? Changes in European youth participation. *Partecipazione e Conflitto*, *9*(3), 748–776. doi:10.1285/i20356609v9i3p748

Graczyk, M., Olszewski, R., Golinski, M., Spychala, M., Szafranski, M., Weber, G. W., & Miadowicz, M. (2022). Human resources optimization with MARS and ANN: Innovation geolocation model for generation Z. *Journal of Industrial and Management Optimization*, *18*(6), 4093–4110. doi:10.3934/jimo.2021149

Greena Team. (2014). Waste based biofuels, waste based feedstock. Retrieved from http://www.greenea.com/wp-content/uploads/2016/07/10.-HVO-market.pdf

Gregolin, G. C., Gregolin, M. R. P., Triches, R. M., & Zonin, W. J. (2019). Desenvolvimento: do unicamente econômico ao sustentável multidimensional. [Development: from the only economic to the multidimensional sustainable.] *Rev. Eletrônica de Humanidades do Curso de Ciências Sociais da UNIFAP*, *12*(3), 51–64.

Grievson, O., Holloway, T., & Johnson, B. (Eds.). (2022). *A Strategic Digital Transformation for the Water Industry*. IWA Publishing., doi:10.2166/9781789063400

Grimm, R., Fox, C., Baines, S., & Albertson, K. (2013). Social innovation, an answer to contemporary societal challenges? Locating the concept in theory and practice. *Innovation (Abingdon)*, *26*(4), 436–455. doi:10.1080/13511610.2013.848163

Grönlund, Å., & Horan, T. A. (2005). Introducing e-Gov: History, Definitions, and Issues. *Communications of the Association for Information Systems*, *15*, 39.

Grossman, G. M., & Krueger, A. B. (1995). Economic growth and the environment. *The Quarterly Journal of Economics*, *110*(2), 353–377. doi:10.2307/2118443

Guerra, M. P., Silveira, V., Reis, M. S., & Schneider, L. (2002). Exploração, manejo e conservação da Araucária (Araucaria angustifolia). [Exploration, management and conservation of Araucaria (Araucaria angustifolia).] In L. L. Simões, & C. F. Lino (Orgs.). Sustentável Mata Atlântica: a exploração de seus recursos florestais [Sustainable Atlantic forest: the exploitation of its forest resources] (pp. 55-80). São Paulo: SENAC São Paulo.

Guimarães, D. (2019). Sustentabilidade. (Meio sustentável) Retrieved 2022, from https://meiosustentavel.com.br/sustentabilidade/

Gulbenkian. (2022). Fiundação Calouste Gulbenkian. Foresight Portugal 2030- 3 cenários para o futuro de Portugal. Retrieved 2022, from https://gulbenkian.pt/wp-content/uploads/2022/02/FCG_BROCHURA_ForesightPortugal2030_05as.pdf

Gunderson, L. H., & Holling, C. S. (Eds.). (2002). Panarchy; Understanding Transformations in Human and Natural Systems, Island, Washington, DC.

Gunn, G., & Stanley, M. (2018). Harnessing the Flow of Data: Fintech opportunities for ecosystem management (p. 16).

Guo, J., Liu, Z., & Liu, Y. (2016). Key success factors for the launch of government social media platform: Identifying the formation mechanism of continuance intention. *Computers in Human Behavior*, *55*, 750–763.

Guo, P., Tian, W., Li, H., Zhang, G., & Li, J. (2020). Global characteristics and trends of research on construction dust: Based on bibliometric and visualized analysis. *Environmental Science and Pollution Research International*, *27*(30), 37773–37789. doi:10.100711356-020-09723-y PMID:32613507

Gupeng, Z., & Xiangdong, C. (2012). The value of invention patents in China: Country origin and technology field differences. *China Economic Review*, *23*(2), 357–370. doi:10.1016/j.chieco.2012.02.002

Gupta, M. (2018). Blockchain for dummies (2nd IBM limited edition). John Wiley & Sons, Inc.

Gupta, S. (2013). For mobile devices, think Apps, not ads. *Harvard Business Review*.

Hahn, D., Spitzley, D. I., Brumana, M., Ruzzene, A., Bechthold, L., Prügl, R., & Minola, T. (2021). Founding or succeeding? Exploring how family embeddedness shapes the entrepreneurial intentions of the next generation. *Technological Forecasting and Social Change*, *173*, 121182. doi:10.1016/j.techfore.2021.121182

Haleem, A., Javaid, M., Singh, R. P., Suman, R., & Rab, S. (2021). Blockchain technology applications in healthcare: An overview. *International Journal of Intelligent Networks*, *2*, 130–139. doi:10.1016/j.ijin.2021.09.005

Hall, P. (1998). Cities in Civilization: Culture, Innovation and Urban Order (Weidenfeld & Nicolson Ed). Pantheon Book.

Hall, P. (2000). Creative Cities and Economic Development. *Urban Studies, 37*(4), 639 – 649. https://journals.sagepub.com/doi/pdf/10.1080/00420980050003946

Hamza, R., Yan, Z., Muhammad, K., Bellavista, P., & Titouna, F. (2020). A privacy-preserving cryptosystem for IoT E-healthcare. *Information Sciences, 527*, 493–510. doi:10.1016/j.ins.2019.01.070

Hanggi, S., Elbert, P., Butler, T., Cabalzar, U., Teske, S., Bach, C., & Onder, C. (2019). A review of synthetic fuels for passenger vehicles. *Energy Reports, 5*, 555–569. doi:10.1016/j.egyr.2019.04.007

Hannan, M. T., & Freeman, J. (1989). *Organizational Ecology*. Harvard University Press. doi:10.4159/9780674038288

Hannan, M. T., & Freeman, J. H. (1977). The Population Ecology of Organizations. *American Journal of Sociology, 82*(5), 929–963. doi:10.1086/226424

Hansen, D., Shneiderman, B., & Smith, M. A. (2010). *Analyzing social media networks with NodeXL: Insights from a connected world*. Morgan Kaufmann.

Hansen, T. B. (1995). *Measuring the value of culture* (Vol. 1). Cultural Policy.

Haque, A., Fernando, M., & Caputi, P. (2021). Responsible leadership and employee outcomes: A systematic literature review, integration and propositions. *Asia-Pacific Journal of Business Administration, 13*(3), 383–408. doi:10.1108/APJBA-11-2019-0243

Harris, A., Wyn, J., & Younes, S. (2010). Beyond apathetic or activist youth: 'Ordinary' young people and contemporary forms of participation. *Young, 18*(1), 9–32. doi:10.1177/110330880901800103

Hasenheit, M., Gerdes, H., Kiresiewa, Z., & Beekman, V. (2016). Summary report on the social, economic and environmental impacts of the bioeconomy. European Union´s Horizon 2020.

Hasselgren, A., Kralevska, K., Gligoroski, D., Pedersen, S. A., & Faxvaag, A. (2020). Blockchain in healthcare and health sciences—A scoping review. *International Journal of Medical Informatics, 134*, 104040. doi:10.1016/j.ijmedinf.2019.104040 PMID:31865055

Hatimtai, M. H., & Hassan, H. (2018). The relationship between the characteristics of innovation towards the effectiveness of ICT in Malaysia productivity corporation. *Malaysian Journal of Communication, 34*(1), 253–269. doi:10.17576/JKMJC-2018-3401-15

Hecklau, F., Galeitzke, M., Bourgeois, S., & Kohl, H. (2016). Holistic Approach for Human Resource Management In Industry 4.0. *Procedia CIRP, 54*, 1–6. doi:10.1016/j.procir.2016.05.102

Heinrich, A. J., & Million, A. (2016). Young people as city builders: Youth participation in German Municipalities. *disP-The Planning Review, 52*(1), 56-71. Doi:10.1080/02513625.2016.1171049

Hellenic Statistical Authority. (2022), Ελλάς με αριθμούς, Ιανουάριος – Μάρτιος 2022 [Lexical characteristics of Greek language] [Greece in numbers, January - March 2022]. https://www.statistics.gr/documents/20181/17831637/GreeceInFigures_2022Q1_EN.pdf/82613203-ee3c-eb7b-600b-503211d9dc91 (29/3/2022)

Hellström, T. (2007). Dimensions of environmentally sustainable Innovation: The structure of eco-innovation concepts. *Sustainable Development, 15*(3), 148–159. doi:10.1002d.309

Henn, M., & Foard, N. (2014). Social differentiation in young people's political participation: The impact of social and educational factors on youth political engagement in Britain. *Journal of Youth Studies, 17*(3), 360–380. doi:10.1080/13676261.2013.830704

Herczeg, M. (2013). *Municipal waste management in Hungary*. European Environment Agency. https://www.eea.europa.eu/publications/managing-municipal-solid-waste/hungary-municipal-waste-management/view

Herrera, J., De las Heras-Rosas, C., Rodríguez-Fernández, M., & Ciruela-Lorenzo, A. M. (2022). Teleworking: The Link between Worker, Family and Company. *Systems*, *10*(5), 134. doi:10.3390ystems10050134

Herrera, M. E. B. (2015). Creating competitive advantage by institutionalizing corporate social innovation. *Journal of Business Research*, *68*(7), 1468–1474. doi:10.1016/j.jbusres.2015.01.036

Herrero-Jáuregui, C., Arnaiz-Schmitz, C., Reyes, M. F., Telesnicki, M., Agramonte, I., Easdale, M. H., Schmitz, M. F., Aguiar, M., Gómez-Sal, A., & Montes, C. (2018). What do we talk about when we talk about social-ecological systems? A literature review. *Sustainability*, *10*(8), 2950. doi:10.3390u10082950

Hevner, A., & Chatterjee, S. (2004). *Design Science Research in Information Systems*. Springer. doi:10.2307/25148625

Higgins, L. (1999). Applying principles of creativity management to marketing research efforts in high-technology markets. *Industrial Marketing Management*, *28*(3), 305–317. https://tinyurl.com/yms62zxb

Higgins, M., & Morgan, J. (2000). The Role of Creativity in Planning: The "Creative Practitioner". *Planning Practice and Research*, *15*(1–2), 117–127. https://www.tandfonline.com/doi/abs/10.1080/713691881

Ho, Y. S. (2019a). Comments on "A Bibliometric Analysis of Research on Intangible Cultural Heritage Using CiteSpace" by Su et al. (2019). *SAGE Open*, *9*(4), 0–1. doi:10.1177/2158244019894291

Hobson, K. (2003). Thinking habits into action: The role of knowledge and process in questioning household consumption practices. *Local Environment*, *8*(1), 95–112. doi:10.1080/135498303200041359

Ho, E., Clarke, A., & Dougherty, I. (2015). Youth-led social change: Topics, engagement types, organizational types, strategies, and impacts. *Futures*, *67*, 52–62. doi:10.1016/j.futures.2015.01.006

Hofstrand, D. (2009). Ammonia as a Transportation Fuel. AgMRC Renewable Energy Newsletter.

Holling, C. S. (1973). Resilience and Stability of Ecological Systems. *Annual Review of Ecology and Systematics*, *4*(1), 1–23. doi:10.1146/annurev.es.04.110173.000245

Holme, R., & Watts, P. (2001). Making good business sense. *Journal of Corporate Citizenship*, *2001*(2), 17–20. doi:10.9774/GLEAF.4700.2001.su.00005

Holmes, R. M. Jr, Miller, T. J., Hitt, M. A., & Salmador, M. P. (2013). The interrelationships among informal institutions, formal institutions and inward foreign direct investment. *Journal of Management*, *39*(2), 531–566. doi:10.1177/0149206310393503

Hopwood, B., Mellor, M., & O'Brien, G. (2005). Sustainable development: Mapping different approaches. *Sustainable Development*, *13*(1), 38–52. doi:10.1002d.244

Horbach, J., Rammer, C., & Rennings, K. (2012). Determinants of eco-innovations by type of environmental impact - The role of regulatory push/pull, technology push and market pull. *Ecological Economics*, *78*, 112–122. doi:10.1016/j.ecolecon.2012.04.005

Hospers, G. J. (2003). Creative Cities in Europe: Urban Competitiveness in the Knowledge Economy. *Intereconomics*, *38*(5), pp. 260–269. https://www.researchgate.net/publication/47872007_Creative_Cities_in_Europe_Urban_Competitiveness_in_the_Knowledge_Economy

Compilation of References

Hospers, G. J. (2003). Creative cities: Breeding places in the knowledge economy. *Knowledge, Technology & Policy, 16*(3), 143–162. https://www.researchgate.net/publication/240357359_Creative_cities_Breeding_places_in_the_knowledge_economy (3/4/2022)

Hou, J. L., & Yeh, K. H. (2015). Novel authentication schemes for IoT based healthcare systems. *International Journal of Distributed Sensor Networks, 11*(11), 183659. doi:10.1155/2015/183659

Ho, Y. S. (2018). Comment on: "A bibliometric analysis and visualization of medical big data research" Sustainability 2018, 10, 166. *Sustainability (Switzerland), 10*(12), 2017–2018. doi:10.3390u10124851

Ho, Y. S. (2019b). Comments on Research trends of macrophage polarization: A bibliometric analysis. *Chinese Medical Journal, 132*(22), 2772. doi:10.1097/CM9.0000000000000499 PMID:31765362

Ho, Y. S. (2019c). Rebuttal to: Su et al. "The neurotoxicity of nanoparticles: A bibliometric analysis," Vol. 34, pp. 922–929. *Toxicology and Industrial Health, 35*(6), 399–402. doi:10.1177/0748233719850657 PMID:31244406

Hu, A. G. Z., Zhang, P., & Zhao, L. (2017). China as number one? Evidence from China's most recent patenting surge. *Journal of Development Economics, 124*, 107–119. doi:10.1016/j.jdeveco.2016.09.004

Huang, L., Zhou, M., Lv, J., & Chen, K. (2020). Trends in global research in forest carbon sequestration: A bibliometric analysis. *Journal of Cleaner Production, 252*, 1–17. doi:10.1016/j.jclepro.2019.119908

Huang, M., Ding, R., & Xin, C. (2021). Impact of technological innovation and industrial-structure upgrades on ecological efficiency in China in terms of spatial spillover and the threshold effect. *Integrated Environmental Assessment and Management, 17*(4), 852–865. doi:10.1002/ieam.4381 PMID:33325155

Hua, Y., Xie, R., & Su, Y. (2018). Fiscal spending and air pollution in Chinese cities: Identifying composition and technique effects. *China Economic Review, 47*, 156–169. doi:10.1016/j.chieco.2017.09.007

Hungarian Central Statistical Office. (2022). *Központi Statisztikai Hivatal 2020. [Central Statistical Office 2020.]*. KSH. https://www.ksh.hu/?lang=en

Hussain, N., Rigoni, U., & Orij, R. P. (2018). Corporate Governance and Sustainability Performance: Analysis of Triple Bottom Line Performance. *Journal of Business Ethics, 149*(2), 411–432. doi:10.100710551-016-3099-5

Hussien, H. M., Yasin, S. M., Udzir, N. I., Ninggal, M. I. H., & Salman, S. (2021). Blockchain technology in the healthcare industry: Trends and opportunities. *Journal of Industrial Information Integration, 22*, 100217. doi:10.1016/j.jii.2021.100217

Hussien, H. M., Yasin, S. M., Udzir, S. N. I., Zaidan, A. A., & Zaidan, B. B. (2019). A systematic review for enabling of develop a blockchain technology in healthcare application: Taxonomy, substantially analysis, motivations, challenges, recommendations and future direction. *Journal of Medical Systems, 43*(10), 1–35. doi:10.100710916-019-1445-8 PMID:31522262

Hutagalung, L., Purwanto, A., & Prasetya, A. B. (2020). The Five Leadership Style in Time of Pandemic Covid-19 throughout Industrial Revolution 4.0 as compared to Humane Leadership. *International Journal of Social. Policy and Law, 1*(1), 79–87.

Hu, Y., Chen, S., & Wang, J. (2018). Managerial humanistic attention and CSR: Do firm characteristics matter? *Sustainability, 10*(11), 4029. doi:10.3390u10114029

Idrees, S. M., Nowostawski, M., Jameel, R., & Mourya, A. K. (2021). Security aspects of blockchain technology intended for industrial applications. *Electronics (Basel), 10*(8), 951. doi:10.3390/electronics10080951

IEA. (2019). International Energy Agency - The Future of hydrogen - Seizing today's opportunities - Report prepared by the IEA for the G20, Japan. Retrieved 2020, from https://iea.blob.core.windows.net/assets/9e3a3493-b9a6-4b7d-b499-7ca48e357561/The_Future_of_Hydrogen.pdf

IEA. (2021). International Energy Agency - Hydrogen. Retrieved from https://www.iea.org/reports/hydrogen

IEA. (2021). Renewables 2021- Biofuels. (IEA – International Energy Agency) Retrieved 2022, from https://www.iea.org/reports/renewables-2021/biofuels?mode=transport&region=North+America&publication=2021&flow=Consumption&product=Ethanol

IEA. (2021). World Energy Outlook. Retrieved from https://www.iea.org/reports/world-energy-outlook-2021

Iglesias, O., Markovic, S., Bagherzadeh, M., & Singh, J. J. (2020). Co-creation: A Key Link Between Corporate Social Responsibility, Customer Trust, and Customer Loyalty. *Journal of Business Ethics*, *163*(1), 151–166. doi:10.100710551-018-4015-y

ILO. (2020). *Teleworking during the COVID-19 pandemic and beyond: a practical guide*. International Labour Office. Geneva, Switzerland. https://www.ilo.org/travail/whatwedo/publications/WCMS_751232/lang--n/index.htm

Imran, A., Humiyion, M., Arshad, M. U., Saeed, F., Arshad, M. S., Afzaal, M., Imran, M., Usman, I., Ikram, A., Naeem, U., Hussain, M., & Al Jbawi, E. (2022). Extraction, amino acid estimation, and characterization of bioactive constituents from peanut shell through eco-innovative techniques for food application. *International Journal of Food Properties*, *25*(1), 2055–2065. doi:10.1080/10942912.2022.2119999

Instituto Brasileiro de Geografia e Estatística [IBGE]. (2020). *Estimativas populacionais para os municípios e para as Unidades da Federação brasileiros em 01.07.2020. [Population estimates for the municipalities and for the Brazilian Federation Units as of 07.01.2020.]*. IBGE.

Instituto Nacional de Pesquisas Espaciais [INPE]. (2020). *Projeto Prodes Digital: Mapeamento do desmatamento da Amazônia com Imagens de Satélite. [Prodes Digital Project: Mapping deforestation of the Amazon with Satellite Images.]* São José dos Campos: Instituto Nacional de Pesquisas Espaciais.

International Center for Creativity and Sustainable Development (ICCSD). UNESCO. (2019). *Observatory on Creative Cities. The Development of the UNESCO Creative Cities Network (2004-2019)*. UNESCO. https://f2.cri.cn/M00/21/8A/rBABC2BF0MqAWMGXAMBztfzOTqQ662.pdf (01/04/2022)

Irawan, B., & Hidayat, M. N. (2022). Evaluating Local Government Website Using a Synthetic Website Evaluation Model. [IJISM]. *International Journal of Information Science and Management*, *20*(1).

IRENA. (2019). International Renewable Energy Agency - Hydrogen: a renewable energy perspective. Abu Dhabi.

IRENA-International Reneawable Energy Agency. (2014). IRENA- Global bionergy supply and deand projections: A working paper for REmap 2030. Retrieved 2022, from https://www.irena.org/publications/2014/Sep/Global-Bioenergy-Supply-and-Demand-Projections-A-working-paper-for-REmap-2030

Iscan, O., & Naktiyok, A. (2005). Attitudes towards telecommuting: The Turkish case. *Journal of Information Technology*, *20*(1), 52–63. doi:10.1057/palgrave.jit.2000023

Compilation of References

ISO/IEC. (2017). *TS 25011:2017 Information technology — Systems and software Quality Requirements and Evaluation (SQuaRE) — Service quality models*. ISO. https://www.iso.org/standard/35735.html

Jacobs, G. (2006). Communication for commitment in remote technical workforces. *Journal of Communication Management (London)*, *10*(4), 353–370. doi:10.1108/13632540610714809

Jacobson, I., Lawson, H. "., Ng, P.-W., McMahon, P. E., & Goedicke, M. (2019). *The Essentials of Modern Software Engineering: Free the Practices from the Method Prisons*. ACM Books.

Jacobson, I., Spence, I., & Ng, P. (2013). Agile and SEMAT: Perfect partners. *Communications of the ACM*, *11*(9), 1–12.

Jaiswal, K., & Anand, V. (2021). A survey on IoT-based healthcare system: potential applications, issues, and challenges. In *Advances in Biomedical Engineering and Technology* (pp. 459–471). Springer. doi:10.1007/978-981-15-6329-4_38

Janetasari, S. A., & Bokányi, L. (2022). Challenges on creation of sustainable municipal waste and wastewater management in Indonesia using experience of Hungary. *IOP Conference Series: Earth and Environmental Science*, *1017*(1), 012028. doi:10.1088/1755-1315/1017/1/012028

Javaid, M., & Khan, I. H. (2021). Internet of Things (IoT) enabled healthcare helps to take the challenges of COVID-19 Pandemic. *Journal of Oral Biology and Craniofacial Research*, *11*(2), 209–214. doi:10.1016/j.jobcr.2021.01.015 PMID:33665069

JeffersonG. H.TanakaS. YinW. (2013). Environmental Regulation and Industrial Performance: Evidence from Unexpected Externalities in China. doi:10.2139/ssrn.2216220

Jeong, H. J., & Ko, Y. (2016). Configuring an alliance portfolio for eco-friendly innovation in the car industry: Hyundai and Toyota. *Journal of Open Innovation*, *2*(4), 24. doi:10.118640852-016-0050-z

Jiang, Z., Wang, Z., and Lan, X. (2021). How Environmental Regulations Affect Corporate Innovation? the Coupling Mechanism of Mandatory Rules and Voluntary Management. *Technol. Soc.*, *65*, 101575. doi:. 2021.101575 doi:10.1016/j.techsoc

Johannes, W. (2021, February 20). The innovative contribution of multinational enterprises to sustainable development goals. *Journal of Cleaner Production*, *285*, 1–13. doi:10.1016/j.jclepro.2020.125319

JRC Publication Repository. (2021). *Historical Analysis of FCH 2 JU Electrolyser Projects*. (European COmission) Retrieved 2022, from https://publications.jrc.ec.europa.eu/repository/handle/JRC121704

Jumde, A., & du Plessis, J. (2022). Legislated corporate social responsibility (CSR) in India: The law and practicalities of its compliance. *Statute Law Review*, *43*(2), 170–197. doi:10.1093lr/hmaa004

Kameyama, N. (2004). Ética Empresarial [Business ethics]. *Revista Praia Vermelha [Red Beach Magazine]*, *11*, 148–166.

Karakos, H. L. (2015). *Understanding civic engagement among youth in diverse contexts*. Vanderbilt University.

Karmous-Edwards, G., Tomić, S., & Cooper, J. P. (2022). Developing a Unified Definition of Digital Twins. *Journal - American Water Works Association*, *114*(6), 76–78. doi:10.1002/awwa.1946

Karnani, A. (2013). Mandatory CSR in India: A Bad Proposal. *Stanford Social Innovation Review*, *16*, 20.

Kashani, M. H., Madanipour, M., Nikravan, M., Asghari, P., & Mahdipour, E. (2021). A systematic review of IoT in healthcare: Applications, techniques, and trends. *Journal of Network and Computer Applications*, *192*, 103164. doi:10.1016/j.jnca.2021.103164

Kates, R. W., Clark, W. C., Corell, R., Hall, J. M., Jaeger, C. C., Lowe, I., McCarthy, J. J., Schellnhuber, H. J., Bolin, B., Dickson, N. M., Faucheux, S., Gallopin, G. C., Grübler, A., Huntley, B., Jäger, J., Jodha, N. S., Kasperson, R. E., Mabogunje, A., Matson, P., ... Svedin, U. (2001). Sustainability Science. *Science*, *292*(5517), 641–642. doi:10.1126cience.1059386 PMID:11330321

Kaza, S., Yao, L. C., Bhada-Tata, P., & Van Woerden, F. (2018). *What a Waste 2.0: A Global Snapshot of Solid Waste Management to 2050*. World Bank. doi:10.1596/978-1-4648-1329-0

Kelaidi, E. (2020), Έξυπνες Πόλεις, Ψηφιακές Συνεργασίες και Επιχειρηματικότητα [Smart Cities, Digital Collaborations and Entrepreneurship]. [Unpublished master's thesis, National Technical University of Athens, Athens]. https://dspace.lib.ntua.gr/xmlui/bitstream/handle/123456789/51797/diplomatiki_ekelaidi.pdf?sequence=2 (1/4/2022)

Kelliher, C., & Anderson, D. (2010). Doing more with less? Flexible working practices and the intensification of work. *Human Relations*, *63*(1), 83–106. doi:10.1177/0018726709349199

Kes-Erkul, A., & Erkul, R. (2009). Web 2.0 in the Process of e-participation: The Case of Organizing for America and the Obama Administration. *National Center for Digital Government Working Paper Series*, *9*(1), 1-19.

Khan, S., Umer, R., Umer, S., & Naqvi, S. (2021). Antecedents of trust in using social media for E-government services: An empirical study in Pakistan. *Technology in Society*, *64*, 101400. https://doi.org/10.1016/j.techsoc.2020.101400

Kim, H., Koo, K. Y., & Joung, T. (2020). A study on the necessity of integrated evaluation of alternative marine fuels. Journal of International Maritime Safety, Environmental Affairs, and Shipping.

Kim, Y., Russo, S., & Amnå, E. (2017). The longitudinal relation between online and offline political participation among youth at two different developmental stages. *New Media & Society*, *19*(6), 899–917. doi:10.1177/1461444815624181

Kirchherr, J., Reike, D., & Hekkert, M. (2017). Conceptualizing the circular economy: An analysis of 114 definitions. *Resources, Conservation and Recycling*, *127*, 221–232. https://doi.org/10.1016/j.resconrec.2017.09.005

Kirchhoff, B. A. (1994). *Entrepreneurship and Dynamic Capitalism*. Praeger.

Klerk, A. (2016). Chapter 10 - Aviation Turbine Fuels Through the Fischer-Tropsch Process. Biofuels for aviation, pp. 241-259.

Klewitz, J., & Hansen, E. G. (2014). Sustainability-oriented innovation of SMEs: A systematic review. *Journal of Cleaner Production*, *65*, 57–75. doi:10.1016/j.jclepro.2013.07.017

Kline, K. L., Msangi, S., Dale, V. H., Woods, J., Souza, G. M., Osseweijer, P., . . . Mugera, H. K. (2016). Reconciling food security and bioenergy: priorities for action. GCB-Bioenergy- Bioproducts for a sustainable bioeconomy.

Ko, Y.-C., Zigan, K., & Liu, Y.-L. (2021). Carbon capture and storage in South Africa: A technological innovation system with a political economy focus. *Technological Forecasting and Social Change*, *166*, 120633. . techfore. 2021. 120633 doi:10.1016/j

Kobal, F., Agner, T., & Oliveira, A. (2009). Vantagens e desvantagens do teletrabalho: Uma pesquisa de campo em uma multinacional [Advantages and disadvantages of telecommuting: a field survey in a multinational]. XXIX Encontro Nacional de Engenharia de Produção - A Engenharia de Produção e o Desenvolvimento Sustentável: Integrando Tecnologia e Gestão [XXIX National Meetinf of Production Engineering—Production engineering and sustainable Development: Integrating technology and management]. Brasil.

Kobayashi, M. (2021). Introduction. In *Dry Syngas Purification Processes for Coal Gasification Systems* (pp. 1–49). Elsevier. doi:10.1016/B978-0-12-818866-8.00001-X

Kodali, R. K., Swamy, G., & Lakshmi, B. (2015, December). An implementation of IoT for healthcare. In *2015 IEEE Recent Advances in Intelligent Computational Systems* (RAICS) (pp. 411-416). IEEE.

Kolk, A. (2016). The social responsibility of international business: From ethics and the environment to CSR and sustainable development. *Journal of World Business, 51*(1), 23–34. doi:10.1016/j.jwb.2015.08.010

Kon, A. (2018). A inovação nos serviços como instrumento para a Inovação Social: Uma visão integrativa. [Service innovation as a tool for social innovation: an integrative view.]. *Revista de Economia Política, 38*(3), 584–605. doi:10.1590/0101-35172018-2814

Korhonen, J., Honkasalo, A., & Seppälä, J. (2018). Circular economy: The concept and its limitations. *Ecological Economics, 143*, 37–46. doi:10.1016/j.ecolecon.2017.06.041

Kotler, P., & Lee, N. (2005). Best of breed: When it comes to gaining a market edge while supporting a social cause, "corporate social marketing" leads the pack. *Social Marketing Quarterly, 11*(3-4), 91–103. doi:10.1080/15245000500414480

Kousar, S., Sabri, P. S. U., Zafar, M., & Akhtar, A. (2017). Technological factors and adoption of green innovation-moderating role of government intervention: A case of SMEs in Pakistan. *Pakistan Journal of Commerce and Social Sciences, 11*(3), 833–861.

Krauss, S. E. (2018). Towards enhancing youth participation in Muslim-Majority countries: The Case of youth-adult partnership in Malaysia. *Pertanika Journal of Social Science & Humanities, 26*, 165–188.

Kremser, W., Pentland, B. T., & Brunswicker, S. (2017). *The Continuous Transformation of Interdependence in Networks of Routines*. Presented at European Group for Organizational.

Krishnamoorthy, S., Dua, A., & Gupta, S. (2021). Role of emerging technologies in future IoT-driven Healthcare 4.0 technologies: A survey, current challenges and future directions. *Journal of Ambient Intelligence and Humanized Computing*, 1–47.

Kubule, A., Klavenieks, K., Vesere, R., & Blumberga, D. (2019). Towards Efficient Waste Management in Latvia: An Empirical Assessment of Waste Composition. *Environmental and Climate Technologies, 23*(2), 114–130. doi:10.2478/rtuect-2019-0059

Kuhlman, T., & Farrington, J. (2010, November 01). What is sustainability? *Sustainability, 2*(11), 3436–3448. doi:10.3390u2113436

Kumar, T., Ramani, V., Ahmad, I., Braeken, A., Harjula, E., & Ylianttila, M. (2018, September). Blockchain utilization in healthcare: Key requirements and challenges. In *2018 IEEE 20th International conference on e-health networking, applications and services (Healthcom)* (pp. 1-7). IEEE.

Kumari, P., & Kumar, R. (2020). Scientometric Analysis of Computer Science Publications in Journals and Conferences with Publication Patterns. *Journal of Scientometric Research, 9*(1), 54–62. doi:10.5530/jscires.9.1.6

Kunzmann, K. (2004). Culture, Creativity and Spatial Planning. *The Town Planning Review, 75*(4), 383–404. http://www.scholars-on-bilbao.info/fichas/KUNZMANN%20CultureCreativitySpatialPlanningTPR2004.pdf

Kwon, S. A. (2019). The politics of global youth participation. *Journal of Youth Studies, 22*(7), 926–940. doi:10.1080/13676261.2018.1559282

Lala, R. M. (2004). *The creation of wealth: The Tatas from the 19th to the 21st century*. Penguin Books India.

Land Oberösterreich. (2022). *Land Oberösterreich—Entwicklung der Abfallmengen 1990 bis 2020.* [*Province of Upper Austria—Development of waste volumes 1990 to 2020.*] Land Oberösterreich. https://www.land-oberoesterreich.gv.at

Landry, C. (2005). Lineages of the Creative City. In *Creativity and the City: How the creative economy is changing the city,* pp. 42–55. Charles Landry. http://charleslandry.com/panel/wp-content/uploads/downloads/2013/03/Lineages-of-the-Creative-City.pdf (28/4/2022)

Laroiya, C., Saxena, D., & Komalavalli, C. (2020). Applications of blockchain technology. In *Handbook of research on blockchain technology* (pp. 213–243). Academic Press. doi:10.1016/B978-0-12-819816-2.00009-5

Latkovikj, M. T., &, Popovskab, M. B. (2020). *How Millennials, Gen Z, and Technology are Changing the Workplace Design?* Conference: STPIS 2020 Socio-Technical Perspective in IS Development 2020. Grenoble, France.

Latkovikj, M. T., Popovska, M. B., & Popovski, V. (2016). Work Values and Preferences of the New Workforce: HRM Implications for Macedonian Millennial Generation. *Journal of Advanced Management Science, 4*(4), 312–319. doi:10.12720/joams.4.4.312-319

Lawrence, P. R., & Pirson, M. (2015). Economistic and humanistic narratives of leadership in the age of globality: Toward a renewed Darwinian theory of leadership. *Journal of Business Ethics, 128*(2), 383–394. doi:10.100710551-014-2090-2

Lazaretou, S. (2014). Η έξυπνη οικονομία: πολιτιστικές και δημιουργικές βιομηχανίες στην Ελλάδα. Μπορούν να αποτελέσουν προοπτική διεξόδου από την κρίση. [The smart economy: cultural and creative industries in Greece. Can they be a way out of the crisis?]. Bank of Greece. https://www.bankofgreece.gr/Publications/Paper2014175.pdf

Lazar, N., & Chithra, K. (2021). Comprehensive bibliometric mapping of publication trends in the development of Building Sustainability Assessment Systems. *Environment, Development and Sustainability, 23*(4), 4899–4923. doi:10.100710668-020-00796-w

Lebel, L., Anderies, J. M., Campbell, B., Folke, C., Hatfield-Dodds, S., Hughes, T. P., & Wilson, J. (2006). Governance and the Capacity to Manage Resilience in Regional Social-Ecological Systems. *Ecology and Society, 11*(1), 19. doi:10.5751/ES-01606-110119

Leeming, G., Cunningham, J., & Ainsworth, J. (2019). A ledger of me: Personalizing healthcare using blockchain technology. *Frontiers in medicine, 6,* 171. doi:10.3389/fmed.2019.00171 PMID:31396516

Lele, S. (2013). Rethinking Sustainable Development. *Current History (New York, N.Y.), 112*(757), 311–316. doi:10.1525/curh.2013.112.757.311

Lettice, F., & Parekh, M. (2010). The social innovation process: Themes, challenges and implications for practice. *International Journal of Technology Management, 51*(1), 139–158. doi:10.1504/IJTM.2010.033133

Levickaite, R. (2010). Generations X, Y, Z: How Social Networks Form the Concept of the World Without Borders (the case of Lithuania). *LIMES: Cultural Regionalistics, 3*(2), 170–183. doi:10.3846/limes.2010.17

Liao, Y. C., & Tsai, K. H. (2019). Innovation intensity, creativity enhancement, and eco-innovation strategy: The roles of customer demand and environmental regulation. *Business Strategy and the Environment, 28*(2), 316–326. doi:10.1002/bse.2232

Li, C.-Z., Crépin, A.-S., & Folke, C. (2018). The Economics of Resilience. *International Review of Environmental and Resource Economics, 11*(4), 309–353. doi:10.1561/101.00000096

Li, J., Cai, J., Khan, F., Rehman, A. U., Balasubramaniam, V., Sun, J., & Venu, P. (2020). A secured framework for sdn-based edge computing in IOT-enabled healthcare system. *IEEE Access : Practical Innovations, Open Solutions, 8,* 135479–135490. doi:10.1109/ACCESS.2020.3011503

Compilation of References

Li, J., Pan, S. Y., Kim, H., Linn, J. H., & Chiang, P. C. (2015). Building green supply chains in eco-industrial parks towards a green economy: Barriers and strategies. *Journal of Environmental Management, 162*, 158–170. doi:10.1016/j.jenvman.2015.07.030 PMID:26241931

Lin, Y. X., Sha, K. C., & Wang, J. (2020). Practical Experience and Inspiration of Foreign Emission Permit System. *Environ. Impact. Assess., 42* (1), 14–16.

Li, N. B. (2019). Environmental Regulation and Corporate Green Technology Innovation - a Conditional Process Analysis. *Inn. Mong. Soc. Sci. Chin. Ed, 40*(6), 109–115. doi:10.14137/j.cnki.issn1003-5281.2019.06.016

Lin, C. Y., & Ho, Y. H. (2010). The influences of environmental uncertainty on corporate green behavior: An empirical study with small and medium-size enterprises. *Social Behavior and Personality, 38*(5), 691–696. doi:10.2224bp.2010.38.5.691

Linden, A., & Carlsson-Kanyama, A. (2003). Environmentally friendly disposal behaviour and local support systems: Lessons from a metropolitan area. *Local Environment, 8*(3), 291–301. doi:10.1080/13549830306664

Lindenberg, S., & Steg, L. (2007). Normative, gain and hedonic goal frames guiding environmental behaviour. *The Journal of Social Issues, 63*(1), 117–137. doi:10.1111/j.1540-4560.2007.00499.x

Linders, D. (2012). From e-government to we-government: Defining a typology for citizen coproduction in the age of social media. *Government Information Quarterly, 29*(4), 446–454.

Lin, H., Zeng, S. X., Ma, H. Y., Qi, G. Y., & Tam, V. W. Y. (2014). Can political capital drive corporate green innovation? Lessons from China. *Journal of Cleaner Production, 64*, 63–72. doi:10.1016/j.jclepro.2013.07.046

Li, S., Song, X., & Wu, H. (2015). Political connection, ownership structure, and corporate philanthropy in China: A strategic-political perspective. *Journal of Business Ethics, 129*(2), 399–411. doi:10.100710551-014-2167-y

Liswani, H., & Scarioni, B. (2021). Tech4Dev - Operational Report. *EPFL.* https://www.epfl.ch/innovation/domains/wp-content/uploads/2022/03/Tech4Dev-2021-Long.pdf

Liu, J., Zhao, M., & Wang, Y. (2020). Impacts of government subsidies and environmental regulations on green process innovation: A nonlinear approach. *Technology in Society, 63*, 101417. doi:10.1016/j.techsoc.2020.101417

Liu, Y., Wang, A., & Wu, Y. (2021). Environmental regulation and green innovation: Evidence from China's new environmental protection law. *Journal of Cleaner Production, 297*, 126698. doi:10.1016/j.jclepro.2021.126698

Liu, Y., Zhu, J., Li, E., Meng, Z., & Song, Y. (2020). Environmental regulation, green technological innovation, and ecoefficiency: The case of Yangtze river economic belt in China. *Technological Forecasting and Social Change, 155*, 1–21. doi:10.1016/j.techfore.2020.119993

Li, Y., Shan, B., Li, B., Liu, X., & Pu, Y. (2021). Literature review on the applications of machine learning and blockchain technology in smart healthcare industry: A bibliometric analysis. *Journal of Healthcare Engineering, 2021*, 2021. doi:10.1155/2021/9739219 PMID:34426765

Li, Z., Liao, G., Wang, Z., & Huang, Z. (2018). Green loan and subsidy for promoting clean production innovation. *Journal of Cleaner Production, 187*, 421–431. doi:10.1016/j.jclepro.2018.03.066

Lodovici, M. S. (2021). *The impact of teleworking and digital work on workers and society.* Study Requested by the EMPL Committee.

López, F. (2002). El análisis de contenido como método de investigación. [Content analysis as a research method.]. *XXI Revista de Educación, 4*, 167–179.

López, P., & Rodríguez, P. (2020). Who is Teleworking and Where from? Exploring the Main Determinants of Telework in Europe. *Sustainability*, *12*(21), 87–97. doi:10.3390u12218797

Luo, Y., Salman, M., & Lu, Z. (2021). Heterogeneous impacts of environmental regulations and foreign direct investment on green innovation across different regions in China. *The Science of the Total Environment*, *759*, 143744. doi:10.1016/j.scitotenv.2020.143744 PMID:33341514

Lupton, P., & Haynes, B. (2000). Teleworking – the perception-reality gap. *Facilities*, *18*(7/8), 323–328. doi:10.1108/02632770010340726

Lynch, T., & Khan, T. (2020). Understanding what sustainability is not – and what it is. *The Ecological Citizen*, *3*(B), 55-65.

Maak, T. (2007). Responsible leadership, stakeholder engagement, and the emergence of social capital. *Journal of Business Ethics*, *74*(4), 329–343. doi:10.100710551-007-9510-5

Maak, T., & Pless, N. M. (2006). Responsible leadership in a stakeholder society–a relational perspective. *Journal of Business Ethics*, *66*(1), 99–115. doi:10.100710551-006-9047-z

Määttä, M., & Aaltonen, S. (2016). Between rights and obligations–rethinking youth participation at the margins. *The International Journal of Sociology and Social Policy*, *36*(3/4), 157–172. doi:10.1108/IJSSP-09-2014-0066

MacArthur, E. (2013). *Towards the circular economy, economic and business rationale for an accelerated transition*. Ellen MacArthur Foundation.

Machado, F. (2016). *Análise e Gestão de Requisitos de Software – Onde nascem os sistemas* [Analysis and Management of Software Requirements – Where systems are born.]. Érica - Saraiva.

Maddison, A. (2008). The West and the Rest in the World Economy: 1000-2030. *World Economy*, *9*(4).

Magro, M. J. (2012). A Review of Social Media Use in E-Government. *Administrative Sciences*, *2*(2), 148–161.

Mainsah, H., Brandtzæg, P. B., & Følstad, A. (2016). Bridging the generational culture gap in youth civic engagement through social media: Lessons learnt from young designers in three civic organizations. *The Journal of Media Innovations*, *3*(1), 23–40. doi:10.5617/jmi.v3i1.2724

Maisey, S. (2022, July 6). Lifestyle. *The National News*. www.thenationalnews.com: https://www.thenationalnews.com/lifestyle/fashion/2022/07/06/dubai-school-becomes-first-in-the-world-to-add-thaely-vegan-shoes-to-its-uniform/

Mallawaarachchi, H., Sandanayake, Y., Karunasena, G., & Liu, C. (2020). Unveiling the conceptual development of industrial symbiosis: Bibliometric analysis. *Journal of Cleaner Production*, *258*, 120618. doi:10.1016/j.jclepro.2020.120618

Mallikarjuna, B., Shrivastava, G., & Sharma, M. (2022). Blockchain technology: A DNN token-based approach in healthcare and COVID-19 to generate extracted data. *Expert Systems: International Journal of Knowledge Engineering and Neural Networks*, *39*(3), e12778. doi:10.1111/exsy.12778 PMID:34511692

Ma, M., Yang, G., Wang, H., Lu, Y., Zhang, B., Cao, X., Peng, D., Du, X., Liu, Y., & Huang, Y. (2019). Ordered distributed nickel sulfide nanoparticles across graphite nanosheets for efficient oxygen evolution reaction electrocatalyst. *International Journal of Hydrogen Energy*, *44*(3), 1544–1554. doi:10.1016/j.ijhydene.2018.11.176

Manisha, K., M., & Kaur, K. (2021). Impact of Agile Scrum Methodology on Team's Productivity and Client Satisfaction – A Case Study. *3rd International Conference on Advances in Computing, Communication Control and Networking (ICAC3N)*, (pp. 1686-1691). 10.1109/ICAC3N53548.2021.9725505

Compilation of References

Marketeer. (2020). *Marketeer.* https://www.marketeer.sapo.pt: https://www.marketeer.sapo.pt/numero-de-portugueses-que-descarregam-apps-gratis-triplica

Marshall, K. (2018). Global education challenges: Exploring religious dimensions. *International Journal of Educational Development, 62*, 184–191. doi:10.1016/j.ijedudev.2018.04.005

Martin, B., & MacDonnell, R. (2012). Is telework effective for organizations? A meta-analysis of empirical research on perceptions of telework and organizational outcomes. *Management Research Review, 35*(7), 602–616. doi:10.1108/01409171211238820

Martinez, F., O'Sullivan, P., Smith, M., & Esposito, M. (2017). Perspectives on the role of business in social innovation. *Journal of Management Development, 36*(5), 36. doi:10.1108/JMD-10-2016-0212

Martínez-Fernández, J., & Banos-González, I. (2021). *An integral approach to address socio-ecological systems sustainability.*

Martínez-Sánchez, A., Pérez-Pérez, M., Vela-Jiménez, M., & Carnicer, P. (2008). Telework adoption, change management and firm performance. *Journal of Organizational Change Management, 21*(1), 7–31. doi:10.1108/09534810810847011

Martins, J. D. D. (2020). Função social e responsabilidade social empresarial: O princípio da solidariedade como marco jurídico-constitucional para uma nova empresa cidadã. [Social function and corporate social responsibility: The principle of solidarity as a legal-constitutional framework for a new citizen enterprise.]. *Revista de Direito Ambiental e Socioambientalismo, 6*(2), 38–52. doi:10.26668/IndexLawJournals/2525-9628/2020.v6i2.7124

Marx, C., Reimann, M., & Diewald, M. (2021). Do Work-Life Measures Really Matter? The Impact of Flexible Working Hours and Home-Based Teleworking in Preventing Voluntary Employee Exits. *Social Sciences, 10*(1), 1–22. doi:10.3390ocsci10010009

Mascheroni, G. (2017). A practice-based approach to online participation: Young people's participatory habitus as a source of diverse online engagement. *International Journal of Communication, 11*, 4630–4651.

Masquietto, C. D., Sacomano Neto, M., & Giuliani, A. C. (2011). Centrality and Density in Interfirm Networks: a Study of an Ethanol Local Productive Arrangement. *Review of Administration and Innovation - RAI, 8*(1), 122–147. doi:10.5773/rai.v8i1.456

Matar, S., & Hatchl, F. (2001). *Ammonia production (Haber Process).* Chem Petrochemical Process.

Mathew, P. S., Pillai, A. S., & Palade, V. (2018). Applications of IoT in healthcare. In *Cognitive Computing for Big Data Systems Over IoT* (pp. 263–288). Springer.

McBee, M. P., & Wilcox, C. (2020). Blockchain technology: Principles and applications in medical imaging. *Journal of Digital Imaging, 33*(3), 726–734. doi:10.100710278-019-00310-3 PMID:31898037

McCaffery, C., Zhu, H., Ahmed, C. S., Canchola, A., Chen, J. Y., Li, C., ... Karavalaskis, G. (2022). Effects of hydrogenated vegetable oil (HVO) and HVO/biodiesel blends on the physicochemical and toxicological properties of emissions from an off-road heavy-duty diesel engine. *Fuel, 323*, 323. doi:10.1016/j.fuel.2022.124283

McClelland, D. C., & Boyatzis, R. E. (1982). Leadership motive pattern and long-term success in management. *The Journal of Applied Psychology, 67*(6), 737–743. doi:10.1037/0021-9010.67.6.737

McCowan, T. (2009). A 'seamless enactment' of citizenship education. *Journal of Philosophy of Education, 43*(1), 85–99. doi:10.1111/j.1467-9752.2009.00669.x

Medellin. Official city portal. (2022). Medellin. https://www.medellin.gov.co/irj/portal/medellin (1/4/)

Mehrad, J., Eftekhar, Z., & Goltaji, M. (2020). Vaccinating users against the hypodermic needle theory of social media: Libraries and improving media literacy. [IJISM]. *International Journal of Information Science and Management*, *18*(1), 17–24.

Meinzen-Dick, R., Di Gregorio, M., & McCarthy, N. (2004). *Methods of Studying Collective Action in Rural Development*. International Food Policy Research Institute. doi:10.1016/j.agsy.2004.07.006

Melé, D. (2016). Understanding humanistic management. *Humanistic Management Journal*, *1*(1), 33–55. doi:10.100741463-016-0011-5

Melles, M. O., & Ricker, C. L. (2017). Youth participation in HIV and sexual and reproductive health decision-making, policies, programs: Perspectives from the field. *International Journal of Adolescence and Youth*, *23*(2), 159–167. doi:10.1080/02673843.2017.1317642

Mello, J. (2007). Managing Telework Programs Effectively. *Employee Responsibilities and Rights Journal*, *19*(4), 247–261. doi:10.100710672-007-9051-1

Mercur, B. ([s.d.]). *Mercur - Desde 1924—O mundo de um jeito bom pra todo o mundo. [Mercur - Since 1924—The world in a good way for everyone.]* Mercur. https://mercur.com.br/

Mergel, I. (2013). A framework for interpreting social media interactions in the public sector. *Government Information Quarterly*, *30*(4), 327–334.

Meschengieser, G. (2022). *Waterfuturism, a new perspective that is here to stay*. IWA. https://iwa-network.org/waterfuturism-a-new-perspective-that-is-here-to-stay/

Mesquida, M. (2021). *Digital Twin in Water Distribution Networks*. Issue February.

Metallo, C., Gesuele, B., Guillamón, M., & Ríos, A. (2020). Determinants of public engagement on municipal Facebook pages. *The Information Society*, *36*(3), 147–159.

Metcalf, L., & Benn, S. (2013). Leadership for sustainability: An evolution of leadership ability. *Journal of Business Ethics*, *112*(3), 369–384. doi:10.100710551-012-1278-6

Metzger, M., Erete, S., Barton, D. L., Desler, M. K., & Lewis, D. A. (2014). The new political voice of young Americans: Online engagement and civic development among first-year college students. *Education, Citizenship and Social Justice*, *10*(1), 55–66. doi:10.1177/1746197914558398

Miller, J. G., Bersoff, D. M., & Harwood, R. L. (1990). Perceptions of social responsibilities in India and in the United States: Moral imperatives or personal decisions? *Journal of Personality and Social Psychology*, *58*(1), 33–47. doi:10.1037/0022-3514.58.1.33 PMID:2308074

Mills, J., Wong-Ellison, C., Werner, W., & Clay, J. (2001). Employer liability for telecommuting employees. *The Cornell Hotel and Restaurant Administration Quarterly*, *42*(5), 48–59. doi:10.1016/S0010-8804(01)80057-4

Minniti, M. (2012). El emprendimiento y el crecimiento económico de las naciones. [Entrepreneurship and the economic growth of nations.]. *Economía Industrial*, *383*, 23–30.

Mirvis, P., Herrera, M. E. B., Googins, B., & Albareda, L. (2016). Corporate social innovation: How firms learn to innovate for the greater good. *Journal of Business Research*, *69*(11), 5014–5021. doi:10.1016/j.jbusres.2016.04.073

Compilation of References

Mitra, N., & Schmidpeter, R. (2017). The why, what and how of the CSR mandate: The India story. In *Corporate Social Responsibility in India* (pp. 1–8). Springer. doi:10.1007/978-3-319-41781-3_1

Mohr, K. A., & Mohr, E. (2017). Understanding Generation Z Students to Promote a Contemporary Learning Environment. *Journal on Empowering Teaching Excellence*, *1*(1), 84–94. doi:10.15142/T3M05T

Mollica, C. (2017). The diversity of identity: Youth participation at the Solomon Islands Truth and Reconciliation Commission. *Australian Journal of International Affairs*, *71*(4), 371–388. doi:10.1080/10357718.2017.1290045

Mondal, A. (2022). *Feet & fine: How vegan leather, biodegradable shoes are trying to save Earth*. MSN. www.msn.com: https://www.msn.com/en-in/news/other/feet-fine-how-vegan-leather-biodegradable-shoes-are-trying-to-save-earth/ar-AASOTG0

Mondejar, M. E., Avtar, R., Diaz, H. L. B., Dubey, R. K., Esteban, J., Gómez-Morales, A., Hallam, B., Mbungu, N. T., Okolo, C. C., Prasad, K. A., She, Q., & Garcia-Segura, S. (2021). Digitalization to achieve sustainable development goals: Steps towards a Smart Green Planet. *The Science of the Total Environment*, *794*(June), 148539. doi:10.1016/j.scitotenv.2021.148539 PMID:34323742

Moneva, J. M., & Martín, E. (2012). Universidad y Desarrollo sostenible: Análisis de la rendición de cuentas de las universidades públicas desde un enfoque de responsabilidad social. [University and Sustainable Development: Analysis of the accountability of public universities from a social responsibility approach.]. *Revista Iberoamericana de Contabilidad de Gestión*, *10*(19), 1–18.

Morciano, D., Scardigno, F., Manuti, A., & Pastore, A. (2016). A theory-based evaluation to improve youth participation in progress: A case study of a youth policy in Italy. *Child and Youth Services*, *37*(4), 304–324. doi:10.1080/0145935X.2015.1125289

Moreira, M., Mourato, S., Rodrigues, C., Silva, S., Guimarães, R., & Chibeles, C. (2021). Building a Digital Twin for the Management of Pressurised Collective Irrigation Systems. In *Proceedings of the 1st International Conference on Water Energy Food and Sustainability (ICoWEFS 2021)* (Vol. 3, Issue ICoWEFS 2021, pp. 785–795). Springer International Publishing. 10.1007/978-3-030-75315-3_83

Moren, E. (1998). Η πολιτική πολιτισμού. [The politics of culture]. In E. Moren & S. Nair, *Μια πολιτική πολιτισμού [Lexical characteristics of Greek language] [A policy of culture]*. Athens Nantes, Official city portal. https://metropole.nantes.fr/

Morvan, Y. (1985). *Filière de production: fondementes d'economie industrielle [Production chain: foundations of industrial economics.]*. Economica.

Mossberger, K., Wu, Y., & Crawford, J. (2013). Connecting citizens and local governments? Social media and interactivity in major U.S. cities. *Government Information Quarterly*, *30*(4), 351–358.

Motesharrei, S., Rivas, J., & Kalnay, E. (2014). Human and nature dynamics (HANDY): Modeling inequality and use of resources in the collapse or sustainability of societies. *Ecological Economics*, *101*, 90–102. doi:10.1016/j.ecolecon.2014.02.014

Motesharrei, S., Rivas, J., Kalnay, E., Asrar, G. R., Busalacchi, A. J., Cahalan, R. F., Cane, M. A., Colwell, R. R., Feng, K., Franklin, R. S., Hubacek, K., Miralles-Wilhelm, F., Miyoshi, T., Ruth, M., Sagdeev, R., Shirmohammadi, A., Shukla, J., Srebric, J., Yakovenko, V. M., & Zeng, N. (2016). Modeling sustainability: Population, inequality, consumption, and bidirectional coupling of the Earth and Human Systems. *National Science Review*, *3*(4), 470–494. doi:10.1093/nsr/nww081 PMID:32747868

Mousavi, S., Bossink, B., & van Vliet, M. (2019). Microfoundations of companies' dynamic capabilities for environmentally sustainable innovation: Case study insights from high-tech innovation in science-based companies. *Business Strategy and the Environment*, *28*(2), 366–387. doi:10.1002/bse.2255

Mulgan, G. (2007). *Gestão de organizações sem fins lucrativos: o desafio da inovação social* [*Management of nonprofit organizations: the challenge of social innovation.*]. Edições Vida Económica.

Müller-Czygan, G., Tarasyuk, V., Wagner, C., & Wimmer, M. (2021). How does digitization succeed in the municipal water sector? The waterexe4.0 meta-study identifies barriers as well as success factors, and reveals expectations for the future. *Energies*, *14*(22), 1–21. doi:10.3390/en14227709

Murphy, K. M., Shleifer, A., & Vishny, R. W. (1993). Why is rent-seeking so costly to growth? *The American Economic Review*, *83*(2), 409–414. https://www.jstor.org/stable/2117699

Naciti, V., Cesaroni, F., & Pulejo, L. (2022). Corporate governance and sustainability: A review of the existing literature. *The Journal of Management and Governance*, *26*(1), 55–74. doi:10.100710997-020-09554-6

Nakamura, B., Williams, P. (US) Inc. (2017). Intelligent Water System: The Path to a SMART Utility. *Joint Knowledge Development Forum (KDF) in Conjunction with the International Society of Automation's (ISA) Water/Wastewater and Automation Controls Symposium*.

Narksompong, J., & Limjirakan, S. (2015). Youth participation in climate change for sustainable engagement. *Review of European, Comparative & International Environmental Law*, *24*(2), 171–181. doi:10.1111/reel.12121

Natale, D. (2011). Complexity and data quality. *Commission UNINFO JTC1/SC7*. *Software Engineering*, 1–4.

Navarro, A., Ruiz, M., De los Ríos, A., & Tirado, P. (2011). Responsabilidad social y administración pública local: un análisis del grado de divulgación de información en Reino Unido e Irlanda. [Social responsibility and local public administration: an analysis of the degree of disclosure of information in the United Kingdom and Ireland.] *In Actas del XVI Congreso AECA*, Granada: Asociación Española de Contabilidad y Administración de Empresas.

Navarro, A., Alcaraz, F. J., & Ortiz, D. (2010). La divulgación de información sobre responsabilidad corporativa en administraciones públicas: Un estudio empírico en gobiernos locales. [The disclosure of information on corporate responsibility in public administrations: an empirical study in local governments.]. *Revista de Contabilidad*, *13*(2), 285–314. doi:10.1016/S1138-4891(10)70019-4

Navarro, A., Tirado, P., Ruiz, M., & De los Ríos, A. (2015). Divulgación de información sobre responsabilidad social de los gobiernos locales europeos: El caso de los países nórdicos. [Dissemination of information on social responsibility of European local governments: The case of the Nordic countries.]. *Gestión y Política Pública*, *24*(1), 229–269.

Neumann, T. (2021). Does it pay for new firms to be green? An empirical analysis of when and how different greening strategies affect the performance of new firms. *Journal of Cleaner Production*, *317*, 128403. doi:10.1016/j.jclepro.2021.128403

Neumeier, S. (2012). Why do Social Innovations in Rural Development Matter and Should They be Considered More Seriously in Rural Development Research? Proposal for a Stronger Focus on Social Innovations in Rural Development Research. *Journal of the European Society for Rural Sociology*, *52*(1), 48–69. doi:10.1111/j.1467-9523.2011.00553.x

Nevado, M. T., Gallardo, D., & Carvalho, L. (2019). Entrepreneurship in a local government: An empirical study of information in the websites of Alentejo region municipalities (Portugal). *Innovar (Universidad Nacional de Colombia)*, *29*(71), 97–112. doi:10.15446/innovar.v29n71.76398

Nicolini, D., & Monteiro, P. (2017). The practice approach: For a praxeology of organizational and management studies. In A. Langley & H. Tsoukas (Eds.), *The Sage Handbook of Process Organization Studies*. Sage.

Compilation of References

Noonan, M., & Glass, J. (2012). The Hard Truth About Telecommuting. *Monthly Labor Review*, U. S. Department of Labor. *Bureau of Labor Statistics.*, *135*, 38–45.

NovaInnovation. (2020). Green syngas from carbon dioxide. Retrieved 2022, from https://novainnovation.unl.pt/2020/01/22/green-syngas-from-carbon-dioxide/

Nwachukwu, P. T., & Kang'ethe, S. M. (2014). An insightful assessment on impact of local economic development programs in South Africa and Nigeria: Linking socio-cooperative connections and youths' participation. *Mediterranean Journal of Social Sciences*, *5*(23), 1621–1621. doi:10.5901/mjss.2014.v5n23p1621

Νέα Αρχιτεκτονική της Αυτοδιοίκησης και της Αποκεντρωμένης Διοίκησης – Πρόγραμμα Καλλικράτης. [New Architecture of Local Government and Decentralized Administration - Kallikratis Program]. http://www.et.gr (22/4/2022)

Object Management Group. (2018). *Essence—Kernel and Language for Software Engineering Methods*. OMG. www.omg.org/spec/Essence/

Oddo, E., & Masi, M. (2021). Roadmap to 2050: The Land-Water-Energy Nexus of Biofuels - Biofuels Technologies. Retrieved 2022, from https://roadmap2050.report/biofuels/biofuels-technologies

OECD. (2015). *G20/OECD Principles of Corporate Governance*. Organisation for Economic Co-operation and Development. doi:10.1787/9789264236882-

OECD. (2018). *OECD Environmental Performance Reviews: Hungary 2018*. Organisation for Economic Co-operation and Development. https://www.oecd-ilibrary.org/environment/hungary-2018_9789264298613-en

OECD. (2021). Biofuels. (OECD-Organisation for Economic Co-operation and Development- FAO Agricultural Outlook 2021-2030) Retrieved 2022, from https://www.oecd-ilibrary.org/sites/89d2ac54-en/index.html?itemId=/content/component/89d2ac54-en

OECD . (2022) Global Plastic Outlook. OECD Library.

Okolie, J. A., Patra, B. R., Mukherjee, A., Nanda, S., Dalai, A. K., & Kozinski, J. A. (2021). Futuristic applications of hydrogen in energy, biorefining, aerospace, pharmaceuticals and metallurgy. Journal of Hydrogen Energy, 46.

Oliveira, G. H. M., & Welch, E. W. (2013). Social media use in local government: Linkage of technology, task, and organizational context. *Government Information Quarterly*, *30*(4), 397–405.

Oller Alonso, M. (2021). *La Responsabilidad Social Corporativa de las empresas del mármol en la comarca del Almanzora, Almería (2019-2021): análisis de sus estrategias de comunicación integral. [The Corporate Social Responsibility of marble companies in the Almanzora region, Almería (2019-2021): analysis of their integral communication strategies.]* [Doctorate, Universidad de Murcia Escuela Internacional de Doctorado].

Onofrei, M., Vatamanu, A. F., & Cigu, E. (2022). The Relationship Between Economic Growth and CO2 Emissions in EU Countries: A Cointegration Analysis. *Frontiers in Environmental Science*, *10*, 934885. doi:10.3389/fenvs.2022.934885

Onyechi, N. J. (2018). Taking their destiny in their hands: Social media, youth participation and the 2015 political campaigns in Nigeria. *African Journalism Studies*, *39*(1), 69–89. doi:10.1080/23743670.2018.1434998

Organisation for Economic Co-operation and Development (OECD). (2010). *Education at a glance 2010: OECD indicators*. OECD.

Organização das Nações Unidas [ONU]. (2018). *Objetivos de Desenvolvimento Sustentável.* [*Production chain: foundations of industrial economics.*] ONU.

Organização das Nações Unidas. (2020). UN government survey 2020. UN. https://publicadministration.un.org/egovkb/en-us/Reports/UN-E-Government-Survey-2020

Orloff, A. S., & Palier, B. (2009). The power of gender perspectives: Feminist influence on policy paradigms, social science, and social politics. *Social Politics, 16*(4), 405–412. doi:10.1093p/jxp021

Orville, H. (2019). The Relationship between Sustainability and Creativity. *Cadmus, Promoting Leadership in Thought that Leads to Action, 4*(1). https://tinyurl.com/2s3hsame

Osamah Siddiqui, I. D. (2018). A review and comparative assessment of direct ammonia fuel cells. *Thermal Science and Engineering Progress, 5,* 568–578. doi:10.1016/j.tsep.2018.02.011

OVAM. (2022a). *Jaarverslag 2021.* OVAM. https://jaarverslag.ovam.be/sites/default/files/2022-05/OVAM_jaarverslag_2021-JB.pdf

OVAM. (2022b). Preventie- en sorteergedrag van de Vlaamse Bevolking. Samenvatting.... [Prevention and sorting behaviour of the Flemish Population. Summary]. OVAM. www.vlaanderen.be. https://www.vlaanderen.be/publicaties/preventie-en-sorteergedrag-van-de-vlaamse-bevolking-samenvatting-kwantitatieve-en-kwalitatieve-bevraging-2021

Pacheco, D. A. de J., ten Caten, C. S., Jung, C. F., Ribeiro, J. L. D., Navas, H. V. G., & Cruz-Machado, V. A. (2017). Eco-innovation determinants in manufacturing SMEs: Systematic review and research directions. *Journal of Cleaner Production, 142,* 2277–2287. doi:10.1016/j.jclepro.2016.11.049

Pache, M., & Nevado, M. T. (2021). Compromiso de los Ayuntamientos Malagueños con la divulgación de información responsable. [Commitment of Malaga City Councils to the dissemination of responsible information.]. *Transinformação, 33.*

Paes-de-Souza, M., Silva, T. N., Pedrozo, E. A., & Souza Filho, T. A. (2011). O Produto Florestal Não Madeirável (PFNM) amazônico açaí nativo: Proposição de uma organização social baseada na lógica de cadeia e rede para potencializar a exploração local. [The Native Amazonian Non-Timber Forest Product (PFNM): proposition of a social organization based on chain and network logic to enhance local exploration.]. *Revista de Administração e Negócios da Amazônia, 3*(2), 44–57.

Pahija, E., Golshan, S., Blais, B., & Boffito, D. C. (2022). Chemical Engineering and Processing - Process Intensification.

Palmer, G. (1992). Earth summit: What went wrong at Rio. *Wash. ULQ, 70,* 1005.

Panebianco, S. (2021). Towards a Human and Humane Approach? The EU Discourse on Migration amidst the Covid-19 Crisis. *The International Spectator, 56*(2), 19–37. doi:10.1080/03932729.2021.1902650

Papadopoulou, E. (2019). *Δημιουργική πόλη και πολιτικές: Θεωρητικές προσεγγίσεις και η περίπτωση της Θεσσαλονίκης* [*Creative city and policies: Theoretical approaches and the case of Thessaloniki*] [Unpublished master's thesis, Aristotle University of Thessaloniki. Thessaloniki.] http://ikee.lib.auth.gr/record/308004/files/PAPADOPOULOUERMIONH_DE.pdf (1/4/2022)

Parry, E., & Urwin, P. (2011). Generational Differences in Work Values: A Review of Theory and Evidence. *International Journal of Management Reviews, 73*(1), 79–96. doi:10.1111/j.1468-2370.2010.00285.x

Compilation of References

Paschalidis, Gr. (2002). Η συμβολή του πολιτισμού στην κοινωνική και οικονομική ανάπτυξη [The contribution of culture to social and economic development]. In Paschalidis, Gr. & Hambouri-Ioannidou Aik., Οι Διαστάσεις των Πολιτιστικών Φαινομένων: Εισαγωγή στον Πολιτισμό [Lexical characteristics of Greek language] [The Dimensions of Cultural Phenomena: Introduction to Culture]. Hellenic Open University. Patras

Pasricha, P., Singh, B., & Verma, P. (2018). Ethical leadership, organic organizational cultures, and corporate social responsibility: An empirical study in social enterprises. *Journal of Business Ethics*, *151*(4), 941–958. doi:10.100710551-017-3568-5

Pedersen, A. N., Borup, M., Brink-Kjær, A., Christiansen, L. E., & Mikkelsen, P. S. (2021). Living and prototyping digital twins for urban water systems: Towards multi-purpose value creation using models and sensors. *Water (Switzerland)*, *13*(5), 592. doi:10.3390/w13050592

Pedrozo, E. A., Silva, T. N., Sato, S. A. S., & Oliveira, N. D. A. (2011). Produtos Florestais Não Madeiráveis (PFNMS): As Filières do Açaí e da Castanha da Amazônia. [Non-Timber Forest Products (PFNMS): the Filières do Açaí and Castanha da Amazônia]. *Revista de Administração e Negócios da Amazônia*, *3*(2), 88–112.

Peffers, T., Tuunanen, T., Rothenberger, M. A., & Chatterjee, S. (2007). A Design Science Research Methodology for Information Systems Research. *Journal of Management Information Systems*, *24*(3), 45–78. doi:10.2753/MIS0742-1222240302

Peng, B., Guo, D., Qiao, H., Yang, Q., Zhang, B., Hayat, T., Alsaedi, A., & Ahmad, B. (2018). Bibliometric and visualized analysis of China's coal research 2000–2015. *Journal of Cleaner Production*, *197*, 1177–1189. doi:10.1016/j.jclepro.2018.06.283

Penzenstadler, B., Raturi, A., Richardson, D., & Tomlinson, B. (2014). Safety, Security, Now Sustainability: The Nonfunctional Requirement for the 21st Century. *IEEE Software*, *31*(3), 40–47. doi:10.1109/MS.2014.22

Percy-Smith, B., McMahon, G., & Thomas, N. (2019). Recognition, inclusion and democracy: Learning from action research with young people. *Educational Action Research*, *27*(3), 347–361. doi:10.1080/09650792.2019.1577149

Pérez-Pérez, M., Sánchez, A., & Carnicer, M. (2003). The organizational implications of human resources managers' perception of teleworking. *Personnel Review*, *32*(6), 733–755. doi:10.1108/00483480310498693

Perincherry V. (2009). A Framework for Evaluating Regional Impacts of Broadband Internet Access: Application to Telecommuting Behavior. doi:10.2139/ssrn.1489377

Perrin, T., & Delvainquière, J. C. (2015). *France/ 1. Historical perspective: cultural policies and instruments*. Compendium: *Cultural Policies and Trends in Europe*. https://tinyurl.com/33j7nkn8 (29/03/2022)

Pesantez, J. E., Alghamdi, F., Sabu, S., Mahinthakumar, G., & Berglund, E. Z. (2021, July). Using a digital twin to explore water infrastructure impacts during the COVID-19 pandemic. *Sustainable Cities and Society*, *77*, 103520. doi:10.1016/j.scs.2021.103520 PMID:34777984

PETA India. (2021, November 8). PETA. www.petaindia.com: https://www.petaindia.com/blog/milind-soman-alia-bhatts-ed-a-mamma-and-sunny-leones-i-am-animal-among-winners-of-peta-in dias-vegan-fashion-awards-2021/

Petrakos, G., & Oikonomou, D. (1999). Διεθνοποίηση και διαρθρωτικές αλλαγές στο Ευρωπαϊκό σύστημα αστικών κέντρων. [Internationalization and structural changes in the European system of urban centers] In D. Economou & G. Petrakos (Eds.), *Η ανάπτυξη των Ελληνικών πόλεων. Διεπιστημονικές προσεγγίσεις αστικής ανάλυσης και πολιτικής* [The development of Greek cities. Interdisciplinary approaches to urban analysis and politics]. (pp. 13–44). University Publications of Thessaly – Gutenberg.

Pettigrew, A., & Whipp, R. (1992). Managing change and corporate performance. In *European industrial restructuring in the 1990s* (pp. 227–265). Palgrave Macmillan. doi:10.1007/978-1-349-12582-1_9

Phillips, W., Lee, H., Ghobadian, A., O'Regan, N., & James, P. (2015). Social Innovation and Social Entrepreneurship: A Systematic Review. *Group & Organization Management*, 40(3), 428–461. doi:10.1177/1059601114560063

Phills, J. A. Jr, Deiglmeier, K., & Miller, D. T. (2008). Rediscovering Social Innovation. *Stanford Social Innovation Review*, 6(4), 34–43. doi:10.48558/GBJY-GJ47

Pirro, A. L., & Róna, D. (2019). Far-right activism in Hungary: Youth participation in Jobbik and its network. *European Societies*, 21(4), 603–626. doi:10.1080/14616696.2018.1494292

Pirson, M. (2017). *Humanistic management: Protecting dignity and promoting well-being*. Cambridge University Press. doi:10.1017/9781316675946

Pirson, M. A., & Lawrence, P. R. (2010). Humanism in business–towards a paradigm shift? *Journal of Business Ethics*, 93(4), 553–565. doi:10.100710551-009-0239-1

Pisters, S. R., Vihinen, H., & Figueiredo, E. (2020). Inner change and sustainability initiatives: Exploring the narratives from eco-villagers through a place-based transformative learning approach. *Sustainability Science*, 15(2), 395–409. doi:10.100711625-019-00775-9

Pohl, M., & Tolhurst, N. (2010). *Responsible business: How to manage a CSR strategy successfully*. John Wiley & Sons.

Pokhrel, N. (2022). *Five Lessons For Digitizing Asia's Water Systems* [Text]. Asian Development Bank. https://blogs.adb.org/blog/five-lessons-digitizing-asia-s-water-systems

Pompeo, W. (2016). *(R)Evolução Digital: Análises e perspectivas das novas tecnologias da informação e comunicação no direito, educação e gestão de negócios [(R)Digital Evolution: Analysis and perspectives of new information and communication technologies in law, education and business management.]*. Fadisma.

Popoli, P. (2016). Social Enterprise and Social Innovation: A Look Beyond Corporate Social Responsibility. Em R. Laratta, Social Enterprise—Context-Dependent Dynamics In A Global Perspective. IntechOpen. doi:10.5772/62980

Porter, M. E., & Kramer, M. R. (2006). Strategy and society: The link between competitive advantage and corporate social responsibility. *Harvard Business Review*, 84(12), 78–92, 163. PMID:17183795

Portugal, 2. (17 de 04 de 2022). *Portugal Inovação social*. Portugal Inovação social. https://inovacaosocial.portugal2020.pt

Poteat, V. P., Calzo, J. P., & Yoshikawa, H. (2018). Gay-Straight Alliance involvement and youths' participation in civic engagement, advocacy, and awareness-raising. *Journal of Applied Developmental Psychology*, 56, 13–20. doi:10.1016/j.appdev.2018.01.001 PMID:29805190

Pounder, P. (2021). Responsible leadership and COVID-19: small Island making big waves in cruise tourism. *International Journal of Public Leadership*.

Prensky, M. (2001). Digital Natives, Digital Immigrants. *On the Horizon*, 9(5), 1–6. doi:10.1108/10748120110424816

Preskill, H., & Beer, T. (2012). *Evaluating social innovation*. Centre for Evaluation Innovation. doi:10.22163/fteval.2012.119

Prieto-Sandoval, V., Jaca, C., & Ormazabal, M. (2018). Towards a consensus on the circular economy. *Journal of Cleaner Production*, 179, 605–615. doi:10.1016/j.jclepro.2017.12.224

Compilation of References

Primăria Municipiului Timişoara. (2022). *Acasa—Primăria Municipiului Timişoara.* [*Home—Timişoara City Hall.*] PMT. https://www.primariatm.ro/

Pub, S. N. W. A. (2020). *Digitalising Water – Sharing Singapore's Experience.* doi:10.2166/9781789061871

Purdue, S., Peterson, H., & Deng, C. (2018). The case for greater youth participation in monitoring and evaluation in international development. *Evaluation Journal of Australasia, 18*(4), 206–221. doi:10.1177/1035719X18804401

PURE- European Renewable Ethanol. (2000). Fuel Blends. Retrieved 2022, from https://www.epure.org/about-ethanol/fuel-market/fuel-blends/

PURON., G., & Rodríguez, M. P. (2016). Financial transparency in Mexican municipalities: An empirical research. En *Proceedings of the 17th International Digital Government Research Conference on Digital Government Research.* ACM.

Quinn, L., & Dalton, M. (2009). Leading for sustainability: implementing the tasks of leadership. *Corporate Governance: The international journal of business in society.*

Radanović, I., & Likić, R. (2018). Opportunities for use of blockchain technology in medicine. *Applied Health Economics and Health Policy, 16*(5), 583–590. doi:10.100740258-018-0412-8 PMID:30022440

Raghuram, S., Wiesenfield, B., & Garud, R. (2003). Technology enabled work: The role of self- efficacy in determining telecommuter adjustment and structuring behavior. *Journal of Vocational Behavior, 63*(2), 180–198. doi:10.1016/S0001-8791(03)00040-X

Rajeev, P. N., & Kalagnanam, S. (2017). India's mandatory CSR policy: Implications and implementation challenges. *International Journal of Business Governance and Ethics, 12*(1), 90–106. doi:10.1504/IJBGE.2017.085240

Rajput, S., & Singh, S. P. (2019). Connecting circular economy and industry 4.0. *International Journal of Information Management, 49*, 98–113. doi:10.1016/j.ijinfomgt.2019.03.002

Raj, T., Chandrasekhar, K., Kumar, A. N., Banu, R. J., Yoon, J.-J., Bhatia, S. K., ... Kim, S.-H. (2022). Recent advances in commercial biorefineries for lignocellulosic ethanol production: Current status, challenges and future perspectives. *Bioresource Technology, 344*, 126292. doi:10.1016/j.biortech.2021.126292 PMID:34748984

Ramírez, Y., & Nembhard, D. (2004). Measuring Knowledge Worker Productivity – a taxonomy. *Journal of Intellectual Capital, 5*(4), 602–628. doi:10.1108/14691930410567040

Rasmussen, E., & Corbett, G. (2008). Why Isn't Teleworking Working? *New Zealand Journal of Employment Relations, 33*(2), 20–32.

Ray, P. P., Dash, D., Salah, K., & Kumar, N. (2020). Blockchain for IoT-based healthcare: Background, consensus, platforms, and use cases. *IEEE Systems Journal, 15*(1), 85–94. doi:10.1109/JSYST.2020.2963840

Rebelo, G. (2004). *Trabalho e privacidade: contributos e desafios para o direito do trabalho.* Editora RH.

Recuero, R., Bastos, M., & Zago, G. (2015). *Análise de redes para mídia social* [*Network analysis for social media.*]. Editora Sulina.

Recytyre. (2022). *Waar naartoe?* [*Where to go?*] Recytyre. https://www.recytyre.be/nl/waar-naartoe

Redclift, M. (1989). The environmental consequences of Latin America's agricultural development: Some thoughts on the Brundtland Commission report. *World Development, 17*(3), 365–377. doi:10.1016/0305-750X(89)90210-6

ReddyK.AgrawalR. (2012). Designing Case Studies from Secondary Sources – A Conceptual Framework. doi:10.2139/ssrn.2167776

Reis, L., Cagica Carvalho, L., Silveira, C., Marques, A., & Russo, N. (2021). *Inovação e Sustentabilidade em TIC*. Silabo.

Rennings, K. (2000). Redefining innovation— Eco-innovation research and the contribution from ecological economics. *Ecological Economics, 32*(2), 319–332. doi:10.1016/S0921-8009(99)00112-3

Rennings, K., Ziegler, A., Ankele, K., & Hoffmann, E. (2006). The influence of different characteristics of the EU environmental management and auditing scheme on technical environmental innovations and economic performance. *Ecological Economics, 57*(1), 45–59. doi:10.1016/j.ecolecon.2005.03.013

RESOLUÇÃO N° 807, DE 23 DE JANEIRO DE 2020. (2020). Imprensa Nacional_ Brasil_Diário oficial da União. Retrieved 2022, from https://www.in.gov.br/web/dou/-/resolucao-n-807-de-23-de-janeiro-de-2020-239635261

RETIM. (2022). *Acasă—RETIM SA*. RETIM. https://retim.ro/

Reynolds, P., Bosma, N., Autio, E., Hunt, S., De Bono, N., Servais, I., Lopez-Garcia, P., & Chin, N. (2005). Global entrepreneurship monitor: Data collection design and implementation 1998–2003. *Small Business Economics, 24*(3), 205–231. doi:10.100711187-005-1980-1

RFA. (2021). World Fuel Ethanol Production by Region. (RFA- Renewable Fuels Association) Retrieved 2022, from https://ethanolrfa.org/markets-and-statistics/annual-ethanol-production

Riedmann, C., Schwarzenhofe, H., Huemer, C., & Reidelshöfer, K. (2021). *Nachhaltig und sicher: Glasrecycling in Österreich*. [*Sustainable and safe: glass recycling in Austria*.]. OIT. https://www.oesterreich-isst-informiert.at/verantwortung/nachhaltig-und-sicher-glasrecycling-in-oesterreich/

Rist, G. (2008). The history of development: From Western origins to global faith (3. ed., 2. impr). Zed Books.

Rist, G. (2007). Development as a buzzword. *Development in Practice, 17*(4–5), 485–491. doi:10.1080/09614520701469328

Robertson, M. (2017). *Dictionary of sustainability*. Taylor & Francis. doi:10.4324/9781315536705

Robinson, M., Kleffner, A., & Bertels, S. (2011). Signaling sustainability leadership: Empirical evidence of the value of DJSI membership. *Journal of Business Ethics, 101*(3), 493–505. doi:10.100710551-011-0735-y

Rockström, J., Steffen, W., Noone, K., Persson, Å., Chapin, F. S. I., Lambin, E., Lenton, T. M., Scheffer, M., Folke, C., Schellnhuber, H. J., Nykvist, B., de Wit, C. A., Hughes, T., van der Leeuw, S., Rodhe, H., Sörlin, S., Snyder, P. K., Costanza, R., Svedin, U., & Foley, J. (2009). Planetary Boundaries: Exploring the Safe Operating Space for Humanity. *Ecology and Society, 14*(2), art32. doi:10.5751/ES-03180-140232

Rodrigues, G., Sarabdeen, J., & Balasubramanian, S. (2016). Factors that influence consumer adoption of e-government services in the UAE: A UTAUT model perspective. *Journal of Internet Commerce, 15*(1), 18–39. doi:10.1080/15332861.2015.1121460

Rodrigues, M., Menezes, I., & Ferreira, P. D. (2018). Effects of socialization on scout youth participation behaviors. *Educação e Pesquisa, 44*, 1–16. doi:10.15901678-4634201844175560

Rodríguez, L. F., & Brown, T. M. (2009). From voice to agency: Guiding principles for participatory action research with youth. *New Directions for Youth Development, 123*(123), 19–34. doi:10.1002/yd.312 PMID:19830799

Rodríguez-Pose, A. (2013). Do Institutions Matter for Regional Development?. *Regional Studies, 47*(7), 1034–1047. [Taylor & Francis Online], doi:10.1080/00343404.2012.748978

Rogers, E. M. (2003). Elements of diffusion. *Diffusion of Innovations, 5*, 1–38.

Romano, B. Y. M., Boatwright, S., Mounce, S., & Nikoloudi, E. (2020). *AI – BASED EVENT MANAGEMENT AT UNITED UTILITIES. 4*, 104–109.

Roque, R., Dasgupta, S., & Costanza-Chock, S. (2016). Children's civic engagement in the scratch online community. *Social Sciences*, *5*(4), 1–17. doi:10.3390ocsci5040055

Rosati, F., & Faria, L. G. D. (2019). Business contribution to the Sustainable Development Agenda: Organizational factors related to early adoption of SDG reporting. *Corporate Social Responsibility and Environmental Management*, *26*(3), 588–597. doi:10.1002/csr.1705

Rubin, K. S. (2013). *Essential Scrum: A Practical Guide to the Most Popular Agile Process*. Adisson-Wesley.

Ruggie, J. (2011). Report of the Special Representative of the Secretary-General on the Issue of Human Rights and Transnational Corporations and other Business Enterprises: Guiding Principles on Business and Human Rights: Implementing the United Nations 'Protect, Respect and Remedy' Framework. *Netherlands Quarterly of Human Rights*, *29*(2), 224–253. doi:10.1177/016934411102900206

Rulli, M. C., Bellomi, D., Cazzoli, A., Carolis, G. D., & D'Orico, P. (2016). The water-land-food nexus of first-generation biofuels. *Scientific Reports*, *6*(22521). PMID:26936679

Rutherford, S. (2007). Green governmentality: Insights and opportunities in the study of nature's rule. *Progress in Human Geography*, *31*(3), 291–307. doi:10.1177/0309132507077080

S. D. G. Watch Europe. (2022). *Wardrobe Change*. SDG Watch Europe. https://www.sdgwatcheurope.org/wardrobe-change/

Sachs, I. (1986). *Ecodesenvolvimento: crescer sem destruir* [Eco-development: grow without destroying.]. Editora Vértice.

Sachs, I. (2007). *Rumo à ecossocieconomia: teoria e prática do desenvolvimento* [Towards eco-economics: theory and practice of development.]. Cortez.

Saeed, H., Malik, H., Bashir, U., Ahmad, A., Riaz, S., Ilyas, M., Bukhari, W. A., & Khan, M. I. A. (2022). Blockchain technology in healthcare: A systematic review. *PLoS One*, *17*(4), e0266462. doi:10.1371/journal.pone.0266462 PMID:35404955

Saha, A., Amin, R., Kunal, S., Vollala, S., & Dwivedi, S. K. (2019). Review on "Blockchain technology based medical healthcare system with privacy issues". *Security and Privacy*, *2*(5), e83. doi:10.1002py2.83

Sahimi, A., Hamizah, N.A., Suandi, T., Ismail, I.A., & Hamzah, S.R. (2018). Profiling youth participation in volunteer activities in Malaysia: understanding the motivational factors influencing participation in volunteer work among malaysian youth. *Pertanika Journal of Social Sciences & Humanities, 26*(T), 49-62.

Said, O., & Tolba, A. (2021). Design and evaluation of large-scale IoT-enabled healthcare architecture. *Applied Sciences (Basel, Switzerland)*, *11*(8), 3623. doi:10.3390/app11083623

Saini, D., & Sengupta, S. S. (2016). Responsibility, ethics, and leadership: An Indian study. *Asian Journal of Business Ethics*, *5*(1), 97–109. doi:10.100713520-016-0058-2

Saini, M., Sengupta, E., Singh, M., Singh, H., & Singh, J. (2022). Sustainable Development Goal for Quality Education (SDG 4): A study on SDG 4 to extract the pattern of association among the indicators of SDG 4 employing a genetic algorithm. *Education and Information Technologies*, 1–39. doi:10.100710639-022-11265-4 PMID:35975216

Sakuda, L. & Vasconcelos, F. (2005). Teletrabalho: Desafios e Perspectivas. *O&S., 12* (33), 39-49.

Saldarriaga, G. (2022). *Medellín continues to advance as one of the cities with the lowest unemployment rate in the country.* News, Economic Development, Portal de Medellin. https://www.medellin.gov.co/irj/portal/medellin?NavigationTarget=contenido/12142-Medellin-sigue-avanzando-como-una-de-las-ciudades-con-menor-tasa-de-desempleo-del-pais#google_translate_element (2/4/2022)

Salimath, M., & Jones, R. III. (2011). Population ecology theory: Implications for sustainability. *Management Decision*, *49*(6), 874–910. doi:10.1108/00251741111143595

Salleh, M. S., Mahbob, N., & Baharudin, N. S. (2017). Overview of "Generation Z" Behavioural Characteristic and Its Effect Towards Hostel Facility. *International Journal of Real Estate Studies*, *11*(2), 59–74.

Sánchez, A., Pérez-Pérez, M., Carnicer, P., & Jiménez, P. (2007). Teleworking and workplace flexibility: A study of impact on firm performance. *Personnel Review*, *36*(1), 42–64. doi:10.1108/00483480710716713

Santos, A. J., Corso, N. M., Martins, G., & Bittencourt, E. (2002). Aspectos produtivos e comerciais do Pinhão no estado do Paraná. *Revista Floresta*, *32*(2), 163–169. doi:10.5380/rf.v32i2.2281

Sarasvathy, S. D. (2003). *Effectuation: Elements of entrepreneurial expertise. The Darden School.* University of Virginia.

Satapathy, J., & Paltasingh, T. (2019). CSR in India: A journey from compassion to commitment. *Asian Journal of Business Ethics*, *8*(2), 225–240. doi:10.100713520-019-00095-2

Saud, S., Chen, S., Haseeb, A., Khan, K., & Imran, M. (2019). The Nexus between Financial Development, Income Level, and Environment in Central and Eastern European Countries: A Perspective on Belt and Road Initiative. *Environmental Science and Pollution Research International*, *26*(16), 16053–16075. doi:10.100711356-019-05004-5 PMID:30968296

Savić, D. (2021). Digital Water Developments and Lessons Learned from Automation in the Car and Aircraft Industries. *Engineering*. doi:10.1016/j.eng.2021.05.013

Schaltegger, S., & Burritt, R. (2018). Business Cases and Corporate Engagement with Sustainability: Differentiating Ethical Motivations. *Journal of Business Ethics*, *147*(2), 241–259. doi:10.100710551-015-2938-0

Schelbe, L., Chanmugam, A., Moses, T., Saltzburg, S., Williams, L. R., & Letendre, J. (2015). Youth participation in qualitative research: Challenges and possibilities. *Qualitative Social Work: Research and Practice*, *14*(4), 504–521. doi:10.1177/1473325014556792

Scherer, A. G., & Voegtlin, C. (2020). Corporate Governance for Responsible Innovation: Approaches to Corporate Governance and Their Implications for Sustainable Development. *The Academy of Management Perspectives*, *34*(2), 182–208. doi:10.5465/amp.2017.0175

Schmidt, J. (2018). *Universidades comunitárias e terceiro setor – Fundamentos comunitaristas da cooperação em políticas públicas. [Community universities and the third sector – Communitarian foundations of cooperation in public policies.]* Santa cruz do sul: Edunisc.

SchmidtC.SchneiderY.SteffenS.StreitzD. (2020). Capital misal-location and innovation. doi:10.2139/ssrn.3489801

Schmidt, F. C., Zanini, R. R., Korzenowski, A. L., Schmidt, R. Junior, & Xavier do Nascimento, K. B. (2018). Evaluation of sustainability practices in small and medium-sized manufacturing enterprises in Southern Brazil. *Sustainability*, *10*(7), 2460. doi:10.3390u10072460

Schroth, H. (2019). Are You Ready for Gen Z in the Workplace? *California Management Review*, *61*(3), 5–18. doi:10.1177/0008125619841006

Compilation of References

Schumpeter, J. A. (1934). *The Theory of Economic Development*. Harvard University Press.

Schusler, T., Krings, A., & Hernández, M. (2019). Integrating youth participation and ecosocial work: New possibilities to advance environmental and social justice. *Journal of Community Practice*, 27(3-4), 460–475. doi:10.1080/10705422.2019.1657537

Schwarcz, L. M., & Starling, H. (2015). *Brasil: uma biografia*. Companhia das Letras.

Scott, A. (2006). Creative cities: Conceptual issues and policy questions. *Journal of Urban Affairs*, 28(1). University of California, Los Angeles. https://escholarship.org/content/qt77m9g2g6/qt77m9g2g6.pdf?t & (3/4/2022)

Sebti, A., Buck, M., Sanzone, L., Liduke, B. B., Sanga, G. M., & Carnevale, F. A. (2019). Child and youth participation in sexual health-related discussions, decisions, and actions in Njombe, Tanzania: A focused ethnography. *Journal of Child Health Care*, 23(3), 370–381. doi:10.1177/1367493518823920 PMID:30669864

Selvaraj, S., & Sundaravaradhan, S. (2020). Challenges and opportunities in IoT healthcare systems: A systematic review. *SN Applied Sciences*, 2(1), 1–8. doi:10.100742452-019-1925-y

Sen, A. (1999). *Development as freedom*. Oxford University Press.

Seyfang, G., & Smith, A. (2007). Grassroots Innovations for Sustainable Development: Towards a New Research and Policy Agenda. *Environmental Politics*, 16(4), 584–603. doi:10.1080/09644010701419121

Seyff, N., Penzenstadler, B., Betz, S., Brooks, I., Oyedeji, S., Porras, J., & Venters, C. (2021). The Elephant in the Room-Educating Practitioners on Software Development for Sustainability. *IEEE/ACM International Workshop on Body of Knowledge for Software Sustainability (BoKSS)* (pp. 25-26). IEEE. 10.1109/BoKSS52540.2021.00017

Seymour, K., Bull, M., Homel, R., & Wright, P. (2017). Making the most of youth development: Evidence-based programs and the role of young people in research. *Queensland Review*, 24(1), 147–162. doi:10.1017/qre.2017.17

Shafiee, M. E., Rasekh, A., Sela, L., & Preis, A. (2020). Streaming Smart Meter Data Integration to Enable Dynamic Demand Assignment for Real-Time Hydraulic Simulation. *Journal of Water Resources Planning and Management*, 146(6), 06020008. doi:10.1061/(ASCE)WR.1943-5452.0001221

Shahnaz, A., Qamar, U., & Khalid, A. (2019). Using blockchain for electronic health records. *IEEE Access : Practical Innovations, Open Solutions*, 7, 147782–147795. doi:10.1109/ACCESS.2019.2946373

Shahrul, N. S., & Normah, M. (2015). Digital version newspaper: Implication towards printed newspaper circulation in Malaysia. *Malaysian Journal of Communication*, 31(2), 687–701.

Shahzad, F., Lu, J., & Fareed, Z. (2019). Does firm life cycle impact corporate risk taking and performance? *Journal of Multinational Financial Management*, 51, 23–44. doi:10.1016/j.mulfin.2019.05.001

Sharma, A., Kaur, S., & Singh, M. (2021). A comprehensive review on blockchain and Internet of Things in healthcare. *Transactions on Emerging Telecommunications Technologies*, 32(10), e4333. doi:10.1002/ett.4333

Sharma, N., Bhushan, B., Kaushik, I., & Debnath, N. C. (2021). Applicability of Blockchain Technology in Healthcare Industry: Applications, Challenges, and Solutions. In *Efficient Data Handling for Massive Internet of Medical Things* (pp. 339–370). Springer. doi:10.1007/978-3-030-66633-0_15

Sharma, R. R. (2019). Evolving a model of sustainable leadership: An ex-post facto research. *Vision (Basel)*, 23(2), 152–169. doi:10.1177/0972262919840216

Shaw, A., Brady, B., McGrath, B., Brennan, M. A., & Dolan, P. (2014). Understanding youth civic engagement: Debates, discourses, and lessons from practice. *Community Development (Columbus, Ohio)*, *45*(4), 300–316. doi:10.1080/15575330.2014.931447

Shiazawa, B. (2020). Ammonia Energy Association. Retrieved from https://www.ammoniaenergy.org/articles/the-cost-of-co2-free-ammonia/

Shiratuddin, N., Hassan, S., Mohd Sani, M. A., Ahmad, M. K., Khalid, K. A., Abdull Rahman, N.L., Abd Rahman, Z.S., & Ahmad, N.S. (2017). Media and youth participation in social and political activities: development of a survey instrument and its critical findings. *Pertanika Journal of Social Sciences & Humanities*, *25*(S), 1-20.

Sika, N. (2018). Civil society and the rise of unconventional modes of youth participation in the MENA. *Middle East Law and Governance*, *10*(3), 237–263. doi:10.1163/18763375-01003002

Silva, J. M. (2015). *Políticas públicas para composição de custos e formação de preços da atividade extrativa da castanha-da-Amazônia*. [Master's thesis, Fundação Universidade Federal de Rondônia].

Silva-Jean, M. (2017). Custos e preços da castanha-da-amazônia nos Estados do Acre e Rondônia. [Costs and prices of the Amazon nut in the states of Acre and Rondônia.] *Custos e @gronegócio on-line*, *13*(2), 421-447.

Silva-Jean, M., Paes-De-Souza, M., Souz-Filho, T. A., Riva, F. R., & Barbosa, C. S. (2022). Public Policies Guarantee for Minimum Prices on Products of Sociobiodiversity (PGPMBio): Composition of the extration cost of Amazonian chestnut in Rondônia and Acre. *Revista de Administração UFSM*, *15*(1), 62–82. doi:10.5902/1983465965906

Silva-Jean, M., Souza, M. P., & Filho, T. A. S. (2020). Cadeia produtiva da Castanha-da-Amazônia nos Estados do Acre e Rondônia. [Production chain of the Amazon Nut in the states of Acre and Rondônia.]. *Brazilian Journal of Development*, *6*(11), 91277–91297. doi:10.34117/bjdv6n11-512

Silva, M. C. (2011). *Sustentabilidade no Terceiro Setor: O desafio de armonizer* [Sustainability in the Third Sector: The challenge of harmonizing.]. Reuna.

Simoes, J. A., & Campos, R. (2017). Digital media, subcultural activity and youth participation: The cases of protest rap and graffiti in Portugal. *Journal of Youth Studies*, *20*(1), 16–31. doi:10.1080/13676261.2016.1166190

Simon, H. (1996). *The sciences of artificial*. MIT PRESS.

Simpson, J., & Taylor, J. R. (2013). *Corporate governance ethics and CSR*. Kogan Page Publishers.

Singh, R., & Singh, A. (2020). Socially Responsible Leadership as a Driver for Sustainable Growth in the World of Electronic Commerce. *E-Business: Issues and Challenges of 21st Century*, 137.

Siokas, G. (2018). Οι Ευφυείς Πόλεις και ο ρόλος της Τοπικής Αυτοδιοίκησης: Θεωρητικό πλαίσιο και ελληνικά [The Intelligent Cities and the Role of Local Government: Theoretical Framework and Greek Examples]. *Ermoupolis Seminar on the Information Society and the Knowledge Economy*. Technical University of Athens. https://www.researchgate.net/publication/349916627_Prosdioristikoi_paragontes_kai_strategikes_sto_schediasmo_mias_Euphyous_Poles (1/5/2022)

Siyal, A. A., Junejo, A. Z., Zawish, M., Ahmed, K., Khalil, A., & Soursou, G. (2019). Applications of blockchain technology in medicine and healthcare: Challenges and future perspectives. *Cryptography*, *3*(1), 3. doi:10.3390/cryptography3010003

Smith, S., & Strawser, M. (2022). Welcome Gen Z to the Workforce. In Atay, A. and Ashlock, M. Z. (eds) Social Media, Technology and New Generations. Lexington Books.

Compilation of References

Sneideriene, A., & Rugine, H. (2019). Theoretical approach on the green technologies development. *Regional Formation and Development Studies, 2*(28), 124–134.

Soares, T. S., Fiedler, N. C., Silva, J. A., & Gasparini Junior, A. J. (2008). Produtos Florestais Não Madeireiros. [Non-Timber Forest Products.] *Revista Científica de Engenharia Florestal, 11*.

Soler-i-Martí, R., & Ferrer-Fons, M. (2015). Youth participation in context: The impact of youth transition regimes on political action strategies in Europe. *The Sociological Review, 63*(2_suppl), 92–117. doi:10.1111/1467-954X.12264

Solomon, R. C. (1993). Business Ethics. In *A Companion to Ethics* (pp. 354–365). Blackwell Publishers.

Solow, R. M. (1991). Sustainability: an economist's perspective.

Somayaji, S. R. K., Alazab, M., Manoj, M. K., Bucchiarone, A., Chowdhary, C. L., & Gadekallu, T. R. (2020, December). A framework for prediction and storage of battery life in iot devices using dnn and blockchain. In *2020 IEEE Globecom Workshops* (pp. 1-6). IEEE.

Sommerville, I. (2016). *Software Engineering* (10th ed.). Pearson.

Song, J. J., Kim, J., Jones, D. R., Baker, J., & Chin, W. W. (2014). Application discoverability and user satisfaction in mobile application stores: An environmental psychology perspective. Decision Support Systems. [Decision Support Systems.]. *Decision Support Systems, 59*, 37–51. doi:10.1016/j.dss.2013.10.004

Sorkhabi, R. (2015). The First Oil Shock. (GeoExPro) Retrieved 2022, from https://www.geoexpro.com/articles/2015/06/the-first-oil-shock

Sorknaes, P., Johannsen, R. M., Korberg, A. D., Nielsen, T. B., Petersen, U. R., & Mathiesen, B. V. (2022). Electrification of the industrial sector in 100% renewable energy scenarios. Energy.

Souza, M. C. G. D. (2009). *Ética no ambiente de trabalho*. Elsevier Editora.

Spreitzer, G. M., & Quinn, R. E. (2001). *A company of leaders: Five disciplines for unleashing the power in your workforce* (Vol. 3). Jossey-Bass.

Srivastava, G., Parizi, R. M., & Dehghantanha, A. (2020). The future of blockchain technology in healthcare internet of things security. *Blockchain cybersecurity, trust and privacy*, 161-184.

Srivastava, A., Jain, P., Hazela, B., Asthana, P., & Rizvi, S. W. A. (2021). Application of Fog Computing, Internet of Things, and Blockchain Technology in Healthcare Industry. In *Fog Computing for Healthcare 4.0 Environments* (pp. 563–591). Springer. doi:10.1007/978-3-030-46197-3_22

Stadt Wien. (2018). *Repair Network Vienna*. Reparatur Netzwerk. https://www.reparaturnetzwerk.at/repair-network-vienna

Staista. (2022). Chemical & Resource- HVO biodiesel production volume worldwide from 2013 to 2020. *Staista*. https://www.statista.com/statistics/1297290/hvo-biodiesel-production-worldwide/

Stancin, H., Mikuleié, H., Wang, X., & Duié, N. (2020). A review on aalternative fules in futiure energy system. *Renewable & Sustainable Energy Reviews, 128*.

Statista. (2021). Biofuel productin form 2000-2020. *Statista*. https://www.statista.com/statistics/274163/global-biofuel-production-in-oil-equivalent/

StEP. (2022). *Organisation*. StEP Initiative. https://www.step-initiative.org/organisation-rev.html

Sterioti, Th. (2016). *Δημιουργική Οικονομία: ο πολιτισμός και η δημιουργικότητα, μοχλοί ανάπτυξης των σύγχρονων πόλεων* [Creative Economy: culture and creativity, levers of development of modern cities] [Unpublished Master Thesis, Hellenic Open University. Athens]. https://apothesis.eap.gr/handle/repo/32067

Steurer, R., Langer, M. E., Konrad, A., & Martinuzzi, A. (2005). Corporations, stakeholders and sustainable development I: A theoretical exploration of business–society relations. *Journal of Business Ethics*, *61*(3), 263–281. doi:10.100710551-005-7054-0

Stoffels, M., & Ziemer, C. (2017). Digitalization in the process industries – Evidence from the German water industry. *Journal of Business Chemistry*, *14*(3), 94–105. doi:10.17879/20249613743

Stratu-Strelet, D., Gil-Gómez, H., Oltra-Badenes, R., & Oltra-Gutierrez, J. V. (2021). Critical factors in the institutionalization of e-participation in e-government in Europe: Technology or leadership? *Technological Forecasting and Social Change*, *164*, 120489.

Strehl, L. (2005). O fator de impacto do ISI e a avaliação da produção científica: Aspectos conceituais e metodológicos [The impact factor of the ISI and the evaluation of scientific production: Conceptual and methodological aspects]. *Ci. Inf.*, *34*(1), 19–27. doi:10.1590/S0100-19652005000100003

Studies: EGOS Le Boterf, G. (1999). *L'ingénierie des compétences*. París: Éditions d'Organisation.

Studio, A. (2022). *Developer Android*. Obtido de https://developer.android.com/studio?gclid=CjwKCAjwqvyFBhB7EiwAER786XUrp2bo8yfAcorObMLYazNeRtNEzEXk63p-qsAe7DvgPaf8HECUARoC2PkQAvD_BwE&gclsrc=aw.ds

Subramaniyaswamy, V., Manogaran, G., Logesh, R., Vijayakumar, V., Chilamkurti, N., Malathi, D., & Senthilselvan, N. (2019). An ontology-driven personalized food recommendation in IoT-based healthcare system. *The Journal of Supercomputing*, *75*(6), 3184–3216. doi:10.100711227-018-2331-8

Sueyoshi, T., & Goto, M. (2012). Efficiency-based rank assessment for electric power industry: A combined use of data envelopment analysis (DEA) and DEA-discriminant analysis (DA). *Energy Economics*, *34*(3), 634–644. doi:10.1016/j.eneco.2011.04.001

Sung, K. (2015). A Review on Upcycling: Current Body of Literature, knowledge gaps and a way forward. Venice Italy, 28-40.

Sun, X., Zhou, X., Chen, Z., & Yang, Y. (2020b). Environmental efficiency of electric power industry, market segmentation and technologi-cal innovation: Empirical evidence from China. *The Science of the Total Environment*, *706*, 135749. doi:10.1016/j.scitotenv.2019.135749 PMID:31940733

SYNGASCHEM. (2022). Clean Coal to Liquids as a Transitional Technology. Syngachem. https://www.syngaschem.com/our-vision/

Taborosi, S., Strukan, E., Postin, J., Konjikusic, M., & Nikolic, M. (2020). Organizational Commitment and Trust at Work by Remote Employees. *Journal of Engineering Management and Competitiveness*, *10* (1), 48-60.

Taklo, S. K., & Tooranloo, H. S. & Shahabaldini parizi, Z. (. (2020). Green Innovation: A Systematic Literature Review. *Journal of Cleaner Production*, *2020*(7). Advance online publication. doi:10.1016/j.jclepro.2020.122474

Tan, K. S., & Eze, U. C. (2008). An empirical study of internet-based ICT adoption among Malaysian SMEs. *Communications of the IBIMA*, *1*, 1–12.

Tanwar, S., Parekh, K., & Evans, R. (2020). Blockchain-based electronic healthcare record system for healthcare 4.0 applications. *Journal of Information Security and Applications*, *50*, 102407. doi:10.1016/j.jisa.2019.102407

Compilation of References

Tao, H., Bhuiyan, M. Z. A., Abdalla, A. N., Hassan, M. M., Zain, J. M., & Hayajneh, T. (2018). Secured data collection with hardware-based ciphers for IoT-based healthcare. *IEEE Internet of Things Journal*, 6(1), 410–420. doi:10.1109/JIOT.2018.2854714

TATA Consultancy Services. (2019). *Corporate Sustainability Report 2018-2019*. TATA Consultancy Services, Mumbai, India (https://www.tcs.com/content/dam/tcs/pdf/discover-tcs/investor-relations/corporate-sustainability/GRI-Sustainability-Report-2018-2019.pdf)

TE- Transport & Enviroment. (2020). *RED II and advanced biofuels - Recommendations about Annex IX of the Renewable Energy Directive and its implementation at national level*.

Techskill Brew. (2021). Blockchain and IoT in smart healthcare. *Medium*. https://medium.com/techskill-brew/blockchain-and-iot-in-smart-healthcare-814287551300

Tejedo-Romero, F., Araujo, J. F. F. E., Tejada, Á., & Ramírez, Y. (2022). E-government mechanisms to enhance the participation of citizens and society: Exploratory analysis through the dimension of municipalities. *Technology in Society*, 70, 101978. https://doi.org/10.1016/j.techsoc.2022.101978

Telemaco, U., Oliveira, T., Alencar, P., & Cowan, D. (2020). A Catalogue of Agile Smells for Agility Assessment. *IEEE Access: Practical Innovations, Open Solutions*, 8, 79239–79259. doi:10.1109/ACCESS.2020.2989106

Tenório, O. (2015). *Responsabilidade Social Empresarial: Teoria e Prática* [*Corporate Social Responsibility: Theory and Practice.*]. FGV.

Thakar, A. T., & Pandya, S. (2017, July). Survey of IoT enables healthcare devices. In *2017 International Conference on Computing Methodologies and Communication (ICCMC)* (pp. 1087-1090). IEEE. 10.1109/ICCMC.2017.8282640

Thatte, A. A. (2007). *Competitive advantage of a firm through supply chain responsiveness and SCM practices*. [Doctoral dissertation, The University of Toledo].

The Ellen MacArthur Foundation. (2022). *What is a circular economy?* Ellen MacArthur Foundation. https://ellenmacarthurfoundation.org/topics/circular-economy-introduction/overview

Themistokleous, S., & Avraamidou, L. (2016). The role of online games in promoting young adults' civic engagement. *Educational Media International*, 53(1), 53–67. doi:10.1080/09523987.2016.1192352

Thew, H. (2018). Youth participation and agency in the United Nations framework convention on climate change. *International Environmental Agreement: Politics, Law and Economics*, 18(3), 369–389. doi:10.100710784-018-9392-2

Thomas, G., & Pring, R. (2004). *Evidence-based practice in education*. Open University Press.

Timsal, A., & Awais, M. (2016). Flexibility or ethical dilemma: An overview of the work from home policies in modern organizations around the world. *Human Resource Management International Digest*, 14(7), 12–15. doi:10.1108/HRMID-03-2016-0027

Tolliver, C., Fujii, H., Keeley, A. R., & Managi, S. (2021). Green innovation and finance in Asia. *Asian Economic Policy Review*, 16(1), 67–87. doi:10.1111/aepr.12320

Tornatzky, L. G., & Klein, K. J. (1982). Innovation characteristics and innovation adoption-implementation: A meta-analysis of findings. *IEEE Transactions on Engineering Management*, 29(1), 28–45. doi:10.1109/TEM.1982.6447463

Torten, R., Reaiche, C., & Caraballo, E. (2016). Teleworking in the New Millennium. *The Journal of Developing Areas. Special Issue on Kuala Lumpur Conference Held in N*, 50(5), 317–326. doi:10.1353/jda.2016.0060

Towoju, O. A. (2021, April). Fuels for automobiles: The Sustainable Future. *Journal of Energy Research and Reviews*, pp. 8-13.

Transclean. (2022). *Transclean*. http://transclean.ro/

Trayush, T., Bathla, R., Saini, S., & Shukla, V. K. (2021, March). IoT in Healthcare: Challenges, Benefits, Applications, and Opportunities. In *2021 International Conference on Advance Computing and Innovative Technologies in Engineering (ICACITE)* (pp. 107-111). IEEE. 10.1109/ICACITE51222.2021.9404583

Tremblay, D. G. (2002). Organização e satisfação no contexto do teletrabalho [Organization and satisfaction in the context of telework]. *ERA – Revista de Administração de Empresas [ERA- Business Administration Magazine]*, *42* (3), 54-65. doi:10.1590/S0034-75902002000300006

Triguero, A., Moreno-Mondéjar, L., & Davia, M. A. (2015). Eco-innovation by small and medium-sized firms in Europe: From end-of-pipe to cleaner technologies. *Innovation (North Sydney, N.S.W.)*, *17*(1), 24–40. doi:10.1080/14479338.2015.1011059

Trilla, J., & Novella, A. M. (2011). Participación, democracia y formación para la ciudadanía: Los consejos de infancia. *Review of Education*, *356*, 26–46.

Tripathi, R., & Kumar, A. (2020). Humanistic leadership in the Tata group: The synergy in personal values, organisational strategy and national cultural ethos. *Cross Cultural & Strategic Management*, *27*(4), 607–626. doi:10.1108/CCSM-01-2020-0025

Truong, Y., Mazloomi, H., & Berrone, P. (2021). Understanding the impact of symbolic and substantive environmental actions on organizational reputation. *Industrial Marketing Management*, *92*, 307–320. doi:10.1016/j.indmarman.2020.05.006

Tsalis, T. A., Malamateniou, K. E., Koulouriotis, D., & Nikolaou, I. E. (2020). New challenges for corporate sustainability reporting: United Nations' 2030 Agenda for sustainable development and the sustainable development goals. *Corporate Social Responsibility and Environmental Management*, *27*(4), 1617–1629. doi:10.1002/csr.1910

Tsekoura, M. (2016a). Debates on youth participation: From citizens in preparation to active social agents. *Revista Katálysis*, *19*(1), 118–125. doi:10.1590/1414-49802016.00100012

Tsekoura, M. (2016b). Spaces for youth participation and youth empowerment: Case studies from the UK and Greece. *Young*, *24*(4), 326–341. doi:10.1177/1103308815618505

Turnbull, S. (1997). Stakeholder Governance: A Cybernetic and Property Rights Analysis. *Corporate Governance*, *5*(1), 11–23. doi:10.1111/1467-8683.00035

Turner, A. (2015). Generation Z: Technology and Social Interest. *Journal of Individual Psychology*, *71*(2), 103–113. doi:10.1353/jip.2015.0021

Tyagi, S., Agarwal, A., & Maheshwari, P. (2016, January). A conceptual framework for IoT-based healthcare system using cloud computing. In *2016 6th International Conference-Cloud System and Big Data Engineering (Confluence)* (pp. 503-507). IEEE.

Tzankova, I., & Cicognani, E. (2019). Youth participation in psychological literature: A semantic analysis of scholarly publications in the PsycInfo Database. *Europe's Journal of Psychology*, *15*(2), 276–291. doi:10.5964/ejop.v15i2.1647 PMID:33574955

U.S. department of energy. (2019). Flexible Fuel Vehicles (FFV). *Energy Efficiency & Renewable Energy*. https://afdc.energy.gov/vehicles/flexible_fuel.html

Compilation of References

Ulitin, A., Mier-Alpaño, J. D., Labarda, M., Juban, N., Mier, A. R., Tucker, J. D., & Chan, P. L. (2022). Youth social innovation during the COVID-19 pandemic in the Philippines: A quantitative and qualitative descriptive analyses from a crowdsourcing open call and online hackathon. *BMJ Innovations*, *0*(3), 1–8. doi:10.1136/bmjinnov-2021-000887

Ulrich, D., Zenger, J., & Smallwood, N. (1999). *Results-based leadership*. Harvard Business Press.

UNCTAD. (2010). *Creative Economy Report 2010*. UNCTAD. https://unctad.org/system/files/official-document/ditctab20103_en.pdf

UNDP. (2020). *Integrated Solutions for Sustainable Development*. United Nations Development Programme: //sdgintegration.undp.org/

UNESCO. (2004). *Creative Cities Network*. UNESCO. https://en.unesco.org/creative-cities/home (5/4/2022)

UNESCO. (2021). *Cities, culture, creativity: leveraging culture and creativity for sustainable urban development and inclusive growth*. World Bank and UNESCO. Washington. (Online). Available at: https://unesdoc.unesco.org/ark:/48223/pf0000377427 (10/4/2022)

UNESCO. Institute for Statistics. (2016). *The globalization of cultural trade: A shift in consumption. International flows of cultural goods and services 2004-2013*. Canada. http://uis.unesco.org/sites/default/files/documents/the-globalisation-of-cultural-trade-a-shift-in-consumption-international-flows-of-cultural-goods-services-2004-2013-en_0.pdf (23/4/2022)

UNFCCC- United Nation Climate Change. (2022). *United Nation Climate Change- What is the Paris Agreement?* UN. https://unfccc.int/process-and-meetings/the-paris-agreement/the-paris-agreement

Ungerer, R. (2013). *Sociedade globalizada e mundo digital [Globalized society and digital world.]*. Artmed.

United Nations. (2015). *Transforming our world: The 2030 agenda for sustainable development*. UN. https://sustainabledevelopment.un.org/content/documents/21252030%20Agenda%20for%20Sustainable%20Development%20web.pdf (4/4/2022)

United Nations. (2020). *E-Government*. UN. https://publicadministration.un.org/egovkb/en-us/about/unegovdd-framework

United Nations. (2022). *Objetivos de Desenvolvimento Sustentável*. UNRIC. https://unric.org/pt/objetivos-de-desenvolvimento-sustentavel/

United Nations. (2022). *Sustainable Development Goals*. UN. https://www.un.org/sustainabledevelopment/cities/ (1/4/2022)

Uršič, M., Dekker, K., & Filipovič Hrast, M., (2014). Spatial organization and youth participation: case of the University of Ljubljana and Tokyo Metropolitan University. *Annales, Series historia et sociología*, *24*(3), 433-450.

Valera-Medina, A., Amer-Hatem, F., Joannon, M., Fernandes, R. X., Glarborg, P., Hashemi, H., & Costa, M. (2021). Review on Ammonia as Potential Fuel: From Synthesis to Economics. *Energy & Fuels*, *35*(9), 6964–7029. doi:10.1021/acs.energyfuels.0c03685

Valera-Medina, A., Xiao, H., Owen-Jones, M., David, W., & Bowen, P. (2018). Ammonia for power. *Progress in Energy and Combustion Science*, *69*, 63–102. doi:10.1016/j.pecs.2018.07.001

Vallance, E. (1995). *Business ethics at work*. Cambridge University Press. doi:10.1017/CBO9781139166461

Vallance, S., Perkins, H. C., & Dixon, J. E. (2011). What is social sustainability? A clarification of concepts. *Geoforum*, *42*(3), 342–348. doi:10.1016/j.geoforum.2011.01.002

Valliere, D., & Peterson, R. (2009). Entrepreneurship and economic growth: Evidence from emerging and developed countries. *Entrepreneurship and Regional Development*, *21*(5-6), 459–480. doi:10.1080/08985620802332723

Valverde-Pérez, B., Johnson, B., Wärff, C., Lumley, D., Torfs, E., Nopens, I., & Townley, L. (2021). *Digital Water in the urban water*. International Water Association.

Van den Berg, L., & Braun, E. (2001). Growth clusters in European cities: An integral approach. *Urban Studies*, *38*(1), 185–205. Sage. https://journals.sagepub.com/doi/pdf/10.1080/00420980124001 (29/4/2022)

van den Bergh, J. C. J. M. (2009). The GDP paradox. *Journal of Economic Psychology*, *30*(2), 117–135. doi:10.1016/j.joep.2008.12.001

Van Haperen, S., Nicholls, W., & Uitermark, J. (2018). Building protest online: Engagement with the digitally networked #not1more protest campaign on Twitter. *Social Movement Studies*, *17*(4), 408–423. doi:10.1080/14742837.2018.1434499

VARAM. (2021). *Minister Plešs: State waste management plan will ensure the development of the sector [Pinister Pless: State waste management plan will ensure the development of the sector.]*. Vides aizsardzības un reģionālās attīstības ministrija [Ministry of Environmental Protection and Regional Development]. https://www.varam.gov.lv/en/article/minister-pless-state-waste-management-plan-will-ensure-development-sector

Varela, J. A., & António, N. S. (2012). *O Bem Comum e a Teoria dos Stakeholders*. [*The Common Good and the Stakeholder Theory*.] ISCTE-IUL: Business Research Unit.

Vázquez, J. L., Lanero, A., Gutiérrez, P., & García, M. P. (2015). Expressive and instrumental motivations explaining youth participation in non-profit voluntary associations: An application in Spain. *International Review on Public and Nonprofit Marketing*, *12*(3), 237–251. doi:10.100712208-015-0128-5

Vega, R., Anderson, A., & Kaplan, S. (2015). A Within-Person Examination of the Effects of Telework. *Journal of Business and Psychology*, *30*(2), 313–323. doi:10.100710869-014-9359-4

Verdejo Espinosa, Á., López, J. L., Mata Mata, F., & Estevez, M. E. (2021). Application of IoT in healthcare: Keys to implementation of the sustainable development goals. *Sensors (Basel)*, *21*(7), 2330. doi:10.339021072330 PMID:33810606

Viaggi, D. (2015). Research and innovation in agriculture: Beyond productivity? *Bio-Based and Applied Economics*, *4*(3), 279–300. doi:10.13128/BAE-17555

Vileou, G. (2015). *Portrait-robot des habitants de l'île de Nantes*. [*Portrait-robot of the inhabitants of the island of Nantes*.] https://datajournalisme2013.hyblab.fr/projets/population/ (28/04/2022)

Vlaanderen. (2022). *Afvalinzameling en sorteren [Waste collection and sorting]*. Vlaanderen. www.vlaanderen.be. https://www.vlaanderen.be/afvalinzameling-en-sorteren

Vlek, C., & Steg, L. (2007). Human behavior and environmental sustainability: Problems, driving forces, and research topics. *The Journal of Social Issues*, *63*(1), 1–19. doi:10.1111/j.1540-4560.2007.00493.x

VOEB. (2022). *Erreichung der EU-Klimaziele: Recyceln statt Deponieren*. [*Achieving EU climate targets: recycling instead of landfilling*.] VOEB. https://www.voeb.at/service/voeb-blog/detail/show-article/erreichung-der-eu-klimaziele-recyceln-statt-deponieren/

Voices of Culture. (2021). Culture and the United Nations Sustainable Development Goals: Challenges and Opportunities. *Brainstorming Report*. European Union. https://voicesofculture.eu/wp-content/uploads/2021/02/VoC-Brainstorming-Report-Culture-and-SDGs.pdf (12/10/2022)

von Ditfurth, H., Weisbord, E., Danielsen, T., Zutari, L. F.-J., Hafemann, A. C., Hima, J., & Oraeki, T. C. (2021). *Digital Water: An overview of the future of digital water from a YWP perspective*. International Water Association. https://iwa-network.org/publications/digital-water-an-overview-of-the-future-of-digital-water-from-a-ywp-perspective/

von Schomberg, R. (2012). Prospects for technology assessment in a framework of responsible research and innovation. Em M. Dusseldorp & R. Beecroft (Orgs.), Technikfolgen abschätzen lehren: Bildungspotenziale transdisziplinärer Methoden (p. 39–61). VS Verlag für Sozialwissenschaften. doi:10.1007/978-3-531-93468-6_2

Waddock, S. (2004). Creating corporate accountability: Foundational principles to make corporate citizenship real. *Journal of Business Ethics*, 50(4), 313–327. doi:10.1023/B:BUSI.0000025080.77652.a3

Waddock, S. (2016). Developing humanistic leadership education. *Humanistic Management Journal*, 1(1), 57–73. doi:10.100741463-016-0003-5

Walker, B., Holling, C. S., Carpenter, S., & Kinzig, A. (2004). Resilience, Adaptability and Transformability in Social–ecological Systems. *Ecology and Society*, 9(2), 5. doi:10.5751/ES-00650-090205

Wang, F. Z., Jiang, T., & Guo, X. C. (2018). Government Quality, Environmental Regulation and Corporate Green Technology Innovation. *Sci. Res. Manage.* 39 (1), 26–33. CNKI:SUN:KYGL.0.2018-01-004.

Wang, H. (2020). IoT based clinical sensor data management and transfer using blockchain technology. *Journal of ISMAC*, 2(03), 154–159. doi:10.36548/jismac.2020.3.003

Wang, K., & Jiang, W. (2021). State ownership and green innovation in China: The con-tingent roles of environmental and organizational factors. *Journal of Cleaner Production*, 314, 128029. doi:10.1016/j.jclepro.2021.128029

Wang, Y., Sun, X., & Guo, X. (2019). Environmental regulation and green productivity growth: Empirical evidence on the Porter Hypothesis from OECD industrial sectors. *Energy Policy*, 132, 611–619. doi:10.1016/j.enpol.2019.06.016

Warren, A. M., Sulaiman, A., & Jaafar, N. I. (2014). Social media effects on fostering online civic engagement and building citizen trust and trust in institutions. *Government Information Quarterly*, 31(2), 291–301. doi:10.1016/j.giq.2013.11.007

Watts, R. J., & Hipólito-Delgado, C. P. (2015). Thinking ourselves to liberation? Advancing sociopolitical action in critical consciousness. *The Urban Review*, 47(5), 847–867. doi:10.100711256-015-0341-x

WCED. (1987). Our Common Future. Oxford University Press.

Weng, M. H., & Lin, C. Y. (2011). Determinants of green innovation adoption for small and medium-size enterprises (SMES). *African Journal of Business Management*, 5(22), 9154–9163.

Wennekers, S., & Thurik, R. (1999). Linking Entrepreneurship and Economic Growth. *Small Business Economics*, 13(1), 27–55. doi:10.1023/A:1008063200484

Westwood, R., & Low, D. R. (2003). The multicultural muse: Culture, creativity, and innovation. *International Journal of Cross Cultural Management*, 3(2), 235–259. doi:10.1177/14705958030032006

Wieringa, R. (2009). *Design Science Methodology For Information Systems And Software Engineering*. Springer.

Wiesenfeld, B., Raghuram, S., & Garud, R. (2001). Organizational identification among virtual workers: The role of need for affiliation and perceived work-based social support. *Journal of Management*, *27*(2), 213–229. doi:10.1177/014920630102700205

Williamson, O. E. (2000). The new institutional economics: Taking stock, looking ahead. Journal of Economic Literature, *38*(3), 595–613. [Crossref], [Web of Science ®], . doi:10.1257/jel.38.3.595

Wilson, D. C., Velis, C., & Cheeseman, C. (2006). Role of informal sector recycling in waste management in developing countries. *Habitat International*, *30*(4), 797–808. https://doi.org/10.1016/j.habitatint.2005.09.005

WKO. (2022a). *Abfallwirtschaft im Betrieb*. [*Waste management in the company*.] WKO. https://www.wko.at/service/umwelt-energie/Abfallwirtschaft_im_Betrieb.html

WKO. (2022b). *Information zur Verpackungsverordnung*. [*Information on the Packaging Ordinance*.] WKO. https://www.wko.at/service/umwelt-energie/information-verpackungsverordnung.html

Wong, M. C., Yee, K. C., & Nøhr, C. (2018). Socio-technical considerations for the use of blockchain technology in healthcare. In *Building Continents of Knowledge in Oceans of Data: The Future of Co-Created eHealth* (pp. 636–640). IOS Press.

Wong, P. K., Ho, Y. P., & Autio, E. (2005). Entrepreneurship, innovation, and economic growth: Evidence from GEM data. *Small Business Economics*, *24*(3), 335–350.

World Commission On Environment And Development [WCED]. (1987). *Our Common Future*. Oxford: Oxford University Press.

Wu, H. Y., Tsai, A., & Wu, H. S. (2019). A hybrid multi-criteria decision analysis approach for environmental performance evaluation: an example of the tft-lcd manufacturers in taiwan. [EEMJ]. *Environmental Engineering and Management Journal*, *18*(3), 597–616. doi:10.30638/eemj.2019.056

Wu, J., Guo, S., Huang, H., Liu, W., & Xiang, Y. (2018). Information and communications technologies for sustainable development goals: State-of-the-art, needs and perspectives. *IEEE Communications Surveys and Tutorials*, *20*(3), 2389–2406. doi:10.1109/COMST.2018.2812301

Wukich, C., & Mergel, I. (2016). Reusing social media information in government. *Government Information Quarterly*, 1–8.

Wurlod, J. D., & Noailly, J. (2018). The impact of green innovation on energy intensity: An empirical analysis for 14 industrial sectors in OECD countries. *Energy Econ*, *71*, 47–61. doi:10.1016/j.eneco.2017.12.012

Wu, T., Wu, F., Redoute, J. M., & Yuce, M. R. (2017). An autonomous wireless body area network implementation towards IoT connected healthcare applications. *IEEE Access : Practical Innovations, Open Solutions*, *5*, 11413–11422. doi:10.1109/ACCESS.2017.2716344

Xavier, A. F., Naveiro, R. M., Aoussat, A., & Reyes, T. (2017). Systematic literature review of eco-innovation models: Opportunities and recommendations for future research. *Journal of Cleaner Production*, *149*, 1278–1302. doi:10.1016/j.jclepro.2017.02.145

Yaeger, K., Martini, M., Rasouli, J., & Costa, A. (2019). Emerging blockchain technology solutions for modern healthcare infrastructure. *Journal of Scientific Innovation in Medicine*, *2*(1), 1. doi:10.29024/jsim.7

Yang, B., Fu, P., Beveridge, A. J., & Qu, Q. (2020). Humanistic leadership in a Chinese context. *Cross Cultural & Strategic Management*, *27*(4), 547–566. doi:10.1108/CCSM-01-2020-0019

Compilation of References

Yang, S., & Feng, N. (2008). A case study of industrial symbiosis: Nanning Sugar Co., Ltd. in China. *Resources, Conservation and Recycling, 52*(5), 813–820. doi:10.1016/j.resconrec.2007.11.008

Yang, Y., Zheng, X., Guo, W., Liu, X., & Chang, V. (2019). Privacy-preserving smart IoT-based healthcare big data storage and self-adaptive access control system. *Information Sciences, 479*, 567–592. doi:10.1016/j.ins.2018.02.005

Yapicioglu, A., & Dincer, I. (2019). A review on clean ammonia as a potential fuel for power generatiors. *Renewable & Sustainable Energy Reviews, 103*, 96–108. doi:10.1016/j.rser.2018.12.023

Yeole, A. S., & Kalbande, D. R. (2016, March). Use of Internet of Things (IoT) in healthcare: A survey. In *Proceedings of the ACM Symposium on Women in Research 2016* (pp. 71-76). ACM. 10.1145/2909067.2909079

Yin, J., Gong, L., & Wang, S. (2018). Large-scale assessment of global green innovation research trends from 1981 to 2016: A bibliometric study. *Journal of Cleaner Production, 197*, 827–841. doi:10.1016/j.jclepro.2018.06.169

Yoon, H. J. (2019). Blockchain technology and healthcare. *Healthcare Informatics Research, 25*(2), 59–60. doi:10.4258/hir.2019.25.2.59 PMID:31131139

Youniss, J., Bales, S., Christmas-Best, V., Diversi, M., Mclaughlin, M., & Silbereisen, R. (2002). Youth civic engagement in the twenty-first century. *Journal of Research on Adolescence, 12*(1), 121–148. doi:10.1111/1532-7795.00027

Yusuf, S., Musa, M. A., Diugwu, I., Adindu, C., & Afeez, B. (2021). A Systematic Literature Review Approach on the Role of Digitalization in Construction Infrastructure and Sustainable City Development in Developing Countries. *ZEMCH International Conference*, (pp. 1075–1093).

Yu, Z., Khan, S. A. R., Ponce, P., & Jabbour, A. B. L. (2022). Factors affecting carbon emissions in emerging economies in the context of a green recovery: Implications for sustainable development goals. *Technological Forecasting and Social Change, 176*, 121417. doi:10.1016/j.techfore.2021.121417

Zailani, S., Iranmanesh, M., Nikbin, D., & Jumadi, H. B. (2014). Determinants and environmental outcome of green technology innovation adoption in the transportation industry in Malaysia. *Asian Journal of Technology Innovation, 22*(2), 286–301. doi:10.1080/19761597.2014.973167

Zaman, R., Jain, T., Samara, G., & Jamali, D. (2022). Corporate Governance Meets Corporate Social Responsibility: Mapping the Interface. *Business & Society, 61*(3), 690–752. doi:10.1177/0007650320973415

Zárate-Rueda, R., Beltrán-Villamizar, Y. I., & Murallas-Sánchez, D. (2021). Social representations of socioenvironmental dynamics in extractive ecosystems and conservation practices with sustainable development: A bibliometric analysis. *Environment, Development and Sustainability, 23*(11), 16428–16453. Advance online publication. doi:10.100710668-021-01358-4

Zeman, P., Honig, V., Kotek, M., Táborsky, J., Obergruber, M., Marik, J., Hartová, V., & Pechout, M. (2019). Hydrotreated Vegetable Oil as a Fuel from Waste Materials. *Catalysts, 9*(4), 337. doi:10.3390/catal9040337

Zhang, D. G., & Lu, Y. Q. (2017). Impact of market segmentation on energy efficiency. *China Popul Resour Environ, 27*(1), 65–72.

Zhang, X., & Lin, W. Y. (2018). Hanging together or not? Impacts of social media use and organizational membership on individual and collective political actions. *International Political Science Review, 39*(2), 273–289. doi:10.1177/0192512116641842

Zhao, X., Ding, X., & Li, L. (2021). Research on Environmental Regulation, Technological Innovation and Green Transformation of Manufacturing Industry in the Yangtze River Economic Belt. *Sustainability, 13*(18), 10005. doi:10.3390u131810005

Zhao, X., Shang, Y., & Song, M. (2020). Industrial structure distortion and urban ecological efficiency from the perspective of green entre-preneurial ecosystems. *Socio-Economic Planning Sciences*, *72*, 100757. doi:10.1016/j.seps.2019.100757

Zhou, K. Z., Gao, G. Y., & Zhao, H. (2017). State Ownership and firm innovation in China: An integrated view of institutional and efficiency logics. *Administrative Science Quarterly*, *62*(2), 375–404. doi:10.1177/0001839216674457

Zhou, P., Ang, B. W., & Poh, K. L. (2008). Measuring environmental performance under different environmental DEA technologies. *Energy Economics*, *30*(1), 1–14. doi:10.1016/j.eneco.2006.05.001

Zhou, W., Chen, J., & Huang, Y. (2019). Co-Citation Analysis and Burst Detection on Financial Bubbles with Scientometrics Approach. *Economic Research Journal*, *32*(1), 2310–2328. doi:10.1080/1331677X.2019.1645716

Zhou, X., & Du, J. (2021). Does environmental regulation induce improved financial development for green technological innovation in China? *Journal of Environmental Management*, *300*, 113685. doi:10.1016/j.jenvman.2021.113685 PMID:34517232

Zink, T., & Geyer, R. (2017). Circular economy rebound. *Journal of Industrial Ecology*, *21*(3), 593–602. doi:10.1111/jiec.12545

Zivnuska, S., Carlson, J., Carlson, D., Harris, R., & Harris, K. (2019). Social Media Addiction and Social Media Reactions: The Implications for Job Performance. *The Journal of Social Psychology*, *159*(6), 745–760. doi:10.1080/00224545.2019.1578725 PMID:30821647

Zubaydi, H. D., Chong, Y. W., Ko, K., Hanshi, S. M., & Karuppayah, S. (2019). A review on the role of blockchain technology in the healthcare domain. *Electronics (Basel)*, *8*(6), 679. doi:10.3390/electronics8060679

Zubkova, V., Strojwas, A., Bielecki, M., Kieush, L., & Koverya, A. (2019). Comparative study of pyrolytic behavior of the biomass wastes originating in the Ukraine and potential application of such biomass. Part 1. Analysis of the course of pyrolysis process and the composition of formed products. *Fuel*, *254*, 115688. doi:10.1016/j.fuel.2019.115688

Zupic, I., & Čater, T. (2015). Bibliometric Methods in Management and Organization. *Organizational Research Methods*, *18*(3), 429–472. doi:10.1177/1094428114562629

About the Contributors

Luísa Cagica Carvalho held a PhD in Management in University of Évora – Portugal. Professor of Management on Department of Economics and Management, Institute Polytecnic of Setubal– Portugal. Guest professor in international universities teaches in courses of master and PhDs programs. Researcher at CEFAGE (Center for Advanced Studies in Management and Economics) University of Evora – Portugal. Author of several publications in national and international journals, books and book chapters.

* * *

Juhi Agarwal is a research scholar in the Department of Management Studies, Indian Institute of Technology, Roorkee. She holds her MBA Degree (2019) from Institute of Management Science, Lucknow University with specialization in Human Resources and Industrial Relations, and has completed her internship project with Tata Consultancy Services, Lucknow (June 2018-August 2018), and Bachelor Degree (2016) from National Post Graduate College, Lucknow University. She has qualified for UGC-NET,JRF in December 2021. Her scholarly interest is in the areas of Corporate Social Responsibility, Leadership, Motivation, Talent Management, Labour Welfare, etc.

Alexandra Anderluh is Senior Researcher at the Carl Ritter von Ghega Institute for Integrated Mobility Research at the St. Pölten University of Applied Sciences, Department Rail Technology & Mobility. She has earned a doctoral degree in Supply Chain Management at WU Vienna. Within the institute she is in charge of numerous research projects in the field of sustainability and resilience in logistics and supply chains as well as circular economy, and she authored various publications.

André Antunes received the BSc degree (2016) and the MSc degree (2022) in Software Engineering from Instituto Politécnico de Setúbal, Portugal. He's currently undergoing a PhD in Computer Science since 2022 in Universidade Nova de Lisboa, Portugal. Since 2016, he has been an invited assistant in web programming subjects at Instituto Politécnico de Setúbal, Portugal. His current research interests include digital systems for efficient management of water distribution networks and serious games, smart environments, adaptive systems, machine learning, and digital twins for therapy with patients with special needs.

Luciana Barbieri da Rosa graduated in Business Administration - Bachelor's Degree from Faculdade de Educação São Luis - FESL (2007) and Full Degree from the Special Teacher Education Graduation Program for Professional Education - PEG (UFSM - 2015). Specialist in Environmental Education from the Federal University of Santa Maria / RS (UFSM-2009) and Specialist in Public Management from the Federal University of Santa Maria (UAB / UFSM - 2015). Master in Business Administration from the Postgraduate Program in Business Administration PPGA (UFSM - 2013). PhD in Business Administration from the Postgraduate Program in Business Administration PPGA (UFSM - 2019). He is currently a teacher trainer at the Federal Institute of Rondônia (UAB / IFRO), a member of the research groups: Ecoinovar (UFSM), Innovation and Technology Management (GEITEC / UNIR) and Innovation and Organizational Sustainability (UFSM). It operates in the areas of: Innovation, Sustainability, Education, Strategy, Marketing and Production.

Gloria Braga Blanco is a permanent teacher in Curriculum Studies and innovation in the Faculty of Education of the University of Oviedo (Spain). She is specialist in action research and innovation processes. She has participated in the coordination of PhD, postgraduate and master programmes. One of her actual research interests has to do with how young people build global citizenship in a word in crisis. She participates in the I+D+I project "The construction of global citizenship with young people: investigating transformative practices with participatory methodologies" (PID2020-114478RB-C22) financed by Spanish Ministry of Science and Innovation (I+D+i, 2020).

João Caetano received the B.S. degree (2013) and the M.S. degree (2018) in Civil Engineering from the Universidade do Algarve, Portugal. He is currently undergoing a Ph.D. in Civil Engineering at Instituto Superior Técnico, Portugal. He is a member of the IPS R&D Centre for Innovation in Science and Technology (INCITE). His main research interests are in multiple criteria decision analysis, decision-making process, infrastructure asset management, and optimization applied to water distribution networks.

Adelina Calvo-Salvador, with a doctorate in Pedagogy (University of Oviedo, Spain), is a senior lecturer in the Department of Education at the University of Cantabria, Spain, in the area of Didactics and School Organization. Her research interests include socio-educational inclusion/exclusion mediated by technology, global education, student voice, gender and education and school improvement. She is a member of ANGEL (Academic Network on Global Education and Learning).

Nelson Carriço holds 5-year graduation in water resources engineering, an MSc. degree in hydraulics and water resources, and a PhD. in civil engineering. He worked for 5 years as a consultant and designer in urban hydraulic engineering, was a research fellow at Instituto Superior Técnico (IST) for 5 years and was visiting adjunct professor in higher education institutions from 2012-2014 at Instituto Superior Politécnico Autónomo (IPA), and from 2014-2018 at the Civil Engineering Department of the Barreiro School of Technology from the Polytechnic Institute of Setúbal (IPS) and in 2018 he has turned adjunct professor at IPS where he is until nowadays. In IPS he participates in different institutional activities, such as member of the Technical-Scientific Council and of the General Council. He has participated in 4 research projects and has coordinated 2 research projects related to the digitalization of urban water and irrigation infrastructures. Has published 15 articles in international journals, 25 national and international conference papers and 3 book chapters. Has also supervised and co-supervised several master's dissertations and currently co-supervises two doctoral students at the IST since IPS as a University of Applied Sciences (or Polytechnic) does not grant doctoral programmes. Also coordinates the IPS's R&D Centre for Innovation in Science and Technology (INCITE) where he organizes several scientific events, and write several proposals to national and international R&D funding. His main research interests are in

About the Contributors

Multicriteria decision analysis (MCDA); Infrastructure asset management (IAM); Hydraulic modelling; Digitalization of urban water systems; Water and Energy Efficiencies and Circular Economy.

Chiranji Lal Chowdhary is an Associate Professor in the School of Information Technology & Engineering at VIT University, where he has been since 2010. He received a B.E. (CSE) from MBM Engineering College at Jodhpur in 2001, and M. Tech. (CSE) from the M.S. Ramaiah Institute of Technology at Bangalore in 2008. He received his PhD in Information Technology and Engineering from the VIT University Vellore in 2017. From 2006 to 2010 he worked at M.S. Ramaiah Institute of Technology in Bangalore, eventually as a Lecturer. His research interests span both computer vision and image processing. Much of his work has been on images, mainly through the application of image processing, computer vision, pattern recognition, machine learning, biometric systems, deep learning, soft computing, and computational intelligence. He has given a few invited talks on medical image processing. Professor Chowdhary is editor/co-editor of 8 books and is the author of over forty articles on computer science. He filed two patents deriving from his research. He was selected in the Stanford University List of Top 2% Scientists Worldwide for 2021. Google Scholar Link: Link: https://scholar.google.com/citations?user=PpJt13oAAAAJ&hl=en.539e3bcd-53f5-4ac9-8246-5540014cd757

Marcos Cohen holds Master's (1998) and Doctoral (2007) degrees in Business Administration from the Pontifical Catholic University of Rio de Janeiro (PUC-Rio). He has been a Strategy professor and researcher on the main staff of the Department of Business Administration (IAG Business School) at PUC-Rio since 2007. Coordinator of the Center for Studies in Sustainable Organizations (NEOS) at the IAG Business School at PUC-Rio. Co-coordinator of the online MBA ESG Management at IAG/PUC-Rio. of Environmental Management Coordinator of the Interdisciplinary Center for the Environment at PUC-Rio (NIMA). Co-leader of the theme "Sustainability-Oriented Strategies" at ESO / EnANPAD and co-leader of the theme "Sustainable Entrepreneurship and Impact Business" at ENGEMA / USP. He develops two lines of research: 1- Strategies for the sustainability of public and private organizations 2- Sustainable entrepreneurship and impact business.

Bruno Ferreira received the B.S. degree (2015) and the M.S. degree (2017) in Civil Engineering from the Instituto Politécnico de Setúbal, Portugal. He's currently undergoing a PhD in Civil Engineering since 2019 in Instituto Superior Técnico, Portugal. Since 2018, he has been an invited assistant in hydraulic-related subjects at the Instituto Politécnico de Setúbal, Portugal. His current research interests include the detection and location of anomalous events in water distribution networks using hydraulic simulation and artificial intelligence techniques, and data and information systems management for infrastructure asset management in water distribution networks.

Aquilina Fueyo Gutiérrez, since 1988 she has developed educational activities as a university professor in the subjects of Educational Technology, Communication and Media Education, Educational, Computing Education, New Technologies for Education, and New Technologies for Education Innovation. She has taken part in official Master´s Degrees and Doctorate Programs about Social Media, Communication and Education. She has a very long experience in the use of e-learning platforms and in the direction of projects based in Information and communication technologies. She has been an active researcher in the areas of Media education and Digital competence, Virtual Learning and Development Education. She has published a lot of articles in indexed journals and chapters in books, and various books on the aforementioned subjects. She is currently the Principal Investigator of the IETIC EVEA Research Group accredited by ANECA and of the R&D Project Building global citizenship with young people. Researching transformative practices with participatory and inclusive methodologies.

Harpreet Kaur is working as a faculty member at Chandigarh University, Mohali, Chandigarh, India.

Celso Machado Jr. is professor of the Postgraduate Programs in Administration - PPGA/USCS, and of the Professional Master's Degree in Innovation in Higher Education in Health PPGES/USCS at the Universidade Municipal de São Caetano do Sul USCS.

Daielly Mantovani is professor at the Management Department in University of Sao Paulo, Brazil. Researcher on Smart and Sustainable Cities.

Maria Carolina Martins-Rodrigues, International PhD in Management, University of Extremadura, Spain. Post-graduate in Knowledge management, University of Belgrano, Buenos Aires, Argentina, and Member of the Research Unit - CinTurs - Research Center of Tourism, Sustainability and Well Being of University of Algarve. Her research interests currently are Tourism; Sustainability, Economia Circular, Green knowledge, Green intellectual Capital, Innovation, Knowledge Management, Entrepreneurship, Social Responsibility, Business Models, Business Intelligence, Business incubators, Smart Cities. She is co-authored over 40 articles and book chapters and published in several scientific journals. She has also organised the Iberian Conference on Entrepreneurship and international Conferences of Sustainability and Innovation. Founder member and vice-president of the direction of Empreend - Portuguese Association for Entrepreneurship. She was teacher in University Aberta (UAb), and University Moderna. In TAX Administration was Tax Administration Technician Advisor acting as coordinator in the International Relations Services Directorate (DSRI) and was Head of the Tax Service.

Fabio de Oliveira Paula is an adjunct professor of the Pontifical Catholic University of Rio de Janeiro (PUC-Rio). Young Scientist of Our State by FAPERJ (2020). PhD in Business Administration at PUC-Rio, in the research line of Strategy (2017). FAPERJ Doctorate-Sandwich scholarship from September 2016 to August 2017, having conducted academic research at Aalto University School of Business (Helsinki, Finland). Master in Business Administration at PUC-Rio (2006). Graduated in Computer Engineering at the same University (2001). Coordinates the Research Center on Entrepreneurship and Innovation (MAGIS) at IAG-PUC Rio. Conducts research and has articles published in national and international academic journals of high impact and in national and international conferences in the fields of Innovation Management, Strategy and Alliance Networks.

Leonilde Reis is a Coordinator Professor with Aggregation at the School of Business and Administration (ESCE) of the Polytechnic Institute of Setúbal (IPS). The activities of teaching in higher education were developed since 1992 in the field of "Information Systems" and focused on undergraduate, master's and doctoral courses. Aggregation in Information Sciences, Fernando Pessoa University; PhD in Systems Information and Technologies, Minho University; Master's in management informatics, Católica University. Author of several publications in national and international journals, books and book chapters.

About the Contributors

Leon Rodrigues is a PhD candidate in Social Sustainability and Development at Universidade Aberta (UAb, Lisbon; 2021-Current), Master in Ecology at the Federal University of Rio Grande do Sul (UFRGS; 2002), graduated in Biological Sciences/Ecology at the University of Santa Cruz do Sul (UNISC; 1999). He currently works as a laboratory technician at the State University of Rio Grande do Sul (UERGS; 2018-Current) and as a guide and environmental educator at Fundação Gaia - Legado Lutzemberger (2015-Current). He has experience in Ecology, with an emphasis on Limnology and Ecosystem Ecology, and Environmental Sciences, with a focus on Environmental Education and sustainable agriculture, working mainly on the following topics: basic and applied ecology, biogeochemistry, environmental education and literacy and environmental licensing. He has teaching experience in higher education, environmental education and environmental licensing.

Carlos Rodríguez-Hoyos, is a senior lecturer at University of Cantabria. He has a Degree in Education and Doctorate awarded by the University of Oviedo. He has been a lecturer in the Department of Education at the University of Cantabria. since 2009 and has collaborated on the Modular Program in Digital Technologies and the Knowledge Society at the National University of Distance Education (UNED) since 2006. His lines of research include: the analysis of e-learning from a Teaching perspective, ICTs and the dynamics of educational and social inclusion and exclusion.

Damini Saini is currently associated with IIM Raipur working as an Assistant Professor in HRM/OB area. She also worked with the University of Lucknow in the management department. Dr. Saini has received her PhD from Faculty of Management Studies (FMS), University of Delhi. She has also been awarded UGC's Junior and Senior Research Fellowship. Dr. Saini has been awarded the Best research paper award at times for example at the International Conference on, Vedic Foundations of Indian Management, organised by VFIM(2013) and PRMEs International Conference "Responsible Management Education, Training and Practice"(2015). Her scholarly interests lie in leadership and ethics, values and spiritual foundations for developing leadership etc. Dr. Damini has contributed to nationally and internationally acclaimed journals and conferences.

Mirian Benair Semedo is a Postgraduate Assistant Professor at the Universidade de Santiago, Cape Verde. Degree in Organizational Communication and Public Relations, Master in Interior Tourism - Education for Sustainability from the Escola Superior de Educação de Coimbra. PhD student in Social Sustainability and Development at Universidade Aberta. Lines of research: Tourism, Public Relations, Communication, Marketing and sustainability.

Maria de Fátima Nunes Serralha completed the Doutoramento in Engenharia Química in 2001 by Universidade de Lisboa and Licenciatura in Engenharia Química in 1994 by Universidade de Lisboa; is Assistant Teacher in Instituto Politécnico de Setúbal since 2016; published 8 articles in journals and others publications. The main research interests are: Optimization process, Catalysis, Energetic Integration, Sustainable Fuels and Circular Economy.

Clara Silveira is a Coordinator Professor at the School of Technology and Management (ESTG) of the Polytechnic of Guarda. Teaching activities, since 1991, in the field of "Software Engineering and Information Systems" to the Computer Science Engineering degree and the MSc in Mobile Computing. She holds a PhD in Electrical and Computer Engineering from the Faculty of Engineering, University of Porto (FEUP); Master in Electrical and Computer Engineering, specializing in Industrial Informatics at FEUP; and completed the Training Program on Public Management (FORGEP) by INA. Director of the Software Engineering and Information Systems field. She has over eight years of experience leading the ESTG management team.

Maria José Sousa (Ph.D. in Management) is a University Professor and a research fellow at ISCTE/ Instituto Universitário de Lisboa. Her research interests currently are public policies, information science, innovation and management issues. She is a best seller author in ICT and People Management and has co-authored over 70 articles and book chapters and published in several scientific journals (e.g. Journal of Business Research, Information Systems Frontiers, European Planning Studies, Systems Research, and Behavioral Science, Computational and Mathematical Organization Theory, Future Generation Computer Systems and others), she has also organized and peer-reviewed international conferences, and is the guest-editor of several Special Issues. She has participated in several European projects of innovation transfer and is also External Expert of COST Association - European Cooperation in Science and Technology and President of the ISO/TC 260 – Human Resources Management, representing Portugal in the International Organization for Standardization.

Antonia Stefanidou is a Ph.D. Candidate in Technical University of Crete in the Department of Production Engineering and Management. She was born in 1978 in Chania, Crete, where she lives. She studied communication at the Department of Journalism and Mass Media of the Aristotle University of Thessaloniki. She holds a master's degree from the Social Sciences Department of Cultural Units Management, Hellenic Open University. She works as a journalist in the Communication and Public Relations Office of the Municipality of Chania.

Omar Vargas-González is Professor and Head of Systems and Computing Department at Tecnologico Nacional de Mexico Campus Ciudad Guzman, professor at Telematic Engineering at Centro Universitario del Sur Universidad de Guadalajara with a master degree in Computer Systems. Has been trained in Innovation and Multidisciplinary Entrepreneurship at Arizona State University (2018) and a Generation of Ecosystems of Innovation, Entrepreneurship and Sustainability for Jalisco course by Harvard University T.H. Chan School of Health. At present conduct research on diverse fields such as Entrepreneurship, Economy, Statistics, Mathematics and Information and Computer Sciences. Has collaborated in the publication of over 20 scientific articles and conducted diverse Innovation and Technological Development projects.

Waleska Yamakawa Zavatti Campos has a Master's in administration from the Pontifical Catholic University of Rio de Janeiro. Graduated in Pedagogy from the University of São Paulo and in Administration from the Federal University of Goiás. Participation in a course with a Capes grant, under the Internationalization Program of the University Institute of Lisbon - ISCTE / IUL, Portugal. MBA in People Management with a degree in Higher Education. Specialization course in Public Management. Specialization course in External Control and Public Governance. External Control Analyst - Specialty in Personnel Management at the Court of Auditors. Member of the research groups Innovation and Technology Management at the Federal University of Rondônia (GEITEC) and Research Center in Entrepreneurship and Innovation (MAGIS / PUC- Rio).

Index

A

Agency 24-25, 27, 34, 36, 160-162, 167-168, 170, 175, 177-178, 204, 219, 290, 328-329, 332-333, 336, 341-342, 345, 348, 350-351
Agile Development 93, 98-99, 111
Agile Software Development 113-114
Alternative Fuels 17-18, 23, 31-32, 39
Artificial Intelligence 44-45, 48, 51

B

Barriers 13, 17, 41, 44-47, 54, 69, 165, 199, 285, 312
Bibliometry 77
Biofuels 17-18, 23-25, 34, 36-39
Biomass 17-18, 23-24, 26-29, 31, 39
Blockchain 51, 53, 56-64, 66-73
Brazilian Municipalities 287

C

CCIs 136, 157
Circular Economy 2, 17-18, 24, 29, 31, 33, 35, 39, 87-88, 92, 159-161, 163, 166, 168-169, 174-175, 177-181, 243-244, 247-249, 252-259
Circular Economy Green Ammonia 17
CiteSpace 74, 76, 78, 89-90
Citizen Participation 285-286, 288-289, 291-292, 297, 300-303, 305, 330
Civic Education 328, 345
Civic Engagement 328-334, 336-351
Co-Creation 221, 228-229, 238, 253, 284, 286, 297, 301-302
Community of Barro Preto 261, 263, 270, 273-274
Composting 163, 165, 167, 171, 180
Co-Productive Interactivity 339, 351
Corporate Governance 10, 215-217, 234, 236-241, 258, 267
Creative City 135-136, 138, 140-141, 146-147, 149-151, 155
Creative Economy 135-138, 140, 146, 151-152, 154-156
Creative Industries 135-138, 141-143, 145, 150-153, 155
Cryptocurrency 56-57, 61
Cybersecurity 44, 46, 51, 56, 60, 62, 72

D

Data 9, 14, 17, 41-47, 49-53, 55, 57-60, 62-64, 66-67, 69, 71-73, 75-77, 89-90, 94, 105-106, 114, 120-121, 123, 130-131, 137, 146, 151, 158, 163, 184, 191, 201-202, 226, 231, 235, 244, 269-276, 278-279, 284-287, 289, 291-295, 297-305, 326-328, 330-331, 334-336, 341-342
Data Transformation 41
Decentralisation 56
Digital Twin 41, 43, 49-51, 53-54

E

Eco-Innovation 74-76, 78-92, 210
Ecology Innovation 74-76
Economic Development 4, 117-118, 127, 130, 139, 141, 144, 146, 154-156, 211, 246, 348
e-Government 118, 127, 130, 284-285, 287-291, 301, 303-305
e-Healthcare 56-57, 70
Entrepreneurship 90, 116-121, 123-131, 154, 201-202, 240, 281
Environmental Innovation 10-11, 74-76, 81, 85-88, 90
Environmental Regulation 1-3, 5-9, 11, 13, 15-16, 87, 91, 235, 238
e-Participation 284-287, 290-291, 299, 303-306
Ethics for sustainability 215-216, 221, 223-224
EUT 157

F

Forest Products 261-263, 267-269, 271, 280-283

G

GDP 118, 123, 138, 158, 166, 235, 241, 269
Generation Z 183-184, 188-190, 192-196, 198-200
Goals 8, 15, 19, 35, 49, 54, 72, 93-95, 102, 107-108, 112, 114, 119, 136, 139, 142, 146, 156-158, 171, 190, 193, 210, 212, 216-221, 223, 228, 233, 246, 253, 267-268, 283, 308, 315
Green Ammonia 17, 39
Green Hydrogen 17-22, 29-30, 35, 39
Green Innovation 2, 4-6, 8-13, 15, 74-76, 85, 88, 91-92
Green Technological Innovation 1-2, 4-9, 11, 13, 16

H

Humanistic Leadership 243, 245, 249, 251-256, 259
Hydrogenated Vegetable Oil 17, 21, 24-25, 27, 37, 39

I

ICT 12, 15, 94-95, 129, 184-186, 188, 192-193, 284, 290-292, 306, 308-309, 312-313, 315, 318, 321, 324, 326
Indicators 5, 8, 44, 77, 79-80, 82, 113, 116-118, 120-121, 129, 139, 187, 227-228, 230, 257
Informal Structures of Participation 338, 351
Information and Communication Technologies 93-94, 114, 118, 184, 188, 306-309, 326-327
Information Divulgation 116
Information Systems 10, 42-43, 47, 53, 94, 114, 126, 304, 308-309, 325-327
Innovation 1-6, 8-16, 25, 63-64, 69, 73, 75-76, 81, 84-88, 90-92, 113, 116-118, 120, 126, 129, 131, 135-138, 140, 149-154, 157, 176, 183, 190, 196, 201-202, 210, 212, 215, 219-221, 226-227, 229, 237-241, 253, 256, 265, 267, 272, 278-279, 291, 304, 309, 311-312, 324-326, 328-329, 336, 342-343, 346, 348, 350
Innovative Strategies 308
International Standards 248, 327
Internet of Things 42, 56-58, 62, 64, 67-73
IoT data 56, 66

K

Karlskrona Manifesto 93-94, 107, 113-114

L

Local Government 116, 122, 129-130, 138, 155-156, 287, 304-305

M

Medellin 135, 137, 146-149, 152-153, 155-157
Mercur 215, 226-235, 239, 241-242
Meta-Participation 337-338, 351
Meta-Research 328, 330-332, 336, 340-342, 351
Mobile Application 93-95, 98, 104-105, 111-113, 307-309, 312-315, 317-319, 321-322, 324, 326
Municipal Waste 30, 159-163, 166-172, 174-175, 177-178, 180

N

Nantes 135, 137, 146, 149-153, 155, 157
Narrative 261, 271, 273-274, 279-280
Network Analysis 79, 284, 293-294, 300, 305-306
New Challenges 120, 171, 183-184, 188-189, 194, 235, 283
New Trends 183
Non-Timber Forest Products 261-263, 267, 269, 271, 280-283

O

OECD 15, 24-26, 37, 80, 82, 84, 88, 158, 166, 179, 203, 205, 213, 219, 239, 257
Online Civic Engagement 332, 334, 345, 350-351
Organizational Commitment 187-188, 190-191, 198, 200, 217, 237, 311
Organizational Performance 200

P

Participation 78, 120, 137, 141, 145-146, 220, 228-229, 269, 277-278, 284-292, 297, 300-303, 305, 328-339, 341-351
Pinhão 261-263, 269-271, 273-274, 277-280, 282
Portugal 17-18, 26, 30, 36, 41, 74, 93-94, 116-117, 120, 126, 128, 130, 159, 169-171, 175, 183, 190, 215, 307-308, 312, 315, 325-326, 349
Productive Chain 261, 269, 274, 277-279
Productivity 12, 15, 77, 80, 82, 87, 92, 113, 120, 183-184, 186-187, 191-195, 198, 200, 210, 217, 253, 262
Public Administration 130, 199, 284-285, 292, 303

Q

QdlC 150, 158

R

Real-Time Information 66
Recover 159-160, 180, 232
Recycle 159, 161, 164-166, 169, 171, 174, 180, 204-205, 211
Reduce 3, 6, 8-9, 11, 23-25, 30-31, 43, 45, 64, 74, 95, 112, 159-161, 168-170, 172, 174, 180, 194, 203-204, 206, 211-212, 230-231, 294, 312
Refuse 159-160, 180
Regift 159-160, 180
Renewable Fuels 18-19, 38-39
Repair 159-160, 174, 179, 181, 203
Requirements Analysis 102, 114, 315, 327
Rethink 159-160, 181, 247, 329, 333

S

Scientometrics 77, 90, 92
SDGs 94-98, 102, 107-109, 111-112, 136, 142, 158, 210, 307-308, 318, 322
SEMAT 93, 98-99, 101, 111, 113
Separately Collected Waste 159-160, 166, 175, 180-181
Small and Medium-sized Enterprises 184, 190, 200
Smart Water 41, 44
Social Entrepreneurship 201-202, 240
Social Innovation 219, 221, 229, 238-241, 256, 309, 311-312, 324-326, 328-329, 336, 342-343, 346, 348, 350
Social Media 52, 129, 190, 195-196, 198-199, 284-293, 295, 300-306, 336, 342, 344, 346-348, 350
Social Network Analysis 284, 300
Social Responsibility 3-4, 9, 14, 124, 128-130, 202, 218, 224, 234, 236-241, 244-246, 248-249, 251-254, 256-257, 283, 311, 326
Social-Ecological System 215, 225, 230
Socio-Ecology 1-3, 6, 8, 10
Software Systems Development 114, 327
Startups 201, 203
Storytelling 271, 279-280
Sustainability 1-3, 8, 10-11, 14, 16-19, 24, 27, 30-31, 35, 40-41, 45, 54, 75, 80-81, 84, 86, 88, 90-95, 107-109, 112-114, 129, 136, 144-146, 155, 160-161, 175, 183, 188, 190, 195, 197, 201-202, 206, 210, 212, 215-221, 223-227, 230, 233-240, 243-244, 246-259, 261-267, 270-271, 273-274, 277-283, 286, 303, 307-308, 311, 318, 321-322, 324, 326-327, 329, 342-343, 345
Sustainability-Oriented Innovation 75-76, 88, 91, 210
Sustainable Development 11, 15, 19, 34, 39-40, 54, 72, 75, 90, 92-96, 98, 105, 107, 112-114, 129, 136, 138, 140, 142, 151-152, 154, 156-158, 160, 168, 171, 197, 209, 211-212, 216, 220, 224, 237-240, 245-246, 256, 259, 263, 267-268, 278, 280-281, 283, 285-286, 308, 324, 349
Sustainable Development Sustainability 93
Sustainable Innovation 75-76, 86, 88, 90-91, 210
Sustainable Urban Development 135, 145, 154, 156
Synthetic Fuels 18, 21, 28-31, 33, 36, 39-40

T

Talent Retention 187-188, 191-193, 200
Tata 243-244, 250-252, 254, 259
Telework 183-188, 191-197, 199
Teleworking 183-188, 190, 193, 195-200
Text Mining 284, 291, 293-294, 297, 299, 303

U

UN 38, 129-130, 136, 140, 156, 158, 179, 216, 246, 278, 285, 287-288, 291, 305, 332, 345
UNCTAD 142-143, 156, 158
UNESCO 136-137, 140, 144-145, 148, 154, 156-158
Upcycling 201, 204-206, 213
Urban Creativity 141
Urban Water Systems 41, 43-44, 54

W

Water 4.0 42
Water Utilities 41-53
Web 2.0 285-288, 304, 306

Y

Young 124, 151, 183-184, 188-189, 192-194, 200, 211, 274, 279, 328-340, 342-351
Young Citizenship 328
Youth Civic Engagement 328, 330-345, 347, 349-351

Recommended Reference Books

IGI Global's reference books are available in three unique pricing formats:
Print Only, E-Book Only, or Print + E-Book.
Shipping fees may apply.
www.igi-global.com

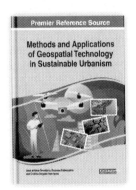

ISBN: 9781799822493
EISBN: 9781799822516
© 2021; 684 pp.
List Price: US$ **195**

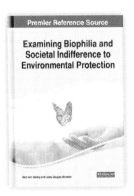

ISBN: 9781799844082
EISBN: 9781799844099
© 2021; 279 pp.
List Price: US$ **195**

ISBN: 9781799884262
EISBN: 9781799884286
© 2021; 461 pp.
List Price: US$ **295**

ISBN: 9781799872504
EISBN: 9781799872528
© 2021; 297 pp.
List Price: US$ **195**

ISBN: 9781799875123
EISBN: 9781799875192
© 2021; 416 pp.
List Price: US$ **295**

ISBN: 9781799867098
EISBN: 9781799867111
© 2021; 405 pp.
List Price: US$ **195**

Do you want to stay current on the latest research trends, product announcements, news, and special offers?
Join IGI Global's mailing list to receive customized recommendations, exclusive discounts, and more.
Sign up at: **www.igi-global.com/newsletters**.

Publisher of Timely, Peer-Reviewed Inclusive Research Since 1988

www.igi-global.com | Sign up at www.igi-global.com/newsletters | facebook.com/igiglobal | twitter.com/igiglobal | linkedin.com/igiglobal

Ensure Quality Research is Introduced to the Academic Community

Become an Evaluator for IGI Global Authored Book Projects

 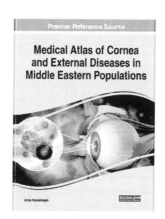 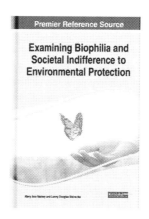 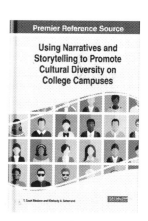

The overall success of an authored book project is dependent on quality and timely manuscript evaluations.

Applications and Inquiries may be sent to:
development@igi-global.com

Applicants must have a doctorate (or equivalent degree) as well as publishing, research, and reviewing experience. Authored Book Evaluators are appointed for one-year terms and are expected to complete at least three evaluations per term. Upon successful completion of this term, evaluators can be considered for an additional term.

If you have a colleague that may be interested in this opportunity, we encourage you to share this information with them.

Easily Identify, Acquire, and Utilize Published Peer-Reviewed Findings in Support of Your Current Research

IGI Global OnDemand

Purchase Individual IGI Global OnDemand Book Chapters and Journal Articles

For More Information:
www.igi-global.com/e-resources/ondemand/

Browse through 150,000+ Articles and Chapters!

Find specific research related to your current studies and projects that have been contributed by international researchers from prestigious institutions, including:

- Accurate and Advanced Search
- Affordably Acquire Research
- Instantly Access Your Content
- Benefit from the InfoSci Platform Features

"It really provides an excellent entry into the research literature of the field. It presents a manageable number of highly relevant sources on topics of interest to a wide range of researchers. The sources are scholarly, but also accessible to 'practitioners'."

- Ms. Lisa Stimatz, MLS, University of North Carolina at Chapel Hill, USA

Interested in Additional Savings?

Subscribe to **IGI Global OnDemand *Plus***

Learn More

Acquire content from over 128,000+ research-focused book chapters and 33,000+ scholarly journal articles for as low as US$ 5 per article/chapter (original retail price for an article/chapter: US$ 37.50).

6,600+ E-BOOKS.
ADVANCED RESEARCH.
INCLUSIVE & ACCESSIBLE.

IGI Global e-Book Collection

- Flexible Purchasing Options (Perpetual, Subscription, EBA, etc.)
- Multi-Year Agreements with No Price Increases Guaranteed
- No Additional Charge for Multi-User Licensing
- No Maintenance, Hosting, or Archiving Fees
- Transformative Open Access Options Available

Request More Information, or Recommend the IGI Global e-Book Collection to Your Institution's Librarian

Among Titles Included in the IGI Global e-Book Collection

 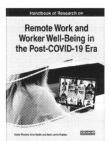  

Research Anthology on Racial Equity, Identity, and Privilege (3 Vols.)
EISBN: 9781668445082
Price: US$ 895

Handbook of Research on Remote Work and Worker Well-Being in the Post-COVID-19 Era
EISBN: 9781799867562
Price: US$ 265

Research Anthology on Big Data Analytics, Architectures, and Applications (4 Vols.)
EISBN: 9781668436639
Price: US$ 1,950

Handbook of Research on Challenging Deficit Thinking for Exceptional Education Improvement
EISBN: 9781799888628
Price: US$ 265

Acquire & Open

When your library acquires an IGI Global e-Book and/or e-Journal Collection, your faculty's published work will be considered for immediate conversion to Open Access *(CC BY License)*, at no additional cost to the library or its faculty *(cost only applies to the e-Collection content being acquired)*, through our popular **Transformative Open Access (Read & Publish) Initiative**.

For More Information or to Request a Free Trial, Contact IGI Global's e-Collections Team: eresources@igi-global.com | 1-866-342-6657 ext. 100 | 717-533-8845 ext. 100